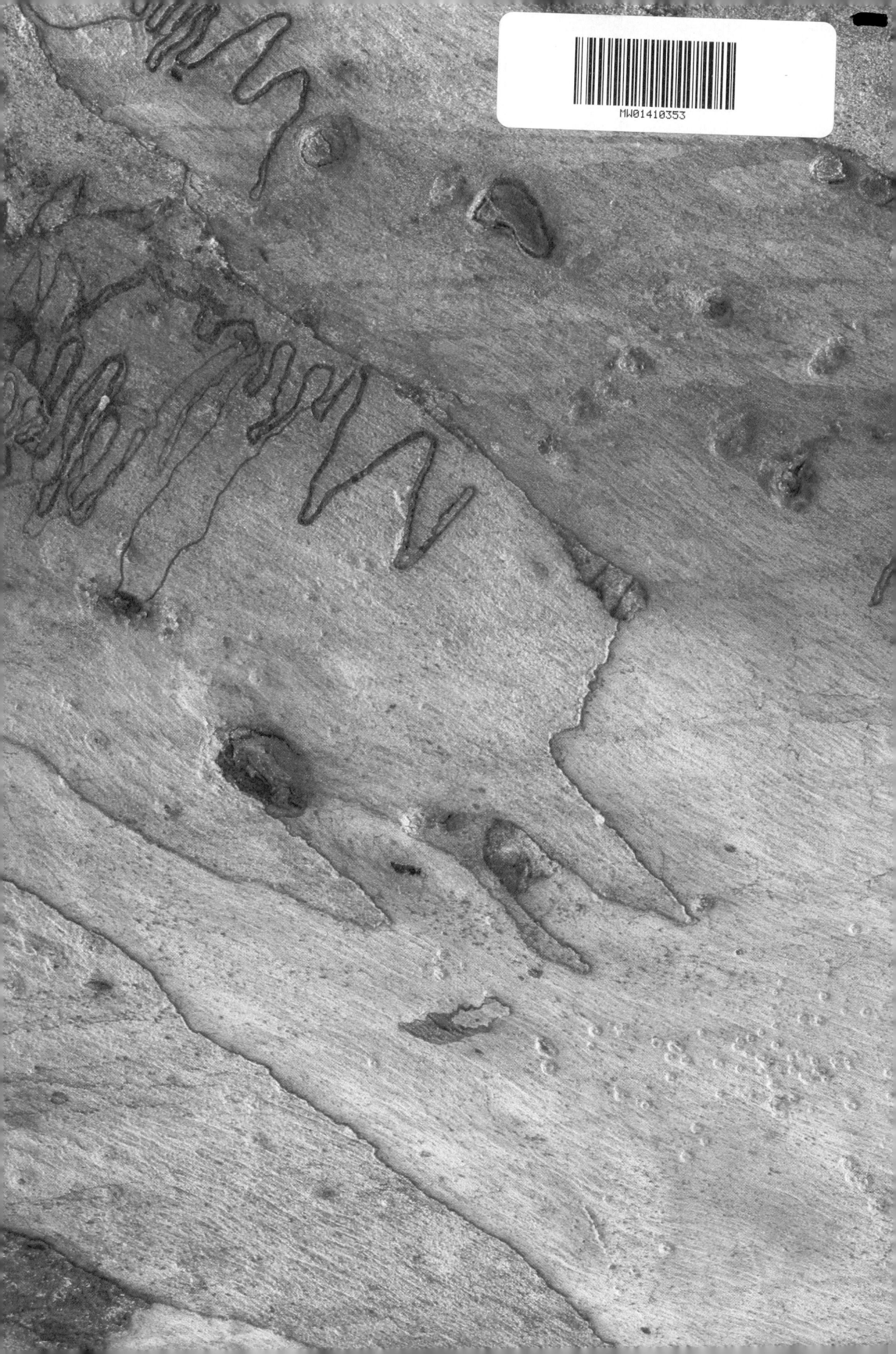

BUSH MEDICINE

A PHARMACOPOEIA OF NATURAL REMEDIES

TIM LOW

AUTHOR OF *BUSH TUCKER*

ANGUS & ROBERTSON

A division of HarperCollins *Publishers*

To Vincent Alexander Orlowski
A dead herbalist

AN ANGUS & ROBERTSON BOOK
An imprint of HarperCollinsPublishers

First published in Australia in 1990 and reprinted in 1991 by
CollinsAngus&Robertson Publishers Pty Limited (ACN 009 913 517)
A division of HarperCollinsPublishers (Australia) Pty Limited
4 Eden Park, 31 Waterloo Road, North Ryde, NSW 2113, Australia

HarperCollinsPublishers (New Zealand) Limited
31 View Road, Glenfield, Auckland 10, New Zealand

HarperCollinsPublishers Limited
77-85 Fulham Palace Road, London W6 8JB, United Kingdom

Copyright © Tim Low 1990

This book is copyright.
Apart from any fair dealing for the purposes of private study,
research, criticism or review, as permitted under the Copyright
Act, no part may be reproduced by any process without written
permission. Inquiries should be addressed to the publishers.

National Library of Australia
Cataloguing-in-Publication data:

Low, Tim, 1956–
Bush medicine: a pharmacopoeia of natural remedies
Bibliography.
Includes index.
ISBN 0 207 16462 2.
1. Medicinal plants — Australia. 2. Materia medica
Vegetable — Australia. 3. Naturopathy — Australia. I. Title.
615.53

Typeset in 11pt Palatino By Best-set Typesetters Ltd.
Printed by Griffin Press
5 4 3 2 1
95 94 93 92 91

CONTENTS

Preface 2
Introduction 3

PEOPLE & PLANTS
Plant Chemistry ....
Aboriginal Medicine 19
Colonial 31
In the European Tradition 43
Weeds as Herbs 55
Belyuen's Changing Ways 61
On the Pituri Trail 69

THE PLANTS
The Plants 77
The Mighty Eucalyptus 83
Aromatic Trees & Shrubs 93
Fragrant Herbs 103
Native Mints 111

THE ANIMALS
Animal Remedies 117
Animal Oils 127

INTERNAL AILMENTS
Ailing Gut 137
Scurvy Stricken 143

Headaches, Colds, Fevers 151
Baby Medicine 157

EXTERNAL AILMENTS
Cuts, Wounds, Sores 163
Animal Stings 169
Snakebite! 175
Miscellaneous Cures 181

DRUGS FOR PLEASURE
Aboriginal Drugs 187
Pituri 195
Bush Beer to Magic Mushrooms .. 201

THE OUTLOOK
Commercial Prospects 209
Conservation 217

APPENDIX
Home Remedies 221
Aboriginal Remedies 226
Colonial Remedies 228

Bibliography 230
Index 233

PREFACE

Plants and people are my areas of interest, not chemistry and physiology, and I do not claim an expert understanding of how bush remedies work. Whether goanna oil, for example, has any special healing power, or whether it is the rubbing that does the trick, I leave for others to decide. Many bush remedies do work, but obviously many do not.

Aromatic plants have been overemphasised in this book because so much is known about them. I find them an especially interesting subject, and hope the reader will too.

This is intended to be a popular book and I have not burdened the reader with intrusive reference citations. Most sources are listed in the bibliography; others may be found in the useful bibliographies in Lassak & McCarthy (1983) and Brock (1988). References to Aboriginal medicine in central Australia, Dampierland, Groote Eylandt and Kakadu are taken respectively from Latz (1982), Smith and Kalotas (1985), Levitt (1981), and Chaloupka and Giuliani (1984). The Northern Territory Pharmacopoeia project, cited several times in the text, published many of its results in *Traditional Bush Medicines*, collated by Andy Barr and listed in the bibliography. The references used to compile the book and accompanying notes have been lodged with the library of the Institute of Aboriginal Studies in Canberra for public access.

Much that has been previously written about Aboriginal medicine is erroneous, and where I have had strong reservations about a reported remedy, I have prefaced it with a qualifier such as "supposedly" or "reportedly".

Many people helped me research this book and I thank the following Aboriginal people for generously sharing their knowledge: Agnes Lippo, Maggie Timber Kwakow, Marjorie Bil Bil, Alice Jarrock and Ruby Yarrowin of Belyuen, Vai Stanton of Darwin, Theresa Ryder of Alice Springs, Grace Macintyre of Bedourie, George Quartpot, Peggy James and Robin Lane of Boulia, Jack Melville of Glengyle, and Lennie Duncan of Cherbourg.

Herbalist Kerry Bone prepared the recipes in the appendix, and provided much helpful information on herb use and chemistry. The staff of the Queensland and Darwin Herbaria identified many plants, and Beth Povinelli facilitated my visit to Belyuen. Judith McKay supplied some remarkable references to colonial medicine and Steve Page accompanied me in the search for pituri. I also thank Owen Foale and the other residents of 2 Lindsay Street, Darwin, and Julian Barry in Alice Springs for their hospitality.

I owe an immense debt to Peter Latz of the Northern Territory Conservation Commission with whom I had long and valuable discussions about Aboriginal medicine in central Australia. I have drawn heavily from his thesis on the subject, and his many ideas, so generously given, have helped shape my own.

The chapter entitled "Pituri" is adapted from an article that appeared in Australian Natural History Magazine in 1987 (vol 22 [6]: 257–260).

▶ The pale turpentine bush (*Beyeria lechenaultii*) is one of very few Aboriginal remedies recorded from southern South Australia. A leaf decoction was taken for fevers. Photographed at Beachport, South Australia.

INTRODUCTION

Three questions have troubled me during the writing of this book: How well did Aboriginal medicine work? How much has it changed in two hundred years? How much can we ever know about past medicinal practices?

The first question is superficially easy to answer. Aboriginal medicine worked wonders. Most Europeans who wrote on the subject, sympathetically or otherwise, confirmed the great power of Aboriginal healing, and many colonists were cured dramatically by native practitioners, as the quoted examples show.

But the real question and one that I cannot happily resolve, is by what means did Aboriginal medicine work? Recent books on the subject emphasise the chemical constituents of the plant remedies, and draw parallels with herbal practices in Asia. I believe that these authors, although well meaning, have distorted the picture.

Herbs were only one part of Aboriginal medicine. They were less important than the craft of magic. Witch doctors, spirits, chants and amulets were potent medicines as were the practices of bloodletting and blistering.

Even when herbs were used they were often applied in seemingly ineffectual ways. North Queensland Aborigines, for example, placed leaves over their eyes to cure headaches, and treated diarrhoea and internal pains with herbal baths. It is difficult to see how such treatments could have worked chemically.

There was also much inconsistency in the use of herbs. In northern Australia especially, different tribes treated ailments with different plants. Although some plant remedies were very widely used, most were not.

Faced with this kind of evidence, I am drawn to the idea that much Aboriginal medicine worked through strength of belief. Probably only about half of the plant remedies were pharmaceutically effective; the rest were placebos. The same proportions may well hold true for Indian and other Third World pharmacopoeias.

Perhaps European medicine, prior to the late nineteenth century, was much the same. It is remarkable, for example, that both Aborigines and Europeans placed so much faith in bleeding, a remedy that both consider useless today. Colonial literature is also full of enthusiastic testimonials for other remedies that seem to have no medical basis.

The power of positive thinking is much stronger than most of us usually acknowledge. The will to survive can make the difference between life and death, between sickness and health. Even in our own scientifically orientated society, willpower is a potent force. New Age devotees swear by the power of crystal healing, Bach flower remedies and shamanism. Aborigines believed strongly in the power of magic and in the unseen. They swore their "witch doctors" could run up trees and fly to the skies. The proudest warrior would calmly lie down and die if a bone was pointed in his direction. Guided by such a powerful belief system, it is no wonder their medicine worked.

◀ The "fleshy rooted geranium" recorded in the nineteenth century as an Aboriginal remedy was probably a native geranium (*Geranium*), but whether it was *G.solanderi* (illustrated) or another species, will never be known.

Colonial doctors had superior treatments for some ailments, such as fevers, but many of their remedies contained mercury and other poisons and were dangerous. Herbal remedies were a popular nineteenth-century alternative for those who feared the doctor's heavy hand. Given the choice, I would rather have entrusted my fate to an Aboriginal healer than to an eighteenth-century doctor.

◆ ◆ ◆

The second question, concerning changes in Aboriginal medicine under European impact, I can answer more confidently. After spending many months reading the literature on the subject, it is obvious that Aboriginal medicine changed dramatically under European influence. The evidence is compelling — the universal adoption of boiling (in billies), the wealth of remedies for non-traditional diseases, the incorporation of introduced plants (and the rabbit) into medicine, and the enormous disparity between nineteenth-century literature on Aboriginal medicine and Aboriginal herbalism as practised today.

▶ Victorian Aborigines may have prescribed the aromatic round-leaved mint-bush (*Prostanthera rotundifolia*) as a remedy for coughs and colds but no record of this survives. Aboriginal medicine in southern Australia remains largely a mystery. Photographed in the Dandenongs, Victoria.

◀ Guardian of the past, George Davis, is the last Yidinyji man well-versed in traditional medicinal lore. Here he prepares a prawn spear from a lawyer vine. Photographed in the rainforest of Cairns.

4   BUSH MEDICINE

▼ Quaker missionary James Backhouse, writing in 1843, observed that Tasmanian Aborigines used fleshy pigface leaves (*Carpobrotus rossii*) as a purgative. The red fruits are edible. Other remedies of Tasmanian Aborigines included muttonbird oil and ashes. Photographed at Beachport, South Australia.

The most obvious change has been one of westernisation. Aborigines have dispensed with the bloodletting, the amulets, and most of the magic. It seems likely that many of their animal-based remedies have also been abandoned.

This leads to the third question. It is one thing to know that change has occurred, but it is difficult to accurately reconstruct what went before. Can we ever be confident of knowing what Aboriginal medicine was like in the past?

Aboriginal medicine in temperate Australia will always remain a mystery. Only scraps of information survive for Tasmania and Victoria (summarised by L. Roth and A. Campbell respectively). Tasmanian Aborigines ate karkalla leaves (*Carpobrotus rossii*) and made a head bandage of stink-wood (*Zieria*) for pains. Few other plant remedies are known. Victorian Aborigines used eucalypt kinos and roots, wattle bark, and a number of herbs — native geranium (*Geranium*), sneezeweed (*Centipeda cunninghamii*), centaury (*Centaurium*), sowthistle (*Sonchus oleraceus*), and native hollyhock (*Lavatera plebeiai*).

From contemporary knowledge of medical practice in the north, one would expect Victorian Aborigines to have used the aromatic leaves of eucalypts, tea trees, mints and mint-bushes. But there is no evidence to support this, presumably because it has been lost. (Surprisingly, not one of the eucalypts used in colonial medicine for its aromatic oil is a recorded Aboriginal remedy!)

Of the five Aboriginal herbs that are listed, it is curious to note that all but one of them (the geranium) were also remedies of colonists. The parallels with early European medicine

▶ Aborigines doted on sowthistle leaves (*Sonchus oleraceus*) as a vegetable, but according to one nineteenth century report, impossible to confirm, the bitter leaves were also given as an emetic, although mixed with other ingredients. Photographed in a vacant allotment in Brisbane.

are striking. Victorian Aborigines supposedly shared the erroneous belief of the English peasant that sowthistle leaves induced sleep (sowthistle is in fact a popular Third World vegetable). They also boiled up centaury leaves for "bilious" headaches and made poultices from native hollyhocks, the Australian equivalents to the marshmallows of Europe.

It is tempting to use these parallels as evidence that Aborigines and Europeans independently discovered the same remedies despite living on opposite sides of the globe. I think the truth, however, is more mundane. Aborigines, in a state of cultural dislocation, probably adopted these "traditional" remedies from missionaries and settlers, and then remembered them as their own.

The animal-based cures known from Victoria are also dubious. Settler James Dawson, who recorded the sowthistle remedy, also reported that difficult sores could be cured by rubbing them with powerful owl fat. But the powerful owl is an uncommon bird, a secretive nocturnal denizen of tall forests. It is difficult to believe that the Aborigines would base a remedy on such an inaccessible source, when their other treatments involved easily obtainable ingredients. Further doubt on this remedy is cast when Dawson describes the owl as laying three eggs in a nest of reeds — powerful owls lay only one or two eggs in a tree hollow.

I quote this example to show the kinds of problems faced in reconstructing the past. Dawson's testimony, the most detailed we have from Victoria, is flawed by error and misinterpretation. The same is probably true of all the old articles written on bush medicine. The colonial literature gives no hint to the scores of herbal remedies reported in recent years from the Northern Territory, and living Aborigines know almost nothing of the bloodletting and much of the magic practised in the past.

Faced with so much discrepancy, I have found it a challenge to write this book. Like other popular writers I have chosen to emphasise the popular plant remedies, to leave out most of the blood and magic, and to downplay the discrepancies. I cannot pretend this is a balanced book; the best I can do is declare its biases.

Some medicinal plants are dangerous, and the author and the publisher accept no responsibility for any mishaps arising from the administering of plants mentioned herein.

The Testimonies of Cured White People

Early this century Mr Crawford of Walcha told of Aborigines curing a man so stricken with rheumatism he could not walk: "The blacks took him in hand, stripped him, laid him out on a sheet of bark, rubbing him with the young leaves and bark of the [stinging tree (*Dendrocnide excelsa*)] pounded up and boiled until it was of the consistency of treacle. It is said that they almost rubbed the man's skin off, but they cured the patient."

In 1868, when she was three, author and poet Dame Mary Gilmore fell seriously ill. Her parents, close friends of the Goulbourn tribes, gave her over to their care after the Aborigines had said she would die on the white man's diet. "Every night," Dame Mary said, "they used to put me in a nest of grass, which was so warm, and yet I had fresh air and the sky above me". Six weeks later she was cured.

Esta Ziviani, who grew up in rainforests near Innisfail in the 1930s, remembers local Aborigines teaching her mother herbal remedies. "One time I was very sick and wasn't getting better so my mother took me to the women and they did a search of me and found a tick embedded right behind my ear. They knew what to look for right away."

Some whites were even cured by Aboriginal magic. Pioneer Edward Stephens told of an interesting case in 1889:

> A friend of mine suffering from a severe catarrh and violent pains in the head, was cured by a native in less than half an hour. The blackfellow placed his patient near the fire. Warming his hands at the fire he would instantly clap them to his patient's ears, rubbing vigorously but not roughly. This he repeated without intermission. Presently he began to blow with his breath on the head and neck of his patient, still heating the palms of his hands and applying them to the ears; then he commenced to hum a sort of song or charm; a peculiar sensation came over my friend, he seemed to hear something "give way," or burst in his head. The cure was complete and no ill effects were felt afterwards.

In 1905 pioneer K. Langloh Parker reported a remarkable example of Bootha, an old Aboriginal "witch doctor" curing a white girl named Adelaide by communing with spirits. Bootha visited the girl in Parker's house, engaged in an eerie dialogue with the dead, and was told the girl had offended the spirit bees by bathing under a taboo tree.

To Parker's surprise, the girl admitted the transgression, and agreed with Bootha she had "bee stings" (pimples) on her back. Bootha entered into another dialogue with the dead spirit Guadgee, and was told the cure: Adelaide should drink only cold water, and during the night Guadgee would come and remove the offending wax planted by the bees in the girl's liver. Concluded Parker, "Adelaide slept that night, looked a better colour the next morning, and rapidly recovered".

Dame Mary Gilmore, poet, socialist, and social activist, was an early campaigner for Aboriginal rights. In *Old Days Old Ways*, she remembers in anger that the Aborigines who restored her health as a child were later raided and killed.

8   PEOPLE & PLANTS

PLANT CHEMISTRY

Most chemicals we get from herbs are actually poisons, produced by the plant to fend off the chomping insects, grazing beasts, hostile fungi and bacteria that threaten its existence. Because plants are rooted to the spot, and cannot fight or flee from predators, they are especially dependent upon chemical defence.

When plant poisons are taken as medicines, they either poison or attack the body in a therapeutic way, or kill health-threatening bacteria (germs) and fungi. They are taken in doses that are irritating or stimulating rather than life-threatening.

The cornerstone chemicals in Aboriginal medicine were the aromatic oils and tannins. The more potent alkaloids and other toxins, although still important, were less widely used.

Aromatic oils give eucalypts, tea trees, native mints and other herbs their strong aromas. Known also as essential or volatile oils (they evaporate readily to produce inhalable vapour), they are unrelated to the mineral oils of industry

◀ Starry flowers of pink fringe myrtle (*Calytrix exstipulata*) sprout among the tiny scale-like leaves, which yield an aromatic oil smelling like liniment. Some tribes crushed the foliage to treat wounds and pains, but most did not use this plant medicinally. Photographed at Adelaide River.

and the fixed plant and animal oils used in frying. All aromatic herbs and spices, and all strong-smelling medicine plants, owe their flavours to these oils.

The aromatic oils found in plants are complex mixtures of hydrocarbons. Many different hydrocarbons may make up the oil of one plant, but one or two compounds usually predominate. The same hydrocarbons are often found in very different plants. Limonene, for example, which helps give oranges their flavour, is also found in various eucalypts, tea trees, lemon grasses (*Cymbopogon*), fringe-myrtle (*Calytrix brownii*), fuchsia bushes (*Eremophila*), and even in peppermint, fennel and pepper. Related compounds give similar flavours to lemons, lemon grasses, lemon-scented tea trees, and lemon-scented eucalypts.

The role of aromatic oils is not completely understood, but it is clear that they reduce the

▶ A cheap source of lemon scent is the lemon-scented tea tree (*Leptospermum petersonii*), a shrub native to northern New South Wales and southern Queensland, and widely cultivated in the Third World for its aromatic oils.

▶ Fragrant pillows of the foliage of desert fuchsia bush (*Eremophila gilesii*) were fashioned by Aborigines stricken with colds and chest complaints. The rich aromatic oils in the leaves, inhaled during sleep, helped the body free itself of phlegm. Photographed at Ayers Rock.

▲ Despite its lemony smell, crushed stems of the lemon grass (*Cymbopogon procerus*) contain aromatic oils that differ from its namesake. Camphene and borneol are its major components, but 27 others have been identified, including limonene, the flavouring found in oranges.

palatability of plants and discourage the growth of bacteria and fungi. The germ-killing properties of aromatic plants have been acknowledged for thousands of years.

Aromatic oils also have general medicinal properties, the strength of which depend upon the precise composition of the oil. They tend to be antiseptic, they stimulate appetite and enhance mood, they relax involuntary muscles, and act as irritants on the respiratory tract and as counter-irritants on the skin.

Specific compounds found in some oils also have very specific effects. The eugenol in cloves and sacred basil (*Ocimum tenuiflorum*), for example, dulls nerves, the menthol in peppermint creates an illusion of cooling, and the oils in certain spices act as irritants.

Aromatic oils are very important to the food-flavouring and perfume industries, and chemists know much about their structure and properties.

◀ Smoking and soaking were the techniques used by desert Aborigines to make medicines from the crimson fuchsia bush (*Eremophila latrobei*). The Arrante treated scabies by making a wash from the leaves, and the Warlpiri smoked their babies in the smouldering leaves. Photographed at Alice Springs.

Colonial botanists and chemists were quick to recognise the value of Australian oils, and in the first half of the twentieth century the oils of many eucalypts, tea trees and others were investigated by them. Eucalypts are now grown throughout the world and their aromatic oils are used in medicine and perfumery. In the 1980s, the Northern Territory Pharmacopoeia project examined the aromatic oils of many Aboriginal remedies. Because most constituents of these oils were well-known compounds, it was possible to deduce much about their therapeutic activity. White cypress leaves (*Callitris glaucophylla*), for example, were found to contain pinene, a well-known antiseptic. Tribes in central Australia make an infusion from the leaves to treat rabies.

In Aboriginal medicine, aromatic plants were popular remedies for coughs, colds and fevers. The plants were prepared so that the aromatic oils could be inhaled deeply into the lungs, or drunk and then exhaled through the lungs (ingested aromatic oils are largely excreted through the lungs). Aromatic oils, and especially the cineol in eucalypts and tea trees, irritate the cells lining the respiratory tract and increase the secretion of mucus. This is especially helpful for dislodging phlegm, and for alleviating painful coughing when caused by an absence of mucus in the throat.

Aromatic oils were also valued as carminatives for relieving colic. When taken internally they relax the involuntary muscles of the alimentary sphincters, facilitating belching and breaking wind. First Fleet Surgeon General John White found the oil of peppermint eucalypts to be valuable in "removing all cholicky complaints"

▶ Pine-leaf geebung fruits (*Persoonia pinifolia*) contain an anti-bacterial principle that is active against golden staph (*Staphylococcus aureus*) and typhoid (*Salmonella typhi*). This geebung was one of a thousand plants tested in the 1940s, and only four others — two geebungs and two sundews — acted against both bacteria.

◀ The liniment tree (*Melaleuca symphyocarpa*) owes its apt name and healing reputation to its heady aromatic oils, which smell like nasal decongestant. This medicinally important tree grows in swampy woodlands in northern Australia. Photographed at Punsand Bay, Cape York.

and peppermint tea is a well-known carminative. Aromatic liqueurs and after-dinner mints serve the same purpose.

Rheumatism and other muscular pains can be relieved by rubbing aromatic oils on to the skin. Because they are mildly irritating, they draw blood to the afflicted area and stimulate the release of anti-inflammatory agents; eucalyptus oil and other liniments work in this way. Aborigines also used tick-weed (*Cleome viscosa*) and desert poplar (*Codonocarpus cotinifolius*) leaves to treat rheumatism. Both plants yield mustard oils, substances renowned for their irritant effect.

Another property of aromatic oils is their ability to kill fungi and bacteria. Camphor, pinene, geraniol and menthol are strongly antiseptic. The leaves of the river red gum (*Eucalyptus camaldulensis*) and the fuchsia bushes (*Eremophila duttonii*, *E. freelingii*) are rich in pinene. When applied to wounds, they suppress germs. The terpinen-4-ol in the tea tree (*Melaleuca alternifolia*) is especially germicidal, and does not irritate sensitive skin. Tea tree oil also contains pinene and some cineol, a blend that is thought to be especially germicidal as each compound attacks different forms of bacteria.

Equally versatile as bush medicine remedies

▶ A slender stalactite of kino hangs from the injured trunk of a rusty gum (*Angophora costata*) in Royal National Park just south of Sydney. The dark red colour shows that this kino is very rich in tannins.

are the tannins, and Australia's flora is rich in tannin-producers. Tannins may occur in the leaves of shrubs and trees, but they are especially concentrated in the inner barks and gum-like kinos of eucalypts and other woodland trees.

Tannins are large molecules that dissolve readily in water. Their extraordinary versatility as medicines is due to a single property — tannins react with proteins to form an inert layer. Tannins are described as being astringent. The word "astringent" comes from the Latin *stringere*, meaning "to bind", and refers to the tightening effect of tannins on internal body membranes.

When a cup of strong black tea is drunk, or a draught of astringent red wine, the tannins in the drink react with the surface of the mouth, forming a cross-linking of the surface proteins and a familiar fuzzy puckering sensation. This sensation is only temporary and the reaction is easily reversed by a body enzyme. When the throat or gums are inflamed, the tannins help contract the tissue and form a protective surface layer. Tannin-rich plants have been used for thousand of years throughout the world to treat gum and throat afflictions. Even in the first century Pliny recognised the value of tannin-rich blackberry leaves for treating "affections of the mouth".

Tannins are also taken to treat diarrhoea. When a strong tannin solution is drunk, the tannins form a surface lining on the intestine, providing protection when diarrhoea is due to irritation of the intestine. A large number of Aboriginal and colonial diarrhoea remedies employed tannin-rich plants.

Tannins are also used for treating burns and abrasions. By bonding with the proteins of damaged surface cells, the tannins form a tight protective layer, which limits contamination and loss of serum. By reacting with the dead proteins, they neutralise a potential source of food for harmful bacteria.

Aborigines used the inner barks of many kinds of trees in medicine, and it is likely that all were rich in tannins. Dark red inner barks were especially popular: the colour of the bark is produced by decomposed, condensed tannins. The Northern Territory Pharmacopoeia project found that some barks were remarkably rich in tannins, especially southern ironwood (*Acacia estrophiolata* — 11.7 per cent), dead finish (*Acacia tetragonophylla* — 8.25 per cent), milky plum (*Persoonia falcata* — 9 per cent) and cocky apple (*Planchonia careya* — 9 per cent). Much higher levels of tannins, up to 36 per cent, were found in eucalypt kinos, which both Aborigines and colonists used medicinally.

▶ Warts are withered by the spines of dead finish (*Acacia tetragonophylla*) inserted around the base of the wart. The spines are actually modified leaf stalks.

▲ Ironwood (*Acacia estrophiolata*) tells by its name that its timber is tough. Aborigines worked it into spears and other tools. A bark decoction served as a remedy for boils. Photographed at Alice Springs.

Trees produce tannins as defence against insects and microbial attack. When a caterpillar bites into a tannin-rich leaf, the liberated tannins react with the enzymes in the insect's saliva and the proteins in the leaf, rendering the leaf indigestible. Tannins in bark and kino provide a similar kind of protection against insect borers, fungi and bacteria. They bond to the proteins of single-celled organisms, rendering them immobile.

Tannins and aromatic oils are gentle medicines. They are not particularly toxic and pose little danger to human health. The same, however, is not true of the alkaloids. These are potent chemicals that represent some of the most toxic substances known.

Well-known alkaloids include strychnine, curarine, morphine, cocaine and nicotine (alkaloid names always end in "ine"). These are strong poisons produced to deter plant eaters. Nicotine, for example, is a powerful insecticide. Alkaloids have a strong effect upon the human central nervous system, and act more specifically and individually than tannins and aromatic oils. The quinine that cures fevers has little in common with the deadly coniine in hemlock or the caffeine in coffee. Alkaloids are much more important in western medicine than aromatic oils or tannins (the latter are no longer used).

Australia's alkaloids were surveyed just after World War II by the CSIRO's Australian Phytochemical Survey. Almost 500 were identified — 200 of them new to science — from more than 4000 plants screened. Little benefit came from all this and no medicine on the market today is

▶ The northern black wattle (*Acacia auriculiformis*) is a pretty street tree in Darwin, and a source of alkaloids. Top End Aborigines used the leaves and pods to relieve pains, and crushed the pods as "bush soap".

▲ Queensland colonial botanist F.M. Bailey thought medicine might one day be made from strychnine berry (*Strychnos lucida*), a northern shrub. Photographed at East Point, Darwin.

attributable to the survey's findings. The survey also overlooked castanospermine, the Moreton Bay chestnut alkaloid (*Castanospermum australe*), which is showing such promise as a cancer cure.

Nonetheless, alkaloids from Australian plants have achieved prominence in western medicine. Important examples include the hyoscine in corkwood (*Duboisia*), the solasodine in kangaroo apples (*Solanum*), and, less significantly, the reserpine in quinine tree (*Alstonia constricta*). Among introduced weeds there are significant alkaloids in Madagascar periwinkle (*Catharathus roseus*), hemlock (*Conium maculatum*), and common thornapple (*Datura stramonium*).

The importance of alkaloids in Aboriginal medicine is difficult to determine. Of 52 medicine plants tested by the Northern Territory Pharmacopoeia project, only four — snakevine (*Tinospora smilacina*), onion lily (*Crinum angustifolium*), dead finish and northern black wattle (*Acacia auriculiformis*) — yielded alkaloids. The first two of these were very important remedies.

Other medicine plants known to contain alkaloids include the strychnine berry (*Strychnos lucida*), white cheesewood (*Alstonia scholaris*), milkwood tree (*Alstonia actinophylla*) and bitter-bark (*Ervatamia orientalis*). Aborigines took few medicines internally, and did not measure out doses (a dangerous practice where alkaloids are concerned), which suggests that alkaloids were not of the greatest importance. They were certainly important as intoxicants: pituri (*Duboisia hopwoodii*) and the native tobaccos (*Nicotiana*) are rich in nicotine and other alkaloids, and rock isotome (*Isotoma petraea*) contains lobeline.

Much is known about alkaloids because they are easy to detect in screening tests. But Australian plants also contain many other medicinal compounds of great potency, most of which remain unidentified. Worth investigating are the corrosive milky latexes of caustic creeper (*Sarcostemma australe*) and spurges (*Euphorbia*), the bitter compounds in fuchsia bushes and striped mint-bush (*Prostanthera striatiflora*), and the constituents of green plum (*Buchanania obovata*), beach convolvulus (*Ipomoea pes-caprae*), great morinda (*Morinda citrifolia*), dysentery bush (*Grewia retusifolia*) and pine-leaf geebung (*Persoonia pinifolia*), among others. It will be many decades before the mysteries of Australia's bush medicines are fully revealed.

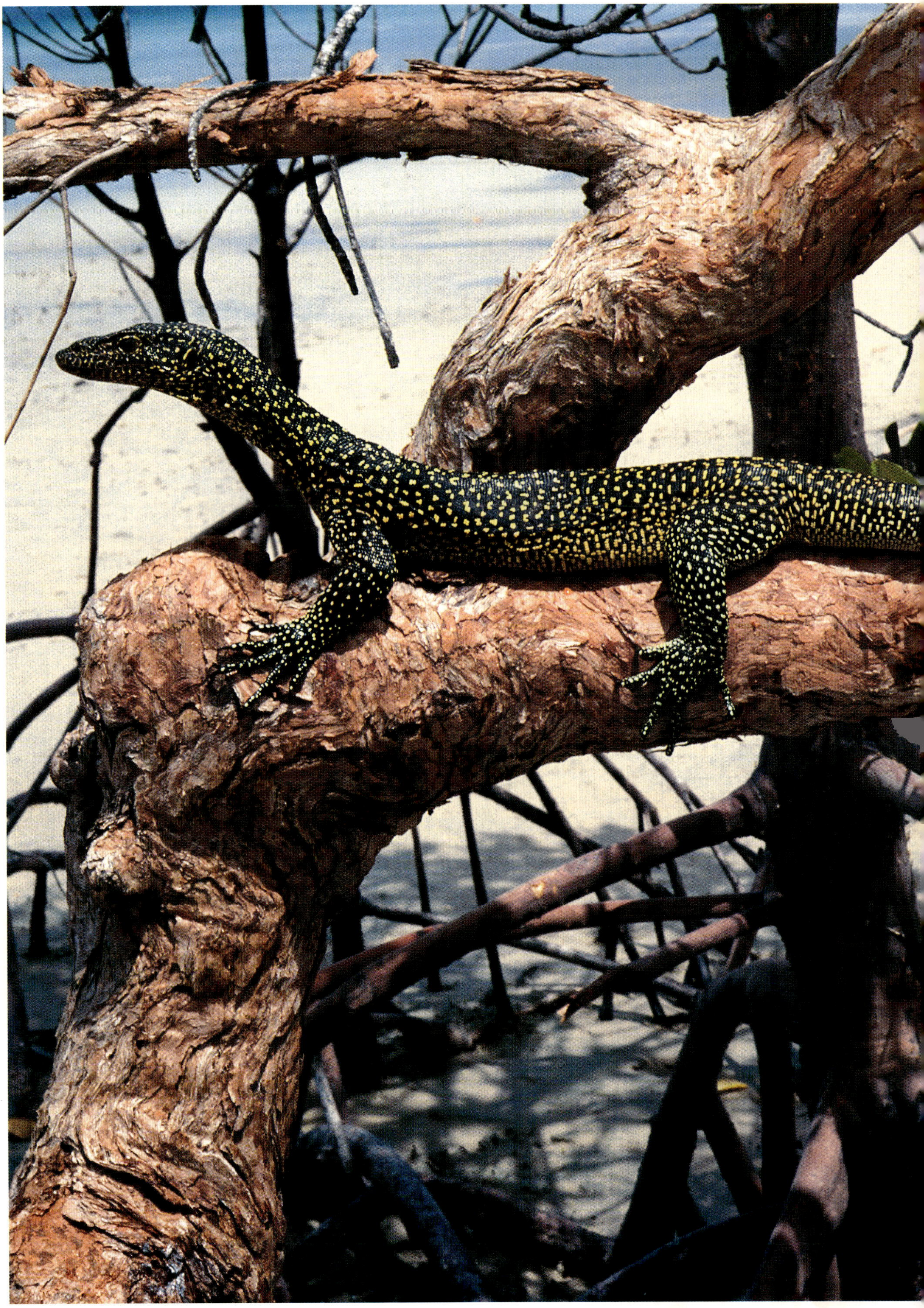

ABORIGINAL MEDICINE

Aborigines traditionally were much healthier than Australians are today. Living in the open in a land largely free from disease, they benefited from a better diet, more exercise, less stress, a more supportive society and a more harmonious world view.

Early colonists were often amazed by the Aborigine's recuperative powers. Disbelieving pioneers wrote of blacks recovering from horrific body wounds where their organs were protruding. The healthy Aboriginal physique promoted rapid healing.

Nonetheless, Aborigines often had need of bush medicines. Sleeping at night by fires meant they sometimes suffered from burns. Strong sunshine and certain foods caused headaches, and eye infections were common. Feasting on sour fruits or rancid meat brought on digestive upsets, and although tooth decay was not a problem, coarse gritty food sometimes wore teeth down to the nerves. Aborigines were also occasionally stung by stingrays, stonefish, jelly-

◀ Bush medicine includes animals as well as plants. Aborigines used ants, cobwebs, worms, and especially oils, taken from animals such as the mangrove monitor (*Varanus indicus*) of northern Australia. Photographed on Cape York.

▲ The lady apple fruit (*Syzygium suborbiculare*) is a Groote Eylandt remedy for chest congestion and colds, but it is no longer prescribed in the traditional way. Missionary Dulcie Levitt observed: "The fruit was originally roasted in hot ashes but is now boiled."

fish and snakes. In the bush there was always a chance of injury, and fighting usually ended in great bruises and gashes.

To deal with such ailments, Aborigines resorted to a range of remedies — wild herbs, animal products, steam baths, clay pills, charcoal and mud, massages, string amulets and secret chants. Some of these remedies had no empirical basis, but it is clear from the accounts of colonists that they worked.

Many of the remedies, of course, did directly heal. Aromatic herbs, tannin-rich inner barks and kinos have well-documented therapeutic effects. Other plants undoubtedly harboured alkaloids or other compounds with pronounced healing effects. Unfortunately, very few native remedies have been tested systematically.

It is important to recognise that Aboriginal remedies varied between tribes. There was no one Aboriginal pharmacopoeia, just as there was no one Aboriginal language. Because many plants share the same properties, and because some treatments had no empirical basis, plant remedies varied much more between nearby tribes than, for example, plant foods.

In trying to understand the nature of Aboriginal medicine, we are faced with the dilemma that most of the knowledge has been lost. Very little is known of medical practice in southern and eastern Australia, where Aboriginal culture was ruthlessly crushed more than a century ago. Aborigines were secretive about their remedies, and even the few sympathetic whites recorded little of value. The nineteenth century English mind was more interested in witch doctors and sorcery, and most colonial books on Aborigines depict their medicine only in these terms. Plant remedies are rarely mentioned.

During the last twenty years, anthropologists have worked in central and north-western Australia to record what is left of Aboriginal medical lore. In Arnhem Land, the Kimberley, and in the deserts of Western and central Australia, there are still Aborigines living who were reared in traditional ways. Their testimony has produced a startling picture of a complex and sophisticated pharmacopoeia, which embraced remedies for all manner of ailments. Whether Aborigines in southern Australia had the same range of plant remedies, it is impossible to say. There is no one left to ask. Plant cures may have been more important in tropical Australia where the chemistry of the plants is more varied. We also cannot discount the possibility of Indonesian influence as Macassan fishermen and Aborigines often lived together during the bêche-de-mer (sea slug) season in the Northern Territory. The beach convolvulus (*Ipomoea pes-caprae*), for example, may be an Indonesian medicine.

Compounding our problems of reconstructing the past are the changes that have taken place in the last two hundred years. Early European settlers unwittingly brought in a gamut of new diseases for which Aborigines had no natural resistance and no traditional remedies. Horrific smallpox plagues swept through Aboriginal Australia, carrying off as many as half the population. We do not know how Aborigines responded to these plagues for they preceded settlement by several decades. We do know that Major Mitchell and other explorers met Aborigines in remote regions disfigured by smallpox scars who told stories of horrific deaths and mass graves. It is likely that in attempting to conquer these scourges, terrified Aborigines abandoned old remedies and experimented with new ones.

The later arrival of influenza, tuberculosis, syphilis and other horrors would have further disrupted traditional Aboriginal medicine as

▼ Tobias Ferguson inhales crushed leaves of the weeping tea tree (*Melaleuca leucadendron*), a well-known Aboriginal remedy for colds. Many traditional treatments are still practised by Aborigines today; others are long forgotten.

did the profound changes in diet and lifestyle imposed by white contact. The diseases afflicting Aborigines today are very different from those they would have endured before white contact. Many early colonists, seeing Aborigines disfigured by diseases they had introduced, thought Aborigines lived short lives of abject misery, in ignorance of any medicinal lore.

A second, more benign change was the introduction last century of the billycan. Almost everywhere in Aboriginal Australia, herbs that once were soaked in water are now boiled over a fire. Aborigines today rarely distinguish this from a traditional practice, although they know the billycan is a white man's innovation. Boiling is much quicker than overnight soaking, but it may destroy some active ingredients and increase the potency in solution of others. In many communities only the very elderly can remember back before billies.

A third change is an apparent decline in the use of non-herbal remedies. Aborigines today rarely, if ever, engage in bloodletting, blood drinking, chants and the tying of healing amulets,

◀ Practically a weed, the medicinal fruit-salad plant (*Pterocaulon sphacelatum*) sprouts throughout the outback on roadsides, rockslides, campsites and wherever else the ground is disturbed by people or fire. Aboriginal remedies were usually made from common and widespread plants. Photographed at Alice Springs.

though these were important remedies in the past. Aborigines were probably discouraged in these practices by early missionaries, and after absorbing western ideas about medicine. Sorcery, however, remains a potent belief, and the casting and removing of spells is still practised.

Aboriginal medicine has also changed in more subtle ways. Several communities now make use of exotic plants, usually claiming these to be traditional remedies. In the Northern Territory medicines are made from the exotic weed called asthma plant (*Euphorbia hirta*); from the African tamarind tree fruit (*Tamarindus indica*), introduced from Indonesia up to three hundred years ago; the Latin American shrub Jerusalem thorn (*Parkinsonia aculeata*); the South American stinking passionfruit vine (*Passiflora foetida* — see "Belyuen's Changing Ways"); the American ringworm shrub (*Senna alata*); the African colocynth (*Citrullus colocynthis*); and the South American billygoat weed (*Ageratum* — see "Weeds as Herbs"). Central Australian Pitjantjatjara chew South American tree tobacco (*Nicotiana glauca*), and use the introduced rabbit in medicine. Of the Jerusalem thorn, which is used in central Australia by the Alyawarra and Anmatjirra, botanist Peter Latz concluded:

Although this plant was probably only introduced into the area less than 100 years ago, I have had informants swear to me that it has always been present in their country. The fact that this plant is a native of the Americas and that it bears the same aboriginal name as that for *Acacia tetragonophylla*, a different plant, only having its thorny nature in common, negates this theory. It is quite possible that the medicinal value of this plant was passed onto the Aboriginals by the people who introduced it into the area, or by early Afghan camel drivers.

◀ Beneath the surface of Aboriginal life lurked the world of spirits, beings who shaped and protected the landscape and its resources. These spirits sometimes inflicted ill-health, which only the witch doctors, by entering the spirit realm, could relieve. Photographed at Kakadu.

▶ An unusual form of traditional treatment was the rubbing of ringworm sores with sandpaper fig leaves (*Ficus opposita*), as demonstrated by Ruby Yarrowin of Belyuen. Some tribes infused these leaves to treat influenza, muscle aches and rashes.

◀ Marjorie Bil Bil of Belyuen near Darwin is used to outsiders asking about the remedies of her people. I was the third botanist to visit in a couple of years. Marjorie shows me a sandpaper fig tree (*Ficus opposita*), used against fungal infections.

24   PEOPLE & PLANTS

◀ A proven cure: the Northern Territory Pharmacopoeia project found that white cypress pine leaves (*Callitris glaucophylla*) contain alphapinene, a well-known antiseptic, which justifies their use as a poultice on rashes. Photographed in Finke Gorge, west of Alice Springs.

The adoption of so many introduced plants into bush medicine suggests the possibility that many of the native remedies would also have changed through time. Considering the number of new diseases brought in by the early settlers, the impact of the billycan, and of white settlement generally, this is hardly surprising. White Australians like to think that Aboriginal culture was static, but it has always been changing and adapting to new circumstances. Anthropologists do not like to admit the extent of recent change, for it diminishes the value of their research.

Throughout Australia, Aborigines believed that serious illness and death were caused by spirits or persons practising sorcery. Even trivial ailments, or accidents such as falling from a tree, were often attributed to malevolence. Aboriginal culture was too rich in meaning to allow the possibility of accidental injury and death and when someone succumbed to misfortune, a man versed in magic (the so-called witch doctor or medicine man) was called in to identify the culprit.

"Witch doctors" were men (rarely women) of great wisdom and stature with immense power. Trained from an early age by their elders and initiated into the deepest of tribal secrets, they were the supreme authorities on matters spiritual. They could visit the skies, witness events from afar, and parry with serpents. Only they could pronounce the cause of serious illness or death, and only they, by performing sacred rites, could effect a cure. Although they were not tricksters, they often treated illness through sleight of hand, removing offending quartz crystals, bones or special stones from the patient's body. Often they would pronounce that nothing could be done, and the victim, losing all hope, would sadly await death.

Medicine men sometimes employed herbs in their rites, but they did not usually practise secular medicine. The healing of trivial non-spiritual complaints, using herbs and other remedies, was practised by all Aborigines, although older women were usually the experts. To ensure success, plants and magic were often prescribed side-by-side.

Plants were prepared as remedies in a number of ways. Leafy branches were often placed over a fire while the patient squatted on top and inhaled the steam. Sprigs of aromatic leaves might be crushed and inhaled, inserted into the nasal septum, or prepared into a pillow on which the patient slept. To make an infusion, leaves or bark were crushed and soaked in water (sometimes for a very long time), which was then drunk, or washed over the body. Ointment was prepared by mixing crushed leaves with animal fat. Other external treatments included rubbing down the patient with crushed seed paste, fruit pulp or animal oil, or dripping milky sap or a gummy solution over them. Most plant medicines were externally applied.

Medicine plants were always common plants. Aborigines carried no medicine kits and had to have remedies that grew at hand when needed. If a preferred herb was unavailable, there was usually a local substitute. In the deserts, the strongest medicines are extraordinarily widespread plants. Fuchsia bushes (*Eremophila*) and bloodwood trees (*Eucalyptus terminalis*) grow everywhere. Lemon grasses (*Cymbopogon*) sprout on every ridge top and jirrpirinypa (*Stemodia viscosa*) around every waterhole. In the Top End many different kinds of large leaves are considered useful for staunching wounds, presumably because cases of profuse bleeding allow little time for searching.

Except for ointments, which were made by mixing crushed leaves with animal fat, medicines were rarely mixed. Very occasionally two plants were used together; there is one record from Victoria (of questionable accuracy) of Aborigines mixing together leeches, kangaroo liver and sowthistle (*Sonchus oleraceus*) for use as an emetic.

Aboriginal medicines were never quantified. There were no measured doses or specific times of treatment. Since most remedies were applied externally there was little risk of overdosing.

Some medicines were known to vary in strength with the seasons. Aromatic lemon grasses had to be picked while green, and toothed ragwort leaves (*Pterocaulon serrulatum*) were strongest after rain. A wet season growth of green plum leaves (*Buchanania obovata*), used as a toothache remedy, was considered much stronger than that available during the dry.

Sometimes herbs were kept for future use.

▲ The chemistry of plants changes with the seasons, and it makes sense that Aborigines consider green plum leaves (*Buchanania obovata*), for example, to be medicinally stronger in the Wet season than the Dry. The leaf veins or stems were inserted in sore teeth. Photographed at Palmerston, near Darwin.

◀ The Northern Territory Pharmacopoeia project says of toothed ragwort (*Pterocaulon serrulatum*): "The new leaves produced after rain are regarded as the best for medicinal preparations. In some places this plant is dried and ground, and stored for use when the fresh leaves are not available."

ABORIGINAL MEDICINE 27

In the Northern Territory dried ground leaves of ragwort, and tufts of lemon-scented grass (*Cymbopogon ambiguus*) were occasionally stored. The drug pituri (*Duboisia hopwoodii*) was dried and traded, and leaves of wild tobaccos (*Nicotiana*) were placed in caves. In Western Australia the rock isotome (*Isotoma petraea*) was stored.

One area of Aboriginal medicine with no obvious western parallel was baby medicine. Newborn babies were steamed or rubbed with oils to render them stronger. Often mothers were also steamed.

Almost any kind of plant, animal or mineral product could serve as a medicine (see "The Plants" and "The Animals"). A notable feature of Aboriginal medicine was the importance placed upon oil as a healing agent, an importance that passed to white colonists, and is reflected today in the continuing popularity of goanna oil.

Earth, mud, sand, and termite dirt were often taken as medicines. In the Channel Country, healing mud for packing wounds was taken from the cold beds of waterholes. In many parts of Australia wounds were dressed with dirt or ash. Arnhem Land Aborigines still eat small balls of white clay and pieces of termite mound to cure diarrhoea and stomach upsets. Clay and termite earth probably share the properties of kaolin, which is the white clay used in western medicine. They may also provide essential nutrients: some termite mounds are extraordinarily rich in iron — as high as two per cent. But whether this can be absorbed through the stomach has yet to be determined.

This chapter has focused only upon Aboriginal medicine, but some mention should also be made of the islanders of Torres Strait, whose medical tradition, as far as we can tell, was very different. A recent article, *Torres Strait Medicines*, compiled by Margaret Norris, listed remedies prepared from coconut oil, sea almond fruit (*Terminalia catappa*), beach convolvulus (*Ipomoea pes-caprae*), frangipanni leaves and sap, chilli leaves, and dugong oil. The use of exotic frangipanni and chilli leaves suggests a very changed tradition, probably influenced by the waves of early missionaries from the South Pacific.

◀ The unripe seed of the sea almond (*Terminalia catappa*) is chewed on Torres Strait islands as a remedy for "itchy throat" and colds. The ripe seed is edible, and tastes like almond. Photographed at Evans Bay, Cape York.

▲ Lemon-scented grass (*Cymbopogon ambiguus*), according to the Northern Territory Pharmacopoeia project, "is regarded as an important and potent medicine. It is one of a few plants which are dried for storage when necessary rather than always used fresh." Photographed at Devil's Marbles, Northern Territory.

ABORIGINAL MEDICINE 29

30   PEOPLE & PLANTS

COLONIAL MEDICINE

Life for the pioneering bushman, though rough and simple, was healthier than life in the towns and diggings. Here infectious disease, aided by intemperance and reckless habits, ran rife. But when sickness did strike the campfire, the ailing bushman, wrapped in his opossum skin rug, had no recourse to the town druggist. He was dependent upon simple remedies, often herbal cures from the bush, either self-invented or learned from sympathetic Aborigines.

A different kind of herbalism was practised by pioneering families who lived on farms and the edges of country towns. Many immigrants, especially the older women, had knowledge of European herbs. When they found similar-looking plants growing on Antipodean soil, they put them to the same uses. Some of these plants, including wild raspberries, centaury and native mints, even came to be prescribed by the more adventurous of country doctors. Interestingly, most of these plants were "herbs" in the literal botanical sense — small plants without woody

◄ Australia's pioneers lived in rough huts made from materials carved from the bush. When they fell ill they also depended on remedies from the bush. Photographed by Richard Daintree at a miners' camp in north Queensland.

▲ Dock was the bushman's answer to nettle stings — the juice eased the itch. According to Mary Gilmore it was also applied as a poultice. Illustrated is a curled dock (*Rumex crispus*) growing at Marrickville in Sydney surrounded by other medicinal weeds — fennel in the foreground and castor oil bean and green cestrum against the fence.

stems. Aboriginal medicine in comparison made much greater use of trees, shrubs and vines.

Another kind of colonial medicine was underway in the capital cities, where highly educated botanists, chemists and doctors investigated the chemical properties of bitter barks, astringent gums and aromatic leaves. These imaginative men were largely motivated, not by the need to treat illness, but by the quest to develop products for commercial development and export — for the good of Australia and the British Empire.

The herbs used by pioneer woodcutters, shepherds, boundary riders, farmers' wives and wandering merchants are now largely forgotten. We can be sure there were many more remedies than the several dozen we have on record today. Fortunately the plants tested by chemists and doctors are better known, for their chemical analyses were published in colonial journals, and their virtues proclaimed in museums and intercolonial exhibitions.

Author Dame Mary Gilmore, who grew up near Goulburn in the 1870s, evocatively describes herbalism on a country farm in her memoirs *Old Days, Old Ways*:

Up by the barns and cow-yards there were nettles for the blood, horehound for coughs and colds, and dock for poultices. But dock was used like horehound and nettles, for beer; sometimes it was wrapped round tough meat with the idea of making it tender. In the thick, unfelled bush above the horse-and-cattle yards were native hop, "sarsaparilla," the bottle-brush flower of the wild honeysuckle, together with geebungs, wild cherry, eucalyptus, wattle, kurrajong, and pine. The wild hop made yeast; the "sarsaparilla" made naughty little boys good by clearing their "over-crowded" blood; the bottle-brush soaked in soft water yielded syrup for sore throats and colds; the wattle-bark the aborigines had taught us to make into a tan lotion for unbroken burns and scalds; the eucalyptus (also native teaching) made vapour in pits, or in bed, for chills and pains; the pine, too, was inhaled ...
Into the big kitchen with its cedar tables black with age, all these came; and what was not simmered in the great cauldron, or brewed in a three legged pot, went into the brick oven; there to soak, there to steep, or there to slowly dry.

Gilmore's backyard pharmacopoeia is a charming blend of European, Aboriginal and original remedies. Her horehound (*Marrubium vulgare*) is the same used medicinally in England; it became a widespread weed in Australia, spread about largely by herbalists. Her nettle (*Urtica*) and dock (*Rumex*) are either European species growing here as weeds, and used in the same ways, or native species of similar looks. Her sarsaparilla (*Smilax glyciphylla*) is a native vine closely related to the true sarsaparilla of the West Indies, although Australian colonists used the leaves, not roots, as a tonic. Native hop

▼ Mary Gilmore's "wild cherry" was the native cherry (*Exocarpos cupressiformis*), a cypress-like tree with tiny edible fruit, popular among colonists. According to Dr Lauterer in 1892, the twigs "prove as good a bitter tonic and astringent as the South American Rhatany". Photographed near Mt Lofty.

◀ The intensely bitter taste of hop bush leaves (*Daviesia latifolia*) persuaded pioneers that it was a wondrous medicine. Joseph Maiden praised the plant as "a useful tonic bitter, and therefore a readily available substitute for gentian for country people." Photographed near Wallangara.

COLONIAL MEDICINE

(*Dodonaea*) is unrelated to English hops, but the similar-looking pods were used to flavour beer. The honeysuckle (*Banksia*) is a traditional Aboriginal food that was adopted as a throat medicine presumably because it tastes syrupy. The wattle bark, eucalypts and perhaps the pine (*Callitris*) were remedies learned from Aborigines.

In the forests away from townships and diggings there were few wild plants in the image of English herbs, and bushmen were obliged to invent their own remedies or learn from the Aborigines. Woodcutters made tonic teas from the more aromatic leaves or the bark of trees they were felling. Other plants were chosen largely at random, as the following anecdote by Matthew Butler, J.P., demonstrates:

▶ Joseph Maiden was scornful of a patent medicine made from sea box (*Alyxia buxifolia*): "wonderful were the virtues claimed for it on the prospectus. I do not doubt that these claims are every bit as valid as those of quack or patent medicines in general." Photographed on Wilson's Promontory.

On the 24th December, 1894, I was sent for to make the will of an old man who was, as he thought, dying from rheumatism. In a fit of abstraction he pulled up the root [of *Tephrosia varians*] **and ate it. Fancying it gave him relief, he pulled more, boiled it and drank the liquor. Within a week there was a marked change in him, and now he is quite well and looks ten years younger . . . I had a slight touch of rheumatism in the leg and tried a decoction of the root with the result that the pain has gone and the stiffness is wearing away.**

There were probably dozens of bush medicines learned from the Aborigines. Among the better known were dysentery bush (*Grewia retusifolia*), caustic bush (*Sarcostemma australe*), sacred basil (*Ocimum tenuiflorum*), western bloodwood (*Eucalyptus terminalis*), and rock fuchsia bush (*Eremophila freelingii*). A handful of remedies were even learned from Chinese settlers.

▶ Dysentery bush (*Grewia retusifolia*), wrote Walter Roth in 1901: "Around Normanton, etc., is used by (both whites and) blacks for dysentery: the leaves are either chewed or made into a decoction." This is one of few remedies taken by both blacks and whites.

◀ Aborigines employed at homesteads sometimes shared their traditional remedies with their employers. The popularity of goanna oil among settlers can probably be attributed to this teaching. Illustrated is Widgee Homestead in southern Queensland.

COLONIAL MEDICINE 35

Sometimes bush remedies were processed, packaged in bottles and hawked from town to town as cure-alls. Goanna oil, made from goanna fat and secret herbs, and eucalyptus oils, were the most famous of these, but there were others, including herbal "cures" for snakebite, "bitters" of quinine tree (*Alstonia constricta*), a "Magic Ophthalmia Cure" prepared from sneezeweed (*Centipeda minima*), "Ti Ta", supposedly a blend of a tree, moss and fern, and "Austral Marine Bitters", made from sea box bark (*Alyxia buxifolia*).

A popular myth, of English origin, was the idea that stinging plants grew beside their cure. Nettle stings were treated with the sap of docks, both in England and Australia, where the two plants often sprout side by side. Bushmen adapted this idea by treating stinging tree stings (*Dendrocnide*) with cunjevoi sap (*Alocasia macrorrhizos*). Stinging trees and cunjevois look like giant nettles and docks, and grow together in rainforest clearings.

The medical investigations of city-based botanists, chemists and doctors were conducted with scientific objectivity, and with some scepticism towards bush nostrums. These learned gentlemen were prey to one romantic notion, however, the idea, as put by colonial botanist Joseph Maiden that "Native Australian drugs will probably be found peculiarly efficacious in the treatment of diseases, or modifications of diseases, which are coextensive with their distribution".

▲ The father of Queensland botany, Frederick Manson Bailey, wrote six volumes on Queensland flora, and an 1881 paper on the medicinal plants of Queensland. Regrettably, he recorded almost nothing of Aboriginal medicine. His portraits always present a severe countenance.

▶ Cunjevoi's (*Alocasia macrorrhizos*) curative powers were praised by Queensland colonial parliamentarian W. Pettigrew, who exhibited specimens before the Royal Society in 1885, claiming that if the leaf was rubbed on stinging tree stings, "the pain will cease and not return again". Photographed at Cunningham's Gap, Queensland.

Said Maiden, "This is an important matter which I have often heard referred to by medical men and others".

But ironically, the plants these men studied the most, the eucalypts, seemed to illustrate the inverse principle. Australia's relative freedom from fevers and agues was attributed to the prevalence of these trees.

Colonial scientists believed that Australia, like every colony, harboured in her forests wonderful new foods, fibres and drugs. The government botanists of the day, Baron Ferdinand von Mueller of Melbourne, Joseph Maiden of Sydney, and Frederick Manson Bailey of Brisbane, helped classify the native flora, and called for investigations of promising plants, or those closely

▼ Bedazzling berries of cunjevoi are apt to irritate the throat and should not be eaten. They were exhibited in Brisbane, along with the root and leaves, as part of an 1885 address on the therapeutic merits of cunjevoi leaves by the honourable W. Pettigrew, M.L.C.

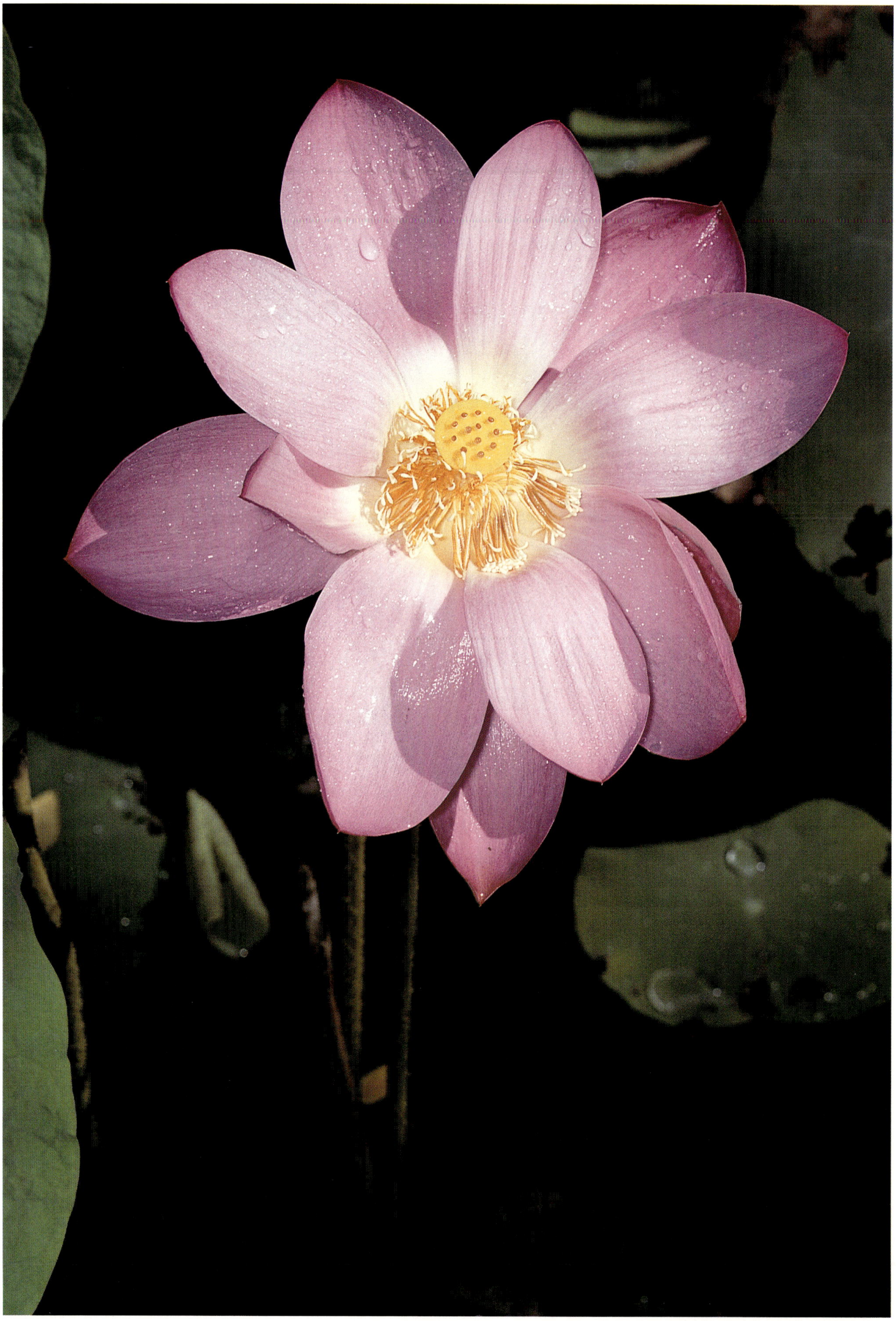

related to overseas medicine plants, such as "Australian ginseng" (*Panax*). Von Mueller, Maiden and Bailey's names appear throughout the literature on plant medicines, alongside those of Melbourne chemist Joseph Bosisto, and Brisbane doctors Thomas and Joseph Bancroft and Joseph Lauterer, who did much to unlock the chemical mysteries of native plants.

Products investigated by these men ranged from the bitter barks of rainforest trees and the toxic leaves of shrubs, to the oozing gum-like kinos of eucalypts and the frightfully toxic leaves and saps of shrubs. Few of these were Aboriginal remedies. Scientists were then unaware of the hundreds of herbs used by Aborigines in central and northern Australia. Bitter barks were eagerly studied for it was known that bitter willow bark (containing natural aspirin) from Europe and cinchona bark (source of quinine) from South America cured fevers.

Another focus of interest was plant poisons. Colonial botanists knew from overseas plants such as hemlock and castor oil bean that medicines were often poisons in sub-lethal doses. Poisonous plant extracts were injected into dogs, cats, rats or frogs, or mixed with sugar and fed to insects, to see what would happen. Much of the colonial literature on medicine is taken up with descriptions of dying animals. Thomas Bancroft, in one paper alone, told of asphyxiated kittens, frogs vomiting for hours on end, convulsing cats, and guinea pigs lying helpless uttering feeble cries.

Although sceptical of Aboriginal medicine, colonial scientists placed great store in the materia medica of India, of which, thanks to the strength of the British Empire, they were well-informed. Wrote Maiden: "much of the knowledge in regard to it is exact, the outcome of intelligent observation and inquiry, and the work of the European practitioner to classify the native drugs is a comparatively easy one". Many Indian medicinal plants also grew in northern Australia, and colonial botanists called for the investigation of these and other plants used in Asian and South Pacific countries; very little came of this.

In retrospect, there was no reason to believe that Indian folk herbalism was any better than Aboriginal medicine, and we may wonder why Maiden placed any faith in Indian claims that cattle bush (*Trichodesma zeylanicum*) treated snakebite, and chaff flower (*Achyranthes aspera*) cured mad dog bites, hydrophobia, and warded off scorpions.

The plant discoveries of scientists were paraded in grand colonial museums. Bailey curated the Museum of Economic Botany in Brisbane, and Maiden helped catalogue the Technological, Industrial, and Sanitary Museum of New South Wales. (Maiden's catalogue was later expanded into his classic book, *The Useful Native Plants of Australia*, published in 1889.) The Technology museum, founded in 1800, displayed everything the colonial mind considered useful, ranging from whalebone and dental appliances

◀ "Our beautiful pink water lily," wrote Bailey in 1881 of the sacred lotus (*Nelumbo nucifera*) "is not without medicinal qualities, for we are told by Endlicher, that the milky viscid juice of the flower-stalks and leaf-stalks is a remedy in India against sickness and diarrhoea."

▶ No place for children, the Brisbane Museum of Economic Botany was a staid display-house of woods, barks and plant powders suitable for commercial exploitation, rather than a hall of recreation. Photographed at the turn of the century.

to hydraulic cement and water-closets. Public health was an important issue in Maiden's time, and many of the papers on plant constituents appeared in Royal Society journals alongside articles on the utilisation and disposal of excreta and the advantages of burning the dead.

Apart from these permanent museums, and live museums in the form of state botanic gardens, Australian medicinal products were exhibited at various intercolonial exhibitions in Australia and Europe. Australian entrepreneurs sent sample medicines, hoping to win markets in Europe. At the Colonial and Indian Exhibition in London in 1886, George Wickham of Brisbane exhibited kangaroo sinews "applicable as Sutures in surgical operations", and Joseph Bancroft displayed nine plant products, including tincture of pituri (*Duboisia hopwoodii*). The Queensland Court of the Centennial International Exhibition of Melbourne in 1888 displayed dugong oil and ointment, essential oils of eucalypts and tea trees, various medicinal barks, and "Natural Mineral Water, from the spouting springs at Helidon".

Despite the optimism of the age, very few of the medicines developed by colonial scientists secured a future in commerce. Eucalypt oils and kino, sandalwood oil, wattle and quinine tree bark, corkwood leaves and dugong oil were sold overseas for a while, before they eventually faded away.

▼ A credulous Joseph Maiden recorded that chaff flower (*Achyranthes aspera*) "is administered in India in cases of dropsy. The seeds are given in hydrophobia, and in cases of snake-bites, as well as in ophthalmia and cutaneous diseases." Photographed on Alpha Rock, Torres Strait.

Today, the legacy of Maiden, von Mueller, Bancroft and friends looks meagre. Australia is a major exporter of corkwood (*Duboisia myoporoides* × *leichhardtii*), a product developed by Thomas Bancroft, but much of the production is controlled by German companies. Eucalypt oils are important in world commerce, but the major suppliers are overseas. Exports of tea tree oil are as yet insignificant. If they were alive today, Maiden, von Mueller and Bancroft would be shocked by Australia's dependence upon overseas drug firms, and appalled by the failure to develop the potential of Australian medicinal plants.

▲ Of cattle bush (*Trichodesma zeylanicum*), Joseph Maiden in 1889 had this to say: "In India this, with other species, is considered diuretic, and one of the cures for the bites of snakes." Photographed in the Olgas after rain.

42   PEOPLE & PLANTS

IN THE EUROPEAN TRADITION

To the first pioneers, Australia was a topsy-turvy land of freakish animals and peculiar plants. In the scraggy forests everything looked unfamiliar, and many homesick colonists pined for the oaks, meadows and songbirds of home.

But as settlement unfolded and colonists ventured farther afield, they began to come across plants they knew. Growing in forest glades were clumps of bracken, wild daisies, and a swag of plants that resembled the medicine herbs of Europe. Immigrants versed in herbalism found that these plants served as remedies, and so arose in Australia a transplanted tradition — a harnessing of European ideas to Australian plants.

The plants used in European-style herbalism were of three kinds. Some were native Australians of close affinity to European herbs. This was possible because Australia's flora was not as topsy turvy as people first thought — many plant genera are shared by Australia and Europe.

Other plants proved to be identical to their

◀ Pink-flowered native raspberry (*Rubus parvifolius*) impressed William Woolls, who in 1867 suggested that this bramble, "the valuable properties of which have been discovered by the old women, may one day claim a place in the Pharmacopoeia." Photographed in New England National Park.

▶ In an 1860 report on the resources of the colony of Victoria, native hollyhock (*Lavatera plebeia*) was mentioned as a plant whose root had been "brought into practical use instead of Althaea [European marshmallow]". Photographed in Palm Valley.

European counterparts. Nineteenth-century botanists, and indeed botanists today, are mystified by the worldwide spread of such plants. We can only guess that seeds were carried from continent to continent long ago by voyaging birds or freak winds.

The third category of medicinal plants are those English herbs that were accidentally brought out on immigrant ships and consequently sprouted in Australia as weeds. By the mid-nineteenth century there were dozens of these.

Probably the earliest of the Australian herbs to be utilised by colonists were the raspberry plants (*Rubus*). There are six native species and at least two were taken as medicines. Leaves of the roseleaf raspberry (*R. rosifolius*), a scrawny scrambler of wet forests, were brewed into a tea, which was drunk for menstrual pains, morning sickness and labour pains. European raspberry (*R. idaeus*), used in the same way in England, contains a uterus-relaxing substance. To treat diarrhoea, country people turned to the pink-flowered native raspberry (*R. parvifolius*), a prickly creeper of open forests. Aborigines at Shoalhaven Bay also treated diarrhoea in this way, drinking a tea of either of these plants or of molucca bramble (*R. moluccanus*).

The European herb heartsease (*Gratiola officinalis*) is replaced in Australia by five native species, two of which — stalked brooklime (*G. pedunculata*) and Austral brooklime (*G. peruviana*) — found their way into "domestic medicine". Joseph Maiden noted in 1889: "A decoction of these plants is used by people in the Braidwood district (New South Wales) for liver complaints with (many say) good results". Brooklimes were also used as tonics, and in western New South Wales, Aborigines supposedly took stalked brooklime as a purgative.

In Victoria, native hollyhock (*Lavatera plebeia*) was used in place of English marshmallow (*Althaea officinalis*) by colonists who could not tell the difference. Colonial hunter Horace Wheelwright declared, "A poultice made of the leaves and stalk of the marsh mallow, which in many places here grows wild, and is the most valuable plant in the bushman's herbal, is an excellent remedy for cuts, bruises, swellings, & c".

Botanist Baron von Mueller (who did know the difference) wrote: "The leaves of our plant can be used for emollient poultices, just like those of the English Marsh-Mallow, and also the roots of ours form a fair substitute for the officinal *Althaea* root."

The European flax plant (*Linum usitatissimum*) is the source of textile flax, which is spun from the stems, and linseed oil, expressed from the seeds. Australia is home to a native flax (*L. marginale*), a very similar herb that Aborigines harvested for its edible seeds and string. European linseed oil was employed as a medicine in England, and according to colonial naturalist William Woolls, native flax could be used in its place: "The seeds may be used for mucilaginous decoctions, for expressing oil, for preparing poultices, and for any other uses for which common linseed is employed." Woolls also claimed that the native trailing speedwell (*Veronica plebeia*) was "very similar in properties" to European speedwells (*Veronica*), used traditionally as diuretics and to promote menstruation.

But the most popular of all substitute herbs, judging by the enthusiasm of antipodean writers, was the native spike centaury (*Centaurium spicatum*). European centaury (*C. erythraea*) was a sacred plant of Celtic Druids and a tonic herb of English peasants, though it is not favoured by herbalists today. Australian spike centaury is almost identical, and it became a surprisingly popular rustic medicine, along with the similar-looking yellow centaury (*Sebaea ovata*), a native species, and European centaury which, brought to Australia, soon ran wild. Many colonists drew attention to these herbs.

Naturalist George Bennett wrote in 1860:

A very pretty diminutive plant grows in the fields about Sydney, bearing a pink flower; it is the Australian Centaury [*C. spicatum*], and so closely resembles the European species, that it might be supposed identical, to the eye of all but a botanist. Like the English species, it varies in height, according to situation and soil, from a few inches to more than a foot. The plant is collected by the colonists, who consider it valuable in cases of dysentery and diarrhoea; it is also useful as a tonic and stomachic, like gentian (to which family it belongs): when required for medicinal purposes, it should be gathered in the spring season.

Reverend Woolls in 1867 told of its adoption by the medical profession:

▲ A delicate herb of the forest floor, mountain gentian (*Gentianella diemensis*) was recommended by Baron von Mueller as a substitute for European gentian, a bitter herb prescribed to aid digestion. Photographed on a grassy knoll in Kinglake National Park near Melbourne.

▲ Martha Nevin, an Aborigine from Coranderrk in Victoria, remembered her mother drinking a bitter infusion of centaury (*Centaurium* — possibly *C.erythraea*, illustrated) for bilious headaches. The whole plant, roots included, was placed in a jug and covered with boiling water.

> A learned physician, not a hundred miles from Parramatta, was so impressed with the efficacy of this little herb, from noticing the use of it amongst certain old women in his neighbourhood, that he was not too proud to adopt their remedy and recommend it to his patients. This was an instance of candour in a great mind which deserves to be recorded, for medical men generally are so wedded to what is popularly called "Doctor's Stuff," that nature with all her endearments appeals to them in vain.

In a later article (1887), Woolls mentioned a Dr Campbell of Gladesville Asylum who "had learned from some country people that a decoction [of native centaury] was highly useful in certain stages of dysentery". Noted Woolls, "It was his practise to collect it in the summer, and to keep little bundles ready for use."

In 1898 botanist Joseph Maiden noted its use in Western Australia as an eczema remedy, and had this to say:

> This plant is useful as a tonic medicine in indigestion, liver complaints, diarrhoea, and dysentery. The whole plant is used, and is pleasantly bitter. It is common enough in grass-land, and appears to be increasing in popularity as a domestic remedy; in fact it is the one native plant remedy known to a great many people, and to my personal knowledge it is used in all parts of New South Wales.

Unfortunately native centaury fell from use in the twentieth century and is no longer harvested today. Professional herbalists import small quantities of European centaury, but it too is a largely forgotten herb. Nonetheless there are people alive today who still remember taking the plant as a childhood tonic. Former Burrendong Gardens director George Althofer recalled his mother brewing a decoction to clear away winter colds: "It was the custom in those days to either get a tonic from the chemist or to make one up. They used various things. Horehound, nettles and the native *Erythraea* [centaury] ... My mother used to use the *Erythraea* by just boiling the whole plant, making a little toddy of it and giving it to us about once a week."

Althofer's nettle was either native scrub nettle (*Urtica incisa*) or the introduced small nettle (*U. urens*), a painfully common weed. Nettle decoctions were probably popular blood tonics in the colonial era, made from either species, and they are still prescribed by herbalists today. Aborigines reportedly boiled scrub nettles to make poultices.

Closely related to centaury is the bitter-rooted European gentian (*Gentiana lutea*), an important tonic medicine. Von Mueller declared that Australia's Alps "enrich us also with a thick-rooted gentian", which was "certainly as valuable" as true gentian. The plant in question, mountain gentian (*Gentianella diemensis*) was apparently ignored by herbalists. Nor was there any attempt to follow von Mueller's suggestion to harvest native orchid tubers for medicinal salep.

One other wild plant used in country medicine cannot be identified, but it was evidently a native plant resembling camomile (*Chamaemelum nobile*). Author Janet Hagger was told of this plant, used to treat biliousness, by an elderly nurse: "You'd go and gather it and make it into a drink — it was very bitter. Some people swore by it for this, that and the other. When the country was opened up to sheep they ate all the camomile and you never heard much more about it."

The plants mentioned so far are all Australian natives that were substituted for European herbs. But at least three kinds of native herb needed no substitution for they are the same species as those found in Europe.

Vervain (*Verbena officinalis*) is a coarse weedy plant, common in paddocks and along roadsides, used by European herbalists to treat nervous disorders and increase the flow of urine. Its presence in Australia is something of a mystery, but it must be considered native for it was collected as early as 1802 in central Queensland by Robert Brown, the botanist who accompanied explorer Matthew Flinders. Today it certainly behaves like a weed, thriving in overgrazed pastures and vacant allotments.

Reverend Woolls, in 1867, told of its use in colonial Australia as a tonic: "It has been used in this country as a decoction for giving tone to the stomach, and arresting the incipient stages of consumption, but whether it really possesses any efficacy in such cases is not known". According to botanist Joseph Maiden it was used by "blacks" in western New South Wales for venereal disease, therefore strengthening the idea that it is a native plant.

▲ A popular peasant potherb in England, the nettle was not used much as a vegetable in Australia, though it found a place in herbal medicine. Illustrated is a native scrub nettle (*Urtica incisa*) growing in a rainforest clearing on Lamington Plateau in southern Queensland.

◄ Colonial herbalist H.P. Rasmussen recommended a tea made from horehound (*Marrubium vulgare*) as "a pleasant bitter and also a good tonic, and if mixed with a little honey and a pinch of cayenne pepper, will give instant relief for cough and asthma."

▶ Dandelion (*Taraxacum officinale*) takes its name from the French *dent de lion* meaning lion's tooth. Whether this refers to the jagged leaves, or to the long white root, no-one knows for sure. Photographed at Bondi Beach, Sydney.

Selfheal (*Prunella vulgaris*) and loosestrife (*Lythrum salicaria*), both of which grow near streams in eucalypt forests, do not behave particularly like weeds, and were probably carried into Australia long ago as seeds stuck to water birds. Both plants were used in Europe as astringents but have fallen from herbal fashion. Queensland botanist F.M. Bailey observed in 1881 that the "common purple Loosestrife of Britain enlivens with its beautiful purple flowers many of our swamps. The whole plant is astringent, and has been recommended in inveterate cases of diarrhoea. It has also been used in tanning."

Herbal practice in Australia was not of course restricted to the use of native plants. Herbalists imported all kinds of dried herbs and herb seeds from overseas. Some of the herbs they grew, such as St John's wort (*Hypericum perforatum*) and horehound cast their seeds into nearby paddocks and wastes and soon ran riot as weeds. Other began appearing on farms and paths from seeds brought in accidentally in potting soil and crop seed. Among these were many familiar

▼ Its name may be a misnomer, for selfheal (*Prunella vulgaris*) is not highly regarded by herbalists today, having largely lost its healing reputation. Cool damp forests are the habitat of this attractive herb, a relative of the mints. Photographed in Morton National Park, west of Wollongong.

backyard weeds such as dandelions (*Taraxacum officinale*) and docks (*Rumex*). Australia soon developed a rural flora rich in European herbs.

The dandelion was sown here last century as a medicine, though most herbalists would have gathered their own from the wild. A popular diuretic and laxative, dandelion was prescribed by herbalist H.P. Rasmussen, along with other herbs, to treat dropsy and epileptic fits.

The most important weed in colonial medicine was horehound (*Marrubium vulgare*). The crinkly bitter-tasting leaves were infused for colds, asthma and other lung complaints, and even prescribed to treat hysteria and Barcoo rot. Colonists drank therapeutic horehound tea, non-alcoholic horehound beer, and even sucked on horehound toffee. Many colonial companies produced horehound health products, gathering their raw herb from the wild. A few companies still brew horehound beer. Pioneers planted the herb in homestead gardens and it spread into nearby sheep paddocks. By 1906 it had become a devastating weed and is now declared noxious in four Australian states.

IN THE EUROPEAN TRADITION

In colonial times physicians "of the highest standing" removed rodent ulcers (sun cancers) by dabbing them with the caustic milky sap of the petty spurge (*Euphorbia peplus*), a common garden weed. Herbalists have known of this corrosive latex since the time of Galen in the second century. In 1976 in the Medical Journal of Australia, dermatologists Weedon and Chick reported that daily application of the sap for five days removed a basal cell carcinoma (the common form of sun cancer with raised surface and pearly edges), leaving no residual scar.

As a treatment for asthma, people turned to the common thornapple (*Datura stramonium*), a poisonous plant supposedly once taken by

▲ In 1890 two men camped near Jerilderie were poisoned by the common thorn-apple (*Datura stramonium*) after drinking a decoction of the plant, believing it to be marshmallow. They recovered in hospital. Photographed beside a stream in the Warrumbungles.

▶ In 1867, William Woolls wrote of scarlet pimpernell (*Anagallis arvensis*): "These pretty little plants were formerly used in mania and hydrophobia, and although perhaps their virtues have been overrated, yet we are assured on good authority that three drachms... are sufficient to kill a dog."

European witches. The leaves were smoked, or boiled for an hour and the steam inhaled. Alkaloids in the leaves dilate the bronchioles, much like a modern bronchodilator. In 1920 Joseph Maiden suggested selling medicinal thornapple cigars:

A rough article could be made for domestic use, or the leaves may be merely smouldered (with or without the aid of saltpetre) and the smoke inhaled. As regards the sale of these medicinal cigars, the revenue authorities offer no objection to them, provided they are made wholy of *Stramonium*, and are not so made up and advertised as to lead the public to believe that they are tobacco.

The wild blackberry (*Rubus fruticosus*), a popular source of fruit for jam-making, also served as a remedy for looseness of the bowels. Rasmussen recommended blackberry tea as a wash for wounds, "as it greatly tends to draw the injured parts together".

Many more weeds from Europe were used as medicines, including, apparently, plantain (*Plantago major*), scarlet pimpernel (*Anagallis arvensis*), cleavers (*Galium aparine*), chickweed (*Stellaria media*) and stinkwort (*Dittrichia graveolens*). Today herbalists still gather weeds. Australian herbal companies employ "wild crafters" to gather many of the herbs they sell. Included in the harvest are such well-known weeds as

► Colonial chemists investigated the constituents of Australian sundews (*Drosera*), but they were not deployed medicinally. Joseph Maiden lamented: "Australia is the land of *Droseras*, but I am not aware they have ever been utilised for medicinal purposes." Photographed at Girraween National Park, Queensland.

▲ In inner-city Sydney, pellitory (*Parietaria judaica*) is one of the most common weeds, found thriving even amid the syringes and squalor of King's Cross. Herbalists gather the sticky leaves (from cleaner locales) as a substitute for European pellitory.

nettles (native and introduced), variegated thistle (*Silybum marianum*), shepherd's-purse (*Capsella bursa-pastoris*), mullein (*Verbascum thapsus*), chickweed, fumitory (*Fumaria officinalis*), horehound, and yellow (curled) dock (*Rumex crispus*).

Some of the wild plants gathered for sale are substitutes, just like the native substitutions of last century. Inkweed (*Phytolacca octandra*) is sold in place of the purgative pokeweed (*P. decandra*), which it closely resembles. The pellitory (*Parietaria judaica*) gathered in Australia is not European pellitory, though many herbalists seem unaware of this. An interesting substitution is the experimental prescription of native sundews (*Drosera*) in place of European *D. rotundifolia*, which has been so overharvested that it is now rare and not available for import.

Kerry Bone of Mediherb, who supplied me with the list of gathered herbs, said that some are difficult to obtain locally, and have to be imported. A special problem is the variegated thistle, a prickly noxious weed with liver-restoring seeds, praised by Kerry as a herb for the modern age. "We use all we can get," he said. "We've tried to get farmers to grow it but they won't. They've got a thing against growing thistles."

▲ Finger-staining berries of inkweed (*Phytolacca octandra*) can be crushed into an inkwell for writing; the roots are prescribed by Australian herbalists for rheumatism. Native birds eat the fruits and spread the seeds. Photographed at Beenleigh, Queensland.

54 PEOPLE & PLANTS

WEEDS AS HERBS

Australia's flora is much richer now than it was two hundred years ago, thanks to the influx of many thousands of exotic weeds. In gardens, paddocks and along roadsides and streams, these foreign invaders flourish at the expense of native plants.

Weeds grow best near human habitation; they have been doing this for thousands of years. During this long association of people and plants, hundreds of weeds have been adopted by peasant cultures into medicine. Dozens of village remedies for every ailment known grow in Australia. For those with the eyes to see, every flush of inner-city weeds is a medicine-chest of pharmaceutical riches.

There are so many such weeds that only a few can be considered here. Of special interest are those that featured in Aboriginal and colonial medicine, and others that are important to modern herbalists or medical research. Most of the remainder have no proven medicinal power.

South American billygoat weed (*Ageratum houstonianum* or *A. conyzoides*), a fluffy-flowered

◀ Queensland botanist F.M. Bailey wrote in 1881: "None of our naturalised plants have spread with greater rapidity over the colony than the castor oil plant, *Ricinis communis*, Linn. and if all accounts are true, few plants are of more value." Photographed beside the Brisbane River.

WEEDS AS HERBS 55

◀ While living at Daintree Mission, "King" Toby of the Bloomfield River tribes told of using mashed billygoat weed (*Ageratum*), a South American weed, to heal wounds. This remedy must be of recent origin. Illustrated is *A. houstanianum*.

weed found in neglected gardens and damp paddocks in eastern Queensland, is especially significant as one of several overseas plants adopted by Aborigines as medicines. On the Bloomfield River and in rainforests near Innisfail, Aborigines crushed the leaves and stems and used them as dressings for sores. Esta Ziviani, who grew up among Aborigines at Erradunga near Innisfail, remembers Aborigines using this plant, which was known locally as "blue top": "They'd crush up the juicy younger leaves, squash it all up to get the juice out and tie it on to any infected cut or scratch," she said.

Esta's family were close friends of the Aborigines and they also adopted this remedy: "Mother was using it on me for cuts and sores and scraped knees. The men used it too. My father had eight or ten men working in the bush for him cutting timber and when they got cuts and scratches that's what they used on them." These weeds are used in exactly the same way by villagers in Nigeria and India; a remarkable achievement for a South American weed.

Another weed used for wounds is the diminutive pennyweed (*Centella asiatica*, formerly *Hydrocotyle asiatica*), a popular healing herb in India, which sprouts in suburban lawns and fields in Australia. It thrives in my back yard. Late last century, among Richmond River pioneers, it was "credited locally as valuable when applied to wounds and sores in the form of a salve or poultice". Brisbane doctors thought very highly of this herb, and its juice was exhibited at the Centennial International Exhibition of Melbourne in 1888.

▶ Pennyweed (*Centella asiatica*), a native garden weed found in eastern Australia, impressed colonial botanist F.M. Bailey, who in 1888 declared: "The medical virtues of this plant have been spoken of by many medical men." Photographed in Brisbane.

Most people have heard of the castor oil plant (*Ricinus communis*), the yielder of purgative castor oil, but few know it runs wild in Australian cities. A sinister-looking weed of urban wastelands, growing two or three metres tall, it throws up tufts of big, grey serrated leaves and spikes of spiny seed-filled pods. In India the plants are farmed and the seeds pressed to extract the oil, an important machinery lubricant (hence "castor" oil). Seeds were once exported to England as domestic laxatives but are now known to be very dangerous, and can kill in doses of two to eight seeds.

In colonial Australia castor oil plant was prescribed experimentally to increase the flow of breast milk, an idea that was probably taken from India. In 1867 Reverend William Woolls reported "two remarkable cases" tested by Dr Pringle, an eminent physician from Parramatta. Noted Woolls:

The one was that of a woman, who from total absence of milk in one breast, and a very limited supply in the other, had lost two children in succession in early infancy. By the application of the Castor Oil leaves for about a week, the effect was truly astonishing, for the evil was remedied, and the woman was enabled to rear her children afterwards.

The second case was that of a delicate lady, who through the same simple application for three days, was rendered capable of supplying the nourishment necessary for twins.

WEEDS AS HERBS

An interesting "cure" for diabetes was reported by an Australian doctor in 1925. Dr W.G. Shellshear of Wallsend claimed a decoction of prickly pear (*Opuntia stricta*) leaf pads "definitely cured" about twenty diabetics. He concluded:

The alleviation of general symptoms due to this condition is really remarkable and from my observations, carried on during the last twelve months, I think that the medical profession ought to thoroughly and scientifically investigate its action. The method of preparation at present being used to extract the substance from the leaf [they are actually flattened stems] is as follows: One pound of leaf. Remove prickles by rubbing with paper. Shred well with a fork. Sprinkle one tablespoonful of bicarbonate of soda; cover with a quart of cold water. Then let the decoction stand all night. Strain through a colander. Strain again through a sieve. Take one wineglassful a day three times a day before meals.
This mixture is unpleasant and slimy to take. If this substance could be extracted more scientifically it could probably be prepared in a way which would make it more pleasant to take. There is not the slightest doubt about its value.

Asthma plant (*Euphorbia hirta*) is another weed that has attracted interest from doctors. A small reddish-leaved herb from Asia, found in bare dirt and pavement cracks in Queensland, it earned a reputation in the late nineteenth century as a cure for asthma. In 1889 Joseph Maiden gave these details of the weed: "The direction usually given by vendors is to simmer one ounce of the dried herb in two quarts of water, and to reduce the liquid to one quart; a wineglassful of this decoction is to be taken three times a day. If the fame of this drug be maintained, doubtless some enterprising pharmacists will present it to the public in a more elegant form."

Medical tests on asthma plant have given conflicting results. The herb appears to dilate the bronchioles but does not always give relief. The herb once earned a place in the British Pharmacopoeia, and doctors have extolled its praises as recently as 1971. Herbalists prescribe it for respiratory complaints and large amounts are imported into Australia from India. Among Queensland and Northern Territory Aborigines this plant, an introduced weed, has earned a reputation as a cure for warts and other ills.

To treat mild forms of coronary heart disease and hypertension, Australian herbalists prescribe a tincture of hawthorn (*Crataegus monogyna*). This prickly shrub, introduced from Europe as a hedge last century, now runs wild in cool farming districts in southern Australia. In numerous experiments conducted overseas, especially in Germany, hawthorn extracts appeared to reduce arterial blood pressure, heart rate, as well as bringing other benefits.

▼ Asthma plant (*Euphorbia hirta*) indicates by its name its function in colonial medicine. Bailey wrote in 1888 that "This common tropical weed is constantly used in the form of tea by persons suffering from asthma, and is said to give instant relief".

▶ The irritating white sap of the asthma plant was applied by Aborigines to remove warts, as demonstrated by Vai Stanton of Darwin, a member of the Koongarunj tribe. Asthma plant is considered an introduced weed, but was so widely used by Aborigines that we may wonder if it is not native to Australia.

▲ Zulus relieve rheumatism by chewing the young leaves of cobbler's pegs (*Bidens pilosa*); elsewhere in the world this weed is taken for dysentery and colds. The young leaves can also be cooked as a vegetable. The small flowers often lack petals. Photographed in Brisbane.

Another weed that is much esteemed by Australian herbalists is the variegated thistle (*Silybum marianum*). Since the time of Dioscorides, two thousand years ago, physicians have valued this thistle as "a friend to the liver". German experiments since the 1960s, resulting in more than a hundred papers, and a 1974 symposium in Cologne helped confirm its liver-restoring function. An extract of the seeds is the only known remedy for the liver destroying toxins of the death cap mushroom (*Amanita phalloides*). Farmers, however, dread this heavily armoured purple-flowered weed, which has been declared noxious by some state governments.

There are many more weeds that are used as medicines somewhere in the world. For example cobbler's pegs (*Bidens pilosa*), from South America, is one of the least loved of garden weeds but is prescribed in different corners of the Third World to treat diarrhoea, coughs, rheumatism and toothache. Chilean green cestrum (*Cestrum parqui*), a smelly shrub with poisonous berries, is used in South America as a cure for just about anything. The plant contains a poisonous alkaloid and a glycoside. Mexican tea (*Chenopodium ambrosioides*) yields oil of chenopodium, and is taken in the Americas to kill intestinal worms. This aromatic weed (it smells like ants) was cultivated in the United States and its de-worming oil sold through Baltimore. The oil is poisonous though, and overdoses have killed both worms and patient.

Weeds from Asia include the garden pest called nutgrass (*Cyperus rotundus*), bearing "nuts" (tubers) prescribed in India, China and South East Asia for digestive complaints, and green amaranth (*Amaranthus viridis*), which is crushed and used as a healing poultice. China and Japan are the homes of the camphor laurel tree (*Cinnamomum camphora*), the original source of camphor (it is now produced synthetically). The tree is proving to be an appalling environmental weed in New South Wales as nothing can eat its camphor-laden leaves.

A number of traditional European herbs grow in Australia as weeds. Dandelions, chickweed, dock and so on are mentioned in an earlier chapter. These are benign herbs, however, compared to the infamous hemlock (*Conium maculatum*). Ancient Greeks mixed it into poison cups for executing political prisoners and it was once prescribed by European doctors as a sedative for nervous disorders. Today it is considered too dangerous to use.

The European daisy (*Bellis perennis*) that sprouts on Melbourne lawns once had a reputation as a wound herb. The noxious shrub gorse (*Ulex europaeus*) supposedly cured jaundice and scarlet fever.

One of the more curious of the European "medicine" weeds is the celery buttercup (*Ranunculus sceleratus*), a pretty yellow-flowered herb found sprouting in bogs in south-eastern Australia. A highly poisonous plant, its irritating juice was used by professional beggars in Europe to raise blisters and open sores to elicit sympathy.

▼ Berries of the hedgerow: In England and Australia handsome hedges of hawthorn (*Crataegus monogyna*) were planted on farms as natural fences. The shiny berries, borne prolifically in summer, were gathered by English peasants for sauces. Photographed near Tenterfield.

60 PEOPLE & PLANTS

BELYUEN'S CHANGING WAYS

Across the harbour from the city of Darwin lies the long, low sweep of the Cox Peninsula. Only six kilometres by sea from the city bustle, the peninsula has escaped the ravages of development and remains largely clothed in its original forest. The only settlement of any size on the peninsula is the Aboriginal community of Belyuen, the home of about two hundred Aborigines, descendants of five tribes who walked in from the south around the turn of the century.

Like any Aboriginal community, Belyuen has its share of older people who remember the bush foods and medicines of long ago. Groups of women go out regularly to gather yams, goannas and other bush treats.

The Belyuen people no longer need bush remedies as the township's medical facilities lie close at hand. But the women take pride in their heritage, and like to remember the traditional remedies, though they are only occasionally practised today.

◄ A sniffing medicine for Belyuen Aborigines, the weeping tea tree (*Melaleuca leucadendron*) also provides a soft papery bark once used for bandaging and bedding, wrapping food, and for making shelters and containers.

▶ Agnes identifies ngarrik (*Croton arnhemicus*), an old-time remedy for headaches, sore joints, cuts and sores. The soft-wooded stems can be fashioned into fire-lighting sticks. Overseas, *Croton* species yield medicinal oils and bark.

▲ A red flush tells of tannins in the inner bark of the ngarrik tree. Agnes specified that only older trees, with pronounced red inner bark, were suitable as remedies. Ngarrik, although a widespread shrub or small tree in northern Australia, has not been recorded as a medicine of other tribes.

One way of keeping traditions alive is by sharing them with outsiders. Belyuen people are used to anthropologists and others coming to ask about the old ways, and when I visited the community, it was easy to arrange a day outing with four of the older women — Agnes Lippo, Maggie Timber Kwakow, Ruby Yarrowin and Alice Djarrug. Beth Povinelli, an anthropologist adopted into the community, accompanied us and spent the day laughing and teasing the women in their own languages.

We bundled into a small truck and sped off along a bush track past groves of cycads, pandanus and ironwood trees. Beth drove, but despite shouts from the women, soon lost the track in a sea of dry speargrass. It was early in the dry season and the coarse grass was thick and tall. The women decided it needed a good burn. While Beth reversed, they walked back along the track flicking matches into the nearby grass. After ominous crackles, great flames burst skyward.

We were driving back slowly when the women decided to stop and show me plants. They leapt from the truck, and before I had time to open my notebook, had handed me the juicy fruits of a vine, *Cynanchum pedunculatum*, and the edible pith of a cabbage palm (*Livistona humilis*). They also hacked open the trunk of a ngarrik tree (*Croton arnhemicus*) to show me its medicinal inner bark.

Ngarrik is a small scrawny tree of tropical woodlands with rough, corky outer bark. Like many medicinal trees (see "Plant Chemistry") it has an inner bark that is dark red. Agnes explained how this was used: "You put it in a billycan over a fire till the water goes red, leave it there till it goes a bit cold, then use it on scabies, sores and swellings. When we don't have medicine in the bush we use that one."

I wanted a photo of the inner bark, but by now the grass nearby was ablaze and I barely had time to check the settings on my camera.

We drove back to a clearing amid cycad and pandanus trees where the women set to with crow bars digging yams (*Dioscorea transversa*). Agnes pointed out edible fruit trees nearby — pandanus, milky plum (*Persoonia falcata*), cocky apple (*Planchonia careya*), and the small dysentery bush (*Grewia retusifolia*). I knew from reading plant lists that all of these species are used medicinally by one or other Northern Territory tribes. Cocky apple and dysentery bush are especially popular remedies. But Agnes and the other women assured me these were only foodplants, not medicine. I noticed Agnes standing near a native lemon grass (*Cymbopogon procerus*), another widely used medicine, and I asked if this was used. It was called "bu", she said. "We crush him, boil him in a billycan, and use him for sores." Ngarrik and bu are both traditional treatments for sores, but ngarrik is evidently the stronger cure. Known to all the women, it was the first medicine they showed me. Bu would not have been identified at all had I not asked about it.

▲ A stowaway from South America, wild passionfruit (*Passiflora foetida*) arrived in the Northern Territory more than a century ago. Many tribes believe it to be a native plant, bestowing upon it Aboriginal names. The sweet aromatic fruit tastes refreshing. Photographed on Cape York.

▲ "Bu" (*Cymbopogon procerus*) is a sweet-smelling lemon grass with a strong reputation for soothing colds and fevers. Photographed at Robin Falls, Northern Territory.

Alice and Ruby showed me the sandpaper fig tree (*Ficus opposita*). Its coarse leaves, armed with silica hairs, are rubbed on ringworm sores, which are then treated with wild passionfruit (*Passiflora foetida*) leaves and fruits. This is one of the more curious of Belyuen treatments, for the wild passionfruit vine is a weed from South America, introduced into the Top End sometime before 1890.

With a bundle of yams, we drove back to Belyuen and collected Marjorie Bil Bil, who told me about baby medicine and the significance of smoking babies over special herb fires, then rubbing them with dugong or emu oil. Marjorie was precise about the effects of the different oils. Both would make a baby stronger, she said, but emu oil would make him run faster (like an emu), while dugong oil quickened the reflexes.

I asked if goanna oil was used for rubbing and Marjorie looked surprised — this was always eaten with the goanna's meat, she said. After thinking over my question, Marjorie said she remembered seeing goanna oil in shops and was vaguely aware of its medicinal reputation. I assured her that emu and dugong oils were just as good.

By now it was late and time to go. The women saw me to the Darwin ferry after offering to take me out again in two days time.

On the second day, after visiting several houses to collect people and equipment, we drove many bumpy miles north of Belyuen to Bagadjat, a small, shady beach fronting a mangrove swamp.

To most Australians a beach overlooking a mangrove mud flat would seem a poor place to picnic, but to the Belyuen people it was ideal. They strode into the mangroves and returned bearing shellfish and mud crabs. By sweeping away a circle of leaves from the clean, white sand, and striking a fire inside a second circle, an instant campsite was made, and one of the crabs was soon cooking. While some of the women went off for more crabs, Marjorie took me to see bush medicines.

The first was wutarr, a small mangrove tree which Marjorie said had poisonous sap. I am not especially interested in poisonous plants and did not bother to take a photo, which seemed to disappoint Marjorie. Beth took me aside and explained that at Belyuen, poisonous plants

◄ To make a quick headache remedy, Maggie roasts a bunch of beach convolvulus leaves gathered from beside the camp. Roasting ruptures the leaf's cell walls, releasing their chemical constituents.

64 PEOPLE & PLANTS

▲ Worth investigating for medicinal compounds is the beach convolvulus (*Ipomoea pes-caprae*), one of Aboriginal Australia's most widely used medicine plants. The Northern Territory Pharmacopoeia project detected saponins in the leaves, but the nature of these has not been determined.

were included under the classification of medicines. They assumed I would be as interested in plants that killed as those that cured.

My second *faux pas* was to ask Marjorie if I could photograph her with bu, the native lemon grass. She had no idea what I was talking about. I repeated what Agnes had said to me, that this was a traditional treatment for sores. Marjorie, in her mid-forties, is younger than the other women and defers to their expertise. We walked back to where Agnes was digging an enormous yam (weighing three kilograms!), and Agnes told Marjorie about this "new" cure. "I might try that one next time," Marjorie said.

Marjorie and I were looking at Nyimara (sandpaper fig) when a raucous pair of black cockatoos flew by. "Where I come from," I said, "people think those birds bring rain."

"Up here," she said, "the women go like this," and placed her hands over her breasts. She said the women believe their breasts will sag if they do not cover them when cockatoos are about. "Even today, some women still do this," she said, smiling.

Back at the campfire, the oldest woman, Maggie Timber, showed me a traditional cure for headache. Taking a bunch of leaves and stems of beach convolvulus (*Ipomoea pes-caprae*), and heating them over a fire, she pressed them to her forehead. There was a fluidity to her motions that showed this was a familiar practice.

We then gathered some eucalyptus bark (*Eucalyptus confertiflora*), which would be burnt to a fine white ash for chewing with tobacco, and drove back to Belyuen. On the way we stopped to look at tea trees (*Melaleuca leucadendron*) — the leaves are a remedy for colds — and to chop down a hollow limb for a didgeridoo. Marjorie invited me to visit Belyuen again in the future, then I thanked the women for an extraordinary day and boarded the ferry to Darwin.

My stay at Belyuen indicated the kind of changes to tradition taking place today in Aboriginal communities. The Belyuen women obviously value their traditional knowledge and draw great pleasure from sharing it, but this knowledge has changed over the years. Two exotic plants, for example, now feature in their pharmacopoeia — the passionfruit vine, and the American ringworm shrub (*Senna alata*). Only a dozen or so native plant medicines are remembered, compared with almost forty on Groote Eylandt to the east. For some ailments, such as the stings of stingrays, no cure is known. (About foodplants, however, the women are much more knowledgeable.)

Some of the Belyuen remedies may have been redeployed over the years. Great morinda (*Morinda citrifolia*) is a well-known remedy for colds and sore throats across much of the Top End. At Belyuen, however, it is a remedy for asthma, along with the meat of flying fox. In all the literature on Aboriginal medicine I can find no other mention of asthma cures, and I suspect that this is a modern ailment, and these are modern cures. Morinda may originally have been a cold remedy. It is also possible that beach convolvulus was once a remedy for marine stings, as it still is in many Top End communities.

The women nearly always concurred about their remedies, but I did witness some discrepancies. Agnes and Alice disagreed about whether tea tree leaves, a traditional treatment for colds, should be boiled or soaked before use (boiling no doubt followed the introduction of the billycan.)

Another discrepancy was in the use of beach convolvulus leaves. When I asked Marjorie to show me how these were applied, she pressed one leaf neatly to her temple. Yet when demonstrating the same remedy, Maggie (the older woman) wound an untidy bundle of leaves and stems around her head. It is easy to see how the details of treatment are progressively lost and distorted when the remedies are rarely used.

The women at Belyuen know that much has been lost. Beth has shown them pictures of trees used for medicines and implements by other communities, and there is an inkling that some of these may once have been used at Belyuen. The women can never recover what has gone, but encouraged by an Australia that is becoming interested in Aboriginal foods and medicines, they will hopefully continue to treasure the knowledge that remains, and act faithfully as custodians to their past.

▶ A remedy remembered from times long past, Maggie holds heated convolvulus leaves to her forehead to show how headaches were relieved.

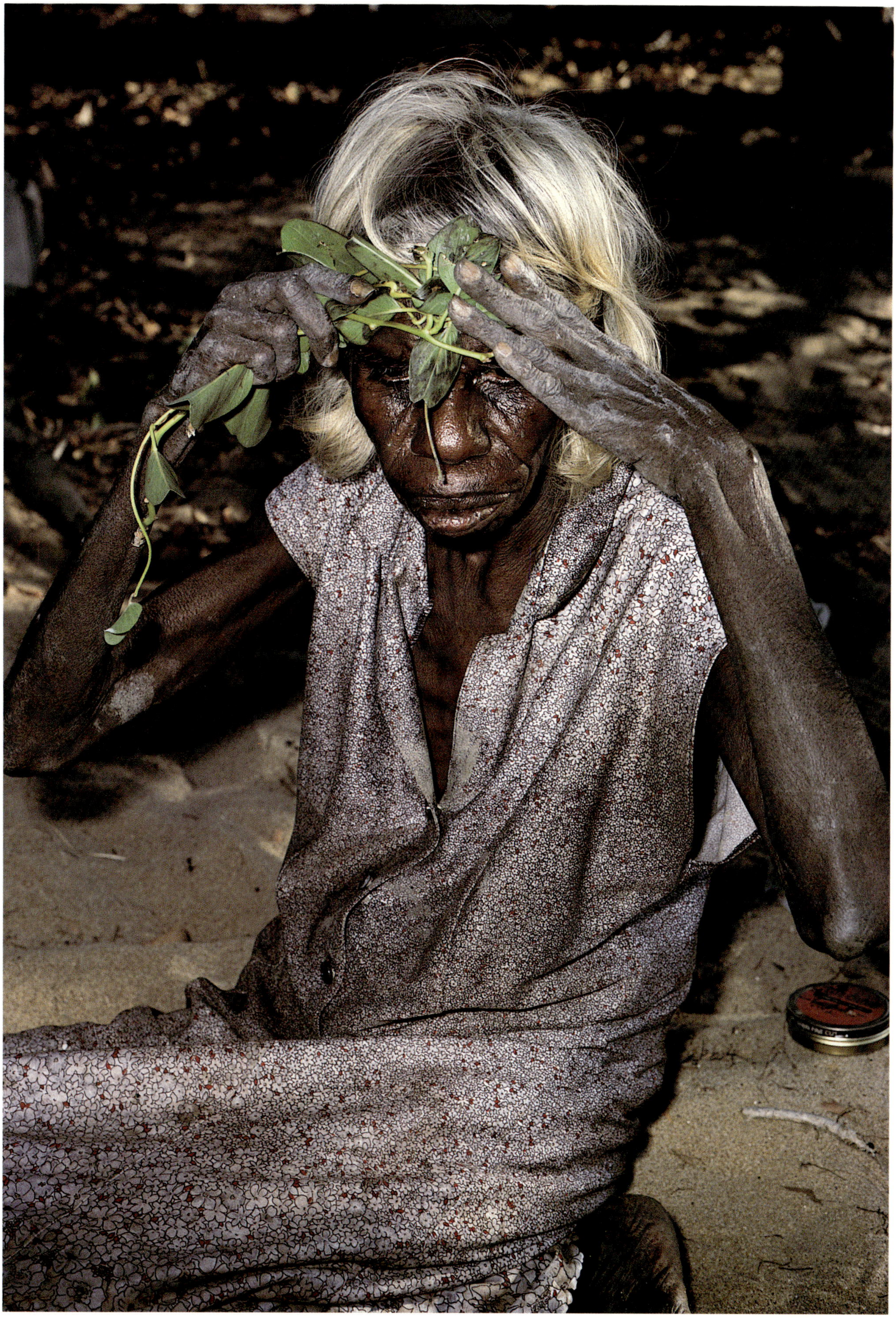

68   PEOPLE & PLANTS

ON THE PITURI TRAIL

◄ Stretching to the horizon and far beyond, this dune south of Bedourie follows the ancient pituri trail, which runs from central Queensland into South Australia. A sandhill wattle (*Acacia ligulata*), one of the plants traditionally chewed with pituri, can be seen as a green patch on the dune crest.

Many generations before Europeans set foot in Australia, great highways crossed the continent from north to south, east to west. Parties of Aborigines trudged along these ancient roads bearing precious items of trade. Seashell ornaments, axe heads and tools were carried thousands of kilometres from their place of manufacture.

The mightiest of the trade highways ran from the ochre mines of the northern Flinders Ranges to Lake Eyre and up the Georgina to the Kalkadoon axe quarries of north-west Queensland. At the very epicentre of this vast trade network lay the coveted dune systems of the north-eastern Simpson Desert, stretching from the Upper Mulligan River to the Upper Toko Range. Here grew the most precious of all trade items in outback Australia, the narcotic shrub pituri (*Duboisia hopwoodii*).

In Queensland pituri is a rare and mysterious plant. Although widespread in the deserts of

▲ According to Joseph Maiden in 1889, the small spurge called caustic weed (*Euphorbia drummondii*) was a remedy of Aborigines in western New South Wales who "use an infusion or decoction of the plant in genital diseases, and use rather strong doses". Photographed at Ayers Rock.

other states, where it was rarely used as a drug, in Queensland it is almost entirely restricted to the inaccessible eastern edge of the Simpson Desert. Botanists have rarely seen it growing. The Queensland Herbarium's vast plant collection boasts a mere five specimens, only one of them collected since 1912, compared with 66 specimens of the related corkwood (*D. myoporoides*).

In February 1989, just before the big drought broke, my friend Steve Page and I journeyed west across the Channel Country to the edge of the Simpson Desert in search of pituri. We were intrigued by the idea that a plant so rare, growing in Australia's harshest desert, could once have been a powerful economic resource. From this small region pituri was traded over 550,000 square kilometres of Australia. I wanted to discover if any Aborigines living in the towns nearby still chewed the drug. Anthropologists generally assume that the use of pituri died with the trade networks long ago. I also hoped to learn if any bush medicines were used by Aborigines or whites in this area.

Our first major stop was Windorah, a small town on the Cooper Creek, which every few years is left stranded when the great channels of the Cooper flood. Windorah is hundreds of kilometres east of where the pituri grows, so I asked in town about bush medicines instead. Paddy Gordon, keeper of the local store, showed me a small spurge (*Euphorbia drummondii*) growing beneath his front steps, and said the sap was used to remove warts. This is one of the better-known of Australian bush remedies. For more information on Aboriginal medicine he suggested I talk to the nurses in Birdsville, and to Jean Smith, the publican at Bedourie, 350 kilometres to the north-west.

Travelling west of Windorah we left the bitumen far behind as the Holden rattled across the baking gibber plains of the Channel Country. At Betoota (population 1) we passed the dusty turn-off to Mooraberrie, the homestead of Alice Duncan-Kemp, who wrote so evocatively of Aboriginal life early this century. Alice often referred to the blacks using pituri, both as a drug, and as a poison thrown into waterholes to stupefy emus, a practise that was forbidden by the station. Alice mistakenly believed that the pituri came from the banks of the Georgina and Mulligan Rivers, an indication that its true location was kept secret.

Three days out of Windorah, Steve and I reached Birdsville, the township that lies beside the Diamantina on the old Aboriginal trade highway. Earlier this century, Aborigines trekked from as far away as Innamincka and Broken Hill, 650 kilometres to the south, to bid for bags of pituri at big markets held here. Today the few Aborigines in town know almost nothing about the drug — or about bush medicines. The town boasts one elderly Aborigine, Linda Crombie, who is versed in traditional ways, but unfortunately she was away during our visit. Locals told me that her sister Clara, who died three years ago, was the last to use pituri. But they could not say how she obtained her supply — the plant grows nowhere near Birdsville. I enquired about bush medicines and the town nurses directed me to the shopkeeper, who suggested I ask Jack Melville, an Aboriginal stockman working at Glengyle Station, 140 kilometres to the north.

Leaving Birdsville, Steve and I turned north along the old trade highway towards Bedourie and Boulia, the two towns closest to the pituri lands. The scalded sand of the Simpson Desert lay beyond the horizon to our west, and treeless plains of gibber crossed by long dunes followed us to the north.

We made camp at a muddy creek close to a stony plain where a colony of bilbies, the endangered outback bandicoot, still survives. Steve discovered he was out of tobacco and I prepared a quick substitute by drying the furry leaves of a native tobacco (*Nicotiana megalosiphon*) growing by the creek. This furnished a rough smoke, but with a pronounced nicotine hit, and Steve was satisfied. In central Australia, where Aboriginal use of tobaccos is well documented, this species was rarely if ever used. Perhaps there is chemical variation between regions.

At this creek, and at other camps, we cooked up pigweed (*Portulaca oleracea*) as an evening vegetable. Early colonists believed this weed prevented scurvy, though recent tests indicate it contains little vitamin C. The flavour is certainly very bland. After dinner we trudged off into the hot night and returned elated having spotted an elusive bilby.

The following day we passed the vast lignum swamps and lagoons that feed Lake Machattie.

▼ The night bloom of the native tobacco (*Nicotiana megalosiphon*). Photographed at Mitchell in southern Queensland.

▲ The native tobacco (*Nicotiana megalosiphon*) thrives on earth degraded by trampling and over-grazing cattle, as here at Sandringham Station just east of the Simpson Desert.

▲ Relieved after two hours spent extricating a dragging muffler, Steve rolls a soothing cigarette of native tobacco (*Nicotiana megalosiphon*). Breakdowns are dangerous on the bleak road from Bedourie to Ethabuka; days go by without passing cars, and temperatures rise above 40° Celsius.

Avocets, stilts, spoonbills and ducks rose from the waters as we passed. A few kilometres further on over a slight rise we came to Glengyle Station, where I asked at the barracks for Jack Melville.

Jack was a gentle old man with rheumy eyes, who told us much about the wild foods of the area. He knew of two bush medicines, a wild "pennyroyal" that grew near water — almost certainly river mint (*Mentha australis*) — and the oil of the goanna, which was good for rubbing in wounds and sores. A fair-sized perenty goanna (*Varanus giganteus*) would yield a tobacco tin full of oil, he said. Jack confirmed that Clara was the last to use pituri, but could not say how she got it. There was certainly none growing locally, he said. That night we camped just south of the station, and with dinner cooked up ruby saltbush leaves (*Enchylaena tomentosa*), a plant used by explorer Sturt as an antiscorbutic (scurvy cure), though I doubt that it contains much vitamin C.

◀ Pigweed (*Portulaca oleracea*), gathered by colonists as a potherb, was believed to be antiscorbutic, though recent tests indicate little vitamin C. A sample analysed by the Department of Defence Support contained only traces of the vitamin. Photographed at St George, southern Queensland.

▶ Dainty button fruits of ruby saltbush (*Enchylaena tomentosa*) provide a tasty treat for birds. Aboriginal children suck on the fruits, and the leaves can be boiled as "bush spinach". Photographed near Goondiwindi, southern Queensland.

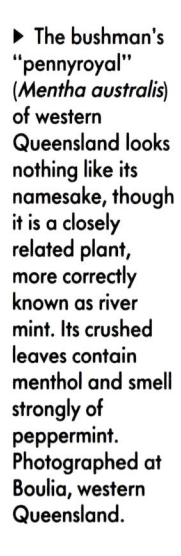

▶ The bushman's "pennyroyal" (*Mentha australis*) of western Queensland looks nothing like its namesake, though it is a closely related plant, more correctly known as river mint. Its crushed leaves contain menthol and smell strongly of peppermint. Photographed at Boulia, western Queensland.

When we arrived in Bedourie the next day we pitched camp by the Georgina River on the edge of town, where I made an interesting find. Wedged in the crotch of a tree was a black basalt axe-head, one that could only have come from the Kalkadoon axe quarries of Mount Isa, 400 kilometres to the north. These sleek axe-heads were traded by the Kalkadoons for pituri at markets north of Boulia. In time, the pituri tribes exchanged some axe-heads, along with pituri, for ochre and other goods traded from the south.

In Bedourie Steve and I sought out the publican, Jean Smith, and asked her about pituri. Like everyone else we had spoken to, she knew nothing. She was certain the few families who made up this tiny country town knew nothing either. But she did know of one medicinal plant, a shrub behind the pub that Aborigines used long ago to treat venereal disease. Unfortunately the shrub had long since died and Jean did not know its name. But she directed us to the home of Gracie Macintyre, a daughter of Linda Crombie, who as a child had travelled with her mother on many desert trips. Gracie knew the shrub, and said she would show us, along with other medicine plants, if we drove out with her the next day. She confirmed that her aunt Clara was the last to use pituri, but like everyone else could not say where it came from.

Gracie took us along a hot sandy road and pointed out the venereal disease plant. It proved to be a fuchsia bush (*Eremophila bignoniiflora*). In central Australia related kinds of fuchsia bush are very important Aboriginal remedies.

more evidence that we were on the pituri trail as Aborigines always chewed their pituri with ash, often made from sandhill wattle leaves. They kept the habit when they switched to white man's tobacco.

Gracie knew of no other medicines, nor did any members of the two other Aboriginal families in town. Nor did any of the white people I spoke to in pubs and stores in the south-west. In this corner of Australia, Aboriginal and pioneer culture is all but forgotten.

Back in the Bedourie pub Jean surprised us by saying her son David might be able to show us pituri. His cattle station Ethabuka took in a stretch of the Simpson Desert near the Upper Mulligan River. This was our first indication that anyone knew what a pituri plant looked like, and where it might grow. We met David that night and he said he knew the pituri plant, and would show us if we came to his station. An Aboriginal stockman, now dead, had shown him the plant years ago.

The road to Ethabuka runs north-west of Bedourie, over baking gibber plains and solitary sand dunes fringed with gidgee. The country

Further on we climbed a sand dune and here Gracie showed me a small spurge (*Euphorbia wheeleri*). She explained: "You boil it up until the leaves go brown, then you throw it out and use the water for sores and scratches". Many kinds of spurge are used by Aborigines in exactly this way, although this species has not been recorded before.

Gracie then took us south of town to see a sandhill wattle (*Acacia ligulata*) growing on the flanks of a high dune. She told us an old Aborigine, Willie Harris, used to chew the ash of the leaves with his tobacco. He also used burnt coolibah bark (*Eucalyptus microtheca*). This was

▲ The spurge (*Euphorbia wheeleri*) was a tribal treatment for sores and wounds, said Gracie, who found me this plant on a sand dune on Kamaran Downs Station, where her brother works as a stockman. Like most spurges, it oozes irritating sap.

*Rarely seen, and never photographed before in Queensland, a pituri shrub (*Duboisia hopwoodii*) perches on a dune in the eastern Simpson Desert on Ethabuka Station. This nondescript desert shrub dominated the lives of Aboriginal tribes living to the east.*

was in the throes of drought, but we passed one green patch where bustards, brolgas and flock pigeons fed. Closer to Ethabuka the land was scalded and many of the scattered gidgees were dead. The edge of the Simpson Desert is harsh country for cattle.

Ethabuka is only about 150 kilometres from Bedourie but we took a full day to get there. The roads are rough and our muffler fell off. David, a man of few words, wasted no time in driving us into the sand-dune country to see what we had come so far for. He took us north, turned east, then we drove up and down and up and down the steep parallel dunes of the Simpson Desert. He pulled to a stop on one high dune crest and there they were: a scatter of scrawny shrubs — pituri!

Steve and I marvelled at the spindly plants, then gaped at the fiery spinifex desert all about us. It was hard to believe that for hundreds of years Aborigines had trekked into such harsh country in search of these plants. Pituri leaves gathered from these dunes had been exported to four Australian states, to be chewed at corroborees thousands of kilometres away.

I asked David if anyone in the area still used pituri. "Clara was the last," he said. "She died three years ago."

"But how did she get her pituri?" I asked.

"I used to send it her," he said, and laughed.

David drove us back through a paddock where a small herb of wild camels galloped away at our approach. His few cattle were surviving on harsh spinifex tops, awaiting rains that had not come for eleven years.

Our last stop was Boulia, a town 217 kilometres north of Bedourie, lying north-east of the pituri fields. Late last century large supplies of pituri were traded into Boulia, according to ethnographer Walter Roth, in exchange for "spears, boomerangs, blankets, nets, and especially red-coloured cloths, ribbons, and handkerchiefs". Boulia is much larger than Bedourie or Birdsville, boasts more than one store, and has a large Aboriginal population, many of them living in "Pituri" Street.

But none of the locals use pituri or know much about it. They cannot get to the desert dunes where it grows. George Quartpot said all the older Aborigines once used it, but police stopped this long ago.

But George was keen to show me his home remedy, a dark liquid made by boiling "dog-

David Smith is probably the only person in outback Queensland who knows what pituri looks like, and without his help we could never have attained our goal. David stands beside a very tall shrub; most were less than two metres tall.

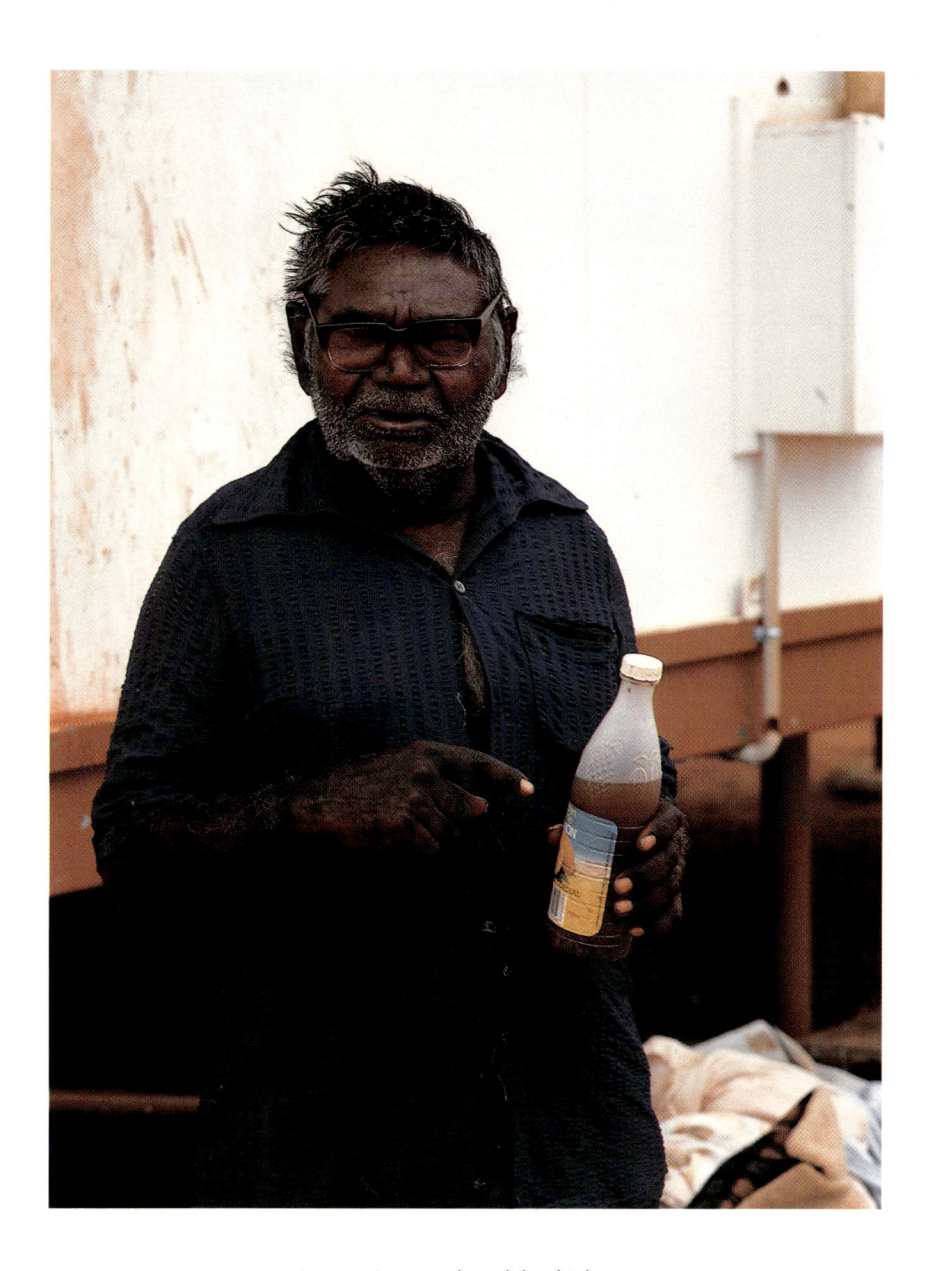

▲ George Quartpot brandishes his home remedy, an elixir of "dogwood" leaves. George told me that when he was once hopping about in pain from stepping on hot coals, he chewed a dogwood leaf, placed it on his foot, and the pain stopped right away.

wood" leaves, which he recommended for pains and burns. "It's like a liniment," he said. "When a haunted place is annoying you, you can't sleep, or dream bad thoughts, this will fix you up. When the people in town are sick they come up here, I give them this." George pointed out the dogwood — it was the fuschia bush Gracie had showed me. I asked how long he had been using it and was surprised to be told he had learnt of it only five years ago, during a visit to relatives in the Northern Territory. This "traditional remedy" was very new in town.

I also spoke to Pĕggy James, another old Aborigine who knew something about bush medicine. She told me she healed up a friend's infected elbow some years ago by sprinkling it with crushed bloodwood gum and applying a bandage. She had remembered this remedy from her childhood (Walter Roth recorded the same treatment at Boulia in 1897.)

Peggy said she knew nothing more about traditional medicine and wandered away. But she returned shortly to tell me of one other remedy she had just remembered: "Wild honey — you put it in sore eyes. You wouldn't think it would do any good, would you? That's all I know. The white people took everything away. They shouldn't have done that, should they?"

76 THE PLANTS

THE PLANTS

Aboriginal medicine presents us with a paradox: up until the 1950s articles written on the subject rarely mentioned plants, yet we now know that more than 200 plant species were used. It appears that Aborigines were very secretive about their remedies, and that those who recorded their ways were insufficiently observant.

But from the wealth of new remedies recorded by anthropologists in recent years, it is now possible to compare plant lists from around Australia and to draw some conclusions about the nature of Aboriginal herbal medicine. Some tentative conclusions are offered in "Aboriginal Medicine"; a survey of the different plant parts used in medicine (both Aboriginal and colonial) is provided here.

Leaves and inner barks were by far the most important of Aboriginal medicines. Aromatic leaves, containing therapeutic essential oils (see "Plant Chemistry") were especially significant, both to Aborigines and colonists. These were usually crushed and inhaled, steamed over fires, infused and drunk, or applied in ointments.

◀ The Kakadu Gagudju roast the hairy tubers of round yam (*Dioscorea bulbifera*) beneath paperbark to create a medicinal smoke, called gundjawuy, which children inhale for strength. On Cape York Peninsula the yam was eaten raw as a contraceptive. Photographed on Cape York.

Colonists often extracted the therapeutic oils in stills. The chapters on Aromatic Trees and Shrubs, Fragrant Herbs, Native Mints and The Mighty Eucalyptus describe many of these remedies in detail. Other kinds of leaves, of large size, were heated and pressed to wounds and sores.

The reddish inner barks of trees are invariably rich in therapeutic tannins, and infusions of these were favoured for many ailments. Outer barks were not much used, except for the powdered ash of desert hakea (*Hakea* — see "Baby Medicine") and honeysuckle oak (*Grevillea juncifolia*), and the fibrous barks of some vines, tied around the body for headaches or as bandages.

Oozing from the trunks of many trees are dark-red, tannin-rich, gum-like kinos, and these were significant medicines of colonists and Aborigines; eucalypt kino was even exported to England. Some paler gums were also used in Aboriginal and colonial medicine, and colonists took sugary manna oozing from gum trees as a laxative.

The milky latex of a number of plants, ranging from herbs to trees, was dripped on to sores and wounds by both Aborigines and colonists (see "Cuts, Wounds, Sores"). Sap from tree orchid stems (*Dendrobium affine, D. canaliculatum, Cymbidium canaliculatum*) was also used in this way.

Roots were rarely used in Aboriginal medicine, apart from the inner root bark of tannin-rich trees. Dysentery bush (*Grewia retusifolia*), Crinum lilies (*Crinum*), pea vine (*Vigna vexillata*), round yam (*Dioscorea bulbifera*), native geranium (*Geranium*) and golden beard grass (*Chrysopogon fallax*) are among the few reported root or tuber remedies.

More than a dozen fleshy fruits were gathered as medicines, including the great morinda (*Morinda citrifolia*), fan flower (*Scaevola sericea*), pandanus (*Pandanus spiralis*), tamarind (*Tamarindus indica*), lady apple (*Syzygium suborbiculare*), stinking passionfruit (*Passiflora foetida*), gardenia (*Gardenia megasperma*), and bitterbark (*Ervatamia orientalis*). Most were applied externally, although morinda fruits were eaten for coughs and colds.

Seeds were a rare item in the Aboriginal medicine chest though the oily seeds of some species were ground into paste and rubbed into the body: beauty leaf (*Calophyllum inophyllum*), woollybutt grass (*Eragrostis eriopoda*), desert walnut (*Owenia reticulata*), weeping pittosporum

◄ Among the few seeds used in Aboriginal medicine are those of beauty leaf (*Calophyllum inophyllum*) as noted by Roth near Townsville in 1903: "Nut broken, kernel triturated with red pigment on stone, mixed with water, and then rubbed all over patient's body, especially where pain is."

▲ Alkaloids and saponins detected in the leaves of *Litsea glutinosa* justify its reputation among Top End Aborigines as a potent medicine, prescribed to relieve nausea, pains, and to soothe sores. This tree of tropical jungles also grows in Asia, where it has a wide reputation as a remedy.

▶ Worth investigating medicinally is the scarlet bracket fungus (*Pycnoporus coccineus*), which the Pintubi and Pitjantjatjara chewed as a remedy. A closely related fungus is prescribed as a medicine in Asia. Photographed at Alice Springs.

(*Pittosporum phylliraeoides*), quandong (*Santalum acuminatum*) and sandalwoods (*Santalum lanceolatum, S. spicatum*). In the Northern Territory young kapok tree seeds (*Cochlospermum fraseri*) were rubbed on to boils.

I can find only a couple of records of Aborigines using flowers in medicine. Nipan nectar (*Capparis lasiantha*) was sipped for colds in Western Australia, and a decoction of kapok tree flowers and bark was drunk for fevers in the Northern Territory.

At least four different fungi featured in bush medicine. Desert Aboriginal babies chewed on scarlet bracket fungus (*Pycnoporus coccineus*) when teething, and puffball spores were used as baby powder. In Arnhem Land the plate fungus (*Phellinus*) was smoked or infused for sore throats, coughs and "bad chest". White people sometimes dressed wounds with "a purple, oval-shaped fungus", probably the puffball *Calvatia lilacina*.

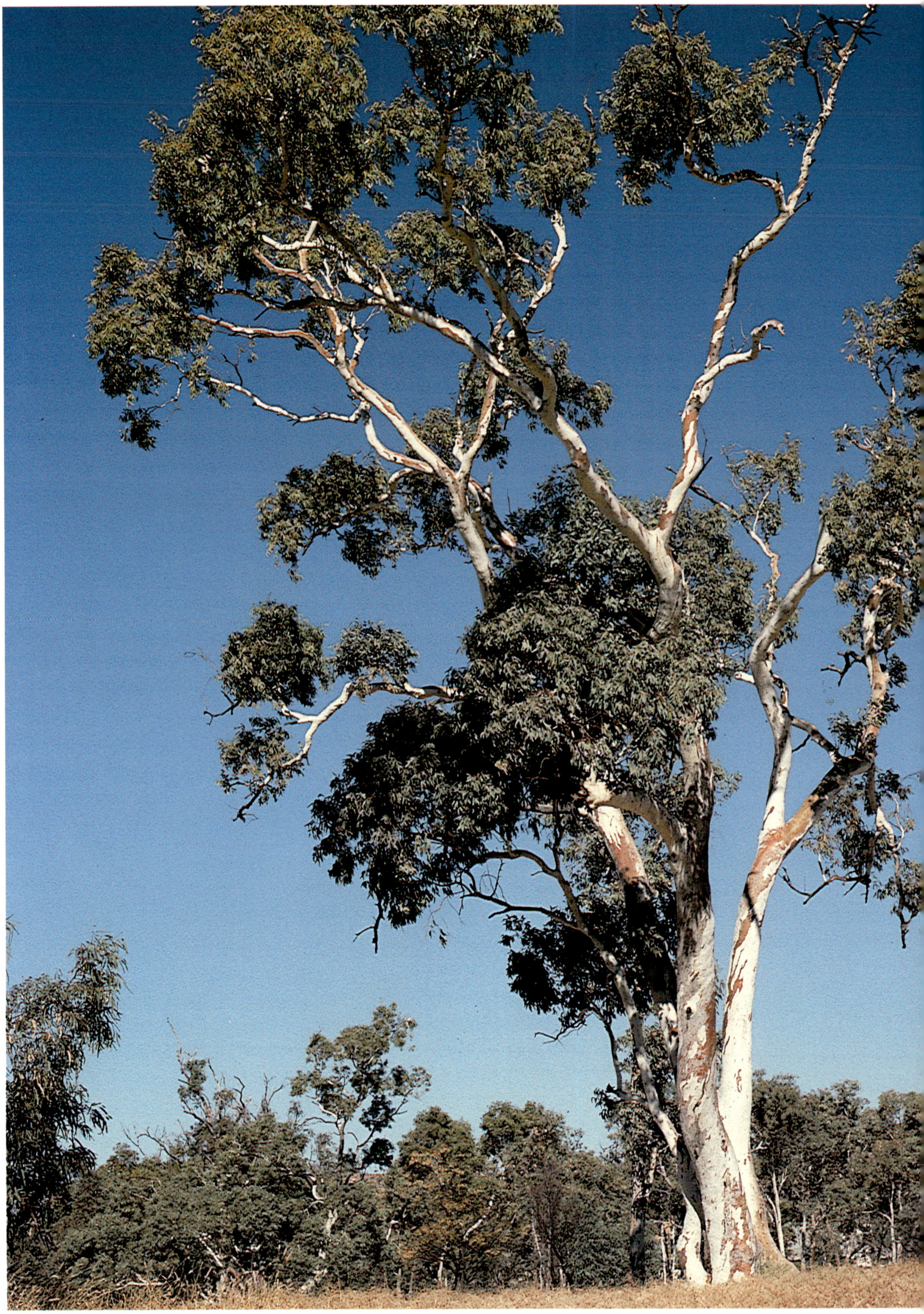

THE MIGHTY EUCALYPTUS

One of the great medical finds of the nineteenth century was the mighty eucalyptus. Throughout the world physicians extolled this tree as nature's answer to life-threatening fevers, dysentery and other ills. Millions of eucalypts, called "fever trees", were planted in Europe, Africa and South America to rid the globe of malaria. Eucalypt remedies for every woe were sold worldwide, and many still sell today.

Australia is home to more than five hundred different eucalypts. The trees have been called "gum" trees ever since the doctors of the First Fleet realised their value as medicines. Surgeon General John White found that "at the heart they are full of veins through which an amazing quality of an astringent red gum issues". This gum proved "very serviceable in an obstinate dysentery that raged at our first landing". Wattle barks were used for wattling hut walls, gum tree gums for healing, and so were Australia's most famous trees named.

First Fleet doctors used any kind of gum tree gum for dysentery, but one species, "the

◀ A stately river red gum (*Eucalyptus camaldulensis*) towers over the Todd riverbed in Alice Springs. Aborigines camp beneath these trees, regarding them as sacred. The burnt bark is chewed with tobacco, and fresh bark, steeped in water, produces a medicinal drink or wash.

peppermint tree" (*Eucalyptus piperita*), impressed White more by the peppermint aroma of its leaves. Distilling the oil, he found it "more efficacious in removing all cholicky complaints than...the English peppermint". White sent a quart of the oil to England for comparison. Eucalypts became so popular as remedies, in Australia and overseas, that it is impossible to list every application known. It is probably safe to say, however, that for every nineteenth-century ailment, a eucalypt remedy was tried.

The active constituents of the trees are easier to define, and fall into four categories. Sticky red gums, oozing from insect injuries in the trunks, known properly as kinos, are rich in astringent tannins and proved effective against dysentery, sores, and inflamed throats and gums. Tannins are also found in the inner barks. Aborigines infused these for similar ailments.

The leaves yield aromatic oils, which doctors (and peddlers of the oil) prescribed for almost anything. The leaves themselves were widely administered in the form of cigarettes, steam baths, sponges, as mattress linings, and to cleanse rooms like bouquets of flowers. Even the living trees, by emitting aromatic oils, were believed to have a powerful disinfecting and antimalarial effect.

A third medicinal product, rutin, which is found in the leaves of certain species, is believed to strengthen blood vessels. Finally, a laxative manna is produced on some gum trees by sap-sucking bugs.

Eucalypts peaked in popularity late in the nineteenth century, and were possibly more important to colonists then than they ever were to Aborigines, who had a choice of many other astringent and aromatic plants.

The eucalypts used by Aborigines included such famous trees as the coolibah (*E. microtheca*), ghost gum (*E. papuana*), river red gum (*E. camaldulensis*), Darwin woollybutt (*E. miniata*), and western bloodwood (*E. terminalis*), all of which were valued for their kinos or astringent inner bark. Aborigines and colonists used kinos in much the same ways: dissolved and gargled for sore throats, swallowed for diarrhoea, or washed on to sores.

In 1889 botanist Joseph Maiden gave a useful summary of eucalypt kino:

Many trees yield their kino in a viscid state on tapping a gum vein in spring or autumn. Exposure to the air usually hardens it almost immediately. As a very general rule, the kinos are collected naturally exuded and hardened on the outside of the bark ... Some of them are used by the settlers for ink and for staining leather black, the process simply consisting in boiling the kino in an iron saucepan.

The British medical journal *Lancet* gave a recipe for *Decoctum Eucalypti gummi*, a mixture of kino and water, taken for diarrhoea in two to four drachm doses. Nosebleeds and throat infections were treated with other preparations. More than 30 eucalypt kinos were tested in Maiden's time,

▶ Baron von Mueller extolled the peppermint eucalypt as "one of the most remarkable and important of all plants in the whole creation" There are now known to be several peppermints, including this, the broad-leaved peppermint (*Eucalyptus dives*), photographed at Kinglake near Melbourne.

◀ Immortalised by painter Albert Namatjirra, the ghost gum (*Eucalyptus papuana*) has come to symbolise outback Australia, although it also grows in coastal woodlands, both in northern Australia and New Guinea, as its Latin name indicates. The kino was a remedy for sores and a bark infusion was drunk for colds. Photographed at Devils Marbles.

▲ Rivers of kino run down a eucalypt trunk in Royal National Park near Sydney. It is not surprising that this substance soon came to the attention of First Fleet doctors, who found it a valuable dysentery remedy.

and one, from the river red gum, remained in the British Pharmacopoeia until at least 1949.

Eucalypt leaves were used by Aborigines in many different ways. They might be steamed for healing vapours, chewed, or crushed and soaked to make a healing wash. The Arnhem Land Yolgnu chewed the leaves of the Darwin stringybark (*E. tetradonta*) for colds or to make poultices for sores and cuts. A leaf infusion was drunk to treat backaches, coughs and diarrhoea.

Compared to the colonists, however, Aborigines seemed almost conservative in their use of gum leaves. Imaginative bushmen and doctors endowed the leaves with the most unlikely qualities, believing they could stimulate hair growth, ward off malaria, cure scurvy, even treat dysentery and cholera in fowls.

Herbalist H.P. Rasmussen advised how to make eucalyptus leaf preparations for earache, malaria, mouth ulcers, haemorrhoids, piles, asthma and skin diseases. Stringybark tea, he claimed in 1890, relieved sore feet, sore throats, and if rubbed in the scalp, prevented hair falling out: "A good healing ointment can be made from the leaves, when boiled together with hogslard and finely strained whilst hot."

He suggested a gum leaf bath as a general tonic: "Pour three gallons of boiling water over four pounds of fresh Eucalyptus Leaves (blue gum), and when drawn for an hour, well covered up, strain, and add to the usual bath. This bath is very invigorating, and greatly strengthens the whole system."

Bushmen chewed gum leaf tips for colds and

▶ The western bloodwood (*Eucalyptus terminalis*) could be renamed the "supermarket tree" judging by the number of items it supplied to Aborigines. The kino was used as glue and medicine, the sap to tan hides, the wood to carve bowls, and the seed capsules for decoration. Galls and lerps on the tree were eaten, bee honey was cut from the limbs, and water was drained from the roots. Photographed in Finke Gorge.

smoked the leaves to treat asthma. Von Mueller recommended bruised blue gum leaves for dressing wounds "to prevent or subdue septic inflammation, especially when no other remedies are at hand".

Colonial doctor Joseph Lauterer, who believed that "Eucalypts have to stand in the first rank" of native remedies, offered handy suggestions for using the leaves. "Small bags of soft washed calico stuffed with recently dried and powdered leaves", he wrote, "can be used like sponges and compresses ... In the bush hut, where diphtheria patients are lying, or where an acute case of bronchitis with troublesome cough is to be soothed, the dry air of the sick-room is much improved, if a big bunch of fresh twigs of Eucalypt is dipped in hot water and then hung up in a corner or on the ceiling."

At Cannes in France a Dr Gimbert dressed wounds with the leaves in place of lint "with very satisfactory results".

Distilled eucalyptus oil came on to the market in 1852 when Melbourne pharmacist Joseph Bosisto set up a still at Dandenong Creek. Bosisto's *Liquor Euc.globuli* was peddled as a malaria remedy and tonic. At first blue gum (*E. globulus*) and various peppermints (grouped under the name *E. amygdalina*) were the main sources, but the industry became largely dependent on mallees such as the Victorian blue mallee (*E. polybractea*) and Kangaroo Island mallee (*E. cneorifolia*). By 1914 there were two hundred stills in Victoria alone.

There are three kinds of commercial oils. Those that are rich in phellandrine and serve as disinfectants and antiseptics, those that contain cineol and act as medicines, and those yielding geranyl acetate or citronella used in the perfume industry.

Miraculous cures were attributed to eucalyptus oil, and many oil-based preparations were peddled by nineteenth-century entrepreneurs. These "cured" such ailments as diabetes, rheumatism, malaria, asthma, headaches, and

◀ Edward Palmer told in 1884 how Darwin stringybark leaves (*Eucalyptus tetradonta*) were taken by Mitchell River Aborigines and "bruised and rubbed in water in a kooliman with the hands till the water is green and thick, when it is drunk for fevers and headache." Photographed at Jabiru, Northern Territory.

▲ Saviour of the nineteenth-century world, according to Baron von Mueller, was the blue gum (*Eucalyptus globulus*). He claimed it had exercised "on regions of the warm temperate zone a greater influence, scenic, industrial and hygienic, than any other single species of arboreous vegetation ever reared anywhere".

stomach fermentation. The following advertising testimonial was typical of its kind:

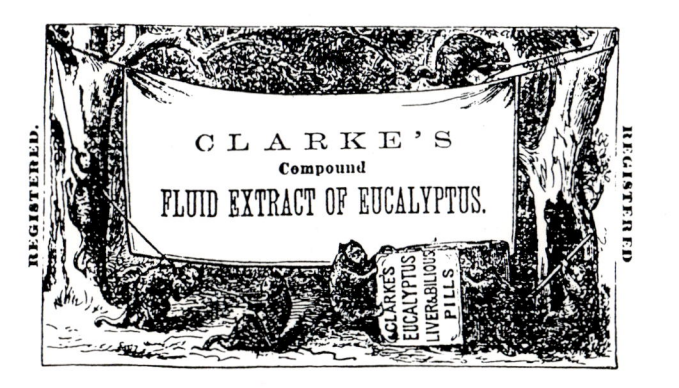

CLARKE'S Compound FLUID EXTRACT OF EUCALYPTUS.

An Engineer of one of the largest sugar manufactories of Queensland says: "My wife suffered martyrdom for TWENTY-FIVE YEARS from Indigestion and its horrible accompaniments. I spent Hundreds of Pounds sterling for advice and medicine, without avail. Three bottles of Clarke's extract of Eucalyptus and two boxes of his Pills *cured* her."

THE MIGHTY EUCALYPTUS 87

The *Brisbane Courier* of 17 July 1892 gave a charming account of a visit by train to an oil factory near Rockhampton:

About twenty-five gentlemen availed themselves of the opportunity of seeing the works. The train reached its destination about 5 o'clock, and about fifty or sixty aborigines, drawn up in a line, greeted it with cheers ... A white man, known by the name of "Dido," has the contract of supplying 5 tons of leaves daily to Mr. Ingham, and he in turn employs the aborigines as leaf pickers ... Mr. Johnson provided an excellent tea in the open air, after which the visitors proceeded to a huge bonfire about a quarter of a mile away in the middle of the scrub. Shortly after 7 o'clock about sixty aborigines, about thirty of each sex, led by Paddy S. Macalister, who was dressed in a long silk coat and tall black hat, marched round the fire, and divesting themselves of all their clothing, executed several dances and sang several songs. The train left Wallaroo at 9 o'clock, arriving in Rockhampton shortly after half-past 12.

Eucalyptus oil was sometimes burned in street lamps, and according to Maiden had "greater illuminating power, pleasant odour, and non-liability to explosion", but was more expensive and rarely used.

Rutin, or vitamin p, a substance also obtained from eucalypts, is believed to strengthen capillaries and increase their permeability. It has been prescribed against haemophilia, bleeding gums, migraine headache and swellings. Many plants around the world produce rutin, but two eucalypts, the red stringybark (*E. macrorhyncha*) and youman's stringybark (*E. youmanii*) are especially rich sources. Rutin from their leaves has been exported to America.

▼ By lining her nest with sprigs of fresh eucalypt leaves, a Pacific baza may be instinctively disinfecting her young. Australia's birds of prey are well known for their habit of bedding their nests with fresh leaves; eucalypts are invariably the leaves chosen. Photographed near Wandoan, southern Queensland.

◄ To the Yolgnu of Arnhem Land, the Darwin woollybutt (*Eucalyptus miniata*) was an old-time remedy for swellings and diarrhoea. Hollow limbs were rendered into didgeridoos, the seeds were eaten, and strips of bark were torn for lighting fires. Photographed near Ubirr Rock, Kakadu.

Manna, another medicine obtained from eucalypts, is a sugary secretion that oozes from the bark or leaves of trees attacked by insects. In Europe, medicinal manna was collected from ash trees, but New South Wales colonists found that copious manna oozed from manna gum (*E. viminalis*) and brittle gum (*E. mannifera*). Manna was often gathered as a treat and occasionally administered as a laxative.

Surgeon P. Cunningham, writing in 1827, was loud in his praises:

Manna is one of the safest, and almost the only pleasant purgative we possess, and it is only its scarcity and high price that have prevented its coming into more general use. Instead of surfeiting yourself with nauseous salts, jalap, and so forth, you have only to sweeten your tea in the morning with manna, or take a paper of it by way of barley-sugar in your pocket, and turn thus the doctor's hitherto nauseous posset into an agreeable *bonne-bouche*.

Cunningham suggested planting brittle gum plantations to supply manna for export to Europe, and calculated an "excellent renumerating price for both the collector and shipper, calculated to call their attention to the procuring of this valuable medicine".

I cannot conclude this account of the eucalypts without some mention of the extraordinary role played by these trees in nineteenth-century sanitation and fever control. Millions of blue gums were planted worldwide to subdue "malarial vapours". This Australian native, known in Europe as the "fever tree", became one of the world's most widely planted trees.

Nineteenth-century doctors in Europe believed that the aromatic oils of pine trees exerted a disinfectant effect on their environs. A Frenchman visiting Australia in 1854, Prosper Ramel, proposed a similar function for the eucalypts, and encouraged plantings in southern Europe and north Africa. The trees seemed to clear away malaria and further plantings followed. Plantations set up by Trappist monks in the Pontine Marshes of Rome were credited with spectacular success against the disease.

A testimonial from C.T. Kingzett in 1888 was typical of the times:

Between Nice and Monaco there existed, until recently, such a very unhealthy district, that the Paris, Lyons, and Mediterranean Railway Company were compelled to change their watchmen who did service at a crossing in the neighbourhood, every few months. Some time since, however, a plantation of the *Eucalyptus globulus* was made there, and now there is said to be no more fever; and the necessity therefore for perpetual change of watchmen (to prevent death) is abolished.

Baron von Mueller ascribed the cleansing power of these trees to four special features: the "copious absorption of humidity from the soil", the "corresponding power of exhalation" of the leaves, the highly antiseptic aromatic oil in the leaves, and finally the "disinfecting action of the dropping foliage on decaying organic matter in the soil".

Australia's near-freedom from malaria was attributed to these wonderful trees. Dr Bosisto calculated that Australia's mallee woodlands alone contained 96,877,440,000 gallons of aromatic oil, "held at one and the same time in a belt of country massed together, over which hot winds pass". In an 1874 article entitled, *Is the Eucalyptus a Fever-destroying Tree*, he concluded that "the whole atmosphere of Australia is more or less affected by the perpetual exhalation of these volatile bodies".

Baron Mueller suggested building sanitoria to benefit from the vapours:

I should assume, that sanitarian dwellings could nowhere on the whole earth be provided for phthisic [tuberculosis] patients more auspiciously and more hopefully, than in mountains clothed with Eucalyptus-forests in extra-tropical Australia and at elevations (varying according to latitude from 1,000 to 3,000 feet), where the slightly rarified air of a very moderate humidity pervaded by Eucalyptus vapor together with the comparative equability of the temperature would ease the respiration greatly.

▲ Site for a sanatorium? Baron von Mueller believed that Australia's mountainous eucalypt forests created abundant ozone, "the most powerfully vitalizing, oxidizing and therefore also chemically and therapeutically disinfecting element in nature's whole range over the globe". Photographed in the Steamers, east of Warwick, southern Queensland.

Other learned gentlemen, such as Dr A.B. Stroud, proposed the planting of eucalypts in gardens "to sanify the atmosphere from those emanations which give rise to epidemic diseases". Joseph Maiden lamented that when forests were cleared for housing estates, "the idea of leaving say one Eucalypt to each allotment for the purpose of desiccating the ground seems never to be thought of." Bosisto warned that "were it not that such happy and benign influences, as those exerted by the *eucalytpus* vegetation, existed around us *independent of ourselves*, we might mourn our fate."

Eucalypt oils have no effect on malaria, a protozoan disease spread by mosquitoes, and we may wonder how so many eminent gentlemen could get it all so wrong. How could Kingzett, for example, so confidently assert of the blue gum: "its fame has been a rapidly increasing one, the evidence in its favour forming an almost unbroken chain down to the present time".

The blue gum is a very fast-growing tree, and its water-hungry roots helped drain the swamps where malarial mosquitoes bred. This alone was apparently the key to its success. A few nineteenth-century doctors did question the merits of this tree, but their objections were loudly overruled.

◀ A favourite of gardeners, the swamp bloodwood (*Eucalyptus ptychocarpa*) of northern Australia has exceptionally attractive flowers. Aborigines pressed the leaves to wounds as instant dressings. Photographed at Robin Falls, Northern Territory.

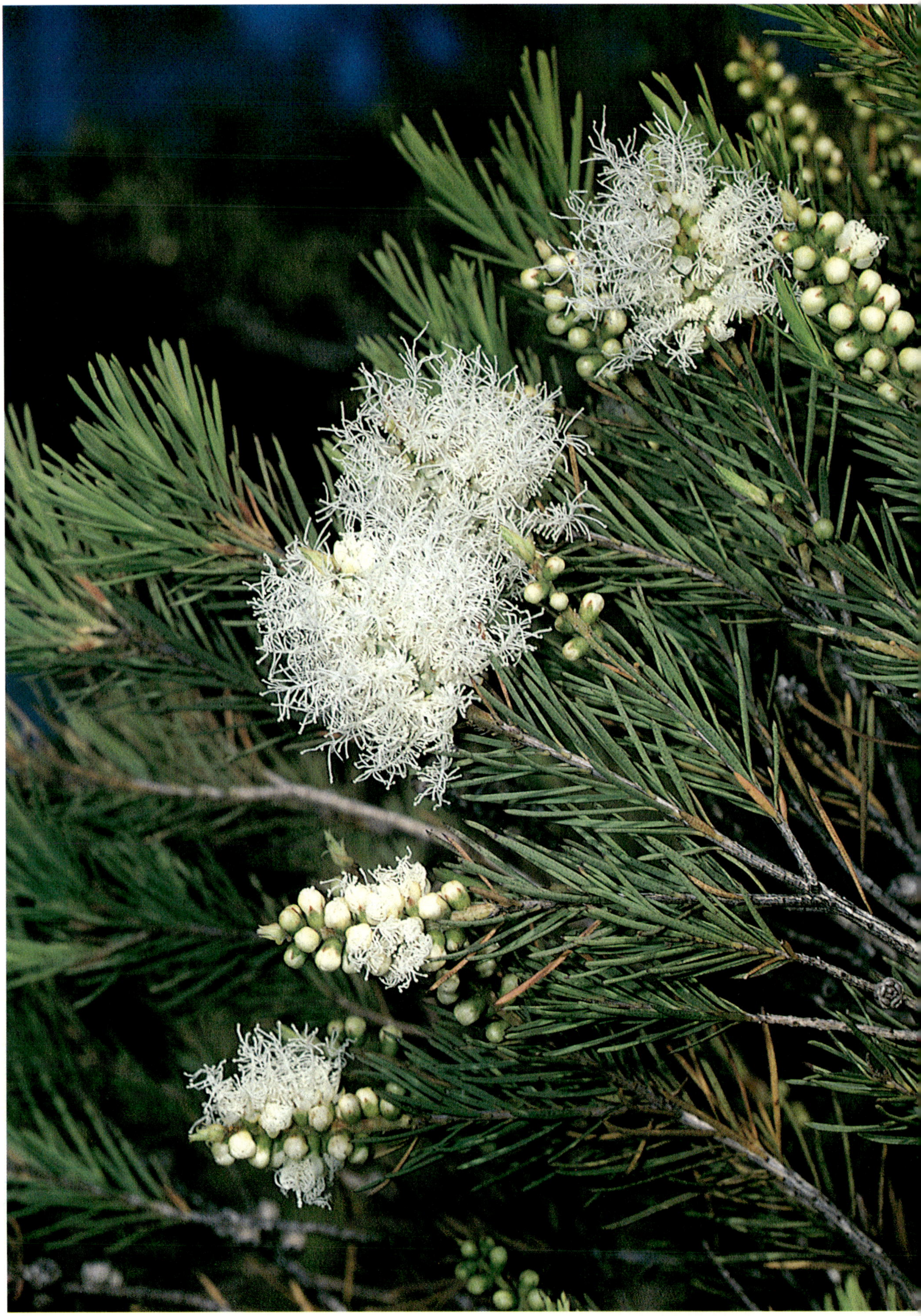

AROMATIC TREES AND SHRUBS

Australia's flora is the most aromatic in the world. The continent is a veritable garden of fragrant eucalypts, tea trees, mint-bushes and fuchsia bushes, which grow over a ground cover of sweet-smelling herbs. Only the pine forests of the Northern Hemisphere are as aromatic, and they harbour fewer species.

Throughout the world aromatic plants have been prominent in medicine, and no more so than in Australia. Eucalypts, tea trees, and other plants were valued remedies of Aborigines and colonists, with a versatility shared by few other plants. Many are still important today.

Aromatic plants owe their healing power to aromatic oils secreted in their leaves, stems or timber. Some aromatic plants, like the cypress pines and eucalypts, also yield medicinal kinos, resins or astringent bark.

Australia's foremost aromatic plants are of course the eucalypts. The country's dominant forest trees, they are described in the previous chapter.

◀ Feathery flowers and slender whorled leaves help identify the tea tree (*Melaleuca alternifolia*), a plant famous medicinally but poorly known in the field. The feathery structures are fused stamens; the tiny petals are largely concealed from view, except in the buds.

▲ Colin and Tobias Ferguson sniff the crushed leaves of the weeping tea tree (*Melaleuca leucadendron*), a traditional Aboriginal remedy for colds and blocked noses. The leaves are rich in cineol. Photographed on Charles Point, Northern Territory.

▲ "Ti-Ta" was hailed as a miraculous panacea by its promoters, but did not remain long on the market.

Next in importance are the tea trees (*Melaleuca*), which are shrubs or trees of soggy soils, often characterised by a shaggy, papery bark. At least a dozen of Australia's 61 species have been used as medicines.

Tea trees are closely related to eucalypts and yield similar oils, often rich in antiseptic cineol. These were not important in colonial medicine, but one of the oils is now becoming popular. *M. alternifolia* is a shrubby tea tree of swamps in northern New South Wales. It yields a non-irritating highly germicidal oil, which is being promoted through health food stores as a cure-all. It cures fungal and bacterial skin infections and effectively cleans away pus. First marketed in the 1920s for dental and surgical use, it was added to machine oils in munitions factories during World War II to reduce infections arising from injuries. In recent years it has been promoted as a natural remedy for home use. Plantations have sprung up in New South Wales and Queensland to meet the sudden demand, and hopefully take pressure off remaining stands of the wild plant.

A kind of tea tree oil was probably an ingredient in the nineteenth-century wonder cure, "Ti Ta", proclaimed by its promoters as "the greatest masterpiece of medical science" and "A discovery far before any of the brilliant theories of Pasteur or Koch". According to the Ti Ta Volatile Oil Company of Brisbane, its product, supposedly an extract of "a Tree, Moss, and Fern Indigenous to Northern Queensland", cured scarlet fever, diphtheria, dropsy, deafness, cancer, and just about anything else. A large advertisement in the *Queenslander* of 1891 listed the beneficiaries of the oil: the colonial secretary of Queensland, the solicitor-general, the railway commissioner, the mayor of Cairns, the editors of the *Hobart Mercury* and Brisbane *Telegraph*, and other persons of prominence. But despite claims that "The leading statesmen of Australia, doctors, and clergy, have given undoubted Testimonials", the product vanished from the market shortly after.

Tea trees had many other uses in colonial medicine. Baron von Mueller advised planting bottlebrush tea tree (*M. squarrosa*) and weeping tea tree (*M. leucadendron*) "for subduing malarian vapours" in swamps. Leaves of broom honey-myrtle (*M. uncinata*) were chewed for catarrh. Therapeutic oil was distilled from swamp paperbark (*M. ericifolia*), river tea tree (*M. bracteata*) and narrow-leaved honey-myrtle (*M. linariifolia*). The oil of weeping tea tree was considered by Dr Bancroft to be a valuable antiseptic inhalant for tuberculosis patients. Cajuput oil, a very important medicine in Europe and Asia, was obtained from Asian stands of cajuput (*M. cajuputi*) and weeping tea tree. Tea tree paperbark was also made into mattresses and baby blankets.

Aborigines used a number of tea trees in medicine. For coughs and colds, leaves were crushed and inhaled or soaked to make an infusion. Leaf washes were applied to pains, sores and burns. Strips of papery bark served as bandages and bedding. Leaves of the liniment tree (*M. symphyocarpa*), cajuput tree, weeping tea tree and broad-leaved tea trees (*M. viridiflora*) have rich medicinal aromas.

"Tea tree" is the name also given to related shrubs of genus *Leptospermum*. The lemon-scented tea tree (*L. petersonii*) is an excellent source of lemon-flavoured citral and citronella for scenting shampoos and toiletries. Plantations are run in Kenya and Guatemala, but in Australia it is grown mainly as an ornamental.

The other trees in Australia with aromatic foliage are the cypress pines. Distant relatives of the true pine trees (*Pinus*), the source of disinfectant pine oil, these native conifers were used by Aborigines and pioneers as medicines.

The white cypress (*Callitris glaucophylla*), a stately tree of rocky ridges and sandy plains, was employed by outback Aborigines as a versatile medicine for colds, sores and other ills. Leaves were smoked over a fire, soaked to make a wash, or mixed with fat to make ointment.

In tropical woodlands the northern cypress (*C. intratropica*) served in different ways. The ash was rubbed on sore chests, a bark infusion was drunk for sore belly and washed on the body for diarrhoea and sores, and the bark and leaves were burned to repel mosquitoes (the bark contains tannins). Vai Stanton of Darwin, a member of the Koongarakunj tribe, told me how she carried in her bag a special splinter of the wood, called a milmil, to prick the pimples of prickly heat, saying it was very effective for relieving the itch. She also remembered her father burning the green branches at camps to chase away mosquitoes.

In eastern Australia the clear resin oozing from

▲ Groote Eylandt Aborigines crush and inhale the fragrant leaves of the liniment tree (*Melaleuca symphyocarpa*) for colds. A liniment is made for aches and sprains. Photographed at Punsand Bay, Cape York.

cypress trunks was gathered by colonists and sold as Australian sandarac, a product with various old-time commercial uses, including the coating of pills and the filling of decayed teeth. In 1907 botanist Joseph Maiden claimed that Australian sandarac was "one of the most valuable of Australian vegetable products", and because the prices paid for it were low, called for an industry based upon child labour: "The collection of Australian sandarac is one of those minor industries which could be readily undertaken by a family of children. As the resin flows from the Cypress Pines it could be accumulated in clean dust-proof tins until a sufficient quantity was obtained to be sold to the local storekeeper, who would again sell to the wholesale chemist, or wholesale oil and colourman of Sydney."

In New South Wales, cypress foliage was steamed by pioneers and inhaled for chills and pains. The burning timber smelt wonderful. Declared Maiden: "There is nothing more delightful in the approach, on a winter evening, to a township where Cypress Pine is used as a fuel. Its delicious perfume is borne on the air for miles, and is often the first intimation that the weary traveller experiences that he is approaching a human habitation, and that his long journey is drawing to a close."

Distantly related to cypress pines is Tasmanian huon pine (*Dacrydium franklini*), once an important timber tree (until it was overexploited) with a heartwood rich in the aromatic methyleugenol. This germicidal oil, distilled from sawdust and waste wood at sawmills, was sold as a remedy for tinea, cuts and wounds. During World War II many Australian essential oils were tested as insect repellents and huon pine wood oil was declared the best.

▶ Incense for babies was made by desert Aborigines burning the aromatic twigs of the white pine (*Callitris glaucophylla*), an important outback medicine tree. The white pine is fire sensitive and grows mainly on rocky ridges beyond the reach of flames, such as here at Finke Gorge.

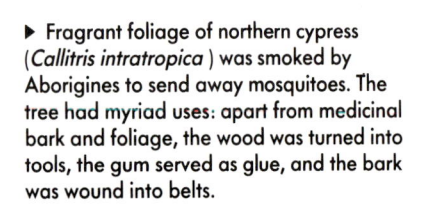

▶ Fragrant foliage of northern cypress (*Callitris intratropica*) was smoked by Aborigines to send away mosquitoes. The tree had myriad uses: apart from medicinal bark and foliage, the wood was turned into tools, the gum served as glue, and the bark was wound into belts.

Of the aromatic shrubs used in medicine, the most important were the outback fuchsia bushes (*Eremophila*). Aborigines still prescribe these for almost any kind of ailment. "Number one medicine", the old people say. Therapeutic activity appears to be due, not only to aromatic oils, but to what are either triterpenes or steroids.

In Australia's outback woodlands, fuchsia bushes are dominant shrubs. Known also as emu- and turkey-bushes, they have large attractive flowers, which are often spotted, and slender leaves with a disinfectant smell. Most are shrubs that grow 1–3 metres high, though a few grow tall enough to be called trees. At least nine species were gathered as medicines, and three of these were especially important — rock fuchsia bush (*E. freelingii*), weeping emu bush or berrigan (*E. longifolia*), and narrow-leaf fuchsia bush (*E. alternifolia*).

Rock fuchsia bush was one of the few Aboriginal remedies in central Australia adopted by whites, who brewed medicinal tea from the fragrant leaves. Aborigines plucked the leaves to make a wash for sores, colds, headaches, diarrhoea and chest pains. This remedy is still popular today. When I joined a bush food outing

AROMATIC TREES AND SHRUBS

◀ The delicately-hued flowers of the rock fuchsia-bush (*Eremophila freelingii*) are among the prettiest in the outback. Aborigines strung them into decorative headbands worn during ceremonies. The shrub may flower at any time of year.

◀ In central Australia Arrante babies and their mothers were smoked over smouldering berrigan leaves (*Eremophila longifolia*). The acrid smoke was believed to strengthen babies and increase milk supply. Photographed south of Alice Springs.

▶ Cyanide-laden leaves of spotted emu bush (*Eremophila maculata*) served as a cold remedy for Aborigines in the Hungerford district of south-western Queensland. Cattle are poisoned by eating this attractive shrub.

98 THE PLANTS

with Theresa Ryder's family near Alice Springs, her mother gathered sprigs of the leaves for home use, telling me it was "nice medicine, bush medicine". The leaves are rich in antiseptic alpha-pinene.

Berrigan is unusual among the fuchsia bushes, in containing no aromatic oils. Infusions of the leaves were applied to sore eyes, boils, headaches and insomnia, and the foliage was an important steaming medicine. Berrigan was widely held to be sacred, and in initiation and funeral rites, served as a lining for graves and shrouding bodies.

Narrow-leaf fuchsia bush served as a remedy for just about anything. Aborigines often dried and carried the leaves with them in case of need. The aromatic oil is mainly fenchone, also found in fennel. Other fuchsia bushes used in medicine include spotted fuchsia bush (*E. maculata*), a cold remedy; purple fuchsia bush (*E. goodwinii*); harlequin fuchsia bush (*E. duttonii*); and terpentine emu bush (*E. sturtii*), used by settlers to thatch meathouses (it was believed to repel flies).

On rocky ranges in central Australia, alongside the rock fuchsia bush, there grows another aromatic medicine plant, the striped mint-bush, or jockeys' caps (*Prostanthera striatiflora*). A particularly pretty shrub, it blooms prolifically in winter and spring, carrying hundreds of showy purple-streaked white flowers that resemble jockeys' caps. The leaves are rich in antiseptic apha-pinene and cineol, and contain either triterpenes or steroids. They were effective remedies for symptoms of colds and influenza. Sometimes stems were thrown into waterholes to poison emus, which would then be eaten. The chemistry of this plant deserves closer investigation.

Other mint-bushes taken as remedies by pioneers include *P. cineolifera*, found around Singleton. Rich in cineol, its crushed leaves were inhaled for influenza, and were said to deter flies. Round-leaved mint-bush (*P. rotundifolia*) was a colonial carminative and subject to a patent application last century to sell an extract as medicine.

Two other genera of aromatic medicine shrubs grow in the Top End of the Northern Territory. Wurrumba (*Pityrodia jamesii* and *P.* species) are slender shrubs with an extraordinary sticky-sweet smell. They grow only on the rugged western escarpment of Arnhem Land in small pockets of soil on sandstone crags. Kakadu Aborigines prepare a leaf decoction to soothe headaches, colds, diarrhoea and wounds. Feeding on the leaves is the shockingly colourful blue and red leichhardt grasshopper (*Petasida ephippigera*), a symbol in dreamtime mythology.

The fringe-myrtles (*Calytrix brownii*, and *C. exstipulata*) grow in the woodlands of Kakadu, and elsewhere in the Top End. They are small-leaved shrubs with medicinal foliage crushed by Aborigines as a liniment for aches and wounds. The leaves have a rich medicinal smell and yield antiseptic alpha-pinene.

▶ Visitors to Kakadu are likely to overlook wurrumba, a slender and nondescript shrub found sprouting on sandstone outcrops, as here at Ubirr Rock. Wurrumba belongs to the same plant family as vervain, a traditional European herb.

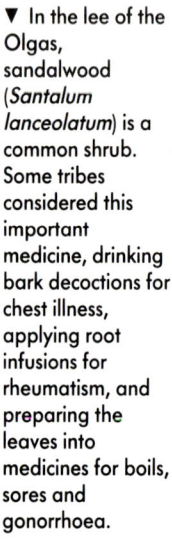

▼ In the lee of the Olgas, sandalwood (*Santalum lanceolatum*) is a common shrub. Some tribes considered this important medicine, drinking bark decoctions for chest illness, applying root infusions for rheumatism, and preparing the leaves into medicines for boils, sores and gonorrhoea.

One other group of aromatic shrubs is worth mentioning — the sandalwoods (*Santalum*). Western sandalwood (*S. spicatum*), a scrawny shrub or small tree, was heavily culled for the aromatic oil in its heartwood. It was exported from Perth to England as a substitute for Asian sandalwood, a remedy for gonorrhoea and other illnesses. In 1876 alone, 7000 tons of wood, worth 70,000 pounds, were shipped from the Swan River. Stands of sandalwood were so heavily cut that the industry almost collapsed in the 1940s. The wood, burnt as incense, and some oil, were exported to eastern Asia as late as the 1950s. Another sandalwood (*S. lanceolatum*) was harvested on a smaller scale and shipped out through Rockhampton and Thursday Island.

Unrelated to these true sandalwoods is the false or bastard sandalwood, known also as budda or budtha (*Eremophila mitchellii*). This is actually one of the fuchsia bushes, although it grows into a substantial tree. The timber yields a unique-smelling aromatic oil, which was once used to flavour soap. Aborigines along the Barwon River valued this tree as a medicine, as reported by K.L. Parker: "For rheumatic pains a fire is made, Budtha twigs laid on it, a little water thrown on them; the ashes raked out, a little more water thrown on then the patient lies on top, his oppossum rug spread over him, and thus his body is steamed."

▼ Pretty but poisonous: Aborigines hurled striped mint-bush foliage (*Prostanthera striatiflora*) into outback waterholes to stupify emus, which were then clubbed and eaten. Photographed near Palm Valley, Northern Territory.

AROMATIC TREES AND SHRUBS

102 THE PLANTS

FRAGRANT HERBS

Herbalism in Europe owes much of its charm to the sweetly scented herbs of the country garden — to the peppermint, tansy, camomile and other herbs favoured as flavourings and remedies. In Aboriginal medicine aromatic herbs were also held in esteem, and an English herbalist would have felt at home with an Aboriginal pharmacopoeia featuring native pennyroyal, mints, basil, lemon grasses and other sweet herbs. These native species are closely related to their namesakes in Europe or Asia.

In Australia as in Europe, the aromatic herbs come from certain plant families, the most prominent of which is the mint family (Lamiaceae). Its European members include horehound, sage, rosemary, thyme, oregano, mints, catnip, and in Australia the native mints, basils, Australian bugle, native coleus, and mint-bushes. The native mints were important remedies to both Aborigines and pioneers, and I have given them a chapter of their own. As the mint-bushes are woody shrubs not herbs, they feature in the previous chapter.

◄ Jirrpirinypa (*Stemodia viscosa*) is so sticky that the leaves and stems often carry a coat of sand grains. The delicate flowers are less than a centimetre wide. Photographed in Palm Valley, central Australia.

The Sacred basil of India (*Ocimum tenuiflorum*), a Hindu holy herb, is an important medicine plant in southern Asia. Surprisingly, this herb also occurs in Australia, growing naturally in acacia woodlands and on ridges in the north. The small green leaves have the delicious aroma of cloves, and Queensland Aborigines gathered them to make an infusion for fevers, dysentery and other woes. Clermont stock inspector W.A. Kearney noted in 1937 that this basil was "reputed to have great medicinal properties being said to have been used by the blacks as a blood tonic, and even at the present used by a good number of whites in tea and claimed to effect a

▼ Writing of the Cloncurry Aborigines in 1884, Edward Palmer recorded all that we know about musk basil (*Basilicum polystachyon*) as a remedy: "Used as a medicine by mixing in water, for fevers &c". Photographed on the banks of Cooper Creek at Windorah.

◀ Of sacred basil (*Ocimum tenuiflorum*), one early writer recorded: "It is said that the blacks used this plant as a cure for dysentery, by soaking the crushed leaves in water and drinking the emulsion. Whether the plant was effective for that purpose the writer knows not." Photographed west of Blackall.

▶ Australian bugle (*Ajuga australis*) sometimes grows so prolifically it imparts to the bush a purple hue. It is occasionally grown in gardens for its blue or purple flowers. Photographed on Mt Coot-tha, Brisbane.

cure taken this way, when afflicted by septic sores on the hand or 'Barcoo Rot' and as a general tonic to the system". Explorer Ludwig Leichhardt added the leaves to his tea, soups and stews. The related musk basil (*Basilicum polystachyon*), a weakly aromatic herb of creek banks in Queensland, was a fever remedy of Cloncurry Aborigines.

English herbalists sometimes prescribe European bugle (*Ajuga reptans*) as a remedy for cuts and bruises, and Aborigines used Australian bugle (*Ajuga australis*) in almost exactly the same way. At Tabulam in northern New South Wales they prepared a decoction of the fresh leaves to bathe sores and boils. Australian bugle is one of eastern Australia's prettiest forest wildflowers; the spikes of flowers are blue or violet.

In north-west Queensland, Aborigines crushed the leaves of the native coleus (*Plectranthus congestus*) in water and drank the liquid to cure internal complaints. Of Queensland's fourteen *Plectranthus* — shrubby herbs, many with sweet-smelling fleshy leaves — this species alone is recorded as a medicine.

After the mint family, the most important aromatic herbs in Australia are the lemon grasses (*Cymbopogon*). There are ten species, found mainly in northern and outback Australia, and

Aborigines used at least five in medicine.

The strongest-smelling, and most potent medicinally, is the outback lemon-scented grass (*C. ambiguus*). On desert rocky outcrops, and along alluvial drainage channels, this grass sprouts in small untidy clumps. Aborigines crushed the leaves, stems and roots between stones, soaked them in water, and applied the fluid as a wash for sores, colds, fevers, muscle cramps and sore eyes. Leaves were sometimes crushed and inhaled for chest complaints. Lemon-scented grass was one of the few Aboriginal remedies that was stored for future use, and one of the few desert remedies taken internally as an infusion, which was drunk sparingly. Botanist Peter Latz, however, warned: "when I tried a decoction of the leaves of this plant one evening I suffered from an almost continuous stream of vivid nightmares throughout the night." Lemon-scented grass contains more than three times the aromatic oils of the other native lemon grasses, and this consists mainly of camphene and borneol. In Warlpiri legend an immortal wallaby of human form once lived upon the leaves.

The silky oilgrass (*C. bombycinus*) grows in open forest and on stony ridges in northern Australia. Although meagerly endowed with aromatic oils (it yields one-thirtieth the oils of lemon-scented grass), this was an important medicine across northern Australia, used much like the lemon-scented grass.

Silky oilgrass is probably one of the lemon grasses remembered by Vai Stanton from her childhood with the Koongarukunj tribe near Katherine. There were two local lemon grasses: one was burnt in bundles to ward off mosquitoes, while the other was used to flavour beef and treat colds. Said Vai, "We'd douse it in boiling water and you'd see the oil droplets coming off. It was good for cleansing the head, for people with colds."

A third widely used lemon grass is *C. procerus* (no common name), a tall stout grass found in northern Australia in open forests and behind beaches wherever the soil is sandy. Like other lemon grasses it is taken as an infusion for fevers, colds and sores. Kakadu Aborigines place the heated plant against a baby's body to give it strength, and to prevent it crying when its mother is away gathering food.

Another kind of lemon grass, silky-heads (*C. obtectus*), is rarely used, for it usually grows near the stronger lemon-scented grass. The Pitjantjatjarra west of Ayers Rock, however, gather it for coughs and colds, and sometimes add its leaves to tea.

The plant family Scrophulariaceae, known affectionately to botanists as "the Scrophs", features some important European herbs — mullein, eyebright, brooklime, speedwell — though none of these is aromatic. But Australia has three strongly scented "Scrophs", which are used in medicines.

Jirrpirinypa (*Stemodia viscosa*) is a sticky herb with pretty bluish flowers, which inhabits river banks and gullies in the outback. Squeezing the leaves releases a heavy, sickly sweet aroma like molasses, which Aborigines inhaled for colds and chest complaints. As the stems are hairy and sticky, they were also used to strain leaves and debris from drinking water.

The rock-haunting *Stemodia lythrifolia*, found across northern Australia, has reportedly been used in the north-west as a headache remedy and bush tobacco. *S. grossa* of Western Australia was a cold, headache and rheumatism remedy.

The delightfully named green crumbweed or rat-tail goosefoot (*Dysphania rhadinostachya*), a

▶ At Kakadu the lemon grass (*Cymbopogon procerus*) known as "yirr", was crushed and steeped in water to make a cleansing fluid, applied to sore eyes, or inhaled or rubbed on the chest for colds and fevers. Photographed at Ubirr Rock.

▶ Agnes Lippo of Belyuen, Darwin, examines "bu" (*Cymbopogon procerus*), an old-time remedy for sores, nowadays rarely if ever used. Elsewhere in northern Australia this grass serves as a cure for colds and fevers.

▼ Aptly-named silky-heads (*Cymbopogon obtectus*), a native lemon grass, is one of outback Australia's more distinctive grasses. These plants, photographed at Alice Springs, are growing beside a sprig of fruit-salad plant (*lower left*), another native medicine.

FRAGRANT HERBS 107

▲ Ailing Aborigines crushed and soaked fragrant leaves of green crumbweed (*Dysphania rhadinostachya*) in water to make a head wash for colds and headaches. This outback herb often sprouts along roadsides, as here at Devil's Marbles in the Northern Territory.

member of the saltbush family (Chenopodiaceae), smells like a bottled cough medicine, and was used to treat colds and headaches. The crested crumbweed (*Chenopodium cristatum*) served as a poultice for swellings and boils.

Finally, the daisy family (*Asteraceae*), the source of such European herbs as tansy, yarrow and camomile, features several native herbs with very unusual smells.

Fruit-salad plant or applebush (*Pterocaulon sphacelatum*) is a strange-looking, strangely named outback plant that truly looks and smells medicinal. Along dry creekbeds, gravelly roadsides and gibber plains, it sprouts fans of fuzzy leaves and heads of dull, pinkish flowers atop flanged stems. The crushed leaves smell sweetly of fruit salad or granny smith apples. South Australian bushmen knew the plant as "horehound", and gathered it to treat colds, or to toss sprigs in the billy to enliven tea.

Central Australian Aborigines used fruit-salad plant in medicine, but found it inferior to toothed ragwort (*P. serrulatum*), a related plant with an even stronger and stranger smell — compared in modern times to Vicks. To treat colds, crushed leaves were soaked in water to make a wash, and mixed with fat as ointment, or rolled into a pillow on which the patient slept. For most efficient inhalation the leaves were inserted into the nasal septum — the hole drilled between the nostrils. Leaves were also infused to treat cuts and sores and chewed as a pituri substitute.

Streptoglossa odora, another outback herb, has an unpleasant smell somewhat like mouse droppings. Along with the sneezeweeds (*Centipeda minima*, *C. thespidioides*), it apparently served as a remedy for colds. Aborigines living in the deserts today remember little about the use of these small herbs, and their true role in the Aboriginal pharmacopoeia may never be known.

◀ Toothed ragwort (*Pterocaulon serrulatum*) is larger than fruit-salad plant (*see page 22*), with flowerheads two to three centimetres tall and a stem ornamented with jagged flaps. This plant, photographed in Palm Valley, is in bud.

110 The Plants

NATIVE MINTS

Among the most charming of Australian herbs are the native mints. In forest glades and along damp billabong banks, these dainty herbs creep and sprout, sending forth from trailing stems, sprigs of sweet-smelling leaves. Ailing Aborigines and bush pioneers, cheered by the soothing scents, gathered armfuls of the herbs as remedies for winter ills.

Australia has six native mints, and at least four of these were taken as medicines. The other two, forest mint (*Mentha laxiflora*) and an unnamed mountain mint, are seldom-seen herbs of cool gullies in south-eastern Australia.

King of the Australian mints is river mint, or native peppermint (*Mentha australis*). One of the most richly aromatic of Australian herbs, its leaves and stems carry a strong peppermint scent. Around rivers and lakes, river mint forms lush clumps up to a metre tall and wide, making it by far Australia's largest mint. It is also the most adaptable, growing across a broad swathe of Australia from the cool forests of Tasmania to the tropical deserts of north-western

◀ A delicate forest creeper with a gentle spearmint scent, forest mint (*Mentha laxiflora*) was possibly one of the now forgotten remedies of Victorian Aborigines. The leaves can be used to flavour drinks. Photographed in the Grampians.

Queensland, sprouting mainly along riverbanks beneath the cooling shade of coolibahs and river red gums. The long slender leaves and supple stems show that this plant is a rheophyte — a plant streamlined to withstand flooding. The pretty flowers are white or purplish, borne in fluffy clusters in the upper leaf axils.

In western Queensland, where Boulia and Cloncurry now stand, river mint was an important Aboriginal remedy for coughs and colds, and was drunk as an infusion. Ethnographer Walter Roth, who recorded this information in 1897, gave the pitta-pitta tribal name as "po-kan-gud-ye". This mint is still remembered in the area today. At Glengyle Station south of Boulia, an old Aboriginal stockman Jack Melville told me of a local bush medicine, an aromatic plant growing near water called "pennyroyal". Walter Roth gave this name to the plant more than 90 years ago.

Along the Narran River in New South Wales, Aborigines held river mint in high esteem. K. Langloh Parker wrote in 1905: "Pennyroyal infused they consider a great blood purifier; they also use a heap as a pillow if suffering from insomnia. It is hard to believe a black ever does suffer from insomnia, yet the cure argues the fact."

South Australian colonists gathered this mint to make tea. By infusing the leaves they rendered a rich and exceptionally refreshing brew — my favourite native tea. Colonists in the east took it as a tonic. River mint may have been the mint tea drunk by explorer Charles Sturt to stay scurvy (and deemed by him unwholesome). The plant grows quickly and is a rich source of menthol. It would be worth cultivating by herbalists as a native alternative to peppermint.

If river mint can be compared with peppermint, then native pennyroyal (*M. satureioides*) is an obvious counterpart to European pennyroyal. Analysis of the aromatic oils in the leaves shows large amounts of pulegone, the dominant oil in true pennyroyal. European pennyroyal has been taken as an abortifacient, and Australia's native species was used illegally in the same way in Queensland in 1921. The plant is also thought to cause abortions in cattle.

Native pennyroyal was a popular colonial tonic. T.W. Shepherd, writing last century in the New South Wales *Medical Gazette,* told how the herb was prepared:

▶ Of native pennyroyal (*Mentha satureioides*), colonial naturalist William Woolls suggested: "a valuable oil may be extracted from the little plant, which can be employed medicinally, as many species of Mint enumerated in the Pharmacopoeia, or chemically in imparting to other ingredients a pleasing flavour and scent."

Strong infusions of the fresh or dried leaves and stems are prepared by boiling in water fifteen minutes, more or less: after straining, the "tea" is sweetened with sugar, and taken warm at bed time q.a.l. It is supposed to be a specific for colds, catarrhs, coughs and a host of other aches and ills peculiar to both our first and second childhood. It is also esteemed by domestic physicians as a useful alterative, a blood purifier, and an invigorator of the whole system.

Colonial women took native pennyroyal, and slender mint, to regulate menstruation. The herbs were strewn on floors and beds to deter blood-sucking insects. According to colonial botanist Joseph Maiden in 1889 "they are very efficient in driving away fleas and bugs". True pennyroyal was used by English peasants in precisely the same way. Aborigines reportedly used the plant for stomach complaints.

Native pennyroyal is a delicate herb, rarely more than a hand-span high, with tiny pink or white flowers borne in whorls of two to eight. In eastern and southern Australia it sprouts in damp hollows near streams. Like other native mints it takes readily to cultivation.

Colonists used slender mint (*M. diemenica*) in much the same ways as native pennyroyal, a plant it closely resembles. (Botanists are unsure of the kinship of these two herbs — intermediates are sometimes found, which suggests the

◀ Whorls of tiny white or purplish flowers in the upper leaf axils distinguish river mint (*Mentha australis*) from other native mints. Aborigines in western New South Wales steamed the foliage to heal the sick. Photographed at Boulia, western Queensland.

two may hybridise.) Slender mint has slightly broader leaves than native pennyroyal, lilac or mauve flowers, and grows close to water in south-eastern Australia. Sprigs of the tiny leaves can be used like spearmint to add zest to foods and drinks.

The fourth medicinal mint is an unnamed species in northern New South Wales and southern Queensland, found growing among grass in open forests. The least water-dependent of the mints, I have seen it growing even on high mountain ridges far from any stream. A tiny herb, it has oval leaves and minute white flowers, which are usually arranged in clusters of three.

Last century this mint was exhibited in Brisbane's Museum of Economic Botany as "Brisbane Pennyroyal". Colonial botanist F.M. Bailey thought it might be worth cultivating, and described it in the museum catalogue as: "a very common plant in Queensland; on hard stony spots it may often be seen to form a dense turf. This herb has long been used as a medicine by bushmen. It yields by distillation a good-flavoured oil, which may be found to possess medicinal properties." Unfortunately Bailey omitted to mention either the ailments treated or the method of preparation. Elsewhere he described the plant as "the common Pennyroyal of Brisbane children".

At Cherbourg Aboriginal settlement near Kingaroy, an elderly man, Lennie Duncan, told me how this herb was once strewn under bedding at

▶ Lennie Duncan, an old stockman living at Cherbourg, southern Queensland, shows me Brisbane pennyroyal, used by him to repel fleas. Lennie also told me of making drinks from green tree ants, and of smoking the leaves of a local unidentified vine, as a tobacco substitute.

◀ In Ferntree Gully behind Melbourne, spearmint (*Mentha spicata*) has escaped from cultivation to become a streamside weed. Aromatic oils in the leaves are widely used to flavour unpalatable medicines.

stock camps to keep away fleas. The herb smells strongly of pennyroyal and was probably an important Aboriginal medicine, but no records of this have survived.

Apart from the native mints, the common mints of Europe now run wild in southern Australia. European pennyroyal, a herbal remedy for menstrual complaints, can now be found sprouting in the muddy banks of the Murray River, far from any town. Peppermint (*M. piperita*), spearmint (*M. spicata*), and apple mint (*M. rotundifolia*), sprout along creeks, gullies and gutters in cool shady places. Colonial pharmacist Joseph Bosisto encouraged the cultivation of peppermint in Victoria for medicinal oil, but despite experimental plantings and extraction of oils, no industry developed.

Two other plants in Australia have been compared to mints. The South American hyptis (*Hyptis suaveolens*), a common weed in the wet tropics, is sometimes called "wild mint". Its highly aromatic leaves have been brewed to make tea, and are said to repel mosquitoes. In Asia they are prescribed to treat skin infections. A sample of leaves tested in the Philippines reportedly yielded menthol.

Around Sydney, the surgeons of the First Fleet found that some of the local gum trees smelled of peppermint, and these became known as "peppermints", a name that still holds today. The oil proved useful in colonial medicine, and today its chief ingredient, piperitone, is widely converted into sythetic menthol.

◀ Reputed to repel mosquitoes, the South American weed called hyptis (*Hyptis suaveolens*) has been employed experimentally in the manufacture of mosquito coils. Photographed at Kakadu in the shelter of a sandstone cliff.

116 THE ANIMALS

ANIMAL REMEDIES

Goannas, ants, cockroaches, and many other things that crawl, swim and fly, were an important part of bush medicine in Australia. Bushmen swore by goanna and emu oil, townspeople doted on leeches, and Aborigines cured with insects, feathers and dung.

The most important animal remedies were the oils. Taken mainly from goannas, emus and dugong, these were considered potent medicines by Aborigines and bushmen throughout Australia. They are the subject of the next chapter.

Next in importance — to Aborigines — were various insects and their nests: bush cockroaches (*Cosmozosteria*), the green tree ant (*Oecophylla smaragdina*) and its woven nest, and the earthen mounds of termites and spinifex ants. Aborigines also made medicines of mammals, birds, fish and marine animals. Remedies of these kinds were probably significant in the past but have now been largely forgotten.

In northern Australia the green tree ant was a popular remedy for coughs and colds. Anyone who has ever travelled in the tropics will have

◄ A baby perenty (*Varanus giganteus*) can grow into a 3.3 metre goanna, which makes it Australia's biggest lizard and the third largest in the world. Large perenties produce plenty of oil, which Aborigines and colonists valued as medicine. The goanna oil symbol looks much like a perenty.

ANIMAL REMEDIES 117

▲ An angry nest of green tree ants (*Oecophylla smaragdina*) looks an unlikely medicine, but the ants were so widely and consistently used by Aborigines as remedies, there can be no doubt they must have worked. Photographed at East Point, Darwin.

unhappy memories of these predatory biting ants, found patrolling the foliage of shrubs and trees, zestfully attacking anyone within reach. By stitching together sprigs of living leaves they weave small suspended nests, and these were much used as medicines. Nests were crushed in water and the milky fluid drunk as a remedy for coughs and colds. Sometimes the ants were crushed in the hands and drunk with water. The formic acid in the ants is thought to promote coughing and to clear the lungs of phlegm.

In Arnhem Land, Aborigines dulled the pain of stonefish, catfish and stingray stings by applying crushed cockroaches. This provided an instant anaesthetic effect. Several different cockroaches, including *Cosmozosteria macula*, were used. The chemical constituents of these insects certainly deserve close examination.

In the central deserts, hairy processionary caterpillars were used as remedies by the Pitjantjatjara. The hairy bags they weave were carefully cleaned, sometimes with mother's milk, and placed as a bandage over burns, which quickly healed. Old Aborigines say these caterpillars were once tools of murder: the victim was punched in the stomach, and upon gasping, the caterpillar's irritating hairs were thrown into his mouth, quickly swelling the throat and choking the victim to death, without leaving signs of any injury.

Pitjantjatjara legends say that when heavy rains pounded the deserts the hairy caterpillars were tossed to the ground and stripped of their hairs. Finding themselves naked and grounded they crawled into the earth and became witchetty grubs.

Outback witchetty grubs were delicious and exceptionally important foods to the Aborigines, but few people realise they were also used as medicines. The ground-up grubs were rubbed over the body to cure ills such as headaches and sore eyes.

Bees, termites and spinifex ants also featured in Aboriginal pharmacy. On Groote Eylandt in the Northen Territory the different parts of the native bee hive (*Trigona hockingsi*) were prescribed in very specific ways. The parts used were the bee bread, called mangkarriya, and two kinds of brood, rich white meningabudangwa marnja and bitter mamurrariyadangwa marnja. Mangkarriya was eaten for diarrhoea, meningabudangwa marnja treated sore eyes,

▶ A tubular tunnel of wax tells of a native beehive (*Trigona*) concealed within the sandstone at Wurdidjileedji, Kakadu. Native bee honey was an Aboriginal remedy for sore eyes at Boulia, and the brood was considered medicinal on Groote Eylandt.

▼ In northern Australia crushed termite mound is a popular diarrhoea remedy. It is also eaten by pregnant women, perhaps acting as a dietary supplement, and is smoked to provide an aromatic vapour for children with colds. This magnetic mound is at Lakefield on Cape York Peninsula.

and mamurrariyadangwa marnja acted as an antiseptic for tinea and sores and a remedy for toothache. (The sweet honey was never used as it was too valuable as food.)

The people of Groote Eylandt are also among those who eat clay from termite mounds. This is a woman's remedy, eaten for diarrhoea or apparently for nutritional defficiency. The clay probably eases diarrhoea by forming a protective lining on the intestine.

Resinous anthills found beneath clumps of spinifex (*Triodia pungens*) served as a medicine for the desert Warlpiri, and were as versatile as the bee hives of Groote Eylandt. According to botanist Peter Latz:

The antbed was burnt or heated until black, ground into a powder and wetted and then either swallowed to cure diarrhoea or rubbed onto babies for undescribed ailments. It was also apparently utilized in some way to increase lactation. The acrid smoke from smouldering antbed was also reputed to have healing properties for sick babies, and the eggs and larvae obtained from the anthill were reputed to be used as a healing liniment.

Other outback medicines included the down of birds, which was used to staunch wounds, and the droppings of zebra finches, taken from abandoned nests and considered a strong medicine. Cooked rat brains were rubbed on babies with colds, and dingo droppings were applied to sores. Aborigines used dung less than colonists, who used cow manure for many ailments.

Introduced rabbits were adopted into medicine, both in central Australia, where the Pitjantjatjarra use the urine, and in Victoria, where rabbit intestines were baked over gum leaves as a cold remedy.

The Aboriginal pharmacopoeia even extended into the sea. Missionary Theodor Webb told of a remarkable treatment practised in Arnhem Land:

◀ Corals are colonies of hundreds (or thousands) of tiny filter-feeding animals, each contributing to construction of the coral skeleton. If corals, such as this brain coral, are exposed during low tide, they secrete a protective slime, which at least one Aboriginal group used as medicine.

A person suffering from [raised sores] goes out on a reef or sand-bank when the tide is low and finds a *yanungani*, sea anemone, which is lying open. He then quickly presses the affected part into the open centre of the creature, whose tentacles immediately contract and cause a certain amount of suction. The member is held there for a considerable time, and when withdrawn the raised exterior is said to have been removed and clean bleeding flesh, which will soon heal over, exposed.

This treatment has its parallel in the European use of leeches to remove blood.

Also in Arnhem Land, coral "phlegm" was rubbed on the body for headaches, colds and flu. This substance, called "garrang" by the Yolgnu, was probably the slime with which coral protects itself during tidal exposure.

At Belyuen near Darwin, Aboriginal community leader Harry Sing told me of a mangrove worm that is used in medicine: "You get this worm mainly in rotten fallen logs. When you boil it the water goes milky. It's good for colds

▲ Bane of Australia, the accursed rabbit has wreaked environmental havoc wherever it has hopped. Aborigines in central Australia value its meat, and even use the urine in medicine, though details of such use are hazy. Photographed in Girraween National Park near Stanthorpe.

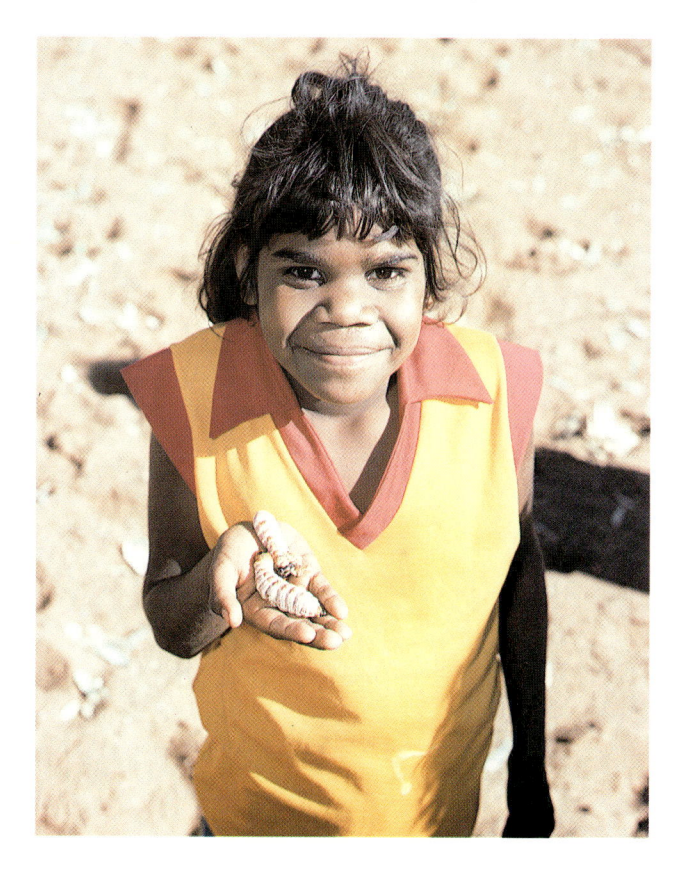

▲ Witchetty grubs (*Xyleutes*) gathered on a picnic bring a smile to Janissa Ryder of Alice Springs. The grubs are still esteemed as delicacies in outback Australia, although their medicinal uses have been largely forgotten.

ANIMAL REMEDIES

and stomach ache. The Tiwi say it is good for breast milk." This same remedy was reported over a hundred years ago by the Northern Territory police inspector, Paul Foelsche: "*Coughs and Colds* are very common complaints among the natives, which they cure by eating a kind of grub found in mangrove trees, and drink the liquid with which the grub is surrounded when in the wood."

On Cape York Peninsula, "the oily mess derived from boiling a young sting-ray's liver", drunk in a baler shell, was considered a cure for constipation, according to colonial ethnographer Walter Roth. On Groote Eylandt powdered cuttlefish was rubbed on the skin for pimples, and then dampened with water. Victorian Aborigines supposedly ate a mixture of leeches, kangaroo liver and sowthistle (*Sonchus oleraceus*) for stomach ache.

Many of the animal cures described here might sound peculiar to white Australian ears. But the remedies are no stranger than those practised by Europeans in the past. The official English pharmacopoeia of the seventeenth century recognised earthworms, woodlice, snails, frogs and toads. Nineteenth-century remedies used roasted mice, raw steak, bacon, eggs and warm manure. Victorian settler J. Dawson noted a similarity between Aboriginal and European use of eels: "For pains in the joints, fresh skins of eels are wrapped round the place, flesh side inwards. The same cure is very common in Scotland for a sprained wrist."

Australian herbalist H.P. Rasmussen recommended cleaning the teeth with cuttlefish: "Cuttle fish bone when finely pulverized makes an excellent tooth powder as it does not scratch or otherwise injure the enamel of the teeth when used in cleaning them. It is an excellent antiseptic and tends to sweeten the breath and is entirely harmless if accidentally swallowed."

Australians in Dawson's time were especially obsessed by the supposedly therapeutic nature of leeches. These slimy blood-sucking worms were prescribed for the treatment of anything from appendicitis to nymphomania, and served as a mainstay of colonial medicine. As late as the 1930s, leeches were being sold in shops in Sydney, and bushmen in western Queensland still use them today.

Pharmacists kept their leeches alive in water-filled casks or jars, for sale at a shilling or more

▶ Contending kangaroos tussle for dominance in a forest clearing near Miles in southern Queensland. Colonial doctors used grey kangaroo sinews as surgical sutures, Aborigines used the fat as an ointment base and the roasted liver as an ingredient in an emetic mixture.

ANIMAL REMEDIES 123

each. Patients took them home in small wooden boxes moistened with damp cotton wool, and applied them to swellings and injuries, especially black eyes, where they eased swelling by drawing off blood. Leeches inject an anticoagulant, and this often caused profuse bleeding after the worm had finished feeding.

Leeches were caught in their thousands in local swamps and streams. A Victorian collector, Horace Wheelwright, recalled in 1861: "we used to catch them by throwing a sheep-skin into the water, and upon taking it up it was covered. We could sell them for a shilling a dozen." Melbourne pharmacist Joseph Bosisto in 1858 gave detailed instructions for cultivating and preservating Murray River medicinal leeches. He concluded, "I see no reason why they should not prove a renumerative article of export from these colonies". Indeed, a cargo of leeches was sent to England in the 1860s, but by this time the English market had declined and the shipment was apparently dumped in the Thames.

Leeches were useless against many of the ailments for which they were prescribed, but

▶ Voracious cane toads (*Bufo marinus*) pose a threat to the Australian bush, and regrettably, few animals can prey upon them. Their bodies are widely used to demonstrate vertebrate anatomy, and hundreds are killed as teaching aids.

◀ Bushmen still deploy leeches today: Ivor Gent, a part-time pig hunter in southern Queensland, told me of treating carbuncles by applying a few leeches caught in a dam. The treatment was very painful.

interestingly, they still serve a role in modern medicine — as a source of nerve and muscle tissue in research.

Today wild animals continue to find a place in modern medicine. Caerulein, a toxin discovered in the skin of Australia's green tree frog (*Litoria caerulea*), and now synthesised artificially, is prescribed throughout the world as a relaxer of the bile duct. The platypus or gastric brooding frog (*Rheobatrichus silus*) may yield a cure for stomach ulcers. It rears its tadpoles in its stomach without digesting them. In the post war period, up until about 1960, cane toads (*Bufo marinus*) were used by Australian laboratories to test for pregnancy in Australian women. A woman's urine was injected into a female toad, and if the toad laid eggs a few hours later, the woman knew she was pregnant. The venom of the mulga snake (*Pseudechis australis*) may provide a treatment for thrombosis and diseases involving blood clotting. The potential of marine animals to supply new medicines has barely been tapped.

126 THE ANIMALS

ANIMAL OILS

Animal oils were among the most revered of Australian bush remedies. Aborigines and colonists ascribed miraculous healing powers to goanna and other oils when rubbed into the skin. Reverence for these oils was a characteristic of bush medicine in Australia and one of the few cultural values that passed from blacks to whites.

Most Aboriginal tribes used oils in healing, but different tribes placed value on different oils. To some tribes goanna oil was supreme, while others held emu or dugong oil to be the best.

Sometimes different oils were used for different purposes. K. Langloh Parker explained in 1905 how the Ualarai of north-western New South Wales used "Iguana [goanna] fat for pains in the head and stiffness anywhere. Porcupine and opossum fats for preserving their hair, fish fat to gloss their skins, emu fat in cold weather to save their skins from chapping."

In western Victoria, settler James Dawson reported on a different range of oils: "Burns are covered with fat. Running sores which are

◀ Snake oil, taken from larger snakes such as this carpet python (*Morelia spilota*), photographed on Lamington Plateau in southern Queensland, was a remedy of some tribes. Mina Rawson wrote in 1919: "Lately I noticed someone praising the virtues of iguana oil for rheumatism. As a matter of fact the blacks will use any animal fat they easily get. It may be iguana to-day and snake or lizard tomorrow."

▶ A sniffing shuffling echidna searches for its dinner of ants in the leaf litter on Black Mountain, Canberra. Aborigines often call this animal "porcupine" (it is unrelated), and in central Australia mix its fat with crushed herbs to make healing ointment.

▲ Fish oil was rubbed over the body to help keep warm and to ward away mosquitoes. Fatty pieces of fish were sometimes placed on children's heads so the melting oil could run onto their bodies. The oil was not generally endowed with medicinal properties. Photographed at Kakadu.

difficult to heal, are rubbed with the fat of the powerful owl, which dries them up quickly. The fat of large grubs is used for anointing the skin of delicate children."

In the deserts of north-western Queensland, wrote Walter Roth, "Iguana-, snake-, or any other kind of fat or grease, mixed more or less with mud and dirt, is used as a dressing for cuts and wounds of all description ... Fat may also be employed as a liniment for rubbing over tired or aching limbs, and in such circumstances affords apparently speedy relief." In this part of Australia children were regularly rubbed "from top to toe" with goanna fat.

Fats and oils were recommended for almost any ill, but were especially favoured for treating burns, wounds, and rheumatic pains. European medicine employed lard in much the same way.

The most popular healing oil, and one that remains famous to this day, was goanna oil. An exceptionally fine oil, it readily penetrates the skin. In western Queensland particularly, Aborigines and bushmen used it as a remedy for every woe. One old prospector, using the oil to treat a broken arm, told pioneer Alice Duncan-Kemp, "Been forty year in the bush; never used nothing but goanna fat".

The oil became famous when Brisbane entrepreneur Joe Marconi set up a goanna oil factory in Brisbane early in the twentieth century. Marconi marketed goanna oil salve, and a liniment comprising a secret mixture of goanna oil, distilled herbs and various aromatic oils including eucalyptus and camphor. Through a flamboyant advertising campaign, which included billboards, railway signs and even nursery rhymes, Marconi claimed cures for everything from rheumatism and dandruff to lumbago and baldness. According to one testimonial a paralysed girl massaged with the salve and oil was walking within a few weeks. Declared her mother: "She is now a healthy, bright, bonny little girlie, running all over the place."

Marconi obtained his oil from shooters in New South Wales and Queensland. The lace monitor (*Varanus varius*) and sand goanna (*V. gouldii*) were no doubt the main species used. One old hunter, James Scanlan, remembers shooting goannas near Gayndah long ago: "I used to cut the yellow fat out of the goanna, put it on sheets of iron in the sun, melt it into a bottle, and post it to Brisbane. It may sound ridiculous, but that oil passes through glass. All the bushmen used it for oiling their guns."

Another shooter, Bill McIntyre, remembered supplying goanna fat, along with their gallstones, to Chinese market farmers in Brisbane during the 1920s. "I was paid two shillings a goanna," he told me. "It was a lot of money during the Depression." McIntyre did not know what the Chinese did with the oil, but he remembers them swallowing the raw gallstones as a medicine.

Marconi's company still exists, but is now owned by Euan Murdoch, who freely admits the oil and salve contain no goanna oil, and have not done for many years. (Goannas are legally protected throughout Australia.) The penetrating agent is oil of wintergreen. Murdoch is now promoting his product in the United States.

Another very popular oil in colonial times was emu oil, which was used by Aborigines and bushmen to treat rheumatism and paralysis. Explorer Ludwig Leichhardt found the oil invaluable during his northern expedition, when at

▲ Rolling away a rock reveals a slumbering spiny-tailed goanna (*Varanus acanthurus*) at Cloncurry in north-west Queensland. Aborigines in this region held goanna oil in great repute and probably contributed to European perceptions of its value.

▲ The showman's shed: the flamboyant Joe Marconi vigorously promoted his salve through nursery rhymes, lecture tours and a goanna salve stall, made of bark slabs, at the Royal National Association Exhibition (the "Ekka") in Brisbane in 1917. Photograph from the J.C. Marconi company archives.

▶ A lace-like body pattern gives the lace monitor (*Varanus varius*) its name, though most people probably know it as the tree goanna because of its habit of fleeing up trees when harassed. Many lace monitors were slaughtered for their oil, and sent to Marconi's Brisbane factory.

times emu drumsticks constituted his staple food: "To obtain the oil, we skinned those parts, and suspended them before a slow fire, and caught the oil in our frying pan; this was of a light yellowish colour, tasteless, and almost free from scent. Several times, when suffering from excessive fatigue, I rubbed it into the skin all over the body, and its slightly exciting properties proved very beneficial. It has always been considered by the white inhabitants of the bush, a good antirheumatic."

Some colonists ascribed the oil with remarkable powers. Police trooper Alexander Tolmer told of a remarkable cure in his 1882 autobiography, modestly titled *Reminiscences of an Adventurous and Chequered Career*. Some weeks after capturing a villain, Tolmer found his arms contracting helplessly towards his shoulders and his fingers becoming rigid:

> In this awful state I sent for Doctors Nash and Woodforde, who were thoroughly struck with amazement, never having seen or heard of a similar case in all their professional experience. After a brief consultation, they agreed to send me an embrocation and other medicine, which, however, had not the least effect, and then suddenly remembering the efficacy of emu oil . . . I got two of the most powerful troopers to rub the oil, one at each arm, two or three times a day, and wonderful to relate, the arms and fingers relaxed and regained their natural position, but very soon again contracted. The rubbing in of the oil, however, was regularly kept up for a week or so, at the expiration of which time I had perfectly recovered, and from that day to this, have never had a relapse.

Testimonials of this kind are still being given today. At Belyuen Aboriginal community near Darwin, Marjorie Bil Bil told me how her son was a near cripple until she rubbed him down with emu oil, effecting a complete cure.

Emus are now farmed by some Aboriginal

◄ Peering from its burrow, a sand goanna surveys the scene at Coral Bay in Western Australia. Explorer Archibold Meston claimed of goanna oil, "The effect on the skin and muscles is particularly beneficial at all times. For athletes it eclipses any other known oil."

▲ Goanna oil and salve are nowadays known to contain no goanna product. The original recipe supposedly consisted of goanna oil mixed with a native herbal remedy for snakebite.

ANIMAL OILS 131

▶ Hunter Horace Wheelwright in 1861 said of the emu: "It is a very fat bird, and when boiled down, emu oil, like the shark's oil among fishermen, is the bushman's universal remedy for rheumatism and other bush complaints." Photographed near Miles in southern Queensland.

groups, and their oil sold as a remedy for rheumatism. The price is high, as emus are now very valuable, having been exterminated from many regions. Demand for oil hastened their decline, as colonist R.N. Breton noted in 1834: "One portion of this singular bird is considered tolerable eating, being somewhat similar to beef, but it is killed more on account of the oil procured from its fat, which is used for burning, preserving leather (it is thought to do so better than any other), and as a remedy for the rheumatism."

Another oil accorded miracle properties was that of the dugong. This gentle-mannered sea mammal paddles in the warmer waters of northern Australia between Shark Bay and Moreton Bay, wherever beds of eelgrass grow. In South East Asia, it is endangered by overhunting. An important food of traditional Aborigines, it is still keenly hunted by islanders in Torres Strait, where numbers are probably dwindling. Aborigines in Moreton Bay valued dugong oil, and impressed its importance upon pioneer Tom Petrie, as recounted by his daughter Constance in 1904:

▶ The dugong of Moreton Bay so impressed Dr Hobbs that in the 1850s he declared: "For miles in every direction are extensive submarine pastures of great luxuriance, affording a never-ending supply of long grass, upon which the herds of dugong feed and fatten like oxen." Illustrated are a cow and calf, and in the background, a loggerhead turtle.

> **The natives were great believers in the curative properties of the dugong. Father has seen sick blacks, unable to walk, apparently in consumption, carried carefully to the mouth of the Brisbane River, and there put into canoes and taken across to Fisherman's Island, to where dugong were being caught. There they would live for some time on the flesh of the dugong, and the oil would be rubbed all over their bodies, and in the end they would return**

quite strong and well. In the early days of Brisbane, my father mentioned how he had seen this for himself to Dr Hobbs, who was greatly interested, and afterwards recommended the use of dugong oil as a remedy similar to cod-liver oil, and this is how it came to be first used medicinally in Queensland.

Thomas Welsby, a settler on Stradbroke Island near Brisbane, told how Hobbs promoted the oil:

By a series of public lectures delivered in Brisbane he demonstrated that the oil was a sovereign remedy for all cases of lung complaint, and he further made use of the cured flesh in the form of bacon, as he found that these valuable properties were more easily assimilated in that form than in any other way. So great was the influence these lectures had upon medical science that they were largely copied into the leading papers of England and commented upon in terms which left upon the mind of the reader the impression that a most valuable remedial agent for consumption had been discovered. As one result from all this, orders were sent from home for large quantities of the oil. A single leading firm of chemists ordered one thousand gallons at three guineas a gallon, and other orders for smaller quantities were received, sufficient to warrant the starting of several people fishing for dugong. For a number of years the business was carried on with great success.

The dugong oil industry collapsed when, as Welsby explained, the dugong hunters "were doing too well. They got lazy and became too fond of the town and its beer. And now and again, when a big shark was caught, in went its liver to the boiling down pot, and the result was a smash-up to the business from which it never recovered."

Dugong oil was exhibited at the International Exhibition of Vienna in 1873, the catalogue of which confirmed Welsby's lament of greed:

> Several years ago [this oil] was introduced in Great Britain and the Continent of Europe: the extent of the demand, and the consequent high price secured, induced unscrupulous persons to adulterate the article with common fish oil, owing to which dugong oil, for a time, went out of favour. This unfortunate circumstance might be remedied, were the English importers to purchase only from persons who are above suspicion, and who are in a position to guarantee that the article is genuine.

The industry never revived, but dugong oil is still used today in Aboriginal and islander communities. In Arnhem Land the oil, along with turtle oil, is rubbed into the skin and hair as a cleanser and moisturiser. This probably dates back to the traditional practice of smearing oil over the skin to deter mosquitoes. In the Torres Strait islands the oil from the tail is stored in jars and drunk in place of vitamins. Said one health worker: "Dugong oil is good for the skin, and is also used as a mixture to stop infections in the body. Usually we take one teaspoon for kids and one tablespoon for adults, three times a day."

Two other kinds of fat also deserve mention: cassowary oil, and fat taken from human bodies. In 1892 Archibold Meston, the Queensland explorer-showman, told of the virtues of cassowary oil: "The cassowary accumulates an immense amount of oil, especially from November to March, and I have obtained six or seven quarts from one bird. The skin is covered by fat and very difficult to preserve. The oil rather increases than prevents rust on articles of steel. It is extremely effective for stiff joints or contracted muscles."

After warfare, Aborigines sometimes rubbed themselves with the fat of their dead adversaries. Police Inspector Paul Foesche wrote of Top End Aborigines in 1882: "Nearly all natives use fat obtained from dead bodies of either their own or other tribes for anointing themselves with, which they believe makes them strong and able to fight well. The fat taken from all parts of the body is mixed with red ochre to prevent it melting away. It is then tied in paper-bark, and in this state is distributed among the men, and very often some is sent to other tribes." Elsewhere in Australia, kidney fat was often stolen symbolically by sorcery.

Animal fats were also valued by Aborigines for keeping the body warm, for deterring biting insects, for disguising body odour when hunting, and for making healing ointments (salves). In central Australia, where oils were not used on their own as medicines, Aborigines mixed animal fats with crushed leaves of aromatic plants to make ointments. The Northern Territory Aboriginal pharmacopoeia tells how the Alyawarra make an ointment from striped mint bush (*Prostanthera striatiflora*):

> Half a cupful of leaves and one cupful of animal fat are heated together until boiling. The green liquid formed is stirred to mix the ingredients thoroughly and allowed to cool. The consistency of the preparation depends upon the source of the fat, and will vary from a thick cream (goanna) to a solid ointment (beef or kangaroo). It is massaged in for colds and general aches and pains, and as a remedy for the crusty sores of infected scabies.

▶ The resplendent cassowary is formidably adorned with brightly coloured skin and a tall bony helmet, used for slicing a swath through the jungle undergrowth. Though the birds feed almost entirely on fruit, their bodies are very oily.

ANIMAL OILS 135

136 INTERNAL AILMENTS

AILING GUT

Traditionally, Aborigines lived by the cycle of the seasons, and the flowers of the forest were their clock. The first flowering of a golden wattle might signal that on the coast the fish were biting or the turtles laying. By following their calender of flowers, Aborigines trekked from highlands to valley, from swamp to river, reaping seasonal riches of fishes and fruits, eggs and possums.

With a change of a season a tribe's diet might alter dramatically from a staple of yams to a mainstay of duck meat or eels. Aborigines needed strong constitutions to adapt to these sudden changes. Occasionally there were windfall feasts, when a whale was beached or a seabird colony discovered. These repasts sometimes brought diarrhoea and indigestion in their wake.

Explorer George Grey told of great whale feasts attended by Aborigines "out of temper from indigestion . . . suffering from a cutaneous disorder by high feeding, and altogether a disgusting spectacle". Victorian R.B. Smyth described a diet of bogong moths bringing on

◀ A calendar tree to Top End tribes, the flowering Darwin woollybutt (*Eucalyptus miniata*) signalled to Groote Eylandt Aborigines that wild bee honey was plentiful. A bark infusion was an Arnhem Land remedy for diarrhoea. Photographed at Robin Falls, Northern Territory.

"violent vomiting and other debilitating effects" until "they became accustomed to its use, and then thrive and fatten exceedingly upon it".

Aborigines needed effective remedies for digestive upsets, and colonial writers witnessed several of these. After successful kangaroo hunts along the Murray River, Aborigines would indulge in what settler E. Stephens described as a "huge tribal feast or gorge lasting for several days", with subsequent ill-effects: "Instances were frequent where those who were uncomfortably full — and that meant an extraordinary distention of the stomach — mutual help would be given and received for the purpose of obtaining relief. One distended native would lie down on his back or stomach while another would roll him from side to side, then mounting the prostrate form would gently tread him into comfort and convalescence."

A similar treatment was observed in 1889 by J.F. Mann: "when a native had eaten more than enough, he would betake himself to the nearest log, across which he would lie stomach downwards, and sometimes sway himself gently to and fro. This position although seemingly an awkward one, appeared to be fraught with ease and relief to the native."

Another settler, Victorian James Dawson, told of an unusual emetic: "When a child gorges itself with food, its mother gathers yellow leeches from underneath dry logs, and bruises them up along with the roasted liver of kangaroo, and sow thistles, and compels it to eat the mess, which is called kallup kallup. It acts as a strong emetic. Adults, when ill from overfeeding, are sometimes induced to take this dose, in ignorance of its composition; and it affects them strongly, but beneficially."

An odd remedy for diarrhoea in western Queensland was witnessed by Walter Roth in 1897. He wrote of "huge clay or mud pills, one or two of which at a time are prescribed for diarrhoea: these pellets are certainly at least twice the size of those to be seen at any pharmaceutist's".

The majority of stomachache remedies, however, were made from plants, and were less dramatic, such as those reported by Dawson: "For indigestion, the small roots of the narrow-leafed gum tree, or the back of the acacia, are infused in hot water, and the liquor drunk as a tonic."

▶ A bunch of dysentery bush fruits beckons forest birds. The fruits have a rich sweet taste. This plant, sprouting in a shady grotto at Kakadu, has much broader leaves than plants growing in open sunshine.

In northern Australia, one Aboriginal remedy for diarrhoea became so popular among bushmen it is known to this day as "dysentery bush" (*Grewia retusifolia*). This small understorey shrub with arching stems and velvety leaves, is also a foodplant, yielding small, bulging, bristly fruits, which give rise to its other, more colourful names — "dog's balls" and "dog's nuts". The fruits are a good source of thiamine. To treat diarrhoea, Aborigines crush and soak the roots, and drink the liquid or chew the astringent leaves. The roots help heal boils and infected sores. Dysentery bush is one of northern Australia's more significant bush remedies, still widely used today. It may repay close chemical examination.

Australian pioneers certainly needed dysentery remedies; their problem was not too much feasting but too few fruits and vegetables, poor hygiene, and too many recycled stews. A number of their remedies were probably learned from Aborigines.

▼ Of sida-retusa (*Sida rhombifolia*), colonial botanist Joseph Maiden wrote: "This herb is largely used by the natives of India in consumption and rheumatism. It is given as an infusion, and is said to promote perspiration; the leaves are used as a poultice for snake-bites."

In Queensland and New South Wales diarrhoea was often treated with sida-retusa or paddy's lucerne (*Sida rhombifolia*). Many elderly people swear by it even today. A diminutive member of the hibiscus family, it sprouts soft, green leaves, which are slimy in taste, tiny yellow hibiscus flowers, and very tough, knee-high stems, which colonists tried to manufacture into string. Though considered to be a native plant in Australia (it also grows widely overseas), sida-retusa has very weedy habits, often sprouting around homesteads and beside paths, to the convenience of ailing colonists. To treat diarrhoea, colonists chewed the leaves or boiled up a decoction; because of its pleasant, slimy taste the plant was sometimes called "jelly leaf". Lismore Aborigines reportedly used the roots to treat indigestion and diarrhoea, and colonists may have adopted what was originally an Aboriginal remedy.

Another treatment probably learned from Aborigines was the chewing of epiphytic

◀ Strings of plump figs hang from the branch of a cluster fig tree (*Ficus racemosa*) at Darwin. Arnhem Land Aborigines savoured the sweet fruits, and soaked scrapings of the inner wood in warm water to make a bath and drink for diarrhoea patients. Medicinal green tree ants can be seen patrolling the branch.

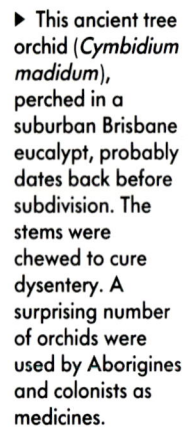

▶ This ancient tree orchid (*Cymbidium madidum*), perched in a suburban Brisbane eucalypt, probably dates back before subdivision. The stems were chewed to cure dysentery. A surprising number of orchids were used by Aborigines and colonists as medicines.

▶ In Tasmania the leaves and twigs of hop goodenia (*Goodenia ovata*) were taken as an infusion to treat diabetes. In damp eucalypt forests in south-eastern Australia, this is one of the most common shrubs. Photographed in eastern Port Phillip Bay, south of Melbourne.

orchids. Aborigines ate the stem bases of tree orchids (*Cymbidium canaliculatum* and *C. madidum*) for dysentery, and settlers apparently followed suit. These orchids, sprouting from hollows in dead trees, store reserves of arrowroot starch in their stems, which colonists extracted as food. C. Hedley of Gladstone told how the stems of black orchid could be grated up and boiled, producing a flour similar to arrowroot: "Delicate children have been reared on this when accidents have cut off from them other supplies. The natives speak of this valuable plant as 'Dampy-ampy', and amongst whites it is known as native arrowroot."

Many of the diarrhoea remedies of Aborigines and colonists were rich in tannins. The best sources are the dark-red, gum-like kinos found oozing from the trunks of eucalypts, angophoras (*Angophora*), and wattles. In 1827 the surgeon P. Cunningham told how wattle "gum" was used successfully in the colony, "in thin mucilage, as a drink in affections of the urinary organs, and dysentery", and suggested an industry based upon their collection: "Something, no doubt, might be made by us of this branch of trade, by encouraging the natives to procure the colonial gums. Even the very idle children, and the hordes of lazy fellows who hate hard work, might for a while obtain here an employment and a livelihood, if any spirited person would pay them well."

The inner barks of many trees are also very rich in tannins. Northern Territory Aborigines soaked the inner bark of cluster fig trees (*Ficus racemosa*), wild orange trees (*Capparis umbonata*), and Darwin woollybutts (*Eucalyptus miniata*), among others.

Leaves can also be a tannin source. In eastern New South Wales, settlers and Aborigines boiled up wild raspberry leaves (*Rubus*) into a therapeutic tea. Chinese tea, made from the leaves of a camellia (*Camellia sinensis*) is very rich in tannins, and if drunk black, is a good remedy for diarrhoea.

An interesting diarrhoea remedy on Groote Eylandt is a drink made from soaked tamarind fruit (*Tamarindus indica*). The tamarind is one of a dozen or so exotic plants to be incorporated into Aboriginal medicine. Originally from Africa, it was brought to the coasts of the Northern Territory some three hundred years ago by Indonesian bêche-de-mer fishermen from Macassar in Sulawesi (Celebes). This tree is so established at Groote Eylandt that Aborigines have adopted two language names for it — Angkayuwaya and Jamba.

Other remedies used to treat digestive complaints included the roots of native asparagus (*Protasparagus racemosus*); the roots of the pea vine (*Vigna vexillata*), a constipation remedy; and the core of the pandanus trunk (*Pandanus spiralis*).

Most Aboriginal remedies for diarrhoea were effective, but some, consisting of baths or the sniffing of leaves, would have been no more than placebos.

However fanciful these treatments, they were no stranger than immigrant ideas about the forest shrub called hop goodenia (*Goodenia ovata*). Around Wollongong this plant was regarded, not so much as a medicine, but as an aperitif. A.G. Hamilton wrote: "*Goodenia ovata* occurs on the edge of the forest [in the Illawara scrubs.] Some of the old Irish people call it 'Hungerweed,' and say that if you walk through it, you will get a good appetite."

▲ Roseleaf raspberry (*Rubus rosifolius*) is also called "thimbleberry" because its edible berries are hollow. Aborigines at Shoalhaven made a decoction to treat diarrhoea. Photographed at Royal National Park, south of Sydney.

142   INTERNAL AILMENTS

SCURVY STRICKEN

Captain Charles Sturt was not a lucky explorer. After becoming stranded in central Australia during drought, he was forced to withdraw suffering severely from scurvy. Under its curse he lost the use of his limbs, and had to be dragged about painfully in a jolting cart.

The Captain learned he was suffering from the disease, caused by severe vitamin C deficiency, in the desert that now bears his name:

> About the beginning of March [1845] I had had occasion to speak to Mr. Browne as to certain indications of disease that were upon me. I had violent headaches, unusual pains in my joints, and a coppery taste in my mouth. These symptoms I attributed to having slept so frequently on the hard ground and in the beds of creeks, and it was only when my mouth became sore, and my gums spongy, that I felt it necessary to trouble Mr. Browne, who at once told me that I was labouring under an attack of scurvy, and I regretted to learn from him that both he and Mr. Poole were similarly affected, but they hoped I had hitherto escaped.

◀ Colonial botanist F.M. Bailey said of false sarsaparilla (*Hardenbergia violacea*): "The roots of this beautiful purple flowered twiner are used by bushmen as a substitute for the true sarsaparilla, which is obtained from a widely different plant. I cannot vouch for any medicinal properties." Photographed on Mt Coot-tha, not far from an old gold mine.

▲ Major Mitchell marvelled at Cooper clover (*Trigonella suavissima*), exclaiming that: "The perfume of this herb, its freshness and flavour, induced me to try it as a vegetable, and we found it to be delicious, tender as spinach, and to preserve a very green colour when boiled."

Poor Mr Poole got worse. On 10 May Sturt recorded that "all his skin along the muscles turned black, and large pieces of spongy flesh hung from the roof of his mouth, which was in such a state that he could hardly eat". Two months later he died.

Sturt himself was lucky to recover, after Mr Browne fed him a "large tureen full" of some small acid berries the Aborigines were eating. Sturt exclaimed that "to the benefit I derived from these berries I attribute my more speedy recovery". The berries were probably wild tomatoes (*Solanum* species), some of which are rich in vitamin C.

Sturt was not the first explorer to be stalked by scurvy, nor the last. The rear party of Burke and Wills' expedition battled the disease, as did Major Mitchell and Matthew Flinders. Mitchell, like other explorers, held erroneous ideas about scurvy, thinking it was contagious, and could be contracted by eating rancid pork. Mitchell guarded against the disease by eating cooper clover (*Trigonella suavissima*). The explorers who triumphed over scurvy were those who regularly dined on wild foods such as Ludwig Leichhardt and Captain James Cook.

When Leichhardt lost rations battling through the thick brigalow scrubs of central Queensland, he was forced to supplement his diet with all kinds of wild fruits, including desert limes (*Eremocitrus glauca*), gums oozing from trees, and tubers pilfered from Aboriginal camps. His men enjoyed hale health.

The astute mariner Captain Cook insisted that

▲ Australia's first native vegetable in cultivation was New Zealand spinach (*Tetragonia tetragonoides*), seen growing here on a beach at Burleigh Heads on the Gold Coast. Joseph Banks took seed back to England where the plant was grown as a summer spinach.

▲ The Australian explorer Charles Sturt made many prodigious journeys, such as his Central Australian Expedition of 1844, on which he was beleaguered by scurvy.

▲ The fruits of wild currant bush (*Leptomeria acida*) were sold last century in Sydney's markets. P. Cunningham in 1827 said they were "strongly acidulous, like the cranberry, and make an excellent preserve when mixed with the raspberry". Photographed at Waterfall near Sydney.

at each landfall the leaves of wild plants be served to his crew. At Botany Bay the men fed on New Zealand spinach (*Tetragonia tetragonoides*), and sea celery (*Apium prostratum*); at Endeavour River they dined on pink purslane (*Sesuvium portulacastrum*). It is likely that none of these plants contains much vitamin C, but the "scurvy grass" (actually a cress *Lepidium oleraceum*) eaten previously in New Zealand was probably an excellent antiscorbutic — cresses usually are — and may have sustained the sailors through their Australian travels.

On return to England Cook was awarded the Copley Medal, the Royal Society's highest honour, for his victory over scurvy. Unfortunately, men continued to die from the disease well into the twentieth century, and the importance of fresh fruits and vegetables was largely overlooked.

The first convict settlement in Australia was severely scourged by scurvy, and wild plants proved to be the only cure. The First Fleet surgeons deemed that two plants were especially efficacious, and these figured prominently in the life of the colony.

Currant bush (*Leptomeria acida*) was no doubt chosen because its tiny green fruits taste like limes, the recommended naval antiscorbutic. The fruits sprout on a broom-like shrub found on the sandstone ridges around Sydney. Surgeon General John White considered them "a good antiscorbutic; but I am sorry to add, that the quantity to be met with is far from sufficient to remove the scurvy". The currants yield modest amounts of vitamin C — about twenty milligrams per hundred grams, compared with fifty milligrams in oranges. Four cupfuls of the fruit *daily* would be needed to cure scurvy, and this amount would rarely have been available.

Nonetheless, currant bush may have saved a

► In 1860 colonial naturalist George Bennett wrote of native sarsaparilla (*Smilax glyciphylla*): "The leaves are sweet when chewed (resembling the taste of liquorice). A decoction of the plant is used as sarsaparilla." Photographed at Burleigh Heads, Queensland.

number of lives. The human cargo aboard the *Neptune*, a convict hulk that limped into Sydney in 1790, was lucky to arrive while the currants were still in season. Conditions aboard the *Neptune* had been appalling, and 164 convicts perished on board. The ship was privately contracted, and its owners profiteered by setting rations at starvation levels. It was said that deaths among the convicts were concealed so that others could share their rations — until the stench of bodies alerted the officers. Once on land the pitiful survivors were restored with special bread rations and supplies of the "acid berry of the country", as Judge Advocate Collins described the fruit.

The other scurvy cure of the colony was the native sarsaparilla vine (*Smilax glyciphylla*), known then as "sweet tea" because of the curious taste of its leaves. A common creeper of gullies around Sydney to this day, it was gathered in large amounts by convicts and troopers. White described the leaves as tasting "sweet, exactly like the liquorice root of the shops. Of this the convicts and soldiers make an infusion which is tolerably pleasant, and serves as no bad succedaneum for tea. Indeed, were it to be met with in greater abundance, it would be found very beneficial to those poor creatures, whose constant diet is salt provisions."

Native sarsaparilla leaves have not been tested for vitamins, but the green berries are known to contain a mere six milligrams of ascorbic acid per hundred grams. At this rate an impossible quantity of leaves would have been needed to cure just one person's scurvy.

As settlement unfolded other vines came into

use as sarsaparillas, either because they vaguely resembled sweet tea or had similar tasting leaves. The substitutes included false sarsaparilla (*Hardenbergia violacea*), native sarsaparilla (*Smilax australis*), climbing lignum (*Muehlenbeckia adpressa* — also called sarsaparilla) and native peas (*Kennedia*).

Colonial botanists scoffed at the use of these plants. In 1898 Joseph Maiden complained of false sarsaparilla: "The roots of this plant are sometimes used by bushmen as a substitute for the true colonial sarsaparilla (*Smilax*), but its virtues are purely imaginary. It is also a common thing, in the spring, in the streets of Sydney, to see persons with large bundles of the leaves on their shoulders, doubtless under the impression that they have the leaves of *Smilax glyciphylla*."

False sarsaparilla was put to one very interesting use. In Victoria, goldminers believed it grew upon lodes of gold. According to Morris: "Old diggers consider the presence of sarsaparilla and the ironbark tree as indicative of the existence of golden wreath below. Whether these can be accepted as indicators in the vegetable kingdom of gold below is questionable, but it is nonetheless a fact that the sarsaparilla and the ironbark tree are common on most of Victoria's gold fields." In New South Wales the blue-flowered herb *Parahebe perfoliata* was sought out for the same reason. Whether anyone became rich by digging beneath these plants, history does not say.

▲ The leaves of yellow wood sorrel (*Oxalis corniculata*) are so sour that Tasmanian colonists baked them in sweet pies in place of barberries. Photographed on the coastal dunes of Moreton Island.

▼ The "scurvy grass" of north-west Queensland pioneers (*Commelina ensifolia*) is not a grass but a close relation of the garden wandering jew. The leaves taste no better than grass, however. Photographed on Cape York.

Native sarsaparilla does not grow west of the Dividing Range, but there were other plants that outback pioneers turned to for scurvy. Settlers in north-west Queensland cooked up a small herb they called "scurvy grass" (*Commelina ensifolia*). In south-western Queensland this name was given to a different plant, a desert mustard, of which pioneer Alice Duncan-Kemp wrote: "Drovers on long trips with cattle gather and boil the leaves like cabbage; its use helps prevent beriberi and scurvy."

On the Victorian goldfields the yellow wood sorrel (*Oxalis corniculata*) was believed to cure scurvy, probably because its tiny leaves have a lemon-juice taste. Explorer John McDougall Stuart depended upon one of the pigfaces (family Aizoaceae) calling it "squash"; it was said to be a "splendid thing" to cure scurvy. Augustus Gregory's exploring party used the young branches of the native grape (*Ampelocissus acetosa*).

Pigweed (*Portulaca oleracea*) in particular was valued as an antiscorbutic. Colonial botanist Baron von Mueller, exploring with Gregory in northern Australia, attributed "the continuance of our health partly to the constant use of this plant". Other antiscorbutics, according to Mueller, included Bower spinach (*Tetragonia implexicoma*), salt lawrencia (*Lawrencia spicata*), native

◆ Old man saltbush (*Atriplex nummularia*), a famous outback shrub, was taken by early settlers to treat scurvy and blood diseases. Some Aboriginal tribes endowed the plant with healing powers. Photographed at Broken Hill.

wood cresses (*Cardamine*), and yellow cress (*Rorippa palustris*). Elsewhere, old man saltbush (*Atriplex nummularia*) was used.

How effective were these plants? Sad to say, hardly any have ever been analysed. In recent years dozens of wild foods have been tested for vitamin C, but the emphasis has been on tropical and desert Aboriginal foods, not on the scurvy vegetables of colonists. Pigweed, which was tested by the Department of Defence Support, yielded only traces of the vitamin. The native mustards and cresses are probably antiscorbutic (plants in the cress/mustard family usually are), but the other plants belong to families not noted for their vitamin C content.

Australia's colonists overlooked the one truly outstanding vitamin C source — the billygoat plum (*Terminalia ferdinandiana*). In 1983 this extraordinary antiscorbutic was "discovered", when its fruit was found to contain a world record amount of vitamin C. A sample of the pulp yielded 3150 milligrams of vitamin C per hundred grams — sixty times the vitamin content of oranges. The plant achieved instant world fame. The fruit had actually been tested three years earlier by the Department of Defence Support, but their sample yielded only 406 milligrams per hundred grams, a lesser but still very impressive figure. The vitamin content of the wild fruits is no doubt very variable.

A scrawny tree of tropical woodlands between Arnhem Land and the Kimberleys, the billygoat plum has huge green deciduous leaves, and sprays of tiny white flowers, which at the beginning of the dry season ripen into small green "plums". Enveloping the stone of each fruit is a layer of pleasantly sour flesh, much loved by Aborigines, and greedily eaten by fruit bats and emus. The tree also oozes edible gum from its trunk, and some tribes soaked the red inner bark as a remedy for sores.

There is now talk of establishing billygoat plum plantations as a vitamin source. The sour fruit is already served as a sauce in Sydney wild food restaurants. Tourist operators in the Top End want the fruit renamed "Kakadu plum", to the chagrin of local Aborigines who have eaten the fruit for thousands of years.

The billygoat plums that made the headlines came from the bush surrounding Belyuen Aboriginal community near Darwin. Chairman Harry Sing told me of local reaction to the news of finding: "We got a shock! We couldn't believe our ears. As kids we used to fill up our pockets with the fruits and put them in billycans and anything. No wonder we never got colds. And to think we used to go to the shop to buy oranges!"

◀ The remarkable billygoat plum tree (*Terminalia ferdinandiana*), also called vitamin C tree and green plum, is a common tree of Top End woodlands, growing here at Robin Falls near Adelaide River. The trees cast off their enormous leaves during the Dry.

150 INTERNAL AILMENTS

HEADACHES, COLDS, FEVERS

One of Australia's most interesting bush medicines grows on sheltered flanks of the Great Dividing Range in wet gullies and fringing rainforests. Headache vine (*Clematis glycinoides*) is a twiner with leaves in threes, cascading starry white flowers, and a pungency that belies its innocent looks.

If you crush a handful of the leaves, and inhale deeply, any headache or nasal congestion is soon left behind as the nose starts smarting, the eyes shed water, and a throbbing head becomes an exploding furnace.

Bushmen and Aborigines swear this vine clears headaches, and a friend of mine who tried it certainly thought it worked. The piercing aroma of the leaves recalls ammonia.

In Papua New Guinea, where this plant also grows, villagers inhale the leaves to treat colds. Also of interest is a related clematis in Africa (*C. hirsutus*), which goes under the same local name and is used in the same way. The chemical

◄ Headache vine (*Clematis glycinoides*) is a famous bushman's remedy for headaches. The plant's popularity, and its popular name, suggest it really works.

HEADACHES, COLDS, FEVERS   151

constituents of this plant would certainly repay investigation.

Aboriginal tribes had a number of herbal remedies for headaches. For example, the crushed leaves of aromatic plants such as liniment tree (*Melaleuca symphyocarpa*) might be inhaled. Tribes in north-western Queensland thought that sniffing any strongly scented plant would help. Different desert tribes made poultices from the leaves of tick-weed (*Cleome viscosa*) or the vine (*Mukia maderaspatana*).

Most headache remedies, however, were non-herbal. Brisbane Aborigines rested the head between hot stones, struck the head with a waddy, or dived under water holding the breath. Other tribes tied ropes around the head or neck. Sometimes blood was drawn or blisters raised. At Kakadu, wrote Chaloupka and Guiliani, "Headache is treated by tying a strip of bark tightly around the head. This is usually that of any species whose inner bark is used for string making, although those that are 'cold', i.e. have running sap are preferred. In the past, when the headache was severe and the binding of the head didn't alleviate the pain, the nose was cut with a stone blade to let the 'bad blood' out." These treatments, largely psychosomatic, were probably very effective.

For treating coughs, colds and sore throats Aborigines had an enormous range of useful herbal remedies. Aromatic plants were especially popular. They were inhaled directly, or slept upon as pillows, prepared into ointments, or drunk as infusions. Lemon grasses (*Cymbopogon*), tea tree (*Melaleuca*), fuchsia bushes (*Eremophila*), ragworts (*Pterocaulon*) and many others were used in this way. Aromatic oils stimulate the cells lining the throat to secrete more lubricating fluid, which eases the irritation causing the cough.

Other cough medicines, such as the inner barks of trees, depended upon their tannin content to provide relief. Still others, including the inner stem of the fan palm (*Livistona humilis*), pandanus leaf bases (*Pandanus spiralis*) and the leaves of the clerodendron (*Clerodendrum floribundum*) worked in less obvious ways.

Pioneers stricken with colds or sore throats turned to many natural remedies. Herbal teas could be brewed from horehound (*Marrubium vulgare*) or fruit salad plant (*Pterocaulon sphacelatum*). Banksia nectar (*Banksia*) made a soothing syrup. To ease sore throats, colonial botanists suggested a gargle of native grapes (*Cissus hypoglauca*), or lozenges made from the sweet astringent bark of blush coondoo (*Planchonella laurifolia*), a rainforest tree. The most favoured bush remedy, however, and one that remains popular to this day, is eucalyptus oil. Those who could not afford the many proprietary products could make do by pouring boiling water over gum leaves to create a soothing vapour. Some bushmen chewed the leaves each day to ward off colds.

Fever was a far more serious malady in Australia, and typhoid in particular cut short many lives last century. Bushmen and botanists scoured the bush for remedies, hoping to discover a substitute for the bitter bark of the South American cinchona tree (*Cinchona succirubra*),

▶ On a desert dune at Windorah a tick weed (*Cleome viscosa*) is oblivious to the scalding heat. Some tribes used this herb as a medicine for rheumatism, headaches, or colds. Near Coen, wrote Walter Roth, "The whole plant is smashed up in the dry, and placed on boils or open sores."

◀ Colonial physicians rightly scorned quinine berry (*Petalostigma pubescens*). Botanist F.M. Bailey noted that it was "often used as an astringent by bushmen. The bitter principle contained in the bark is not considered of value by medical men."

◀ To make a medicine from the fan palm (*Livistona humilis*), Aborigines on Bathurst and Croker islands extract the white base of the growing tip, then pound, soak, and boil it to make a drink taken for sore throats and coughs. Photographed south of Adelaide River, Northern Territory.

the source of quinine. In one bitter-barked, shiny-leaved tree of rainforest edges and inland woodlands they thought they had found it.

Quinine tree (*Alstonia constricta*) is known by country people to this day as quinine, quinine-bush, bitter-bark, fever-bark, or Peruvian bark. This remarkable tally of names shows how pervasive its medicinal reputation has become. A tincture of the bark was taken to treat typhoid and dysentery, and nineteenth-century doctors were loud in their praises. Said Dr Joseph Bancroft, "After fifteen years' experience of the use of Alstonia the writer is of opinion that there is no better or more generally useful tonic". American doctor A.W. Bixby declared, "In typhoid, synochal, and puerperal fevers, where an antiseptic and nerve tonic is demanded, it answers well". And, "I believe it will become a favourite with all who test it".

Quinine bark was exported to England, and during World War II served in place of imported gentian and other bitters in the Australian War Pharmacopoeia. English brewers used it instead of hops. The bark contains a number of interesting alkaloids, including reserpine, which lowers blood pressure and acts as a tranquilliser.

Professional herbalists in Australia have adopted the bark as a remedy for high blood pressure, although this is likely to be banned by the federal government as a health risk in the near future.

Two other trees took their name from quinine, although neither nowadays has any claim to cure fevers. The desert poplar (*Codonocarpus cotinifolius*) or horseradish tree, a plant with a strange mustard smell, was known last century as quinine tree or medicine tree, but its reputation as a fever cure was short-lived. The other tree, called quinine berry (*Petalostigma pubescens*) gained fame as a febrifuge because its hard, orange fruits taste intensely bitter. Bushmen popped them into their tea hoping for protection against malaria. But a colonial naturalist, Reverend Tenison-Woods, was scornful: "[The tree] is usually covered with fruit like a small yellow plum, of eminently nasty taste. This is, I believe, its only claim to be called a 'quinine'." Chemists have since failed to isolate any alkaloids in the plant, suggesting that these sentiments were well-founded.

Another dubious fever cure was the north Queensland Leichhardt tree (*Nauclea orientalis*), a majestic tree of riverbanks and rainforest edges. Botanist Joseph Maiden stated, "it has a bitter bark, long known to bushmen, and used by them occasionally in the form of an infusion for ague, in default of quinine, but with no good results, as far as I could ascertain."

Eucalyptus oil, the colonial cure-all, was also believed to prevent and cure even the most serious of fevers. Extraordinary faith was placed in its powers, as shown, for example, by a testimonial from Mr Graham Mitchell in *The Queenslander* of 28 July 1888: "A few drops taken each morning is a grand preventer of malarial fever, and why not in typhoid, that fell destroyer now in our midst? I had to treat myself for fever and dysentery, and I attribute my recovery in the last stage solely to Bosisto's Eucalyptus Oil."

Nineteenth-century doctors believed that rotting vegetation, particularly in swamps, gave out malaria-producing vapours, and that eucalypts were among the few trees able to destroy this "miasmatic influence". In Europe and north Africa huge plantations of blue gums (*Eucalyptus globulus*) were planted wherever malaria threatened, with apparently beneficial results. This remarkable programme of public hygiene is described in "The Mighty Eucalyptus".

▲ Bland petal salad can be made from the succulent flowers of the kapok tree (*Cochlospermum fraseri*), which also has sappy edible roots. This deciduous tree with its showy yellow flowers is one of the Top End's most distinctive plants. Photographed at Kakadu.

▲ The name is a misnomer, for quinine tree (*Alstonia constricta*) affords no protection against malaria, though an extract of its bark does lower blood pressure. It grows in Queensland and northern New South Wales along rainforest margins and on sandy inland soils, as here at Augathella.

Aborigines had many herbal remedies for fevers and most were the same as their treatments for colds. Fevers were probably rare before the coming of the white man. The Ngarinman of the Northern Territory treat fever by drinking a brew of boiled kapok-tree (*Cochlospermum fraseri*) bark and flowers, one of the very few Australian remedies that employs flowers. The Yidinyji of north Queensland "cured" influenza, a white man's disease, by diving deep into cold water and lying there until the ears "clicked", thus clearing the head.

156   INTERNAL AILMENTS

BABY MEDICINE

Most Aboriginal medicines were known to all the members of the tribe, but the treatments related to birth and contraception were women's secrets. Men knew the treatments existed, but not the ways in which they worked or the plants used.

In most of Aboriginal Australia a baby's birth was attended by special medicinal rites, which have no parallel in western medicine. Immediately after birth, for example, the woman was likely to be smoked over a fire of steaming herbs. Special plants would be thrown on to a small fire and the mother would squat on top; often the fire was lit inside a small pit. Berrigan (*Eremophila longifolia*), various wattles (*Acacia dictyophleba*, *A. lysiphloia*, *A. pruinocarpa*), spike rush (*Eleocharis dulcis*) and silky oilgrass (*Cymbopogon bombycinus*) were among the special plants used. Smoking or steaming of a baby to make it strong was an equally important rite. In the desert, the same kinds of wattles were used. At Kakadu, newborn babies were welcomed into the world over steaming ironwood leaves

◀ Nobbly nuts of pandanus (*Pandanus spiralis*) were a favoured food of northern Aborigines, who cracked the kernels to extract the tiny seeds. Different parts of the plant were prescribed as medicines for colds, sores, diarrhoea, toothache, and other ills. Photographed at Charles Point, Northern Territory.

(*Erythrophleum chlorostachys*); corpses departed in the same way.

Marjorie Bil Bil of Belyuen near Darwin told me how her five-month-old baby was steamed with the foliage of the broad-leaved native cherry (*Exocarpos latifolius*). Marjorie's mother collected the foliage the day before, and directed Marjorie to collect a bundle of small firewood. Very early the next morning, they kindled a small fire. Said Marjorie: "When the flame came up we held the baby by the arms and legs over the fire, first turning her face down, and then away. We did this for about half an hour."

"Did the baby cry?" I asked.

"Of course," Marjorie said. "It cried a lot. It had smoke in its eyes."

Babies were often rubbed with oil, which was usually said to make them stronger. In northwest Queensland, noted Roth in 1897, "the infant is smeared from top to toe with iguana fat, which is renovated continuously during its early years". At Kakadu a one-month-old baby was made strong by pressing heated ironwood leaves to its eyes, mouth, elbows, wrists, fingers, backbone, hips, genitals, knees and ankles. When old enough to turn on its side, a second rite was undertaken, both to make the baby strong and to stop it crying when its mother was away from camp. The baby was dabbed with heated lemon grass (*Cymbopogon procerus*) and made to inhale the fumes of roasting round yams (*Dioscorea bulbifera*).

Sick babies might be treated with special babies' remedies. The desert Warlpiri rubbed a sick baby with burnt antbed taken from resinous anthills. A baby suffering from diarrhoea was smoked over a fire of smouldering wattle leaves (*Acacia adsurgens* or *A. ancistrocarpa*). In Victoria, a delicate baby was rubbed down with grub fat.

To induce lactation, any number of treatments were tried by different tribes. In some outback areas the breasts were steamed above berrigan foliage or old man saltbush (*Atriplex nummularia*). In Victoria the breasts were bathed in lime-water made by burning mussel shells and "dissolving" them in water. Dulcie Levitt tells of the technique practised at Yirrkala in Arnhem Land:

Old Pandanus fruit was cooked on hot stones in an earth oven. When really hot, wet grass from the river was put on the hot stones and fruit. Steam came up, and the woman sat on it, or bent over it, to allow the steam to come up around the breasts to induce a flow of milk. She covered herself with paper-bark to keep the steam in; blankets or clothing are now used but the steam is made in the same way. This was usually practised by women who had to feed a dead relation's baby, but was also used to improve the flow of milk if the mother of a newborn baby had insufficient milk.

Several of the plants taken to bring on milk have copius milky sap, and perhaps worked by analogy more than anything else. Crushed mussel shells probably produced a milky liquid as well. This is similar to the belief of the English peasant that sows eat milky sowthistles to

▶ Marjorie Bil Bil of Belyuen shows me a native cherry tree (*Exocarpos latifolius*), in the smoke of which she strengthened her baby. Some tribes burnt the leaves of this tree as an insect repellant, or treated sores with a leaf solution. The tree produces edible fruit.

increase lactation. But it is worth noting that the milky plants used by Aborigines — milkwood tree (*Alstonia actinophylla*), caustic bush (*Sarcostemma australe*), and spurges (*Euphorbia australis* and *E. coghlanii*) — have a very irritating latex, and this may stimulate the breasts to secrete.

Beastfeeding Aboriginal women sometimes suffered from sore nipples. In the Northern Territory deserts, relief was obtained by applying the black ash from the powdered bark of desert hakeas (*Hakea eyreana* and *H. suberea*). This extraordinary ash is as fine as talcum powder, and Aborigines applied it to sore lips, tongues, and the sore gums of children. Hakeas were important trees to desert Aborigines as they also bore edible seeds and honey-laden flowers from which drinks were made. Increase ceremonies were performed to ensure supply of honey.

If a baby had teething pains, in place of teething rusks, desert Aborigines prescribed the spongy plates of scarlet bracket fungus (*Pycnoporus coccineus*). These were not edible, but had a soft and tough texture, ideal for exercising tender teething gums. Another fungus, a West Australia puffball, had spores that served as baby powder.

If a baby became too hot, desert Aborigines laid it upon a bed of horseradish tree foliage (*Codonocarpus cotinifolius*). The water in the large fleshy leaves evaporates quickly, creating a cooling effect like a canvas waterbag.

Bearing too many babies is a liability in a difficult environment, and it is not surprising that women in some tribes claimed knowledge of "contraceptive" plants. Today little is known of such treatments, and this is a realm of Aboriginal medicine that remains largely a mystery. We do know that such treatments were believed to bring permanent sterility, and they were only taken after a woman had borne enough children.

Anthropologist Donald Thomson gave a rare account of contraception as practised on Cape York Peninsula: "The men, as usual, in all matters pertaining to women, such for example as childbirth, generally disclaim any first-hand knowledge of this medicine, but they freely admit that a '*keni* belong woman', is used, and declare that they would be angry if they found their women using it." The remedies in question were the round yam (*Dioscorea bulbifera*) and matchbox bean (*Entada phaseoloides*). Both are important foods, but unless carefully prepared by washing and cooking, they are highly poisonous. Women using them as contraceptives merely roasted them or ate them raw. Said Thomson: "I was informed that this *keni* is taken in the early morning on an empty stomach, after which the woman lies down, refraining from drinking throughout the day, until sundown.

▲ Clinging to a cliff, a desert hakea (*Hakea suberea*) beckons birds with its bunches of bushy blossoms. Photographed in Finke Gorge, Northern Territory.

The old women declare that once a woman had taken this medicine she would never have a child." It may be significant that related yams in Asia and America provide the raw material for most of the world's contraceptive pills.

Other reports of contraceptive plants lie scattered through the literature. In Arnhem Land a woman was smoked over the burning bark of the ironwood tree or denhamia (*Denhamia obscura*). At Cooktown she used the Cape York lily (*Curcuma australasica* — details unknown) or drank a decoction of leaves and roots of the mapoon shrub (*Morinda reticulata*). In Western Australia she reportedly ate the ground seeds of the vine called doubah (*Leichhardtia australis*), and at Babinda she ate tree orchid seeds (*Cymbidium madidum*). Contraceptive plants have also been alluded to in central Australia, but the main authority on desert plant use, botanist Peter Latz, has been unable to confirm their identity — such plants are not talked of openly, especially to white men.

Did these plants work? In 1975 two CSIRO scientists tested 80 supposed anti-fertility plants, including round yam and matchbox bean, by feeding them to female rats. The rats kept on multiplying, showing that in this situation at

◀ Mysterious mapoon (*Morinda reticulata*), a poisonous shrub known to concentrate selenium, was reportedly taken by Aborigines to induce sterility. A decoction of leaves and roots was supposedly drunk by women near Cooktown. Photographed at Cape York.

▶ Delectable doubah (*Leichhardtia australis*), an edible outback vine, has a pod tasting like baby zucchini, as well as edible shoots, leaves, flowers and roots. The dried ground seeds were supposedly eaten by women of one tribe as a contraceptive. Photographed near Alice Springs.

◀ A colourful contraceptive was the Cape York lily (*Curcuma australasica*), supposedly used by Aborigines at Cooktown. All that is known of this forgotten treatment comes from a comment by botanist Len Webb: "apparently used externally, but method not clear."

least the plants did not work. Unfortunately, only a handful of the plants tested were Aboriginal remedies, as the majority were from New Caledonia and other Pacific islands.

The most detailed descriptions of Aboriginal contraceptives come from missionary Dulcie Levitt in 1975, writing from Groote Eylandt in the Northern Territory. Of the five treatments she records, those involving quinine berry (*Petalostigma pubescens*), pandanus nuts, cypress gum (*Callitris intratropica*) and broadleaved native cherry sound vaguely plausible. But the other two, requiring the drinking of rainwater and the eating of iron filings from a rusty axe, suggest that at least in this part of the world, contraception did not work.

BABY MEDICINE 161

162　EXTERNAL AILMENTS

CUTS, WOUNDS, SORES

For suppurating sores, one of the best healing agents in the Australian outback is the caustic bush (*Sarcostemma australe*), also known as caustic vine and milk bush. A singularly eerie plant, it consists of hundreds of leafless, finger-like stalks, which ooze a copious milk sap when broken. Aborigines dabbed this latex over sores and wounds, letting it dry to form a protective covering — an instant bush bandaid. This sap is so milky that in many desert tongues the plant's name translates as "milk-milk". The Western Arrante regard "ipatja-ipatja" as a feminine symbol and apply it to women's breasts during rituals. At Kalumburu in the Kimberleys the sap is believed to bring on lactation.

Outback pioneers held caustic bush in high esteem. Stockman Max Koch noted in 1898 that "Bushmen use the milk to heal sores", and in a glowing testimonial to botanist Joseph Maiden, a settler from north-west New South Wales claimed: "The caustic plant is really

◀ Unfazed by fires, cycads (*Cycas armstrongii*) thrive in woodlands where Aborigines still burn the land; the fires stimulate seeding. Arnhem Land Aborigines chopped up male flower stalks, mixed them with warmed urine, and applied them to spear wounds. Photographed at Charles Point near Darwin.

▶ On a rocky ridge near Alice Springs the sinking sun strikes the tangled stems of a caustic bush (*Sarcostemma austrae*). This strange plant is very vulnerable to fire and usually grows among rocks, on open plains, in scrubs, or other fire refuges. It is poisonous to cattle.

▲ Bloodwood kino is potent medicine. Desert plant man Peter Latz writes: "Mixed with a little water, it was applied to sore eyes and lips, wounds, burns and sores. A small amount of the liquid was sometimes drunk for sore throats."

very valuable, and will cure warts and corns very quickly".

An enormous variety of plants were used in Aboriginal Australia to treat sores and wounds. Various crushed or heated leaves, bark decoctions, and ointments were applied to cuts, boils, sores and warts. Caustic bush was probably one of the most powerful of these, judging by its wide reputation among both black and white Australians. The latex may contain a caustic property that destroys bacteria and cleans away damaged tissue.

Another famous desert cure, used both by Aborigines and settlers, is the western bloodwood (*Eucalyptus terminalis*), one of the common outback eucalypts. A dark red gum or kino oozing from wounds on the trunk was mixed with water and applied to sores, wounds and burns, forming a protective plastic film.

Jeannie Scollay, a white mother of two from Alice Springs, is full of enthusiasm for this remedy. "It's fantastic for nappy rash," she told me. "It stings a bit when it goes on, but that goes away after a few seconds. But it's drying as well as healing. It just works more effectively than all the expensive creams I was using. But it stains permanently," she said, holding aloft a brown-streaked nappy. "A permanent dark stain."

Other wound-healing eucalypts include the famous ghost gum (*E. papuana*) and the bloodwood (*E. dichromophloia*). The kinos of these and other eucalypts are exceptionally rich in tannins, astringent substances that promote rapid healing by forming a patina over wounds (see "Plant Chemistry"). Colonial doctors prescribed various eucalypt kinos to staunch bleeding, and the kino of river red gum (*E. camaldulensis*) was even exported to Britain as a medicine.

Tannins are also found in inner barks, not

only on eucalypts, but in wattles and many other Australian trees. Aborigines infused the barks in water to prepare a healing wash, the liquid turning red in the presence of tannin by-products. As the Northern Territory Pharmacopoeia notes, in respect to ironwood (*Erythrophleum chlorostachys*): "The bark showing the darkest colour is chosen, as this gives the strongest preparation". Tannin-rich barks used by Aborigines included those of the cocky apple (*Planchonia careya*), native capers (*Capparis*), native bauhinia (*Lysiphyllum cunninghamii*) and markura (*Owenia vernicosa*).

Tannin plants were among the main wound-healing herbs in the Aboriginal medicine chest. Also very important were aromatic plants, which contained antiseptic aromatic oils. Aborigines sometimes bound chewed gum leaves to wounds, or soaked them in water as a wash, and these treatments may have combined the actions of tannins and aromatic oils.

In central Australia the Arrante and Anmatjirra climb into the rocky hills to gather the aromatic leaves of the white cypress pine (*Callitris glaucophylla*), which when crushed and boiled serve as a wash for sores, rashes and scabies. The leaves contain alphapinene, a highly antiseptic oil. Other aromatic plants used in the Northern Territory for wounds and sores include lemon-scented grass (*Cymbopogon ambiguus*), fringe-myrtle (*Calytrix exstipulata*), wurrumbu (*Pityrodia jamesii*), jirrpirinypa (*Stemodia viscosa*), and toothed ragwort (*Pterocaulon serrulatum*).

Colonists treated wounds with eucalyptus leaves and distilled oil. Herbalist H.P. Rasmussen declared: "Eucalyptus oil is an excellent remedy for stopping the bleeding, and an ointment made by boiling two pounds of blue gum leaves in a pound of hogs lard, will heal most wounds."

A third category of wound-healing plants are those with a toxic milky sap such as the caustic bush. The sap may "digest" bacteria and infected tissue. One example is the milkwood tree (*Alstonia actinophylla*) of northern Australia, as described by Dulcie Levitt from Groote Eylandt: "The thick, corky bark was chipped with a piece of stick until the milky sap oozed out. This was collected on a twig and applied to small sores. Care was taken not to get it on the fingers because if it got into the eyes it could cause blindness."

▲ Milkwood tree (*Alstonia actinophylla*), according to police inspector Paul Foelsche writing in 1881, produces latex that "possess extraordinary healing properties. It is applied to the wound with the finger, and is very sticky." Photographed near Darwin.

◀ A gum-leaf bandage was made by Top End Aborigines by plucking the foliage of the swamp bloodwood (*Eucalyptus ptychocarpa*), a eucalypt with striking red flowers. Aromatic oils seeping from the leaf would have an antiseptic effect. Photographed at Robin Falls, Northern Territory.

▲ Stinging penises after circumcision were soothed with the heated leaves of the golden guinea flower (*Dillenia alata*) on Groote Eylandt. This tropical tree also bears edible seeds and has a soft trunk suitable for canoe-making.

▼ The grotesque fruit of the great morinda (*Morinda citrifolia*) has a smell to match — compared to camembert cheese in a urinal. Nonetheless it is a good source of vitamin C, and a popular food of fruit bats, which spread about the seeds of this tropical seashore shrub.

Crowned sandpaper fig (*Ficus coronata*) and the milk creeper (*Secamone elliptica*) of northern Australia are further examples. Latex-oozing plants were often used to remove warts.

Most of the other wound-healing plants are not easily grouped by their properties. They include plants as diverse as a mistletoe (*Decaisnina brittenii*), a potent medicine according to Kakadu Aborigines; several tree orchids; lilies; a grass; and herbs such as Australian bugle (*Ajuga australis*). Many of these plants, such as billygoat weed (*Ageratum*), are described in other chapters.

One of the strongest wound herbs, by reputation, is the northern Australian dysentery bush

▲ The mistletoe (*Decaisnina brittenii*) found hanging from tea trees is a favourite remedy of the Kakadu Gagadju. A solution of mistletoe was washed onto sores and wounds. Photographed at Yellow Waters, Kakadu.

(*Grewia retusifolia*). Its roots probably contain a lot of mucilage, for when boiled in a little water they produce a thin jelly, which is considered very potent when applied to boils and sores.

Some northern communities treated sores by applying crushed fruit pulp such as that from great morinda (*Morinda citrifolia*), sea fan flower (*Scaevola sericea*), bitterbark (*Ervatamia orientalis*) and pandanus (*Pandanus spiralis*).

On Groote Eylandt the coast spinifex (*Spinifex longifolius*) of local beaches is hammered and soaked to prepare a wash for sores and burns. Dulcie Levitt tells how a man on a camping holiday tried this treatment on his daughter: "He washed his daughter with clean water, applied the mixture with a clean rag, and waited to see what would happen. The liquid was applied on a Thursday, by the following Sunday all traces of the sores had gone."

Sometimes wild leaves were used, not to make decoctions, but as dressings to staunch the flow of blood. A leaf, heated over a fire, was pressed on a wound until it stuck. The very large leaves of the cotton tree (*Hibiscus tilaceus*), green plum (*Buchanania obovata*), lady apple (*Syzygium suborbiculare*) and swamp bloodwood (*Eucalyptus ptychocarpa*) were used in this way.

Most wound-healing herbs were used in a very general way, against sores, boils, rashes and cuts, but a select few were considered by particular tribes to be very specific. The cycad (*Cycas armstrongii*) and the golden guinea tree (*Dillenia alata*) were prescribed in the Northern Territory only for spear or circumcision wounds. One north Queensland tribe applied a decoction of long yam (*Dioscorea transversa*) to skin cancer. Native nutmeg resin (*Myristica insipida*) and sandpaper fig (*Ficus opposita*) latex or leaves, served only against ringworm. A yellow box-wood infusion (*Planchonella pohlmaniana*) was specific against sores. The spines of dead finish (*Acacia tetragonophylla*), inserted into warts, made them fall away.

Some tribes used mud or ashes to dress wounds and there are many accounts of Aborigines surviving frightful wounds with no other treatment. In *Our Channel Country*, Alice Duncan-Kemp gave one example from south-west Queensland. "The man had been cutting pegs for his net when the axe slipped and cut his thigh to the bone," she wrote. "Some gins were quickly on the scene with lumps of black medicinal mud which they had scooped from the bed of a waterhole. They packed the wound with thick wet mud and leaves."

Brisbane pioneer Tom Petrie told how an Aborigine called Kebi survived a partial disembowelment after a fight in Queen Street: "... the man pushed the protruding parts in, and holding them so with both hands, walked off to camp, which was near the present Roma Street station. There he had to lie on his back, and the blacks put very fine charcoal and ashes in the wound, and that was all the doctoring he got. He had to keep on his back for a long time, but in the end recovered all right, though the wound left a very large scar."

168 EXTERNAL AILMENTS

ANIMAL STINGS

Australia is the land of things that sting. The world's deadliest snake (western taipan), the most poisonous spider (funnelweb), the deadliest jellyfish (box-jellyfish), the most venomous fish (stonefish) and the world's only poisonous octopuses (blue-ringed) all live in Australia.

Even so, Australia must be reckoned a very safe land. Our snakes are timid and shy, and our deadly sea creatures are rarely seen. Deaths from stings and bites are statistically insignificant.

Tribal Aborigines, however, had more to fear from poisonous animals than we do today. Living naked in forests and by the sea, they ran the gauntlet of snakes, jellyfish, stonefish and other stingers.

Yet surprisingly, the Aboriginal pharmacopoeia contained few remedies for stings. There are a number of remedies for snake bites, but only a handful for marine stingers, and fewer still for spiders and scorpions. This is perhaps because most animal toxins are molecularly very complex, and cannot be countered by simple remedies such as herbs. Even western medicine

◀ The aptly-named fan flower (*Scaevola sericea*) has delicate flowers shaped like fans or outstretched hands. These are followed by small white fruits, which on Groote Eylandt were squeezed onto the skin, along with the crushed stems, to soothe stings. The heated leaves were placed on swollen joints.

ANIMAL STINGS 169

has failed to find a natural or synthetic cure for snake and spider bites, and hospitals throughout the world rely upon enormously expensive antivenoms produced in the blood of venom-injected horses.

Inland Aborigines did not need to worry much about snakebites or stings — rare events in the outback — and I can find reference to only a couple of desert remedies for stings. Nearly all the recorded treatments are from the tropical north, where stingrays, stonefish, and jellyfish pose a threat. There were two remedies of particular interest.

The beach convolvulus (*Ipomoea pes-caprae*) or coast morning glory, is a very common creeper of beaches and dunes in northern and eastern Australia, as far south as Sydney. Like many seashore plants its seeds are spread by ocean currents, and it has a vast overseas distribution ranging from India to Hawaii and Brazil. Hungry Aborigines sometimes ate the bitter fibrous taproots as famine fare, but the plant was much more important as a medicine. The Northern Territory Pharmacopoeia project recorded twelve communities using the heated leaves of beach convolvulus as a medicine, either for the stings of stingrays, stonefish and green tree ants, or to treat boils, sores, pains, swellings and headaches. Leaves are heated over hot sand or coals and pressed against the infected part. Interestingly, the leaves have also been used in similar ways in New Guinea, Torres Strait and South East Asia.

Tests on the plant have been inconclusive. Two American studies found no evidence of pharmacological effects, but a third study from Thailand, reported in the journal *Nature* in 1970, isolated an active principle. According to the author S. Wasuwat: "In the light of the use of the plant by Thai fishermen as an antidote to jelly-fish stings and as an antipruritic agent [stops itching], I examined various crude extracts and have found that there is an active principle (IPA) that is antagonistic to jelly-fish poison and is also mildly antihistaminic. This is of interest because there are no known specific antidotes for jelly-fish poison ..."

Another very interesting remedy for stonefish and stingray stings is the application of bush cockroaches (*Cosmozosteria*). In Arnhem Land the insects are heated over a fire and their body juices dripped on to the sting, or the cockroach

▲ Cleft leaves, shaped like a goat's footprint, give beach convolvulus (*Ipomoea pes-caprae*) its Latin name — *pes-caprae*, which translates as "foot of goat". The sun-loving funnel flowers are like those of morning glories, which are closely related plants. Photographed at Caloundra, Queensland.

is crushed and rubbed over the sting. A sensation of numbness is soon felt. The chemical constitution of these insects deserves close examination.

The other sting remedies reported by anthropologists were apparently used by only one or two tribes, which casts some doubt on their efficacy. At Millingimbi in the Northern Territory the inner bark of the grey mangrove (*Avicennia marina*) is rubbed on stingray and stonefish stings. The Bardi of Dampierland, Western Australia, prescribe white berry bark (*Flueggea virosa*) for catfish stings, and cocky apple roots (*Planchonia careya*) and bindjud (*Tephrosia crocera*) for itchy bites. At Bingal Bay in north Queensland the crinum lily (*Crinum pedunculatum*) was crushed and applied to marine stings; this plant contains an alkaloid, lycorine, which may dull pain. The chewed leaves of broad-leaf hop bush (*Dodonaea viscosa*), which were used as a remedy at Iron Range in north Queensland, may also be mildly anaesthetic, judging by its wide reputation as a medicine, and its use in Peru as a cocaine substitute. Other plant parts used against stings include the white fruits and young stems of the fan flower (*Scaevola sericea*), the heated leaves of the peanut tree (*Sterculia quadrifida*), and the root of snakevine (*Tinospora smilacina*). Some tribes also rubbed hot sand on stings for relief.

▲ Tempting treats: sweet fruits of white berry (*Flueggia virosa*) are a popular snack food of coastal Aborigines in the tropics; the shrub is also found in Asia and Africa. Dampierland Aborigines infuse the bark and roots to treat cuts, sores, itches and catfish stings. Photographed on Cape York.

Sometimes stinging animals themselves were applied as the remedy, and a similar principle is applied in homeopathy today. On Groote Eylandt the stinging cup moth caterpillar was squashed and applied to its own irritation, and in parts of Arnhem Land scorpions were used in the same way. In Torres Strait the tail of the black spinefoot was rubbed over its sting. Death adder bites in north Queensland were similarly treated.

In eastern Australia white Australians have treated the stings of bluebottles (Portuguese men-of-war) and sandflies with the juice of the pigface (*Carpobrotus glaucescens*). The chemistry of Australian pigfaces has not been studied, but South African species contain mesembrine, a weakly anaesthetic alkaloid. The sticky leaf juice of both Australian and African plants has been applied to burns and scalds, and Australian pigface was used by colonists for dysentery. Pigface is probably one of the medicines of Aborigines in eastern Australia that has been completely forgotten.

Ciguatera poisoning is caused by eating tropical fish that have accumulated algal poisons. In the early 1970s, Dr Pat Hanush, a Queensland doctor, was medical superintendent at Cooktown Hospital when three men were admitted with severe poisoning after eating mackerel. Dr Hanush had heard that Thursday Islanders treat the poisoning by drinking coconut milk mixed with fruit juice of the great morinda (*Morinda citrifolia*), a seashore tree with a strong medicinal reputation. She squeezed juice from one of the fruits, mixed it with water (she had no coconut) and gave it to the worst of the patients, whose legs were shaky. In the morning his legs were better and he had recovered more than the untreated patients. Said Dr Hanush, "All I can say is, it seemed to work."

Mosquitoes are the bane of the Australian bush, and though their bite is not severe they can make life unbearable. Aborigines and colonists often deployed smoky fires to keep them at bay. Mina Rawson, who settled near Maryborough in 1880, told how her family coped: "To some extent we had become immune, but all the same we never went into the bush without protecting arms and legs with oiled rags ... And not a day or night during the months of January and February were we without tins of burning tea-tree bark, and later cow-dung. All our mosquito nets were of book muslin, and we had one big room-like net, under which we had meals when they were bad."

Aborigines drove away mosquitoes by burning leaves of special trees, including sandalwood (*Santalum lanceolatum*), ironwood (*Erythrophleum chlorostachys*), native cherry (*Exocarpos latifolius*) and wattle-flowered paperbark (*Melaleuca acacioides*).

Violet Stanton, of the Koongarukunj tribe near Katherine, told me how her father used to burn aromatic cypress pine wood (*Callitris intratropica*): "We had a 44 gallon drum pierced with ten or twelve holes to let in the oxygen. He threw in green branches and this was put in the middle of where we were sleeping." Highly aromatic lemon grass (*Cymbopogon*) was burnt when pine was unavailable.

Aborigines also repelled mosquitoes by coating their bodies with fat, and sometimes clay. Fat also helped keep the body warm.

The most unusual pest control measure practised in Australia was the shepherd's trick for ridding a hut of fleas. As writer Dame Mary Gilmore describes: "There was that other use of a young lamb: to put it in the bed of an empty hut to gather the fleas before you put your own blankets down. We have camped out when travelling, rather than face such a hut when we could not find a lamb, young and tender and milk-smelly enough to entice the intruders. The lamb was put in the bed and the yoe allowed to lie by the fire."

▶ The tasty peanut-flavoured seeds of the peanut tree (*Sterculia quadrifida*), framed inside a bright red pod, attract birds. On Elcho Island the heated leaves are placed on marine stings and insect bites.

◄ Despite what old bushwackers may say, Australia's scorpions are not deadly — though they can deliver a painful sting. Scorpions sting with the spine at the tip of the "tail". Photographed on Kangaroo Island, South Australia.

▲ On a Bribie Island beach, a pigface plant (*Carpobrotus glaucescens*) flaunts its purple flower. Pigface flowers have no true petals; modified stamens take their place. The fruit is salty-sweet and edible.

ANIMAL STINGS 173

174 EXTERNAL AILMENTS

SNAKEBITE!

Colonial Australians were terrified of snakes and they spread fantastic tales of the fates that befell the bitten. A death adder's victim, it was said, had scarcely time to cry "Lord have mercy upon me", before he fell down a lifeless corpse. After such a bite, recorded third-hand by R.N. Breton in 1834, "the blood gushed from his [the victim's] eyes, nose, mouth, and ears. Immediate decomposition commenced, and in a very short space of time the body was in such a state that it was with difficulty removed to where the grave had been dug." Breton thought worse of certain African snakes, whose victim's form "melts at once into a mass of putrefaction". Such tales, of course, are nonsense.

Other colonists swore that death adders had legs and could sting from both ends, that snakebite victims died at sundown, and that glass bottles left carelessly at camp lured snakes. Macabre snake stories are still told today. Bushmen at Hughenden swear that some snakebites bloat the victim so badly he can barely be squeezed into the coffin.

With such a prevailing dread of snakes, it is not surprising that colonial snakebite "cures"

◀ Coiled to strike, a startled death adder (*Acanthophis praelongis*) on Cape York flattens its body and arches its neck. Such displays terrified early colonists, who invented fanciful tales of reptilian prowess. This adder is only of finger thickness.

were gruesome and destructive. Scarification, amputation, burning with hot coals, injection of liquid ammonia, and firing of gunpowder were among the horrific and useless treatments. Colonial hunter Horace Wheelwright, writing from Victoria in 1861, told a typical tale: "One man I knew was bitten in the finger by a whipsnake, when putting up a fence. He coolly laid his finger on the post, and chopped it off with his axe, and thus probably saved his life. My remedy, if I had been bitten, would have been to cut the wound till it bled well, and put on it a charge of powder, and flash it off."

Today, the whip snake in question (*Rhinoplocephalus flagellum*) is considered about as dangerous as a honeybee.

Of the many colonial snakebite "cures" a few consisted of herbal nostrums. Most famous among these was Underwood's Snake Antidote, once sold in the colonies at ten shillings a bottle. During the latter half of the nineteenth century many shepherds and farmers in Tasmania and Victoria kept phials of the murky liquid close at hand.

Underwood was said to have discovered his remedy by watching a battle between a goanna and a snake, or in one version, between a shingleback lizard and a snake. The bitten lizard supposedly recovered after staggering off to eat a special plant from which Underwood made his remedy. His secret ingredient was variously said to be the juice of fern roots, bracken, dock, or some other plant "on which we tread in this country every day of our lives". George Henwood, a salesman for Underwood (who died of snakebite), claimed on his deathbed that the remedy was an infusion of ipecacuanha (a South American medicinal shrub) and ammonia. Underwood himself reportedly died of snakebite, although his supporters claimed he was too drunk to apply the remedy properly. Doctors of the day pronounced his antidote useless.

Underwood was one of a number of snake handlers who travelled from town to town like showmen, all peddling antidotes supposedly learned from goannas. (Another was Lyn Vane, whose herbal remedy was adopted by Joe Marconi as an ingredient in his famous goanna oil liniment.) These men probably acquired some immunity to venom by surviving successive bites, for we can be sure their remedies did not work, and that the snake-goanna encounter was

▶ Underwood's antidote? Bracken (*Pteridium esculentum*) was rumoured to have been an ingredient in the showman's remedy. Aborigines reputedly rubbed bracken stems on insect bites for relief. Photographed on Moreton Island, southern Queensland.

a fiction. Banjo Paterson immortalised these men in his ballad "Johnson's Antidote".

The doctors of the day would have been even more sceptical of such "cures" had they known what we know today about snake venoms. The very large molecules cannot be neutralised by any known natural or synthetic product. All around the world antivenoms are produced painstakingly by injecting horses with ever-increasing doses of snake venom and by isolating the antibodies produced in their blood.

Different snakes have very different venoms requiring different antivenoms.

In the late nineteenth century botanist Joseph Maiden recorded with some scepticism a supposed herbal remedy from northern New South Wales. The plant in question was white root (*Lobelia purpurascens*), a tiny creeping herb with toothed leaves, which is found today as a garden weed. Macleay River Aborigines supposedly made a decoction of the green herb to treat bitten animals, presumably domestic stock.

Maiden wrote of this herb in the Agricultural Gazette of New South Wales in 1894:

> A correspondent from the Bellinger writes a blood-curdling yarn about a black snake and an iguana having a fight. Black snake bites iguana — *iguana* feeling unwell scampers off for a mouthful of *Lobelia purpurascens*, and comes back to continue his encounter with the snake. Another informant tells me that the effect of the *Lobelia* on the iguana is to make him vomit up the snake-poison.
>
> I have had a number of letters and conversations, from which I understand that this belief in [white root] as an antidote to snake-poison is very generally held on the northern rivers. Now I want to know what is the basis for this opinion. Can anybody give me specific instances of men or animals who have recovered from the bite of a poisonous snake after treatment with *Lobelia purpurascens*?
>
> The plant undoubtedly contains an active principle, but will it act in the direction suggested? Of course we have to be nearly as cautious in dealing with snake-stories as in dealing with snakes themselves. I hope, therefore, my readers will thoroughly satisfy themselves as to the genuineness of any "cures" they may refer to.

Maiden's scepticism, commendable for its time, proved justified. No one came forward with evidence. White root attracted no further interest from the scientific community and is remembered today only as a difficult-to-eradicate weed of crops.

Aborigines around Australia had many different treatments for snakebite, none of which would have worked. Some were very similar to the ligatures and incisions practised by western doctors until recently. Others involved burning, blistering and profuse bleeding. Some tribes knew there was no remedy and did not even bother trying.

◀ Farmers spray 2,4-D to rout white root (*Lobelia purpurascens*), a dainty native herb, which once had a reputation as a snakebite cure. White root grows happily on shaded forest floors, as here on Mt Coot-tha near Brisbane, or in open fields and gardens.

EXTRACT FROM JOHNSON'S ANTIDOTE
A.B. "BANJO" PATERSON

Loafing once beside the river, while he thought his heart would break,
There he saw a big goanna, fighting with a tiger snake,
In and out they rolled and wriggled, bit each other, heart and soul,
Till the valiant old goanna swallowed his opponent whole.
Breathless, Johnson sat and watched him, saw him struggle up the bank,
Saw him nibbling at the branches of some bushes, green and rank;
Saw him, happy and contented, lick his lips, as off he crept,
While the bulging in his stomach showed where his opponent slept.
Then a cheer of exultation burst aloud from Johnson's throat;
"Luck at last," said he, "I've struck it! 'tis the famous antidote.

"Here it is, the Grand Elixir, greatest blessing ever known,
Twenty thousand men in India die each year of snakes alone.
Think of all the foreign nations, negro, chow, and blackamoor,
Saved from sudden expiration, by my wondrous snakebite cure.
It will bring me fame and fortune! In the happy days to be,
Men of every clime and nation will be round to gaze on me —
Scientific men in thousands, men of mark and men of note,
Rushing down the Mooki River, after Johnson's antidote.
It will cure *delirium tremens*, when the patient's eyeballs stare
At imaginary spiders, snakes which really are not there.
When he thinks he sees them wriggle, when he thinks he sees them bloat,
It will cure him just to think of Johnson's Snakebite Antidote."

From *The Collected Verse of A.B. Paterson*
Courtesy Collins/Angus & Robertson Publishers.

▲ Aborigines eat Leichhardt tree fruit (*Navdea orientalis*), and according to W. E. Roth writing in 1903: "The Koko-minni break up the bark: if eaten, it produces vomiting, and is employed for curing 'sore belly' and certain snake-bites."

▲ Alluring native grapes (*Ampelocissus acetosa*) have a pleasant sweet taste. North Queensland Aborigines applied the juice as a snakebite remedy. The woody tubers were also eaten. Photographed on Horn Island, Torres Strait.

In north-eastern Arnhem Land a bitten person captured the snake, bound its head to a stick and carried it back to camp. The snake was watched until it passed the blood of its victim, then released. If the snake was killed, its victim would also die. Snakebite deaths were attributed to sorcery, and this was confirmed by blowing smoke over the corpse to see if the shadowy form of a snake appeared.

A number of snakebite remedies were herbal. At Somerset, on Cape York, Ellis Rowan recorded that a brown snake's bite was treated by pounding the leaves of the white apple (*Syzygium*), putting some on the wound and drinking a decoction of the rest. In the case of death adder "they rub a portion of the creature's inside on the wound, and this they say always cures it, though they are sick and weak for some days after". Elsewhere in north Queensland the juice of the native grape vine (*Ampelocissus acetosa*) was used. The death adder in question is the northern death adder (*Acanthophis praelongus*), a dwarf species, and its venom is probably not strong. (A colleague, Klaus Uhlenhut, suffered no ill-effects after a bite from a nothern death adder.)

Other plants used as remedies in different parts of Australia included snakevine (*Tinospora smilacina*), beach convolvulus (*Ipomoea pescaprae*), caustic weed (*Euphorbia drummondii*), coolibah inner bark (*Eucalyptus microtheca*), cunjevoi (*Alocasia macrorrhizos*), Leichhardt tree (*Nauclea orientalis*), and in Western Australia, nipan (*Capparis lasiantha*) and native hops (*Dodonaea lanceolata*).

A north Queensland doctor, G.C. Morrissey, tried a herbal remedy that had been taught to him in the 1930s by an old Chinese man, who used it in many snakebite cases "affecting dogs, aborigines, and Chinese". The remedy was a dried herb (*Hedyotis galioides*), steeped in rum, and given in doses as generous as the patient could stand. Wrote Morrissey, "I prepared a stock of this, using whisky instead of rum". He plied the remedy on a man bitten by a brown snake, with apparent success, but two other patients, small children bitten by death adders, were less fortunate; they both died.

▲ Snakevine (*Tinospora smilacina*) snakes over branches and twines the tallest trees, and was probably named for this habit, not because of its use in Western Australia as a snakebite remedy. It was also one of the Northern Territory's most important general remedies. Photographed in Darwin.

SNAKEBITE! 179

180 EXTERNAL AILMENTS

MISCELLANEOUS CURES

Every book on bush medicines needs a "miscellaneous cures" chapter to store away all the odd remedies that fit nowhere else, such as the Aboriginal soaps, colonial pet cures, and diabetes remedies. It is also the place to note those plants that are recorded as "Aboriginal medicines", but about which nothing further is known. I also briefly mention here three common ailments — toothache, eye infections and rheumatism — for which there were many Aboriginal remedies and that really warrant chapters of their own.

In 1959 anthropologists J.B. Cleland and Norman Tindale recorded that the Arrante of Haasts Bluff considered the boobialla tree (*Myoporum acuminatum*) a very strong medicine. The plant was steeped in water heated wth hot stones to make a medicinal head wash or placed over a fire to produce medicinal steam. Unfortunately the anthropologists did not say what ailments the plant cured. Their informants probably did not know.

There are many outback medicine plants in

◄ Bountiful boobialla fruits (*Myoporum acuminatum*) tempt birds to this elegant outback shrub. Aboriginal children eat the fruits and their elders gather white gum oozing from the stems as glue. Photographed at Broken Hill.

the "don't know" category. Golden beard grass (*Chrysopogon fallax*), and loose-flowered rattle-pod (*Crotalaria eremaea*) are examples. As well, there are plants that are known today as remedies for general pains or illness, which in the past would have been used more specifically. Strychnine tree (*Strychnos lucida*) is probably one example and witchetty bush (*Acacia kempeana*) another. The clash with white culture has obliterated the memory of these. First to go after cultural collision were the intricate details of preparation, followed by the ambit of ailments treated. Finally one is left only with the vague notion that a particular plant was used as a medicine.

Several plants have been listed as Aboriginal remedies for venereal diseases, but here again we face changed circumstances as white men brought these diseases in. Aborigines may have responded by inventing new remedies, or by adapting those already used for sores. An arduous syphilis treatment practised in central Australia, suggesting desperation, was recorded in 1893 by F.H. Wells. The sufferer was "taken to the edge of a waterhole and buried in the mud up to the navel for fourteen days, during which time he is fed, and a shelter is erected to protect him from the weather. At the expiration of that time he is dug out, and it is said generally completely cured." Pioneers treated sphilis with a decoction of rice flower bark (*Pimelea*), and some Aborigines treated gonorrhoea with pounded quandong leaves (*Santalum acuminatum*).

The many treatments for toothache may also reflect change. Traditional Aborigines ate so

▶ Source of strychnine: intensely bitter strychnine berries (*Strychnos lucida*) warn that this tree produces the deadly alkaloid. Aborigines hurled the leaves and fruits into pools to poison fish, which were then eaten. A related *Strychnos* yields curarine, the infamous Amazonian arrow poison. Photographed near Darwin.

▶ The outback Anmatjirra tribe chewed the leaves of witchetty bush (*Acacia kempeana*) as a remedy and prepared the inner bark and roots, but for what ailments is no longer known. The characteristic flowers show that this is a wattle.

182   EXTERNAL AILMENTS

little sugar that tooth decay was probably unknown until the coming of the white man's tucker. But nerves exposed by grinding down gritty and fibrous foods may have caused much pain and many remedies are known from northern Australia. A painful cavity could be plugged with sticky eucalypt gum, with the inner bark of coast she-oak (*Casuarina equisetifolia*) or burned with a twig of pemphis (*Pemphis acidula*). Alternatively, the nearby cheek might be poked with a burning stick, or the mouth numbed by chewing a supplejack stem (*Flagellaria indica*) or sucking a wad of bull's head (*Tribulus cistoides*). In northern Australia a very popular remedy was a heated twig or chewed wood or stem of green plum (*Buchanania obovata*) placed against the

▶ Flaxleaf riceflower (*Pimelea linifolia*) may be the species of *Pimelea* spoken of in 1879 by the Reverend Tenison-Woods: "I have heard it confidently asserted by bushmen, that a decoction of the bark was a remedy for syphylitic symptoms." Photographed in Royal National Park, near Sydney.

◀ According to Walter Roth, the Nggerikudi tribe of north Queensland kept bullshead (*Tribulus cistoides*) "in between the gums and cheek, along the base of a bad tooth, to relieve pain in tooth-ache." Photographed at Casuarina beach near Darwin.

▶ Pemphis (*Pemphis acidula*) was a toothache remedy of Groote Eylandt Aborigines. Noted D. Levitt: "A twig from this tree was sharpened, put in hot sand for 2 to 3 minutes to heat, then it was inserted into the hollow of the tooth and kept there until the pain was gone." Photographed on Alpha Rock, Torres Strait.

tooth. The leaves of this tree contain triterpenes or steroids and deserve close investigation.

Eye infections were a very common malady of Aborigines, judging by the number of remedies they used. Kakadu Aborigines, for example, had six. Infections may have increased dramatically after cattle and sheep were put upon the land, which would have increased the number of flies, although William Dampier, the first outsider to describe Aborigines, spoke of flies and infected eyes in 1699.

Aboriginal remedies ranged from the leaf sap of the regal birdflower (*Crotalaria cunninghamii*) and a root decoction of ironwood (*Acacia estrophiolata*) to the juice of fan flower fruit (*Scaevola sericea*) and a wash of lemon grass (*Cymbopogon*). Other remedies were made from barks, gums (kinos) and witchetty grubs. Theresa Ryder of Alice Springs, an Arrante woman, told me how the gum of the **western** bloodwood (*Eucalyptus terminalis*) could be used: "Put him in water, boil it, and when it's a bit cold you can use him for eyes."

Eye infections, called ophthalmia or sandy blight, were also a serious problem for colonists, who were often temporarily blinded. A very popular remedy was made by boiling the leaves of a small native herb, spreading sneezeweed (*Centipeda minima*). Dr Louis Jockel of Richmond told how he cured a patient with this plant:

> A case came under my notice a few days ago of a drover who was suffering from a severe form of purulent ophthalmia, contracted up the country.
> I made an infusion of the plant according to directions, and the first local application seemed to have an almost magical effect. The man expressed himself as relieved at once of the intense smarting which he had previously suffered. He had got on so well that in two days he was able to start back up country again, and could hardly express his gratitude for the very great relief afforded.

The herb was at one time packed in tins and sold as Magic Ophthalmia Cure.

Aborigines had a great number of remedies for rheumatism and this again may indicate a lifestyle changed by white contact. Aching limbs were steamed over herbs or rubbed with plant decoctions or oils. Plant remedies included the beach bean (*Canavalia rosea*), konkerberry (*Carissa lanceolata*) and tick-weed (*Cleome viscosa*). A very interesting remedy was to flog the affected parts with nettles (*Urtica incisa*) — a traditional European remedy — or stinging tree bark or leaves (*Dendrocnide excelsa*).

Colonists bedding down in chilly shepherds' huts and miners' camps also succumbed to rheumatism. Rubbing down with goanna and emu oil were popular remedies, which were probably learned from Aborigines. In north Queensland the wild bean plant (*Tephrosia varians*) seemed to provide relief. Colonial doctor Joseph Lauterer reported, "I had a slight touch of rheumatism in the leg and tried a decoction of the root, with the result that the pain has gone and the stiffness is wearing away."

An interesting remedy for back pain was reported by the Royal Society of Queensland in 1885. A man was cured of lingering pain by a poultice of cunjevoi leaves (*Alocasia macrorrhizos*); the stems of this plant also cured ulcerated sores. Aborigines had a number of herbal remedies for pains, such as northern black wattle (*Acacia auriculiformis*), but it is usually unclear what kind of pains these remedies were treating.

For the treatment of earaches, North Queensland Aborigines dripped a decoction of river mangrove leaves (*Aegiceras corniculatum*) or lemon grass into the ear. In Arnhem Land the pulp of the lady apple (*Syzygium suborbiculare*) was squeezed into it. Herbalist H.P. Rasmussen advised syringing the ear with a mild decoction of blue gum leaves (*Eucalyptus globulus*) mixed with gum myrrh.

Other miscellaneous colonial treatments included using the yellow resin that oozed from

◀ Billybuttons (*Helichrysum apiculatum*), can reportedly be taken to expel worms. Outback Aborigines decorated their bodies with the feathery seed capsules. Photographed at Alice Springs.

the trunk of the hoop pine (*Araucaria cunninghamii*) as a tincture for kidney complaint. Wrote Joseph Maiden: "A gentleman says he finds it gives great relief in very aggravated cases, when three or four doses are usually sufficient". Hopbush (*Dodonaea*) and the wildflower yellow buttons (*Helichrysum apiculatum*) were taken to expel worms. In Tasmania, diabetes was treated with an infusion of leaves and twigs of hop goodenia (*Goodenia ovata*).

Colonists sometimes practised medicine on their livestock and pets. A correspondent to the Sydney *Town and Country Journal* told how to cure cholera and dysentery in chooks and parrots by feeding them powdered gum leaves. In New South Wales horses were fed boronia (*Boronia rhomboidea*) and black cypress pine foliage (*Callitris endlicheri*) with their fodder to expel worms. Sweet pittosporum gum (*Pittosporum undulatum*) healed a badly wounded dog. Bancroft tested out new plant extracts on pets, but his experiments were cruel, and often ended in the death of the animals.

Sometimes wild plants were used as bandages or other medicinal aids. Aborigines sponged sores with pads of tea tree bark (*Melaleuca*) and bound wounds with mat-rush leaves (*Lomandra longifolia*) or strips of stems or bark. Tobacco smoking pipes were made from the stems of fan flower (*Scaevola sericea*) and gardenia (*Gardenia megasperma*) and many other plants. Colonists made walking sticks from the stems of the rainforest walking stick palm (*Linospadix monostachya*), and these were exported to England last century.

Aborigines made instant bush soap by crushing the leaves or pods of saponin-rich plants such as strap wattle (*Acacia holosericea*), soap bush (*A. pellita*) and soap tree (*Alphitonia excelsa*). Kakadu Aborigines call the cleansing lather "andjana", and use the same name for beer, because it too froths up.

◀ Source of snuff? Baron von Mueller suggested last century making snuff from sneezeweed (*Centipeda minima*), a small damp-loving herb with a piercing smell, highly regarded in pioneering times as a medicine. Photographed in the Olgas.

186   DRUGS FOR PLEASURE

ABORIGINAL DRUGS

Cape York Aborigines were cigarette smokers long before white men came to Australia. Their tobacco was traded in from Torres Strait and smoked, either in paper-bark "cigarettes", or in bamboo or wooden pipes.

In Arnhem Land, Aborigines were probably addicted to the tobacco they obtained each year from Macassan fishermen from Sulawesi. The coarse, twisted tobacco was smoked in wooden pipes, in hollow bird bones, seashells or crab claws. Both in Arnhem Land and north Queensland the use of tobacco, and nicotine addiction, may date back hundreds of years.

The subject of Aboriginal drug use includes much that is puzzling and much that is bizarre. The use of tobacco, although curious, is well-documented, but many of the drugs used by Aborigines remain mysterious and enigmatic.

No one has satisfactorily explained why Aborigines over much of outback Australia traded so eagerly for pituri from north-west Queensland, when the same plant grew on their own tribal lands. Nor are the strange practices associated

◀ Killer of sheep, the native thorn-apple (*Datura leichhardtii*), a shrubby outback herb, contains lethal doses of hyoscyamine and related alkaloids. Hallucinogenic hyoscyamine is also found in corkwood (*Duboisia myoporoides*). Photographed in an Alice Springs riverbed.

with stinging trees and cocky apples easy to understand. As many of the recorded drug plants are poorly known, it seems likely there must have been other narcotic plants that have been forgotten altogether.

Pituri (*Duboisia hopwoodii*) and native tobaccos (*Nicotiana* species) were by far the most widely used drugs. Rich in nicotine and related alkaloids, their leaves were chewed by tribes throughout the outback. Pituri was so culturally and economically significant it is described later in a chapter of its own.

Captain Cook was the first Englishman to see Aborigines using a "drug", if that is what it was. At Endeavour River he observed of the "Indians" that "several of them held leaves of some sort constantly in their mouths, as a European does tobacco, and an East Indian betel: we never saw the plant . . ." No one has since been able to identify this plant.

Aboriginal drugs were mostly chewed, usually with the ash of certain shrubs and trees which enhanced their effects. Smoking pipes were a foreign innovation and were only used in the north.

Aborigines had little interest in hallucinogens. As far as we know they did not use the native relatives of the goldtop mushroom, nor the rainforest tree called argara (*Galbulimima belgraveana*), a hallucinogen of New Guinea tribesmen. They carefully avoided the hallucinogenic native thornapple (*Datura leichhardtii*), which is considered a "cheeky bugger" plant by Aborigines today.

One of the many puzzles about Aboriginal drug use is its skewed distribution. Nearly all the accounts of narcotics are from central or northern Australia, and the plants concerned are mainly subtropical or tropical. If drugs were widely used in the south of the continent, nearly all evidence of this was swept away with the white invasion.

At least two drugs, however, were used in the south. One was the root and bark of the bitter quandong (*Santalum murrayanum*), prepared by Aborigines at Lake Boga in Victoria as a stupefying drink. An intoxicating root used by South Australian Aborigines and mentioned by early settler George Angas, may also be bitter quandong, although another possibility is the coast tobacco (*Nicotiana maritima*), a herb with a bitter-tasting, and probably alkaloid-bearing root.

The other drug used in the south was alcohol. Many Australians think that traditional Aborigines knew nothing of alcohol, but there is plenty of evidence to the contrary. Brews were fermented in many parts of Australia from

▲ The cocky apple (*Planchonia careya*) has toxic bark and roots, which Aborigines hurled into pools to poison fish, which could then be eaten with impunity. The inner bark, roots and leaves were important remedies for sores and wounds. Photographed in Darwin.

◀ Mystery drug? No-one is certain whether coast tobacco (*Nicotiana maritima*) was taken by southern Aborigines as a narcotic. This rare herb grows only on coastal headlands and gullies in eastern South Australia. Photographed for the first time, at Hallet's Cove near Adelaide.

nectars, saps, honeypot ants, fruits and tree gum. Most of these brews were weaker than beer, and relatively few tribes made them, which explains how the misconception arose.

The strongest alcohol was brewed in Tasmania from the sap of the cider gum (*Eucalyptus gunnii*). Aborigines gouged holes in the trunk to collect the intoxicating liquid, and white settlers learned to follow suit. According to one writer, "At Christmas time, in 1826, the Lake Arthur blacks indulged in a great eucalyptus cider orgy".

Another strong brew was made in southwestern Australia, where nectar-laden banksia blossoms were fermented in water-filled bark troughs. According to Walter Roth: "The effect of drinking this 'mead' in quantity was exhilarating, producing excessive volubility".

From northern Australia there are some very strange reports of drug use. In 1901 ethnographer Walter Roth reported the following: "At certain of the corroborees on the lower Tully River some of the blacks will chew, and spit out again, the leaves of the 'stinging tree' [*Dendrocnide*]. The immediate effect is apparently a condition of frenzy, in which the individual may take violent action on his mates, or perhaps more commonly produce in himself a grossly disgusting perversion of the alimentary functions which enables him to eat human excreta."

Equally bizarre is a note attached to an old decaying specimen of cocky apple (*Planchonia careya*) held by the Queensland Museum, saying "used by Aboriginal women as a narcotic resulting in the death of many infants". Cocky apple is a well-known edible and medicinal plant, and no one seems able to explain what this statement means. "Narcotic" here most likely means "sleep inducing", but it is difficult to believe Aboriginal mothers would pacify their children with a dangerous drug.

▲ A bush tobacco of West Australian Aborigines, the aromatic herb *Stemodia lythrifolia* was also infused by some tribes to make a wash for headaches. This herb has blue flowers and grows near rocks. Photographed at Nourlangie Rock, Kakadu.

An intriguing note from a nineteenth-century missionary on the Daly River reported that dried leaves of a stinking arum (*Amorphophallus*) were smoked for an anaesthetic effect. "A short smoke makes one sleepy; if he smoke too long he will not awaken. While so sleeping he is, they say, unconscious of pain."

Other northern drugs, about which very little is known, include the nectar of a yellow-flowered tree, which was drunk by the Yidinyji south of Cairns to produce stupor, and native clerodendron (*Clerodendron floribundum*) and *Stemodia lythrifolia*, two plants chewed with ashes as

▲ Mulga woodland (*Acacia aneura*) in the lee of Ayers Rock is a rich resource for the Pitjantjatjara, supplying edible seed, galls, and lerp, and timber for tools. The twigs are often burnt to supply ash for chewing with native tobaccos.

stimulants. Lilly pilly leaves (*Syzygium* species) were smoked at Cape York, cattle bush (*Trichodesma zeylanicum*) served as "bush tobacco" in Arnhem Land, and near Gladstone the leaves of bitter bush (*Adriana glabrata*) were dried and smoked in place of white man's tobacco. It is not always easy to decide if such plants were traditional drugs or newly adopted tobacco substitutes, taken in desperation to allay nicotine craving.

In central Australia, drug-taking held exceptional importance, and the pituri and native tobaccos of desert tribes were especially significant. Addicted Aborigines valued these plants above all desert products, and traded them over hundreds and sometimes thousands of kilometres.

Australia's sixteen native tobaccos, close relatives of commercial tobacco, are found throughout the outback, but Aborigines used only the few varieties found in rocky ranges in the Northern Territory and Western Australia. The plants contain nicotine and similar alkaloids and compare in strength with European chewing tobacco. East of the Simpson Desert these favoured tobaccos do not grow, and pituri, a shrub of desert dunes, was chewed instead. Although pituri grows throughout the outback, it was almost never used in the Northern Territory or Western Australia, except to poison game in waterholes. Unfortunately the word "pituri"

Latz lists the main tobaccos, in order of preference, as *Nicotiana gossei*, *N. excelsior*, *N. rosulata* and *N. benthamii*. The Pitjantjatjarra also chew the leaves of an introduced tobacco, the South American tree tobacco (*N. glauca*).

Tobaccos were always chewed with the ash of special shrubs or trees, which enhanced the uptake of alkaloids. Each tribe claimed that only certain trees could be used, and yet each tribe used different trees. What all the trees had in common was a wood that burned to a fine lump-free ash.

When wild tobaccos were unavailable, substitutes such as ragworts (*Pterocaulon serrulatum*, *P. sphacelatum*), sneezeweeds (*Centipeda*), and wild tomato (*Solanum ellipticum*) were sometimes chewed in their place. Among the Alyawarra and western Arrante the hairy goodenia (*Goodenia lunata*), a very common desert herb with yellow flowers, was an oft-used substitute. Theresa Ryder of Alice Springs remembers the old people sending her out to gather and grind

▲ Insects do not chew the leaves of the native tobacco (*Nicotiana excelsior*) — they contain too much nicotine, a potent insecticide. This herb grows in caves and along gullies in central Australia, as here in the Olgas.

has been adopted as a name for the native tobaccos, which creates much confusion.

Botanist Peter Latz told of the value of wild tobacco to desert tribes:

> Known stands of high quality pituri were carefully husbanded, the plants were rarely uprooted and only the older more potent leaves were collected. Excess quantities were sometimes stored in caves until needed as it is only in good seasons that these plants were available all year round. Nowadays seeds are sometimes spread around to ensure a future crop but it is possible that this is a recent innovation not practised in the past. It is interesting that even though most desert languages are quite economic in their use of words, there is a quite large range of terms used for various parts of these plants and their products.

it: "You had to crush the green leaves on a flat stone with another stone, and mix it with ash. It gives you a headache the first time you chew it. Then you get used to it." Colonial botanist F.M. Bailey told of a goodenia used by travelling Aborigines to pacify their children, which was probably this species.

On rock outcrops in the centre there grows a small plant which was used, not so much as a substitute for tobacco but as a strengthener. Rock isotome (*Isotoma petraea*) is a lilac-flowered plant with a sap so acrid it can temporarily blind. Small amounts were sometimes mixed with wild tobacco. Latz noted: "When I questioned an informant on the use of this plant, she said that 'weak pituri was put "near" it to make it strong'. Possibly just throwing the pituri plants onto this plant would enable some juices from the bruised rock isotome to seep onto the pituri and boost its narcotic effect."

At least two other plants, both very dangerous, have been taken by Aborigines as drugs. The rainforest tree called corkwood (*Duboisia myoporoides*) is rich in the alkaloid hyoscine, which in southern Queensland is extracted commercially for making medicines. (See "Commercial Prospects"). According to one nineteenth-century report, "The aborigines make holes in the trunk and put some fluid in them, which, when drunk on the following morning, produces stupor." Judging by the known effects of hyoscine, this was probably the most powerfully narcotic of Aboriginal drugs.

In the desert north of Kalgoorlie, older Aborigines use sparingly the kite leaf poison bush (*Gastrolobium laytonii* — see Parker, 1980) as a drug. The plant is rich in sodium fluoroacetate, an extremely potent poison chemically equivalent to the dingo poison 10-80. Stock cannot be grazed in country where this plant is known to grow.

In recent years a new drug, kava, has come to Aboriginal communities in northern Australia. It is a muddy-flavoured mildly stupefying drink, which is prepared from the ground root of a pepper bush (*Piper methysticum*). An important traditional drink in Fiji, Samoa, and Vanuatu, it was introduced to Australia by missionaries from those countries, and now has a local following.

▲ To the Gunwinggu of Arnhem Land, cattle bush (*Trichodesma zeylanicum*) serves as a bush tobacco. Aborigines elsewhere do not use it as a drug, although one west Australian tribe place the boiled plant on sores. Photographed in the Olgas.

▲ Addicted Alyawarra Aborigines, unable to obtain native tobaccos, turned to the hairy goodenia (*Goodenia lunata*) as a substitute. The plants were also hurled into waterholes to poison game coming to drink. Photographed north of Alice Springs.

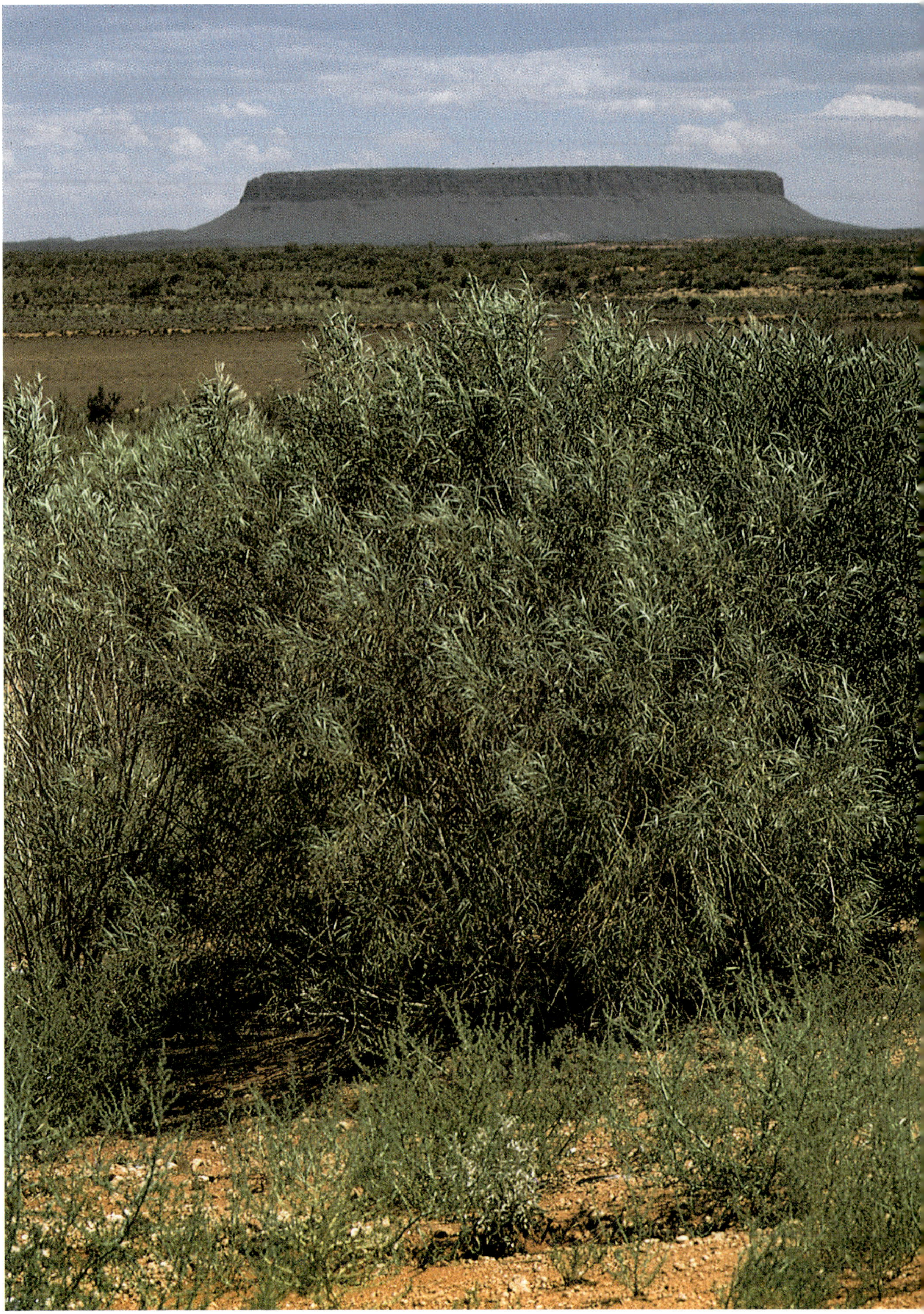

PITURI

Burke and Wills endured frightful hardship during their last wretched weeks in central Australia. Local Aborigines helped ease their pain, plying them with fish, nardoo cakes, "nice fat rats", and the "stuff they call bedgery or petcherry", described in Wills' diary as having a "highly intoxicating effect when chewed, even in small quantities".

Burke and Wills died only weeks later, the first white men ever known to take the Aboriginal drug pituri. Wills' journal does not say whether they enjoyed the drug, but the sole survivor of the expedition, Mr King, certainly did; while living with the Aborigines he chewed pituri to dull his hunger and pain.

Of the drugs used by Aborigines, pituri has attained a mystique all of its own. Traded over hundreds of kilometres, zealously hoarded and preserved, it figured prominently in the lives of dozens of tribes over much of central Australia.

Pituri is the dried leaves and stems of an outback shrub, *Duboisia hopwoodii*, named after Mr Hopwood of Echuca, a patron of the fateful Burke and Wills expedition. A species of the

◀ On the road to Uluru, pituri shrubs (*Duboisia hopwoodii*) sprout on a gravelly verge, while Mt Connor looms in the distance. In this part of Australia, pituri was used to drug emus, and was not taken by people.

alkaloid-rich Solanaceae family, the pituri plant is closely related to the tobaccos (*Nicotiana*), deadly nightshade (*Atropa belladonna*), and the hallucinatory thornapples (*Datura*).

Colonial pharmacists were quick to investigate pituri, hoping its alkaloids might prove useful in medicine. They must have been disappointed to discover the active constituent was nicotine, the alkaloid in cigarettes.

Pituri was often dismissed by whites as mere chewing tobacco, a description that belittled its immense versatility as a drug. Chewed pituri quids were passed from mouth to mouth to promote mirth during talks and feasts. The drug raised stamina on long desert treks: one Aboriginal boy supposedly walked 260 kilometres in two days on no other sustenance. Pituri was said to inspire courage during warfare, and to enable men to "fire-walk" on hot stones. It also made men voluble: the desert traveller Mr Gilmour once encountered an old man who "refused to have anything to say or do until he had chewed the pituri, after which he rose and harangued in grand style, ordering the explorers to leave the place".

Cigarette smoking hardly produces effects like these, and anthropologists in recent years have been forced to reassess the significance of nicotine as a drug. In the Americas, Indians used very potent tobaccos as mind-altering drugs, and the pituri wads chewed by Aborigines were also very strong. Their nicotine content, already two or three times that of cigarettes, was further enhanced by adding highly alkaline ash, which

▶ Tiny black berries and purple-streaked flowers distinguish pituri from similar-looking outback shrubs. The flowers have five joined petals, a characteristic of the tomato family (Solanaceae). Photographed near Ayers Rock.

▶ Desirable ashes were prepared by heating and then burning the leaves of the sandhill wattle (*Acacia ligulata*). The foliage produces a lump-free ash, ideal for chewing with native drugs. Photographed at Bedourie, western Queensland.

196   DRUGS FOR PLEASURE

released the nicotine from bondage with acids and enhanced its uptake by the body. In such doses nicotine has a depressant effect, and the user attains a trance-like state, immune to pain.

The ash chewed with pituri came from the leaves, or sometimes the wood, of certain trees, mostly acacias. Especially important were sandhill wattle (*Acacia ligulata*) and black sally wattle (*A. salicina*), which burned to a fine clean powder. When pituri was unavailable, it is known that at least one tribe chewed the foliage of the tropical speedwell (*Evolvulus alsinoides*) as a substitute.

Pituri was widely traded by Aborigines, and herein lies a mystery. The centre of pituri trade was the north-eastern edge of the Simpson Desert in far western Queensland, beyond the towns of Boulia and Bedourie. Pituri from this area was traded throughout a region of 550,000 square kilometres, passing along trade routes running south to Lake Eyre, north to Cloncurry, and west into the Northern Territory. Yet pituri is very widespread in the outback, and already grew in many of these places. Why was it traded?

A traditional view was that Aborigines living beyond western Queensland simply did not grasp the connection between the precious stashes of dried pituri chips, traded in special woven bags, and the nondescript shrub growing on their own tribal land. A more modern view is that Simpson Desert pituri is safer. Analysis of pituri plants from around Australia has shown considerable chemical variation. Queensland and some Western Australian plants have nicotine as their main alkaloid, but plants from the Northern Territory contain nor-nicotine, a more potent and more toxic alkaloid. Pituri from outside Queensland may have been considered unsafe or unpleasant.

Supporting this idea are the few tests done on wild tobaccos. In most of the Northern Territory pituri was eschewed in favour of wild tobaccos, close relatives of the tobacco used commercially. The most popular wild tobacco (*Nicotiana gossei*) contains only nicotine, while some of the other species, used only as substitutes, contain nor-nicotine and other alkaloids. Northern Territory Aborigines used pituri only to capture prey. They hurled the leaves into waterholes to drug emus and make them easy to club.

Another possibility is that tradition played a part. Native tobaccos grow only on rocky ranges in central Australia, west of the flat deserts where pituri was traded. Perhaps during Aboriginal colonisation of the outback, two drug cultures arose: one that was centred around the central ranges and used native tobaccos, and another in the Lake Eyre basin to the east that used pituri.

The tribes who controlled the pituri trade have long since passed away. Pastoralists with their cattle, and the arrival of the overland telegraph, destroyed the local culture. The best record of the pituri trade was kept by Protector of Aborigines George Aiston, although his account was not published until 1937, long after the trade had ended. Aiston declared that the curing of pituri, achieved by drying the leaves in sand ovens, was the monopoly of local pituri tribes. He observed that "any child born to the horde or tribe who belonged to the pitcheri Moora automatically succeeded to all rights and privileges in the distribution". Only when the men were so old that their beards showed grey were they taught the special drying techniques.

The drug changed hands at unique markets along the trade routes where as many as five hundred eager Aborigines assembled. Bartering would begin at the first camp that was met after leaving the pituri grounds. Aiston wrote:

After everybody had rested and fed, one of the party would throw down a bag in front of the assembled camp; anyone who wished to buy would throw down, perhaps a couple of boomerangs, perhaps

◀ The native tobacco (*Nicotiana excelsior*) is nowadays known as "pituri", while true pituri is often called the emu poison bush. Both plants contain similar alkaloids. Photographed in the Olgas after rain.

▶ Asian herbalists last century prescribed tropical speedwell as a remedy for fever and dysentery. Many varieties of this herb grow in Australia and Asia. This creeping form growing on Horn Island in Torres Strait has mildly narcotic leaves.

> a grinding mill, or whatever he could spare; the
> pitcheri seller would leave his bag until something
> that he wanted was offered; this he would accept by
> picking it up and the buyer would then pick up the
> bag of pitcheri. Perhaps another member of the
> pitcheri party would see something in the goods
> offered and would throw down another bag; if the
> buyers were not satisfied they would pick up their
> offerings, and if the seller was not satisfied he
> would pick up his bag of pitcheri. The camps near
> the pitcheri grounds never became big markets
> because the pitcheri was more valuable the farther
> away it was traded. The near camps were only used
> to get enough utensils and weapons for use when
> travelling to the more profitable markets.

A few white people also used pituri. Anthropologist Pamela Watson interviewed an early settler who remembered a hotel in western Queensland that served whisky spiked with the drug as a knockout drop. Bushmen sometimes smoked the leaves when their tobacco ran out, and during the 1890s, the drug was supplied to the Chinese community in Sydney as an opium substitute.

Aborigines probably became addicted to pituri. The ethnographer Walter Roth wrote in 1901 that "blacks will usually give anything they possess for it — from their women downwards". He noted that among Aborigines "there appears to be as great a craving for pituri as amongst Europeans for alcohol, a fact which is put into practical and economic use by drovers, station managers and others". Traditionally the drug was the preserve of older men, but by the turn of the century, women were using it as well. After chewing, the wad was stowed behind the ear, where additional nicotine may have been absorbed through the sensitive skin.

Pituri was largely supplanted by the white man's tobacco (and perhaps to some extent by opium). By 1900 the two were becoming interchangeable — there are reports of pituri smoked in pipes and of tobacco quids that were mixed with acacia ash.

Pituri remains a popular drug in central Australia today, but the plant now used is one of the wild tobaccos. The word "pituri" has been misapplied since at least the 1920s, creating confusion among anthropologists that continues to this day.

The pituri is a forgotten plant, the solace of a forgotten people, remembered only in the annals of explorers, ethnographers, and the pioneers who settled the Channel country east of where it grows. Pioneer Alice Duncan-Kemp made many fleeting references to Aboriginal use of the drug, and none is more poignant than the following:

> In a lonely little humpy among some sandhills, a
> blind hermit, outcast from his tribe, lived on nardoo
> flour and pituri. He had been implicated in stealing
> and murdering another buck's gin. The penalty was
> worse than death itself. Tried by a tribunal of the
> dead gin's relatives, he was sentenced to have his
> eyes burnt out.
> Banished for ever from his friends, he was fed
> by a faithful gin who, braving severe punishment
> herself, donned Kaditcha shoes and crept out to
> supply him with food. This went on for three years,
> when he died.

200   DRUGS FOR PLEASURE

BUSH BEER TO MAGIC MUSHROOMS

◀ Leafy lather can be made by crushing and frothing the mashed leaves of foam-bark (*Jagera pseudorhus*). Even the bark is soapy, as the name suggests. This remarkable tree, once used to add head to beer, grows on rainforest edges north of Taree.

Last century, Australian doctors tried to obtain opium and cocaine from native plants. The native bristle poppy (*Papaver aculeatum*) was tested for opium, but though it proved poisonous to frogs, it contained no opiates. A native rainforest shrub (*Erythroxylum australe*), closely related to South America's coca tree (*E. coca*), yielded not cocaine but coco-tannic acid. An extract of this plant, labelled "a powerful astringent", was exhibited optimistically at the Colonial and Indian Exhibition of London in 1886.

Opium and cocaine were sought by doctors for noble reasons — as local substitutes for these expensive imported medicines. But there were others who investigated Australia's native flora more indulgently, in the quest for pleasure-seeking drugs.

Alcohol was the popular drug of early pioneers, and wild plants were widely used to make it. In Tasmania, pioneers learned from Aborigines that holes chopped in cider gum trees

▶ Banksia blooms (*Banksia intergrifolia*) dripping nectar were a favourite food of Aborigines, who sometimes brewed alcohol by soaking the cones in water. Colonists, who called the trees "honeysuckles", cut the tough wood into bullock yokes and knees of boats. Photographed on Moreton Island.

(*Eucalyptus gunnii*) would fill with fermenting sap and provide a heady brew. Settlers in New South Wales made wine from scrub cherry (*Syzygium australe*), a rainforest lilly pilly with a fruit chemically similar to the grape. Proof spirits were made experimentally from the sugary pith of grasstress (*Xanthorrhoea*), and there was a proposal to make alcohol from the prickly pear (*Opuntia stricta*), probably to help control this insidious pest. During World War I beer was frothed with the saponin-rich bark of the foambark tree (*Jagera pseudorhus*). Bird collector George Caley even suggested making mead from fermenting banksia nectar. Caley did not know that Aborigines already used this method in Western Australia.

To give beer its bitter flavour, wild plants were sometimes substituted for hops. In 1796 beer in convict Sydney was "bittered" by the leaves of the Cape Gooseberry (*Physalis peruviana*). Colonists used bitter she-oak apples (cones of *Allocasuarina*) and hop bush (*Daviesia latifolia*) leaves, and colonial botanist Thomas Bancroft suggested using bitter Leichhardt tree bark (*Nauclea orientalis*).

But the most popular of the hop substitutes were the shrubs that became known as native hops, or hop-bushes (*Dodonaea* species). Australia has about sixty species, found all over the continent as understorey shrubs, though only a few of those were used for beer. The flattened, often bitter-tasting pods look like hop fruits, and it was for this reason, rather than for any special property, that they became so well known. One species, the broad-leaf hopbush (*D. viscosa*), grows throughout much of the

202   DRUGS FOR PLEASURE

▶ Fragrant jelly can be made from the spongy fruits of scrub cherry (*Syzygium australe*), a small tree of Queensland and New South Wales rainforests. Though suitable for wine-making, the fruits taste quite unlike grapes.

▶ In 1889 Joseph Maiden wrote of native hops (*Dodonaea*): "In the early days of settlement the fruits of these trees were extensively used, yeast and beer of excellent quality being prepared from them. They are still so used to a small extent." Illustrated is *D. triquetra* at Royal National Park, near Sydney.

◀ Grasstrees (*Xanthorrhoea*) supplied First Fleet doctors with a "yellow balsam" (resin) administered for dysentery, though it was scarcer than eucalypt kino, and according to Surgeon-General John White, "nor do I think that its medicinal virtues are by any means so powerful". Photographed atop Mt Mitchell, east of Warwick.

BUSH BEER TO MAGIC MUSHROOMS 203

▲ Corkwood (*Duboisia myoporoides*), a nauseous narcotic, is so dangerous that, according to one colonial writer: "branches of the tree, when hung up in a close room, have had the effect of producing giddiness and vomiting in delicate persons."

world, and was used as a medicine in many countries including South Africa, India and Aboriginal Australia. In Peru the leaves were chewed as a stimulant like coca leaves, which they were also used to adulterate. Aborigines chewed the leaves for toothache, and applied them to stings, which suggests they may have a pain-deadening effect.

The true hop plant (*Humulus lupulus*) was cultivated in Victoria and Tasmania from very early times, and colonial botanist Baron von Mueller hoped it would run wild "on river-banks and in forest-valleys". It did not. He complained that many hop substitutes were "objectionable or deleterious".

Hops have no alcoholic or narcotic properties, despite the colonial belief that hop pillows were good for sleeplessness. What is interesting about this plant is its very close relationship to the hemp or marijuana plant (*Cannabis sativa*). Indeed, *Cannabis* and *Humulus* are the only genera in the tiny plant family Cannabaceae, which has only a handful of species worldwide (no one is sure exactly how many — botanists disagree about whether there are one, two or three different hemps). Colonists did not smoke marijuana, although the plant was occasionally grown in gardens as an ornament, and von Mueller recommended it as a companion plant around edges of cultivated fields to ward off insects.

As substitutes for tobacco, bushmen sometimes turned to pituri (*Duboisia hopwoodii*) and the native tobaccos (*Nicotiana*) of outback Aborigines; some colonists even aquired a taste for smoking gum leaves. Eucalyptus cigarettes were marketed with the slogan "Take a whiff of the gum forests into your home".

The most potent drug taken in the colonial era was probably the hallucinogenic thornapple (*Datura stramonium*), an introduced weed. Dr James Stuart claimed in 1840 that Norfolk Island convicts ate the seeds to invoke temporary or permanent insanity. This was probably preferable to penal life on the island. A fearsome weed with spiky pods, thornapple often sprouts around farms on fertile soils; it has a long history of use as a drug and poison. North American Indians reportedly hallucinated on the plant during initiation ceremonies, and witches were said to have used it in Europe. The plant contains the alkaloids hyoscine (scopolamine) and hyoscyamine. Colonial herbalists mixed the leaves into cigarettes to be smoked by asthma sufferers — its effects are milder when smoked.

◄ Forbidding thorns and a rank aroma identify the common thornapple (*Datura stramonium*), an infamous narcotic plant often found as a weed on farms. The whole plant is poisonous. Photographed on St Helena Island, Moreton Bay.

Thornapple provides a link between colonial and recent Australian drug use, for it was one of the narcotics rediscovered by Australian youth during the early 1970s. The nasty-smelling leaves were cooked as soup or brewed as "tea". A datura "trip" was usually anything but pleasurable, and few availed themselves of the opportunity more than once. Symptoms include extreme disorientation, unquenchable thirst and temporary blindness. The garden shrub angel's trumpet (*Brugmansia candida*) contains similar alkaloids and has been used in the same way.

Thornapple is a member of the pharmacologically active tomato family (Solanaceae), along with such potent drugs and poisons as tobacco, pituri, henbane and deadly nightshade. Corkwood (*Duboisia myoporoides*), the Australian rainforest tree grown commercially, contains similar alkaloids and has occasionally been taken by Australian drug-users, as "bush tea". One acquaintance who smoked a couple of the leaves was unable to walk properly, and became extremely dry in the mouth. He was never tempted to try the drug again. In July 1987 a Wollongong man died after eating the leaves. His death from cardiac arrest must be at least partly attributed to a weak heart, for the alkaloids in corkwood do not usually act in this way.

Thornapple and corkwood have always been fringe drugs and neither has ever attained the mystique reserved for hallucinogenic mushrooms. Since the early 1970s, "magic mushrooms", along with marijuana and LSD, have exerted a profound influence on Australia's alternative movement.

Australia is home to a variety of psychoactive mushrooms. In southern states the fly agaric (*Amanita muscaria*) can be found sprouting beneath introduced trees. This is the familiar fairy tale mushroom with the spotted scarlet cap. Mediaeval housewives supposedly placed pieces of its flesh in saucers to poison flies, hence its name. In northern and eastern Siberia the caps were devoured as hallucinogens, inducing states of violent euphoria. As the active constituents are passed in the urine unchanged, the urine of the drugged was sometimes collected and drunk to prolong the revelry. In Italy and Eastern Europe this species is eaten as a harmless vegetable after careful preparation (but mistakes do happen). Australian drug takers rarely bother with this species; its effects are apparently not desirable.

The fungi favoured by aficionados are the infamous gold tops (*Psilocybe cubensis*) and blue meanies (*Copelandia cyanescens*), nondescript mushrooms found sprouting in cow dung in paddocks after rain. Their possession is strictly prohibited. The active ingredient, psilocybin, is indicated by a blue stain that appears when the caps or stalks are bruised. In effect it resembles LSD. The user experiences a profound shift of perceptions, seeing life as if for the first time. Said one user: "It's like you've always looked at a picture through a frame and you take the frame away. You're aware of a new totality — but it's more of an abyss, a void, than something you see."

Vivid hallucinations are often a feature of the mushroom trip. Sounds seem dissociated, colours throb, faces leer from patterns in curtains and carpets. The images are mostly grotesque distortions of seen objects, but sometimes they have no foundation in reality. One user told of going to get a glass of water, "but there were all these pigs scattered about the kitchen with their throats cut and blood dripping everywhere. But I was so thirsty I just had to step over them to get the water. I felt they were a projection of something unpleasant in me."

Within the alternative movement the mushroom experience has helped reinforce anti-materialist values and a sense of oneness with the universe and other people. Art and music styles have been influenced. But many people, particularly those with shaky self-confidence, find the experience profoundly disturbing, and some appear to suffer permanent ill-effects. Even those who enjoy the experience may complain that the profound revelations bring no lasting meaning.

Because they grow only in cow dung, gold tops and blue meanies are presumably introduced species. But Australia is also home to several native *Psilocybe*, which can be found sprouting on tree stumps and wallaby dung in forests, and some of these are hallucinogenic. In 1972, *P. collybioides* was recorded botanically in Australia for the first time from specimens collected by police in a fridge in Tasmania. The dried crushed fungi were being packed into gelatine capsules for illicit sale in Hobart.

Among garden plants and weeds there are a number of species reputed to be narcotic, and some of these have been tried by drug-users overseas, but the plants appear to be dangerous and are best not mentioned here. One of these supposed drugs, however, is so obviously harmless that I have no reservations about announcing its name — it is the common garden lettuce (*Lactuca sativa*). In times past Europeans believed that the lettuce, and the related weeds, prickly lettuce (*L. serriola*) and sowthistle (*Sonchus oleraceus*), contained a sleep-inducing drug called lactarium. In 1876 Baron von Mueller proclaimed that lettuce was "not without value, especially as a sedative, for medicinal purposes". The belief was probably inspired by the similarity between opium latex and the bitter white sap produced by old lettuces and sowthistles. Drug users in the 1970s revived this charming idea, and a product called "lettuce opium" was pushed upon the market, hailed by its promoters as the perfect drug — it could not be proscribed without also banning the lettuce.

▲ An illegal harvest of gold tops, the infamous "magic mushrooms" of the sixties counterculture. Users made soup of the mushrooms, served them fried on toast, or ate them raw.

◀ Fairytale fungus: since the nineteenth century the fly agaric (*Amanita muscaria*) has featured in pictures found in children's books, such as the original Alice in Wonderland illustrations. The fungus is strongly narcotic. Photographed at Stirling in the Adelaide Hills.

The new drug never caught on, presumably because it did not work.

Nonetheless, the idea of sedative plant foods was very real to the nineteenth-century mind, and pioneers in Australia sometimes claimed dangerous encounters with such plants. South Australian stockman Max Koch told a curious tale about sandalwood fruits (*Santalum lanceolatum*) in 1898: "A friend of mine asserts that on a hot summer's day he was driving a flock of sheep in company with a black boy, and met with a tree loaded with ripe [sandalwood] fruit. Both ate a large quantity of it and fell asleep. Awaking, they drove the sheep further on, and met with another tree; ate some more fruit with the same result. My friend is of opinion that the berries contain narcotic properties."

Sandalwood is a well-known Aboriginal food-plant, and is certainly not narcotic. The same is probably true of the "orchid" nectar described by Queensland pioneer Mina Rawson.

Mrs Rawson found the orchid growing in mosquito-infested tea tree scrubs near Maryborough where she battled to eke out a living late last century: "It was a creamy white in colour, a feathery flower, but in the early morning you could get half a cup of thin white honey. It had a very pleasant flavour, rather like cocoanut, but it used to make me sleepy. On one occasion my boys and another who was on a visit to us shared about three tablespoonfuls on their bread, and all three of them went to sleep for over an hour, so of course I barred it altogether."

BUSH BEER TO MAGIC MUSHROOMS   **207**

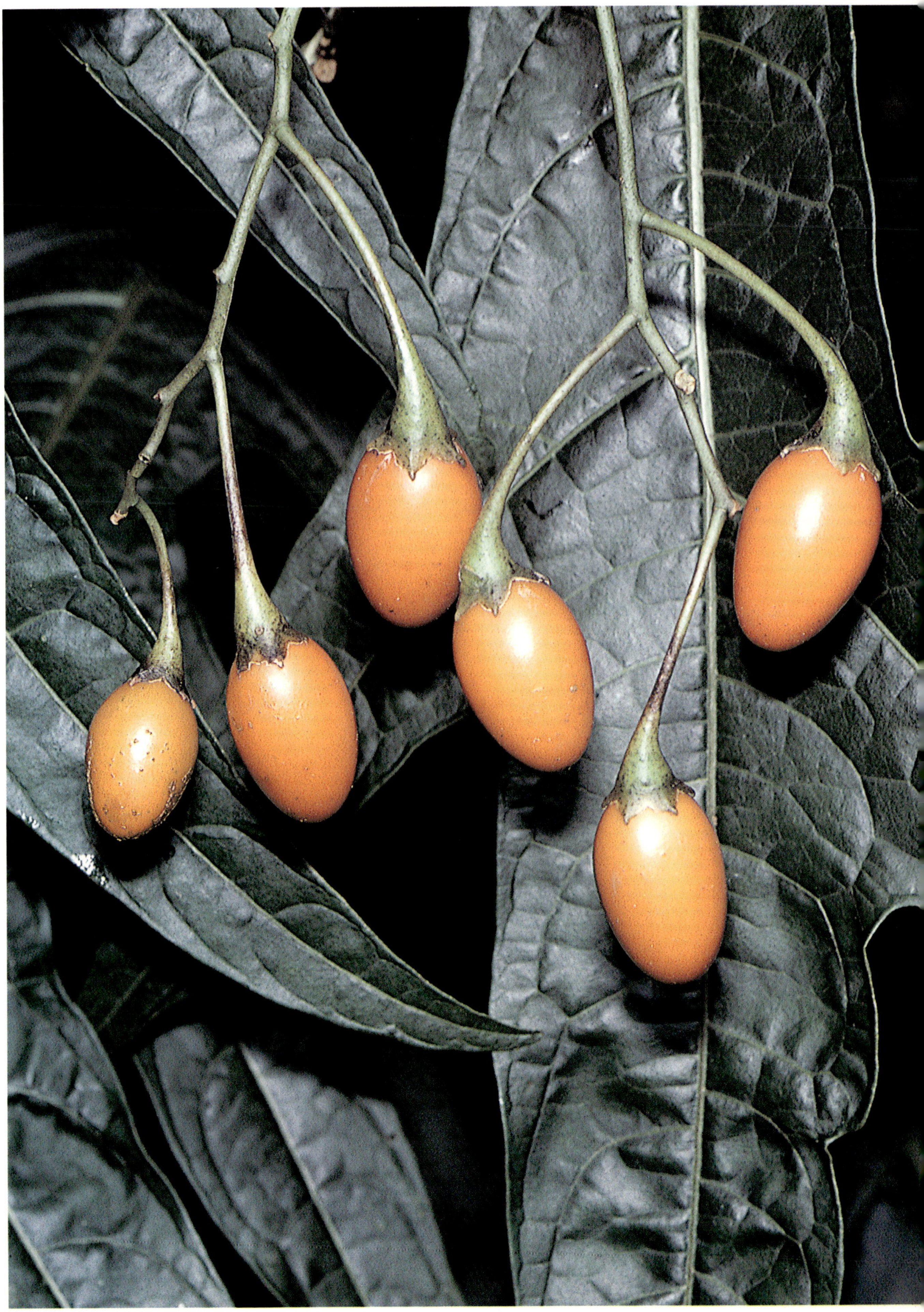

COMMERCIAL PROSPECTS

Australian medicinal plants have made their mark on the world. Tens of millions of Soviet women depend upon kangaroo apples for contraception, and a related rainforest tree called corkwood supplies about half the world's hyoscine, used in travel sickness pills. Eucalyptus oil is internationally renowned, and tea tree oil, distilled from plantations in northern New South Wales, is finding markets overseas. One of our rainforest trees may hold the key to curing AIDS.

Plant medicines were among the first products ever to be exported from Australia. First Fleet surgeons were quick to send England samples of eucalyptus oil, eucalypt kino and native sarsaparilla leaves. During the nineteenth century dugong oil, sandalwood oil and eucalyptus oils were important exports.

But trade in these products dwindled in the twentieth century and Australia today is only a modest exporter of medicines. Our most important pharmaceutical plants, the eucalypts, kangaroo apples and corkwoods, are grown mainly

◄ Framed by dark foliage, a bunch of kangaroo apples (*Solanum aviculare*) attracts birds in a rainforest clearing high on Lamington Plateau. This common shrub is found in eastern Australia, New Zealand, New Guinea and New Caledonia. Colonists believed the plant contained nicotine.

by overseas interests. It is a sad story of opportunities lost.

The native plants we benefit from the least are our kangaroo apples, one of the world's major sources of steroids. The synthesis of steroids in the 1940s helped spawn the sexual revolution of the 1970s. Steroids gave us the contraceptive pill; not to mention the corticosteroids used to treat asthma, allergies and arthritis; and the sex hormones used for menopausal disorders, infertility and impotence.

The use of steroids in medicine is an elegant idea. Certain plant chemicals, the steroidal saponins, so closely resemble human hormones that they can be converted into synthetic equivalents — "the Pill" being the most famous example.

Steroids were first produced commercially from two yams (*Dioscorea composita* and *D. mexicana*) growing wild in Mexico. By the 1970s the yams were a declining resource: their habitat was diminishing, cultivation was proving impractical and the Mexican Government imposed steep price increases. A world search for alternatives began some decades ago, spearheaded by Russian and Hungarian scientists. They developed a highly successful steroid industry, providing the contraceptive requirements for all eastern Europe, based upon Australian rainforest plants.

The kangaroo apples (*Solanum* species) are soft-leaved shrubs of cool, open forests and rainforest edges in eastern and southern Australia and New Zealand. They have pretty, purple flowers and succulent berries that are edible, though not very tasty. All four species produce steroidal saponins in their leaves, stems and fruits, but *S. aviculare* and *S. laciniatum* give the best yields, and only these species are used in steroid manufacture.

The saponin in kangaroo apples is solasodine, an alkaloid very similar to the diosgenin in yams, and easily converted to steroids. (Without conversion the saponins have no birth-control

▲ Back in 1790 leaves of native sarsaparilla (*Smilax glyciphylla*) were exported from Sydney to China by a steward on the *Lady Juliana*. Native sarsaparilla is closely related to the true West Indian sarsaparilla (*S. ornata*), which has medicinal roots. Photographed at Burleigh Heads, Gold Coast.

◄ Soviet contraceptive pills are produced from saponins in the leaves of the kangaroo apple (*Solanum laciniatum*), a shrub of woodlands and shady forests in south-eastern Australia. Aborigines ate the berries. Photographed at Beachport, South Australia.

effect, and cannot be used as herbal contraceptives, except perhaps as abortifacients.) The Soviet Union and Hungary maintain vast kangaroo apple plantations, mainly of *S. laciniatum*, and the plants have also attracted serious interest in Cuba, Japan, India and Egypt. But although the kangaroo apples are native to Australia, there is no local steroid industry, and Australia's contraceptive pills are imported.

One Australian company manufactures derivatives of solasodine, but not for contraception. Dr Bill Cham's *Curaderm* cream, made in Brisbane from a weed, apple of sodom (*Solanum hermannii*), and likely to be produced in future from *S. aviculare*, is sold as a rub-on treatment for sun spots and skin cancers (basal cell carcinomas). Dr Cham claims a high success rate, but the Queensland Cancer Fund is sceptical and the cream remains controversial. Dr Cham believes his formula may also prove effective against internal cancers, and his colleague, Dr Brian Daunter, has modified the solasodine molecule to produce a new mechanism for delivery of anti-cancer drugs which he says will make them more site specific.

Kangaroo apples are members of the remarkable tomato family (Solanaceae), the source of many of Australia's medicines (not to mention foods and narcotic drugs). Among the most important of these are the rainforest trees called corkwood (*Duboisia myoporoides*).

Dr Joseph Bancroft of Brisbane was the first to investigate a corkwood extract (*D. myoporoides*) in 1877, testing its pupil-dilating effects on pets, and he introduced it to the medical world as a useful tool in eye surgery. Other nineteenth-century doctors found it reduced night sweating in tuberculosis patients and the pains of internal inflammations. Corkwood leaves were exported to Europe, earning prices of up to a shilling a pound last century. The leaves contain a number of alkaloids, but the most important is hyoscine (scopalomine), which is used to treat motion sickness, stomach disorders and the side effects of cancer therapy. During World War II, Australian hyoscine was in great demand among allied troops for treating travel sickness and shell shock, and 7000 ounces were exported.

Wartime hyoscine was obtained from wild trees. *D. leichhardtii*, which grows only in a few inland scrubs and woodlands west of Brisbane, was so heavily cut there were fears it would become extinct. *D. myoporoides* is a more widespread tree of rainforest clearings, but its alkaloid content is variable and it was not favoured

▲ A high-yielding hybrid of the two corkwoods (*Duboisia myoporoides* and *D. leichhardtii*), grown in southern Queensland, yields more hyoscine than any plant known. The flowers and fruits resemble those of pituri, a closely related plant.

COMMERCIAL PROSPECTS

for harvesting. (Some colonies of this tree contain nicotine in place of hyoscine; Aborigines once used it as a drug.)

Today the hyoscine harvest is obtained from high-yielding hybrids of the two species grown on plantations in the Burnett district near Kingaroy. In 1989 Australia exported 500 tons of the dried and powdered leaves to West Germany and Switzerland, representing a large share of the world market. Phytex Australia, a Sydney company, extracts the drug locally. Hyoscine is produced overseas from other plants, such as thornapples (*Datura*), but none yields as much as the Queensland hybrid. Unfortunately, German and Swiss companies are establishing their own plantations in Queensland, and Australian earnings from this drug are declining.

The blackthorn or sweet bursaria tree (*Bursaria spinosa*) of Australian forests and woodlands is the source of a curious substance called aesculin. If you soak blackthorn leaves in a bowl, aesculin will turn the water bright blue. Add vinegar or lemon juice, and the water will turn red; caustic soda will make it blue again. Aesculin not only behaves like litmus paper, it also absorbs ultraviolet light. It has been used in sunburn creams and tinted windows.

Aesculin is extracted mainly from the European horse-chestnut tree, but during World War II some was exported from Australia for use by fighter pilots. The substance is now prescribed in medicine to treat haemorrhoids by strengthening blood capillaries. In 1980 Phytex Australia began exporting aesculin to Europe from blackthorn trees culled around Sydney.

By 1983 the industry had become uneconomical due to an overpriced Australian dollar and competition from East European horse-chestnut aesculin. Hopefully the industry may revive in the future.

The medicines I have mentioned so far are not unique to Australia. Solasodine, hyoscine and aesculin are all commercially extracted overseas from European, Asian and American plants. Of special interest, then, are those medicinal products that are unique to Australia. The eucalyptus oils should probably be considered among these. Although their major ingredients — cineol, phellandrine, and other aromatic oils — are found in other plants, the combinations in which they occur are not. Unfortunately, the production of eucalyptus oils cannot be considered an Australian success story, as most of the production occurs in China, Portugal and South Africa.

The tea tree *Melaleuca alternifolia* of northern New South Wales has recently created interest as an oil producer. The germicidal properties of this oil have been known for decades, but the oil has only just been promoted to the general

▲ The forest red gum (*Eucalyptus tereticornis*) is one of three Australian eucalypts grown in Chinese plantations (with *E. globulus* and *E. citriodora*) for its oil. China leads the world in oil production. Photographed on Mt Coot-tha.

◄ Blackthorn (*Bursaria spinosa*) dyes water deep blue, and colonists believed the intense blue of the Lake at Mt Gambier was due to blackthorns growing on its flanks. Photographed on Wilson's Promontory.

► "Nature's best kept secret", the brochures say, referring to the antiseptic oil produced by the tea tree (*Melaleuca alternifolia*). All of Australia's sixty or so tea trees produce medicinal oil, but none has the healing properties of this species. Photographed at Girraween.

COMMERCIAL PROSPECTS

public as a natural remedy — "the complete first aid kit in a bottle". There are now about fifteen plantations between Kempsey and Rockhampton, with most centred around Lismore and Grafton. Christopher Dean of Thursday Plantations has successfully marketed the oil through health food stores in the United States, Canada, New Zealand, Britain and parts of Europe. He hopes to win mainstream pharmaceutical acceptance for the oil in America and Europe.

If Australia has a major future as a drug producer, it lies in our rainforests where the most important plant medicines of modern times, the alkaloids, are mainly found. This became obvious in Australia after World War II when the CSIRO conducted its famous Phytochemical Survey of Australian Plants.

The project was born out of post-war sentiments of Australian self-reliance, and its aim was to identify Australian plants containing alkaloids, which were then a major focus of medical research. At first Len Webb, the survey botanist, tested plants at random, but he soon found that rainforest plants contained the alkaloids — five hundred were discovered altogether.

To understand why alkaloids proliferate in rainforest, Len Webb suggested that alkaloid production be considered "luxury metabolism". All plants manufacture poisons to repel insects and other plant eaters, and alkaloids are among the most potent of these. But alkaloid production costs a plant precious reserves of nitrogen, which is notoriously lacking in Australian soils. Only in rainforests are there enough nutrients to permit production of copious alkaloids.

This helps explain why conservationists often invoke medicine in the quest to preserve rainforests. There can be no doubt that Queensland's rainforests harbour dozens of medicinally valuable alkaloids and other compounds, which only await discovery.

▶ For hanging stores of starch, colonial entrepreneurs turned to the Moreton Bay chestnut (*Castanospermum australe*). Maiden recorded that "As an experiment, a chemist at Lismore once made 40lb. of starch from the beans, which he sold at 4d. per lb."

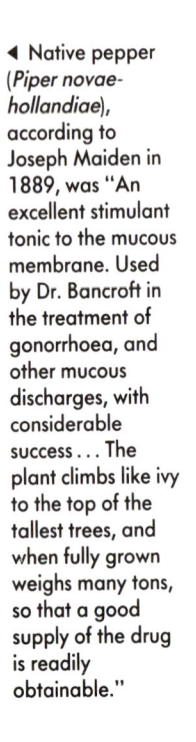

◀ Native pepper (*Piper novae-hollandiae*), according to Joseph Maiden in 1889, was "An excellent stimulant tonic to the mucous membrane. Used by Dr. Bancroft in the treatment of gonorrhoea, and other mucous discharges, with considerable success... The plant climbs like ivy to the top of the tallest trees, and when fully grown weighs many tons, so that a good supply of the drug is readily obtainable."

At present, the kangaroo apples and corkwoods are the only rainforest plants that are cultivated as medicines, but scientists hope these may soon be joined by plants that cure cancer. A number of Queensland rainforest plants have shown promise in this direction.

The native pepper vine (*Piper novae-hollandiae*) is one of these; in Australian tests the plant showed positive activity against lung cancer in mice. Another possibility is coast tylophora (*Tylophora crebriflora*), a rare vine of rainforest margins. One of its two alkaloids, tylocrebrine, retards leukaemia in mice and trials have reached the clinical stage in America. Another promising candidate is scrub yellowwood (*Bauerella simplicifolia*), a nondescript rainforest tree with alkaloids in its corky bark, one of which, acronycine, acts against a wide range of tumours.

By far the best prospect, however, and a candidate in the war against AIDS, is the well-known rainforest tree, the Moreton Bay chestnut or black bean (*Castanospermum australe*). Found on rainforest riverbanks north of Coffs Harbour, this handsome tree is a well-known timber and ornamental tree, easily recognised by its striking orange-red pea flowers and "English" appearance.

In autumn, Moreton bay chestnuts bear crops of enormous seeds, rich in starch, which Aborigines gathered for food. The seeds are highly poisonous but tribes found ways of leaching away the toxins by grating, soaking and baking the seeds. One of the toxins they removed, the alkaloid castanospermine, is showing extraordinary promise as a cancer and AIDS cure. In American experiments castanospermine was found to interrupt metabolism in cancer cells, forcing them to return to normal. It acts synergistically with AZT, the drug used to treat AIDS. This has raised hopes that it may form part of

▲ Parrots flock to the big pea flowers of the Moreton Bay chestnut, a tree often cultivated in parks for its generous shade, attractive flowers, and English appearance.

a multi-drug attack upon the disease, which is the approach favoured by researchers. Castanospermine inhibits viral disorders and is also a potent anti-inflammatory agent. In Australian research it was found to inhibit allergic encephalomyelitis. Its future in medicine looks very promising.

There is one other Australian plant that may have a future in medicine but I cannot tell you what it is because no one will tell me its name. The plant came to notice in 1989 when Griffith University's Dr Ron Quinn found that a crude bark extract was a painkiller with twice the potency of morphine. The bark is a traditional painkiller of Kimberley Aborigines, and its identity is being kept secret while tests continue. Dr Quinn is hoping for research funds to identify the active constituent, which he believes could be up to two hundred times as potent as morphine. There is a slim hope that this Australian plant may yield a non-addictive painkiller, which would be a safe alternative to the opiates.

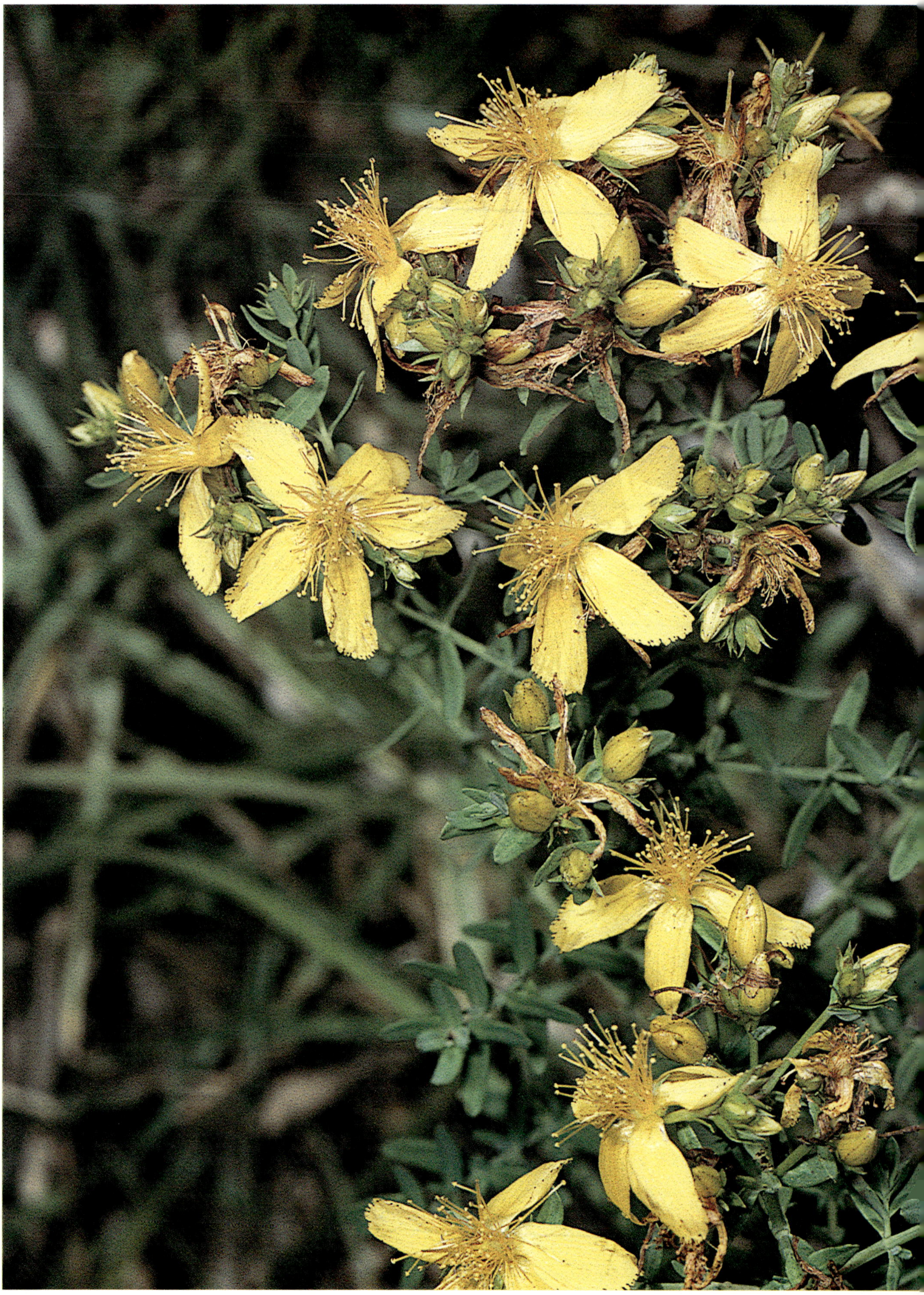

CONSERVATION

St John's wort (*Hypericum perforatum*) is a yellow-flowered herb of woodlands and meadows in Europe. The crushed flowers drip red juice, and English peasants believed it was good for the blood. Throughout Europe the herb was prescribed as a sedative and astringent.

In the 1880s, during Australia's gold-rush boom, a German woman at Bright in Victoria imported St John's wort seeds to grow as a medicine. The herb overran her garden and soon spread into the nearby racetrack where it was dubbed "racecourse weed". By the 1890s it had infested the hills along both sides of the Ovens Valley, and as goldminers moved out seeking new diggings it went with them, mainly as chaff for their horses. Plants appeared at Bendigo, Clunes, Newstead, Gippsland, and in New South Wales, South Australia, Western Australia and Tasmania. By 1917 more than 200,000 acres of land were overrun in Victoria alone. St John's wort is now recognised as one of southern Australia's most serious noxious weeds. It is poisonous to stock and outcompetes other plants. It is still spreading.

Horehound (*Marrubium vulgare*), a bitter herb

◀ If cattle eat poisonous hay, made by drying St John's wort (*Hypericum perforatum*), their skin falls away. They suffer mental depression and sensitivity to water, sometimes drowning when driven across streams.

with white flowers, was also planted by colonists. A popular remedy for coughs and colds, it was grown in homestead gardens. When farms were abandoned, it spread into nearby paddocks, the burrs being carried about on sheep. Explorer Major Mitchell was surprised to find horehound thriving at remote sheep stations in New South Wales in 1845: "I was assured by Mr. Parkinson, a gentleman in charge of these stations, that this plant springs up at all sheep and cattle stations throughout the colony, a remarkable fact". Horehound is now a declared noxious weed in four states.

The explosive spread of these herbs may have startled Mitchell and his contemporaries, but it does not surprise us today. St John's wort and horehound were brought to a land that was free of their pests and predators, and where the native vegetation was unused to cattle, sheep and human disturbance. The medicinal constituents of these herbs help render them distasteful to grazers.

Many of Europe's traditional healing herbs have spread to Australia as weeds. Hemlock (*Conium maculatum*), hawthorn (*Crataegus monogyna*), fennel (*Foeniculum vulgare*), blackberry (*Rubus fruticosus*) and variegated thistle (*Silybum marianum*) have all been declared noxious, and the castor oil bean (*Ricinus communis*), thornapple (*Datura stramonium*), and camphor laurel tree (*Cinnamomum camphora*) are serious weeds. Control of these pests has cost Australian agriculture many millions of dollars. New medicinal plants continue to become established. Comfrey (*Symphytum xuplandium*) has recently joined the weed list of Tasmania.

The spread of exotic plants (and animals) is one of the gravest threats facing the Australian bush. Kakadu, for example, is seriously threatened by two feral ornamental plants — mimosa (*Mimosa pigra*) and salvinia (*Salvinia molesta*). Conservationists and governments have not yet faced up to this problem. It is still legal to import into the country any ornamental or medicinal plant not known to be a weed, or closely related to a weed, and each year at least ten new plants run wild. Unless Australia and other countries halt this trend, in a thousand years the world will become dominated by homogenous eco-

▼ Feral fennel (*Foeniculum vulgare*) thrives along railway lines in Melbourne and Sydney, for example, here at Marrickville. Weed expert Charles Lamp noted that it "is so common on the railway reserves in metropolitan Melbourne that it is surely a candidate for the city's floral emblem".

◀ Sometimes imported into Australia by herbalists, and widely grown in herb gardens, borage (*Borago officinalis*) has become established in the wild in Australia's southern states. "Officinale" implies medicinal use.

▶ Baron von Mueller declared in 1876 that the blackberry (*Rubus fruticosus*) "deserves to be naturalised on the rivulets of our ranges". He is credited with spreading about the seeds of this now noxious weed.

systems of "superior" species such as castor oil beans, thornapples, horehound, cobblers pegs, cane toads, house mice and rabbits.

No one from the nineteenth century would have been more shocked by the wanton spread of weeds than that august gentleman, Baron Ferdinand Jakob Heinrich von Mueller. One of Australia's greatest botanists, von Mueller is remembered for having discovered more than two thousand new plants, for developing a classification of the Australian flora, and for prominently advocating the economic development of native plants.

But von Mueller was also active in the Victorian Acclimatization Society, a movement committed to a course of genteel madness. As Eric Rolls has said, "there was never a body of eminent men so foolishly, so vigorously, and so disastrously wrong". The acclimatisers believed that to ensure future prosperity, Australia's forests should be stocked with useful animals and plants from around the world. Acclimatisers argued for the release of llamas into the Australian Alps, boa constrictors into the scrubs to eat snakes, and monkeys in the eucalypt forests to amuse wayfarers. Von Mueller, as director of the Melbourne Botanic Gardens, was responsible for introducing new plants for cultivation and release into the bush.

In his book *Select Extratropical Plants*, he advocated the naturalisation of many medicinal plants. Mexican jalap (*Ipomoea purga*) was entitled to a trial in warm woodlands, he declared, and Brazilian ipecacuanha (*Cephaelis ipecacuanha*) would grow in the forests of East Gippsland. Of European squill (*Urginea maritima*) he recommended that "settlers living near the coast might encourage its dissemination, and thus obtain the bulbs as drug from natural localities". Meadow saffron (*Colchicum autumnale*), an important gout remedy, he had introduced, "with a view to its naturalization on moist meadows in our ranges".

Von Mueller's charming vision of Australia as a great garden, her meadows and woods stocked with free-growing foods and medicines, could never have materialised. Ecosystems cannot be tinkered with in this way. Attempts to "improve" natural environments always fail. One of von Mueller's favourite introductions, the blackberry, is now Victoria's worst weed. We can be thankful that his plans for squill, jalap and ipecacuanha never came to pass.

Von Mueller and his colleagues saw the Australian bush as a resource to be exploited, managed and manipulated for human ends. While we may laugh at the absurdity of the acclimatisers' ideas, their philosophy of exploitation is still held by men in power today. Unless Australians can come to accept nature's intrinsic right to exist, undisturbed and unexploited, we will continue to degrade this fragile land.

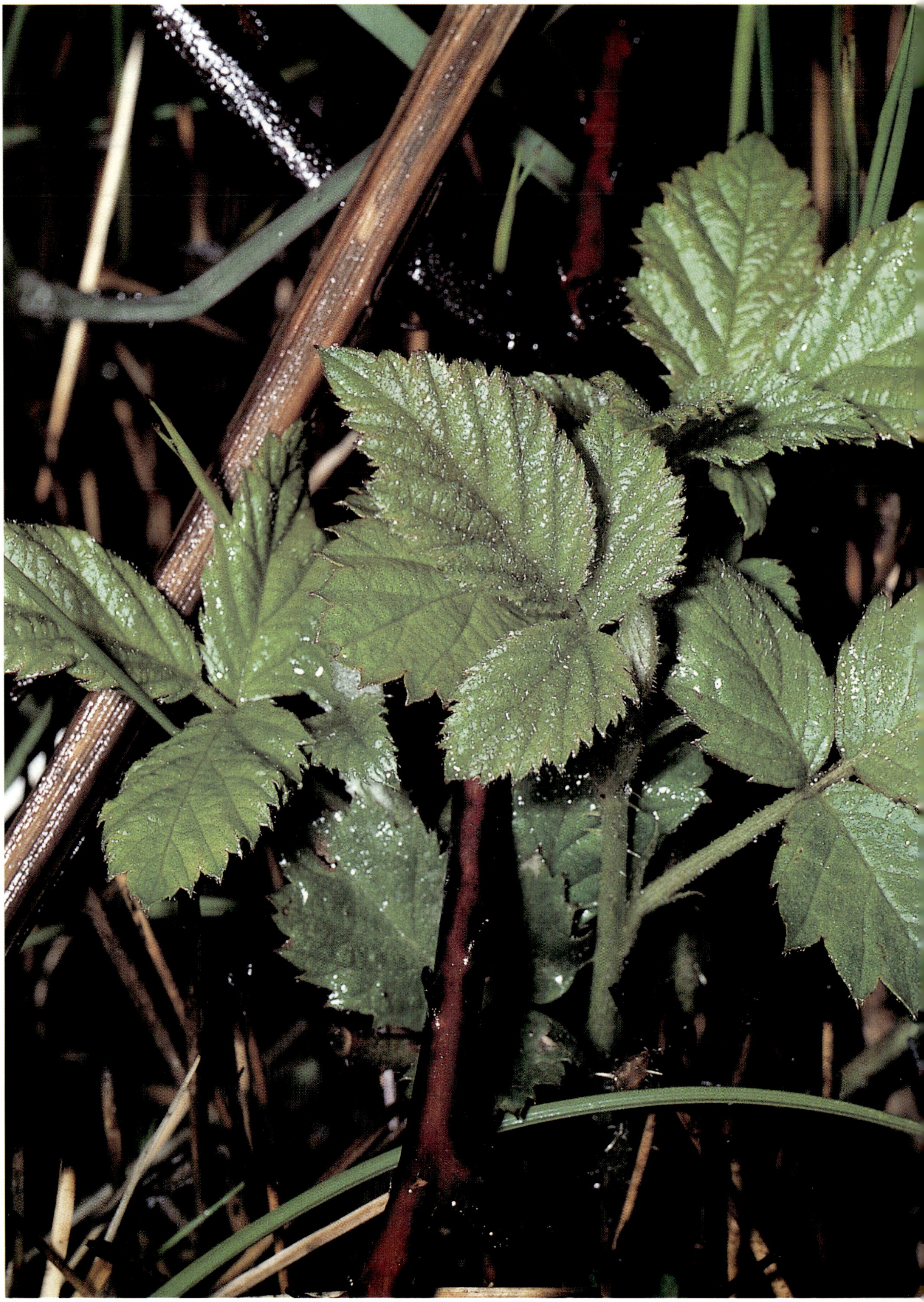

HOME REMEDIES

◀ Foliage for infusing: leaves of the blackberry (*Rubus fruticosus*) can be brewed to make an astringent tea for spongy gums and runny bowels. Two thousand years ago Pliny commended the plant for having "a drying and astringent property". Photographed south of Stanthorpe.

▶ Food, fibre and medicine have been harvested from the native hollyhock (*Lavatera plebeia*), also known as "native marshmallow", an Australian member of the hibiscus family. Aborigines reportedly ate the roots and spun string from the stems; colonists employed the roots and leaves in medicine. Photographed at Broken Hill.

Half the fun of herbalism is the opportunity it provides to gather and prepare your own remedies. Medicinal herbs grow wild in every vacant allotment, in every patch of bush. The following tips for preparing and healing with herbs have been provided by Queensland professional herbalist Kerry Bone. Most of these plants are well-known European herbs that can be found growing in Australia as weeds; also included are a few native herbs. The recipes belong to the European tradition, but it should be possible to substitute Aboriginal herbs. Please remember to take care with wild plants; herbs can be dangerous if not used wisely.

HOME REMEDIES 221

HERBS

These are usually taken internally as infusions, decoctions or tinctures. A single dose should be taken three times a day for chronic complaints, and up to six times a day for acute ills. To make an infusion, place 6–12 tablespoons of dried herb (or five times the weight of the fresh herb) in a teapot and add 6 cups of boiling water. Leave to stand for five or ten minutes. (Note: Infusions need to be much stronger than herbal teas.)

To make a decoction, add the same amounts of herb and unheated water to an open saucepan, bring to the boil, and simmer for at least ten minutes. Replace any evaporated water if the levels fall too low. Infusions are usually made from leaves and other soft material, while decoctions are made from hard barks and roots. For a single dose, drink 100–200 millilitres (about half to three-quarters of a cupful). Prepare the doses fresh each day. When experimenting with a new herb, start with a low dose, in case you react adversely to the herb.

Tinctures are made by soaking the dried herb in alcohol and water, a process called maceration. Vodka and brandy are suitable for home use. Take 50 grams of dried herb and soak it in a quarter of a litre of alcohol, in a sealed container for two to three weeks. Shake thoroughly every day. The herb will swell to fill most of the liquid. To recover the tincture, wrap the herb in coarse calico and press over a bowl (if available, use a small wine press). For a single dose, take a teaspoonful diluted in a quarter of a cup of water. Tinctures can also be diluted and applied to local wounds and sores, either directly as a lotion, or soaked in cloth as a compress.

A better way of applying a herb externally is by making an ointment. There are several methods, but the easiest is to soak the herb in cooking oil, in the same proportions given for the tincture. After three weeks press out the oil, heat it very gently, and thicken with beeswax and petroleum jelly in the following proportions: 55 grams of oil, 15 grams of beeswax, 30 grams of petroleum jelly. (Take care with beeswax: it is highly flammable and should not be overheated.) Pour into a jar while still warm and leave to set.

DRYING

Leaves and flowers may easily be dried at room temperature. Spread them in a thin layer on cardboard in a warm, dry room away from light. Use a heater if the room is cold. Gently rearrange the leaves twice each day for about a week, until they are crisp and crush between the fingers. Stalks should snap cleanly. Dry the roots and bark in a thin layer in the oven at about 40°C. Store the dried herbs in an airtight jar away from direct sunlight.

RECIPES

BLACKBERRY LEAVES

Gather the leaves just before or during early flowering. Rich in astringent tannins, they can be taken internally as an infusion, and either drunk for diarrhoea, or gargled for swollen, bleeding gums and sore throat.

RASPBERRY LEAVES

Raspberry leaves have been used for many years in England to ease the pain of childbirth, and the native raspberry (*Rubus parvifolius*), identified by its pink flowers, and small leaflets with white undersides, is claimed to have similar properties. Take an infusion throughout pregnancy except during the first three months. Start with two doses a day and increase to four or five in the last week. Raspberry leaves are thought to build healthy uterine muscle and to help contractions near birth.

NATIVE MARSHMALLOW

Native marshmallow (*Lavatera plebeia*) and other native members of the hibiscus family (Malvaceae) produce plenty of mucilage, a slimy soothing substance. Eat the fresh flowers or make a petal infusion for digestive, respiratory, and urinary tract inflammations. The flowers of cultivated hibiscuses may also be used; they have an obvious slimy taste.

NATIVE PENNYROYAL

Like European pennyroyal, the native pennyroyals (*Mentha satureioides* and *M.* species) contain aromatic oils that are rich in pulegone, an insect repellant. To drive away flies, rub the body down with a pennyroyal tincture. Add the tincture to baths when bathing pets, and place fresh sprigs in their bedding. Replace these as soon as they lose their strong smell.

▲ Delectable native raspberries (*Rubus parvifolius*) are a food of bushwalker and birds, and the leaves once served in colonial medicine. This scrawny scrambler grows in woodlands and forests between Adelaide and Rockhampton. Photographed in New England National Park.

HOME REMEDIES   223

River Mint

Like peppermint, river mint (*Mentha australis*) contains cooling menthol, and may be drunk as a tea for relieving abdominal pains caused by wind. Judging by the taste of the leaves they also contain pulegone, a mildly irritating oil, so the tea should not be drunk in excess. For tips on identifying mints, see "Native Mints".

Sacred Basil

In India, sacred basil (*Ocimum tenuiflorum*) is regarded almost as a cure-all. Take an infusion for colds, bronchitis, diarrhoea, and apply it externally as an antiseptic to acne.

▲ A holy Hindu herb, sacred basil (*Ocimum tenuiflorum*) was prescribed in India as a "Stimulant, diaphoretic and expectorant" according to colonial botanist Joseph Maiden. Photographed at Mt Grey west of Blackall, central Queensland.

Bronchial Formula

Mix the dried leaves of horehound (*Marrubium vulgare*), eucalyptus (*Eucalyptus*) and asthma plant (*Euphorbia hirta*) in a ratio of 2:2:1. Choose a eucalypt with a strong eucalyptus smell. The mixture may be drunk as an infusion (but it is more effective as a tincture) in single doses of a teaspoonful, taken three times daily, or six times daily if the bronchitis is severe. This blend combines an expectorant (mucus expeller), antiseptic and bronchial dilator.

▲ Milky markings on the leaves of the variegated thistle (*Silybum marianum*) were reputed by legend to be outpourings from the breasts of the Virgin Mary, hence *marianum*, from "Mary". The seedling leaves, young stems, roots and buds, were eaten by European peasants. Photographed near Warwick.

Variegated Thistle

In Europe the seeds of this noxious weed are highly regarded as a liver remedy. It should be harvested when the plant has dried and the seed head has opened. Wear gloves to extract the dark brown or black ripe seeds; reject any that are pale. The seeds should be dried in a thin layer in the oven at 35°C as they are fatty and apt to go mouldy. Make a decoction of 3–5 heaped teaspoons of seeds in 300–400 millilitres of water. Drink half a cupful three times a day for at least a fortnight as a liver tonic and for liver repair after viral infections, insecticide contact and alcohol overuse.

Variegated thistle (*Silybum marianum*) is identified by its purple flowers, tall spineless stems, and dark green, shiny leaves with paler markings.

HAWTHORN

German research suggests that hawthorn (*Crataegus monogyna*) strengthens heart function. Regular intake may help older people keep their heart healthy. Harvest the leaves in early spring while the plant is still in flower, and prepare an infusion, which should be drunk three times a day. The flowers as well as the leaves may be taken.

Hawthorn is a large deciduous shrub or small tree, often grown in hedges in cool climates. It can be identified by its thorny branches, lobed leaves, white, pink or red flowers, and shiny red berries with yellowish flesh. Do not confuse it with cotoneasters, which have unlobed leaves and lack thorns.

▲ "Pixie pears" are how English peasants described the dainty fruits of hawthorn (*Crataegus monogyna*), an ingredient in traditional sauces and jellies. According to folk legend the spiky stems were woven into Christ's crown of thorns. Photographed on a roadside near Tenterfield.

ST JOHN'S WORT

Harvest St John's wort (*Hypericum perforatum*) while the plant is flowering, then dry carefully and remove the larger stems. Make an ointment for cuts, bruises and rashes, and drink an infusion for nerve inflammations (neuralgia). This herb has bright yellow flowers with many stamens, slender, blunt-tipped leaves in pairs, and two tiny ridges along the stems. Preparations from this herb should have a rich, red colour from the pigment hypericin.

THORNAPPLE

Common thornapple (*Datura stramonium*) contains alkaloids that dilate the bronchioles, and it was a popular asthma remedy of generations past. Although no longer recommended today, it is worth remembering during an emergency; it could save a life. Take enough of the coarsely powdered leaves to cover a twenty cent piece (adult dose) and smoke them as a cigarette. Any other kind of preparation from this weed can be very dangerous. Thornapple is illustrated on pages 50 and 204–205.

PETTY SPURGE

An easy way to remove warts is to apply the milky sap of the garden weed petty spurge (*Euphorbia peplus*). First sand the wart to remove the cornified layer of dead skin, then protect the surrounding skin with vaseline or petroleum jelly. Dab the wart liberally with the sap, and repeat the treatment every second day for eight days.

Petty spurge can be recognised by its copious milky sap and its pale green leaves, which are up to 2.5 centimetres long. They are produced singly along the main stem and in pairs on the upper stalks.

▲ Of petty spurge (*Euphorbia peplus*), an English naturalist Anne Pratt in 1891 wrote: "It has powerful medicinal properties, and is used by country people as a caustic for the bite of a viper, and sometimes rubbed on the skin behind the ear as a cure for toothache." Photographed in Brisbane.

Aboriginal Remedies

These tables present a sample of remedies, and only the more important ailments.

Headache

Red ash (*Alphitonia excelsa*)	Bathe with crushed leaves in water
Tick-weed (*Cleome viscosa*)	Mashed plant applied
Headache vine (*Clematis glycinoides*)	Crushed leaves inhaled
Rock fuchsia bush (*Eremophila freelingii*)	Leaf decoction drunk
Liniment tree (*Melaleuca symphyocarpa*)	Crushed leaves rubbed on head
Tamarind (*Tamarindus indica*)	Fruit pulp rubbed on head
Snakevine (*Tinospora smilacina*)	Mashed stems wound around head

Coughs, colds

Lemon grasses (*Cymbopogon*)	Decoction drunk or applied as wash
Fuchsia bushes (*Eremophila*)	Decoction drunk
Tea trees (*Melaleuca*)	Crushed leaves inhaled
River mint (*Mentha australis*)	Decoction drunk
Great morinda (*Morinda citrifolia*)	Ripe fruit eaten
Toothed ragwort (*Pterocaulon serrulatum*)	Crushed leaves inhaled
Jirrpirinypa (*Stemodia viscosa*)	Crushed leaves inhaled

Fevers

Turpentine bush (*Beyeria lechenaultii*)	Leaf decoction taken
Kapok tree (*Cochlospermum fraseri*)	Bark and flower decoction drunk
Lemon grasses (*Cymbopogon*)	External wash of boiled leaves
River red gum (*Eucalyptus camaldulensis*)	Steamed leaves inhaled
Darwin stringybark (*Eucalyptus tetradonta*)	Bark infusion drunk
Tea tree (*Melaleuca viridiflora*)	Bath of crushed leaves in water

Diarrhoea

Lemon grasses (*Cymbopogon*)	Decoction drunk
Eucalypt bark (*Eucalyptus*)	Infusion drunk
Cluster fig (*Ficus racemosa*)	Bark infusion drunk
Dysentery bush (*Grewia retusifolia*)	Root infusion drunk
Sacred basil (*Ocimum tenuiflorum*)	Leaf infusion drunk
Pandanus (*Pandanus spiralis*)	Inner core of trunk eaten
Native raspberries (*Rubus*)	Decoction drunk

Wounds

Billygoat weed (*Ageratum*)	Crushed plant applied
Tree orchid (*Dendrobium affine*)	Bulb sap dabbed on cuts
Spike rush (*Eleocharis dulcis*)	Decaying plant bound to wounds
Paperbark tea trees (*Melaleuca*)	Bark wrapped as a bandage
Cocky apple (*Planchonia careya*)	Bark infusion poured into wounds

SORES

Australian bugle (*Ajuga australis*)	Bathe in infusion of bruised plant
Milkwood tree (*Alstonia actinophylla*)	Sap applied to sores
Green plum (*Buchanania obovata*)	Pounded leaves rubbed on ringworm
Onion lily (*Crinum angustifolium*)	Bulb infusion washed on sores
Rock fuchsia bush (*Eremophila freelingii*)	Sores washed with leaf decoction
Garlpu-nganing (*Hypoestes floribunda*)	Plant mashed and applied
Caustic bush (*Sarcostemma australe*)	Sap dabbed on sores

ACHES AND PAINS

Northern black wattle (*Acacia auriculiformis*)	Root decoction applied
Beach bean (*Canavalia rosea*)	Mashed root infusion rubbed on
Rock fuchsia bush (*Eremophila freelingii*)	Wash with leaf decoction
Beaty leaf (*Calophyllum inophyllum*)	Rub with crushed nut and ochre

STINGS

Nipan (*Capparis lasiantha*)	Whole plant infusion applied
Native hop (*Dodonaea viscosa*)	Chewed leaves bound to sting
Beach convolvulus (*Ipomoea pes-caprae*)	Heated leaf applied
Snakevine (*Tinospora smilacina*)	Root poultice applied
Peanut tree (*Sterculia quadrifida*)	Heated leaves pressed on stings

RHEUMATISM

Blackwood (*Acacia melanoxylon*)	Bathe with bark infusion
Konkerberry (*Carissa lanceolata*)	Oily sap rubbed as liniment
Beach bean (*Canavalia rosea*)	Mashed root infusion rubbed in
Tick-weed (*Cleome viscosa*)	Leaves applied
Stinging tree (*Dendrocnide moroides*)	Boiled leaves and bark rubbed in
Nettle (*Urtica*)	Patient beaten with leaves

SORE EYES

Ironwood (*Acacia estrophiolata*)	Root decoction administered
Green plum (*Buchanania obovata*)	Infusion of inner bark applied
Regal birdflower (*Crotalaria cunninghamii*)	Sap or leaf decoction given
Emu apple (*Owenia acidula*)	Wood decoction applied
Fan flower (*Scaevola sericea*)	Fruit juice applied

SORE EARS

River mangrove (*Aegiceras corniculatum*)	Leaf decoction applied
Lemon grass (*Cymbopogon*)	Root decoction poured into ears
Native hop (*Dodonaea viscosa*)	Boiled root juice applied
Lady apple (*Syzygium suborbiculare*)	Fruit pulp applied

TOOTHACHE

Green plum (*Buchanania obovata*)	Tooth plugged with shredded wood
Coast she-oak (*Casuarina equisetifolia*)	Preparation from inner bark to tooth
Denhamia (*Denhamia obscura*)	Tooth plugged with inner bark
Supplejack (*Flagellaria indica*)	Benumbing stem chewed
Pemphis (*Pemphis acidula*)	Burning twig applied
Quinine berry (*Petalostigma pubescens*)	Fruits held in mouth

Colonial Remedies

Coughs and colds

Banksia (*Banksia*)	Nectar syrup drunk
Blue gum (*Eucalyptus globulus*)	Leaves scalded and steam inhaled
Stringybark (*Eucalyptus*)	Tea for sore throats
Native grape (*Cissus hypoglauca*)	Fruit jelly soothed throat
Horehound (*Marrubium vulgare*)	Leaf tea drunk

Asthma

Thornapple (*Datura stramonium*)	Dried leaves smoked
Asthma plant (*Euphorbia hirta*)	Leaf tea drunk three times daily
Blue gum (*Eucalyptus globulus*)	Leaves smoked or decoction drunk

Fever

Quinine tree (*Alstonia constricta*)	Bark tincture drunk
Hop bush (*Daviesia latifolia*)	Preparation from leaves
Blue gum (*Eucalyptus globulus*)	Tree "disperses malarial vapours"
Leichhardt tree (*Nauclea orientalis*)	Wood decoction drunk
Quinine berry (*Petalostigma pubescens*)	Fruits soaked in tea

Diarrhoea and dysentery

Wattles (*Acacia*)	Bark infusion drunk
Sea box (*Alyxia buxifolia*)	Decoction drunk
Centaury (*Centaurium*)	Decoction drunk
Caustic weed (*Euphorbia drummondii*)	Infusion drunk
Tree orchids (*Cymbidium*)	Starchy stems chewed
Eucalypts (*Eucalyptus*)	Kino (gum) solution drunk
Native raspberry (*Rubus parvifolius*)	Leaf infusion drunk
Sida-retusa (*Sida rhombifolia*)	Leaf tips chewed

Tonics

Southern sassafras (*Atherosperma moschatum*)	Drink made from bark
Centaury (*Centaurium*)	Leaf decoction drunk
Hop bush (*Daviesia latifolia*)	Bitter leaves chewed
Yellow sassafras (*Doryphora sassafras*)	Bark infusion drunk
Native pennyroyal (*Mentha satureoides*)	Leaf tea drunk
River mint (*Mentha australis*)	Leaf tea drunk
Native sarsaparilla (*Smilax glyciphylla*)	Leaf tea drunk
Vervain (*Verbena officinalis*)	Infusion of leaves

Wounds and sores

Pennyweed (*Centella asiatica*)	Leaf poultice applied
Bear's ear (*Cymbonotus lawsonianus*)	Leaf ointment for wounds
Golden orchid (*Dendrobium discolor*)	Stem poultice applied
Caustic weed (*Euphorbia drummondii*)	Sap applied to sores and warts

Eucalypts (*Eucalyptus*)	Leaf poultices applied
Native hollyhock (*Lavatera plebeia*)	Leaf poultice applied
Caustic creeper (*Sarcostemma australe*)	Sap dripped on sores

STINGS

Cunjevoi (*Alocasia macrorrhizos*)	Cut leaf stalk pressed to nettle stings
Pigface (*Carpobrotus glaucescens*)	Leaf juice applied to seaside stings
Crinum lily (*Crinum pedunculatum*)	Crushed plant rubbed on marine stings

RHEUMATISM AND PAINS

Cunjevoi (*Alocasia macrorrhizos*)	Leaf poultice for back pain
Blue gum (*Eucalyptus globulus*)	Oil rubbed on rheumatism
Wild bean plant (*Galactia varians*)	Root decoction drunk for rheumatism

MISCELLANEOUS

Spreading sneezeweed (*Centipeda minima*)	Infected eyes bathed in infusion
Headache vine (*Clematis glycinoides*)	Crushed leaves inhaled for headache
Eucalypts (*Eucalyptus*)	Leaves and oil prescribed for everything
Hops goodenia (*Goodenia ovata*)	Leaf and twig infusion drunk for diabetes
Brooklime (*Gratiola*)	Decoction drunk for liver complaints
Brisbane pennyroyal (*Mentha*)	Preparation drunk to procure abortion
Castor oil bean (*Ricinis communis*)	Leaf poultice for lactation

PET MEDICINE

Scarlet pimpernel (*Anagallis arvensis*)	Juice cleaned dog's eyes
Boronia (*Boronia rhomboidea*)	Fed to horses to expel worms
Eucalyptus (*Eucalyptus*)	Powdered leaves feed to chooks for cholera
Sweet pittosporum (*Pittosporum undulatum*)	Gum healed dog's wound

BIBLIOGRAPHY

Aiston, G. The Aboriginal Narcotic Pitcheri. *Oceania* 1937: 7, 372–377.

Althofer, G. quoted by Lassak & McCarthy, 1983.

Bailey, F.M. Medicinal Plants of Queensland. *Proceedings of the Linnaean Society of New South Wales* 5, 1881: 1–29.

Bailey, F.M. *A Sketch of the Economic Plants of Queensland.* Government Printer, Brisbane, 1888.

Bailey, F.M. *The Queensland Flora.* 6 vols. H.J. Diddams, Brisbane, 1899–1902.

Bancroft, T.L. Preliminary notes on some new Poisonous Plants discovered on the Johnstone River, North Queensland. *Journal & Proceedings of the Royal Society of New South Wales* 20, 1886: 69–71.

Barnard, C. The Duboisias of Australia. *Economic Botany* 6, 1952: 3–17.

Barnes, C.S. and J.R. Price. An Examination of Some Reputed Antifertility Plants. *Lloydia* 38(2), 1975: 135–140.

Barr, A. (ed.) *Traditional Bush Medicines: An Aboriginal Pharmacopoeia.* Greenhouse Publications, Melbourne, 1988.

Bennett, G. *Gatherings of a Naturalist in Australasia.* John Van Voorst, London, 1860.

Bosisto, J. On the Hirudo Australis. *Transactions of the Philosophical Institute of Victoria* 3, 1859: 18–22.

Bosisto, J. Is the Eucalyptus a Fever-destroying Tree? *Transactions of the Royal Society of Victoria* 12, 1874: 10–23.

Bradley, V., D.J. Collins, P.G. Crabbe, F.W. Eastwood, M.C. Irvine, J.M. Swan, and D.E. Symon. A Survey of Australian *Solanum* Plants for Potentially Useful Sources of Solasodine. *Australian Journal of Botany* 26, 1978: 723–754.

Brand, J.C., A.S. Truswell, A. Lee and V. Cherikoff. An outstanding Food Source of Vitamin C. *Lancet* 2, 1982: 873.

Breton, W.H. *Excursions in New South Wales, Western Australia, and Van Diemen's Land.* Bentley, London, 1833.

Brock, J. *Top End Native Plants.* John Brock, Darwin, 1988.

Burkhill, I.H. *A Dictionary of the Economic Products of the Malay Peninsula.* Vols 1 & 2. Ministry of Agriculture and Co-operatives, Kuala Lumpur, 1966.

Butler, M. quoted by Lauterer, 1897.

Campbell, A. "Pharmacy" of Victorian Aborigines. *Australian Journal of Pharmacy* 54(648), 1973–74: 894–900.

Cartwright, L. *A Commonsense Guide to Medicinal Plants.* Angus & Robertson, Sydney, 1985.

Chaloupka, G. and P. Giuliani. *Gundulk Abel Gundalg, Mayali Flora.* Unpublished manuscript, Northern Territory Museum of Arts and Sciences, Darwin, 1984.

Cleland, J.B. and N.B. Tindale. The Native Names and Uses of Plants at Haast Bluff, Central Australia. *Transactions of the Royal Society of South Australia* 82, 1959: 123–140.

Collins, D. *An Account of the English Colony in New South Wales.* Cadell & Davies, London, 1798.

Collins, D.J., F.W. Eastwood and J.M. Swan. A Steroid Industry for Australia? *Search* 7(9), 1976: 378–383.

Covacevich, J., T. Irvine and G. Davis. *A Rainforest Pharmacopoeia: Five Thousand Years of Effective Medicine.* In Pearn, J., 1988.

Crawford (1925) quoted by Maiden, 1925.

Crawford, I.M. *Traditional Aboriginal Plant Resources in the Kalumburu Area.* Western Australian Museum, Perth, 1982.

Cribb, A.B. and J.W. Cribb. *Wild Medicine in Australia.* Collins, Sydney, 1981.

Crowther, W.E. Mr. Charles Underwood and his Antidote, with some Observations on Snake Bite in Tasmania. *The Medical Journal of Australia* Vol 1, 1956, 83–90.

Cunningham, P. *Two Years in New South Wales.* H. Colburn, London, 1827.

Dawson, J. *Australian Aborigines.* George Robertson, Melbourne, 1881.

Djaypila and Yalwidika Medicinal Plants on Elcho Island. *Aboriginal Health Worker* 3(2), 1979: 50–56.

Duncan-Kemp, A. *Our Sandhill Country.* Angus & Robertson, Sydney, 1933.

Elkin, A.P. *Aboriginal Men of High Degree.* Australasian Publishing Co., Sydney, 1945.

Everist, S.L. *Poisonous Plants of Australia.* Angus & Robertson, Sydney, 1974.

Farnsworth, N.R. and D.D. Soejarto. Potential Consequence of Plant Extinction in the United States on the Current and Future Availability of Prescription Drugs. *Economic Botany* 39(3), 1985: 231–240.

Foelsche, P. Notes on the Aborigines of North Australia. *Transactions, Proceedings and Report of the Royal Society of Australia* 5, 1882: 1–18.

Gildemeister, E. and Fr. Hoffmann. *The Volatile Oils.* 3 vols. Longmans, Green, London, 1913–22.

Gilmore, M. *Old Days, Old Ways.* Angus & Robertson, Sydney, 1934.

Gregory, H. *Goanna: The Genuine Bush Remedy.* J.C. Marconi, Brisbane, 1985.

Grieve, M. *A Modern Herbal.* Jonathan Cape, London, 1931.

Griffin, W.J. Duboisias of Australia. *Pharmacy International* 6, 1985: 305–308.

Guenther, E. *The Essential Oils.* 6 vols. Van Nostrand, New York, 1949–52.

Hagger, J. *Australian Colonial Medicine.* Rigby, Adelaide, 1979.

Hedley, C. Uses of some Queensland Plants. *Proceedings of the Royal Society of Queensland* 5(1), 1889: 10–13.

Isaacs, J. *Bush Food.* Weldons, Sydney, 1987.

Jockel, L.C. (1886) Quoted by Maiden, 1889.

Kingzett, C.T. *Nature's Hygiene: A Systematic Manual of Natural Hygiene.* Bailliere, Tindall and Cox, London, 1888.

Koch, M. A List of Plants collected on Mt. Lyndhurst Run, S. Australia. *Transactions of the Royal Society of South Australia* 22, 1898: 101–118.

Lassak, E.V. and T. McCarthy. *Australian Medicinal Plants.* Methuen Australia, Sydney, 1983.

Latz, P.K. *Bushfires and Bushtucker: Aborigines and Plants in Central Australia.* M.A. Thesis, University of New England, 1982.

Lauterer, J. Native Medicinal Plants of Queensland. *Proceedings of the Royal Society of Queensland* 10, 1894: 97–101.

Lauterer, J. New Native Medicinal Plants of Queensland. *Proceedings of the Royal Society of Queensland* 12, 1897: 92–94.

Leichhardt, L. *Journal of an Overland Expedition in Australia, from Moreton Bay to Port Essington, a distance of upwards of 3000 miles, during the years 1844–1845.* T. & W. Boone, London, 1847.

Levitt, D. *Plants and People: Aboriginal Uses of Plants on Groote Eylandt.* Australian Institute of Aboriginal Studies, Canberra, 1981.

Low, T. *Wild Herbs of Australia & New Zealand.* Angus & Robertson, Sydney, 1985.

Low, T. Explorers and Poisonous Plants (in) J. Covacevich, P. Davie and J. Pearn (eds) *Toxic Plants & Animals: A Guide for Australia.* Queensland Museum, Brisbane, 1987.

Low, T. *Wild Food Plants of Australia.* Angus & Robertson, Sydney, 1988.

Low, T. *Bush Tucker.* Angus & Robertson, Sydney, 1989.

Mackillop D. Anthropological Notes on the Aboriginal Tribes of the Daly River, North Australia. *Transactions of the Royal Society of South Australia* 17(1), 1892: 254–264.

Maiden, J.H. *The Useful Native Plants of Australia, (including Tasmania).* Trubner and Co., London, 1889.

Maiden, J.H. Hop-Bush as a Medicine. *Agricultural Gazette of New South Wales* 5, 1894: 143.

Maiden, J.H. Australian Sandarach. *Agricultural Gazette of New South Wales* 5, 1894: 301–305.

Maiden, J.H. Botanical Notes. A reputed antidote to snakebite. *Agricultural Gazette of New South Wales* 5, 1894: 473.

Maiden, J.H. Indigenous Vegetable Drugs. *Agricultural Gazette of New South Wales.* 1898. Part 1. 9.40–53, Part 2. 9.131–141.

Maiden, J.H. *The Forest Flora of New South Wales.* 8 vols. Government Printer, Sydney, 1902–25.

Maiden, J.H. *The Weeds of New South Wales. Part 1.* Government Printer, Sydney, 1920.

Markham N.T. and J.C. Noble. Essential Oil. in Noble, J.C. and R.A. Bradstock (eds) *Mediterranean Landscapes in Australia: Mallee Ecosystems and their Management.* CSIRO, Melbourne, 1989.

Meston, A. On the Australian Cassowary. *Proceedings of the Royal Society of Queensland* 10, 1894: 59–64.

Morrissey, G.C. (1945) quoted by Webb, 1948.

Mueller, F. von *Select Extra-tropical Plants readily eligible for Industrial Culture or Naturalisation.* Government Printer, Sydney, 1881.

Norris, M. Torres Strait Medicines. *Aboriginal Health Worker* 8(1), 1984: 14–18.

O'Connell, J.F., P.K. Latz and P. Barnett. Traditional and Modern Plant Use Among the Alyawara of Central Australia. *Economic Botany* 37, 1983: 80–109.

Parker, A. An Ethnobotany of the Western Desert, Leonora, Western Australia, 1977. *Australian Institute of Aboriginal Studies Newsletter* 13, 1980: 37–42.

Parker, K.L. *The Euahlayi Tribe.* Archibald Constable, London, 1905.

Pearn, J. (edited) *Pioneer Medicine in Australia.* Amphion Press, Brisbane, 1988.

Peterson, N. Aboriginal uses of Australian Solanaceae in J.G. Hawkes, R.N. Lester and A.D. Skelding (eds) *The biology and taxonomy of the Solanaeceae.* Linnean Society Symposium. Series 7. Academic Press, London, 1979.

Petrie, C.C. *Tom Petrie's Reminiscences of Early Queensland.* Watson, Ferguson & Co., Brisbane, 1904.

Pettigrew, W. On the Curative Properties of the *Cunjevoi Colocasia macrorrhiza, Schott. Proceedings of the Royal Society of Queensland* 2, 1885: 211–213.

Rasmussen, H.P. *The Natural Doctor.* Marcus and Andrew, Sydney, 1890.

Rawson, L. *Australian Enquiry Book.* Kangaroo Press, Sydney, 1984.

Reid, E.J. and T.J. Betts. Records of Western Australian Plants Used by Aboriginals as Medicinal Agents. *Planta Medica* 36, 1979: 164–173.

Reid, J. (ed.) *Body, Land & Spirit: Health and Healing in Aboriginal Society.* University of Queensland Press, Brisbane, 1982.

Richardson, L.R. A Contribution to the History of the Australian Medicinal Leech. *Australian Zoologist* 15(3), 1970: 395–399.

Roth, H.L. *The Aborigines of Tasmania.* F. King & Sons, Halifax, 1890.

Roth, W.E. *Ethnological Studies among the North-West-Central Queensland Aborigines.* Government Printer, Brisbane, 1897.

Roth, W.E. Food: Its search, Capture and Preparation. *North Queensland Ethnography* 3, 1901: 1–31.

Scarlett, N., N. White and J. Reid "Bush Medicines": The Pharmacopoeia of the Yolngu of Arnhem Land. In Reid (1982).

Shellshear, W.G. Treatment of Diabetes by "insulin" and Prickly Pear. *The Medical Journal of Australia.* March 1925, 329.

Shellshear, W.G. The Prickly Pear and Diabetes. *The Medical Journal of Australia.* February 1926.

Shepherd, T.W. (1871–2) quoted by Lassak & McCarthy, 1983.

Smith, R. Alkaloids from Native Flora: Commercial Production Old and New. *Chemistry in Australia* 56, 1989: 350–352.

Smyth, R.B. *Aborigines of Victoria.* 2 vols. Government Printer, Melbourne, 1878.

Stephens, E. The Aborigines of Australia. *Journal of the Royal Society of New South Wales* 23, 1889: 476–502.

Sturt, C. *Narrative of an Expedition into Central Australia.* T. & W. Boone, London, 1849.

Tepper, J.G. Probable Curative Properties of Melaleuca uncinata. *Transactions of the Royal Society of South Australia* 3, 1880: 174–175.

Thomson, D.F. The Hero Cult, Initiation and Totemism on Cape York. *The Royal Anthropological Institute of Great Britain and Ireland* 63, 1933: 453–537.

Thomson, D.F. Notes on the Smoking-pipes of North Queensland and the Northern Territory of Australia. *Man* 39(76), 1939: 81–91.

Tolmer, A. *Reminiscences of an Adventurous and Chequered Career at Home and at the Antipodes.* Sampson, Low, Marston, Searle, Rivington, London, 1882.

Turner, F. Botany of North-Western New South Wales. *Proceedings of the Linnaean Society of New South Wales* 30, 1905: 32–90.

Wasuwat, S. Extract of *Ipomoea pes-caprae* (Convolvulaceae) antagonistic to Histamine and Jelly-fish Poison. *Nature* 225, 1970: 758.

Watson, P. This Precious Foliage: A Study of the Aboriginal Psycho-active Drug *Pituri*. *Oceania Monograph* 26. University of Sydney: Sydney, 1983.

Webb, L.J. Guide to the Medicinal and Poisonous Plants of Queensland. *Bulletin 232* Council for Scientific and Industrial Research, Melbourne, 1948.

Webb, L.J. Some New Records of Medicinal Plants used by the Aborigines of Tropical Queensland and New Guinea. *Proceedings of the Royal Society of Queensland* 71(6), 1960: 103–110.

Webb, L.J. The Use of Plant Medicines and Poisons by Australian Aborigines. *Mankind* 7, 1969: 137–146.

Weedon, D. and J. Chick Home Treatment of Basal Cell Carcinoma. *The Medical Journal of Australia* 1, 1976: 928.

Wells, F.H. The Habits, Customs, and Ceremonies of the Aboriginals on the Diamantina, Herbert, and Eleanor Rivers, in east central Australia. *Report of the Fifth Meeting of the Australasian Association for the Advancement of Science*, 1893: 515–522.

Welsby T. *The Collected Works*. Jacaranda, Brisbane, 1967.

Wheelwright, H.W. *Bush Wanderings of a Naturalist*. Routledge, Warne, & Routledge, London, 1861.

White, J. *Journal of a Voyage to New South Wales*. Debrett, London, 1790.

Wightman, G.M. and N.D. Smith. Ethnobotany, Vegetation and Floristics of Milingimbi, Northern Australia. *Northern Territory Botanical Bulletin* 6, 1989: 1–37.

Woolls, W. *A Contribution to the Flora of Australia*. F. White, Sydney, 1867.

Woolls, W. Plants of New South Wales having Medicinal Properties. *Victorian Naturalist* 4, 1887: 103–105.

Ziviani, E. Personal communication with the author, 1989.

Index

Scientific Names

Acacia 228
    adsurgens 158
    ancistrocarpa 158
    aneura 191
    auriculiformis 16, 17, 185, 227
    dictyophleba 157
    estrophiolata 14, 15, 184, 227
    holosericea 185
    kempeana 182
    ligulata 69, 73, 196, 197
    lysiphloia 157
    melanoxylon 227
    pellita 185
    pruinocarpa 157
    salicina 197
    tetragonophylla 14, 15, 23, 167
Acanthophis praelongis 175, 179
Achyranthes aspera 39, 40
Adriana glabrata 190
Aegiceras corniculatum 185, 227
Ageratum 23, 55, 56, 166, 226
    conyzoides 55
    houstanianum 55, 56
Aizoaceae 147
Ajuga australis 104, 105, 166, 227
Ajuga reptans 105
Allocasuarina 202
Alocasia macrorrhizos 36, 179, 185, 229
Alphitonia excelsa 185, 226
Alstonia actinophylla 17, 159, 165, 227
Alstonia constricta 17, 36, 153, 155, 228
Alstonia scholaris 17
Althaea officinalis 44
Alyxia buxifolia 34, 36, 228
Amanita muscaria 205, 207
Amanita phalloides 59
Amaranthus viridis 59
Amorphophallus 189
Ampellocissus acetosa 147, 179
Anagallis arvensis 50, 51, 229
Angophora 140
    costata 14
Apium prostratum 145
Araucaria cunninghamii 185
Asteraceae 109
Atherosperma moschatum 228
Atriplex nummularia 148, 158
Atropa belladonna 196
Avicennia marina 170

Banksia 34, 152, 228
    integrifolia 202
Basilicum polystachyon 104, 105
Baurella simplicifolia 215
Bellis perennis 59
Beyeria lechenaultii 2, 226
Bidens pilosa 59
Borago officinalis 219
Boronia rhomboidea 185, 229
Brugmansia candida 205
Buchanania obovata 17, 26, 167, 183, 227
Bufo marinus 124, 125
Bursaria spinosa 212

Callitris 34, 96
    endlicheri 185
    glaucophylla 12, 25, 95, 96, 165
    intratropica 95, 96, 161, 172
Calophyllum inophyllum 78, 79, 227
Calvatia lilacina 81
Calytrix exstipulata 9, 100, 165
Calytrix brownii 10, 100
Camellia sinensis 141
Canavalia rosea 185, 227
Cannabaceae 204
Cannabis sativa 204

Capparis 165
    lasiantha 80–81, 179, 227
    umbonata 141
Capsella bursa-pastoris 52
Cardamine 148
Carissa lanceolata 185, 227
Carpobrotus glaucescens 172, 173, 229
Carpobrotus rossii 5
Castanospermum australe 17, 214, 215
Casuarina equisetifolia 183, 227
Catharathus roseus 17
Centaurium 5, 228
    erythraea 45, 46
    spicatum 45
Centella asiatica 56, 228
Centipeda 191
    cunninghamii 5
    minima 36, 109, 184, 185, 229
    thespidioides 109
Cephaelis ipecacuanha 219
Cestrum parqui 59
Chamaemelum nobile 47
Chenopodiaceae 109
Chenopodium ambrosioides 59
Chenopodium cristatum 109
Chrysopogon fallax 78, 182
Cinchona succirubra 152
Cinnamomum camphora 59, 218
Cissus hypoglauca 152, 228
Citrullus colocynthis 23
Clematis glycinoides 151, 226, 229
Clematis hirsutus 151
Cleome viscosa 13, 152, 185, 226, 227
Clerodendron floribundum 152, 189, 190
Cochlospermum fraseri 80, 81, 154, 155, 226
Codonocarpus cotinifolius 13, 154, 159
Colchicum autumnale 219
Commelina ensifolia 147
Conium maculatum 17, 59, 218
Copelandia cyanescens 206
Cosmozosteria 117, 170
    macula 118
Crataegus monogyna 58, 59, 218, 225
Crinum 78
    angustifolium 17, 227
    pedunculatum 171, 229
Crotalaria eremaea 182
Crotalaria cunninghamii 184, 227
Croton arnhemicus 62, 63
Curcuma australasica 159, 161
Cycas armstrongii 163, 167
Cymbidium 228
    canaliculatum 78, 140
    madidum 140, 159
Cymbonotus lawsonianus 228
Cymbopogon 10, 26, 105, 152, 172, 184, 226, 227
    ambiguus 28, 29, 106, 165
    bombycinus 106, 157
    obtectus 106, 107
    procerus 10, 63, 106, 107, 158
Cynanchum pedunculatum 63
Cyperus rotundus 59

Dacrydium franklini 96
Datura 196, 212
    leichhardtii 187, 188
    stramonium 17, 50, 51, 204, 205, 218, 225, 228
Daviesia latifolia 33, 202, 228
Decaisnina brittenii 166, 167
Dendrobium affine 78, 226
Dendrobium canaliculatum 78
Dendrobium discolor 228
Dendrocnide 36, 189
    excelsa 7, 185
    moroides 227

Denhamia obscura 159, 227
Dillenia alata 166, 167
Dioscorea bulbifera 77, 78, 158, 159
Dioscorea composita 210
Dioscorea mexicana 210
Dioscorea transversa 63, 167
Dittrichia graveolens 51
Dodonaea 34, 185, 202, 203
    lanceolata 179
    triquetra 203
    viscosa 171, 202, 227
Doryphora sassafras 228
Drosera 52, 53
    rotundifolia 53
Duboisia 17
    hopwoodii 17, 28, 40, 69, 74, 188, 190, 195, 199, 204
    leichhardtii 211
    myoporoides 70, 187, 192, 204, 205, 211
    myoporoides x leichhardtii 41, 211
Dysphania rhadinostachya 106, 108

Eleocharis dulcis 157, 226
Enchylaena tomentosa 71, 72
Entada phaseoloides 159
Eragrostis eriopoda 78
Eremocitrus glauca 144
Eremophila 10, 26, 97, 152
    alternifolia 97
    bignoniiflora 72
    duttonii 13, 99
    freelingii 13, 34, 97, 98, 226, 227
    goodwinii 99
    latrobei 11
    longifolia 97, 98, 99, 157
    maculata 98, 99
    mitchellii 101
    sturtii 99
Ervatamia orientalis 17, 78, 167
Erythraea 46
Erythrophleum chlorostachys 157–158, 165, 172
Erythroxylum australe 201
Erythroxylon coca 201
Eucalyptus 91, 224, 226, 228, 229
    amygdalina 86
    camaldulensis 13, 83, 84, 164, 226
    citriodora 213
    cneorifolia 86
    confertiflora 66
    dichromophloia 164
    dives 84
    globulus 86, 87, 90, 154, 185, 213, 228, 229
    gunnii 189, 202
    macrorhyncha 88
    mannifera 89
    microtheca 73, 84, 179
    miniata 84, 89, 137, 141
    papuana 84, 85, 164
    piperita 84
    polybractea 86
    ptychocarpa 91, 166, 167
    tereticornis 213
    terminalis 26, 34, 84, 86, 164, 184
    tetradonta 85, 87, 226
    viminalis 89
    youmanii 88
Euphorbia 17
    australis 159
    coghlanii 159
    drummondii 70, 179, 228
    hirta 23, 58, 224, 228
    peplus 50, 225
    wheeleri 73
Evolvulus alsinoides 197
Exocarpos cupressiformis 33
Exocarpos latifolius 158, 161, 172

Ficus coronata 166
Ficus opposita 24, 63, 167
Ficus racemosa 140, 141, 226
Flagellaria indica 183, 227
Flueggea virosa 170, 171
Foeniculum vulgare 218
Fumaria officinalis 52

Galactia varians 229
Galbulimima belgraveana 188
Galium aparine 51
Gardenia megasperma 78, 185
Gastrolobium laytonii 192
Gentiana lutea 47
Gentianella diemensis 45, 47
Geranium 3, 5, 78
    solanderi 3
Goodenia lunata 191, 193
Goodenia ovata 141, 185, 229
Gratiola 229
    officinalis 44
    pedunculata 44
    peruviana 44
Grevillea juncifolia 78
Grewia retusifolia 17, 34, 63, 78, 138, 167, 226

Hakea 78
    eyreana 159
    suberea 159
Hardenbergia violacea 143, 147
Hedyotis galioides 179
Helichrysum apiculatum 185
Hibiscus tilaceus 167
Humulus lupulus 204
Hydrocotyle asiatica 56
Hypericum perforatum 48, 217, 225
Hypoestis floribunda 227
Hyptis suaveolens 115

Ipomoea pes-caprae 17, 20, 29, 65, 66, 170, 179, 227
Ipomoea purga 219
Isotoma petraea 17, 28, 192

Jagera pseudorhus 201, 202

Kennedia 147

Lactuca sativa 206
Lactuca serriola 206
Lamiaceae 103
Lavatera plebeia 5, 44, 221, 223, 229
Lawrencia spicata 147
Leichhardtia australis 159, 160
Lepidium oleraceum 145
Leptomeria acida 145-146
Leptospermum 95
    petersonii 10, 95
Linospadix monostachya 185
Linum marginale 45
Linum usitatissimum 45
Litoria caerulea 125
Litsea glutinosa 79
Livistona humilis 63, 152, 153
Lobelia purpurascens 177
Lomandra longifolia 185
Lysiphyllum cunninghamii 165
Lythrum salicaria 48

Marrubium vulgare 32, 47, 152, 217, 224, 228
Melaleuca 152, 185, 226
    acacioides 172
    alternifolia 13, 93, 94, 213
    bracteata 95
    cajuputi 95
    ericifolia 95
    eucadendron 21, 61, 66, 94, 95
    linariifolia 95
    squarrosa 94
    symphyocarpa 12, 95, 152, 226
    uncinata 94
    viridiflora 95, 226
Mentha 229
    australis 71, 72, 111, 114, 224, 226, 228

    diemenica 113
    laxiflora 111
    piperita 115
    rotundifolia 115
    satureioides 112, 223, 228
    spicata 115
Mimosa pigra 218
Morelia spilota 127
Morinda citrifolia 17, 66, 78, 166, 167, 172, 226
Morinda reticulata 159, 160
Muehlenbeckia adpressa 147
Myoporum acuminatum 181
Myristica insipida 167

Nauclea orientalis 154, 178, 179, 202, 228
Nelumbo nucifera 39
Nicotiana 17, 28, 188, 190-191, 196, 204
    benthamii 191
    excelsior 191, 198
    glauca 23, 191
    gossei 191, 198
    maritima 188, 189
    megalosiphon 70-71
    rosulata 191

Ocimum tenuiflorum 11, 34, 104, 224, 226
Oecophylla smaragdina 117-118, 170
Opuntia stricta 58, 202
Owenia acidula 227
Owenia reticulata 78
Owenia vernicosa 165
Oxalis corniculata 147

Panax 39
Pandanus spiralis 78, 141, 152, 157, 167, 226
Papaver aculeatum 201
Parahebe perfoliata 147
Parietaria judaica 52
Parkinsonia aculeata 23
Passiflora foetida 23, 63, 78
Pemphis acidula 183, 184, 227
Persoonia falcata 14, 63
Persoonia pinifolia 12, 17
Petalostigma pubescens 152, 154, 161, 227, 228
Petasida ephippigera 100
Phellinus 81
Physalis peruviana 202
Phytolacca decandra 52
Phytolacca octandra 52, 53
Pimelea 182, 183
    linifolia 183
Pinus 95
Piper methysticum 192
Piper novae-hollandiae 214
Pittosporum phylliraeoides 80
    undulatum 185, 229
Pityrodia jamesii 100, 165
Pityrodia species 100
Planchonella laurifolia 152
Planchonella pohlmaniana 167
Planchonia careya 14, 63, 165, 171, 188, 189, 226
Plantago major 51
Plectranthus congestus 105
Portulaca oleracea 71, 72, 147
Prostanthera cineolifera 100
Prostanthera rotundifolia 4, 100
Prostanthera striatiflora 17, 99, 101, 134
Protasparagus racemosa 141
Prunella vulgaris 48, 49
Pseudechis australis 125
Psilocybe 206
Psilocybe collybioides 206
Psilocybe cubensis 206
Pteridium esculentum 176
Pterocaulon 152
    serrulatum 26, 27, 109, 165, 191, 226
    sphacelatum 23, 109, 152, 191
Pycnoporus coccineus 80, 81, 159

Ranunculus sceleratus 59
Rheobatrichus silus 125
Rhinoplocephalus flagellum 176
Ricinis communis 55, 57, 218, 229
Rorippa palustris 148

Rubus 44, 141, 226
    fruticosus 51, 218, 219
    idaeus 44
    moluccanus 44
    parvifolius 43, 44, 223, 228
    rosifolius 44, 141
Rumex 32, 49
Rumex crispus 32, 52

Salvinia molesta 218
Santalum 101
    acuminatum 80, 182
    lanceolatum 80, 100, 101, 172, 207
    murrayanum 188
    spicatum 80, 101
Sarcostemma australe 17, 34, 159, 163-164, 227, 229
Scaevola sericea 78, 167, 169, 171, 184, 185, 227
Scrophulariaceae 106
Sebaea ovata 45
Secamone elliptica 166
Senna alata 23, 66
Sesuvium portulacastrum 145
Sida rhombifolia 139, 228
Silybum marianum 52, 59, 218, 224
Smilax 147
    australis 147
    glyciphylla 32, 146, 147, 210, 228
Solanaceae 196, 205, 211
Solanum 17
    aviculare 209, 210, 211
    ellipticum 191
    hermannii 211
    laciniatum 210, 211
Sonchus oleraceus 5, 6, 26, 122, 206
Spinifex longifolius 167
Stellaria media 51
Stemodia grossa 106
Stemodia lythrifolia 106, 189
Stemodia viscosa 26, 103, 106, 165, 226
Sterculia quadrifida 171, 172, 227
Streptoglossa odora 109
Strychnos lucida 16, 17, 182
Symphytum xuplandicum 218
Syzygium 179, 190
    australe 202, 203
    suborbiculare 20, 78, 167, 185, 227

Tamarindus indica 23, 78, 141, 226
Taraxacum officinale 48, 49
Tephrosia crocera 171
Tephrosia varians 34, 185
Terminalia catappa 29
Terminalia ferdinandiana 148, 149
Tetragonia tetragonoides 144, 145
Tetragonia implexicoma 147
Tinospora smilacina 17, 171, 179, 226, 227
Tribulus cistoides 183, 184
Trichodesma zeylanicum 39, 41, 190, 192
Trigona 119
    hockingsi 118
Trigonella suavissima 144
Triodia pungens 120
Tylophora crebriflora 214

Ulex europaeus 59
Urginea maritima 219
Urtica 32, 227
    incisa 46, 47, 185
    urens 46

Varanus acanthurus 129
Varanus giganteus 71, 117
Varanus gouldii 129, 131
Varanus indicus 19
Varanus varius 129, 130
Verbascum thapsus 52
Verbena officinalis 47, 228
Veronica 45
    plebeia 45
Vigna vexillata 78, 141

Xanthorrhoea 202, 203

Zieria 5

INDEX

COMMON NAMES

adder 179
    death 171, 175, 179
aesculin 212
agaric, fly 205, 207
AIDS 209, 215
alcohol 188–189
alkaloid 16–17, 214–215
almond, sea 29
amaranth, green 59
anemone, sea 121
angel's trumpet 205
angophora 140
ant, green tree 117–118, 170
    honeypot 189
anthill 120
apple, cocky 14, 63, 165, 171, 188, 189, 226
    emu 227
    kangaroo 17, 209, 210–211, 214
    lady 20, 78, 167, 185, 227
    white 179
applebush 109
apple mint 115
apple of sodom 211
argara 188
aromatic oils 9, 93–101, 129, 165
arum, stinking 189
ash 167
    red 226
asparagus, native 141
asthma plant 23, 58, 224, 228
Austral brooklime 44
Austral Marine Bitters 36
Australian bugle 104, 105, 166, 227

balsam 203
banksia 152, 189, 202, 228
basil, musk 104, 105
    sacred 11, 34, 104, 224, 226
bastard sandalwood 101
bauhinia, native 165
baza, Pacific 88
beach bean 185, 227
beach convolvulus 17, 20, 29, 64, 65, 66, 170, 179, 227
bean, beach 185, 227
    black 215
    castor oil 32, 39, 55, 57, 218, 229
    matchbox 159
    wild 185
bean plant, wild 229
bear's ear 228
beard grass, golden 78, 182
beauty leaf 78, 79, 227
bee, native 118, 119
beef oil 134
bellis perennis 59
berrigan 97, 98, 99, 157
berry, quinine 152, 154, 161, 227, 228
    strychnine 16, 17
    white 170, 171
billygoat plum 148–149
billygoat weed 23, 55, 56, 166, 226
bindjud 171
birdflower, regal 184, 227
bitter bush 190
bitter quandong 188
bitter-bark 17
bitterbark 78, 167
bitters 36
black bean 215
black cypress pine 185
black orchid 140
black sally wattle 197
black wattle, northern 16, 17, 185, 227
blackberry 51, 218, 219, 223
blackthorn 212
blackwood 227

bloodwood 26, 75, 164
    western 34, 84, 86, 164, 184
    swamp 91, 166, 167
blue gum 85, 86, 87, 89–91, 154, 185, 228, 229
blue mallee 86
blue meanies 206
blush coondoo 152
boobialla tree 181
borage 219
boronia 185, 229
bottlebrush tea tree 94
bower spinach 147
box, sea 34, 36, 228
boxwood, yellow 167
bracken 176
bracket fungus, scarlet 80, 81, 159
bramble, molucca 44
Brisbane pennyroyal 114, 229
bristle poppy 201
brittle gum 89
broad-leaf hop bush 171, 202
broad-leaved native cherry 158, 161
broad-leaved peppermint 84
broad-leaved tea tree 95
brooklime 229
    stalked 44
    Austral 44
broom honey-myrtle 94
budda 101
budtha 101
bugle 105
    Australian 104, 105, 166, 227
bull's head 183, 184
bursaria, sweet 212
bush cockroach 117, 118, 170
bush tobacco 190
bush, bitter 190
    broad-leaf hop 171, 202
    cattle 39, 41, 190, 192
    caustic 34, 159, 163–164, 227
    currant 145–146
    desert fuchsia 10
    dysentery 17, 34, 63, 78, 138, 166, 226
    emu 97–98
    fuchsia 17, 26, 72, 75, 97–99, 152
    harlequin fuchsia 99
    hop 33, 185, 202, 228
    kite leaf poison 192
    mint- 4, 5, 100
    narrow-leaf fuchsia 97, 99
    pale turpentine 2
    pepper 192
    purple fuchsia 99
    rock fuchsia 34, 97–99, 226, 227
    round-leaved mint- 4, 100
    soap 185
    spotted fuchsia 99
    striped mint- 17, 99–100, 101, 134
    turkey 97
    turpentine 226
    terpentine emu 99
    weeping emu 97
    witchetty 182
buttercup, celery 59
buttons, yellow 185

cabbage palm 63
cajuput oil 95
cajuput tree 95
camomile 47, 109
camphor laurel 59, 218
cane toad 124, 125
Cape gooseberry 202
Cape York lily 159, 161
caper, native 165
carpet python 127
cassowary oil 134

castor oil bean 32, 39, 55, 57, 218, 229
caterpillar, processionary 118
cattle bush 39, 41, 190, 192
caustic bush 34, 159, 163–164, 227
caustic creeper 17, 229
caustic weed 70, 179, 228
celery, sea 145
celery buttercup 59
centaury 5, 6, 31, 228
    European 45–6
    native spike 45–6
    yellow 45
cestrum, green 32, 59
chaff flower 39, 40
cheesewood, white 17
cherry, broad-leaved native 158, 161
    native 33, 172
    scrub 202, 203
chestnut, Moreton Bay 17, 214, 215
chickweed 51, 52, 59
chilli 29
cider gum 189, 201
cinchona tree 152
clay 28, 138
cleavers 51
clerodendron 152, 189, 190
climbing lignum 147
clover, cooper 144
cluster fig 140, 141, 226
coast she-oak 183, 227
coast spinifex 167
coast tobacco 188, 189
coast tylophora 214
cobbler's pegs 59
coca tree 201
cocaine 201
cockroach 117, 118
    bush 170
cocky apple 14, 63, 165, 171, 188, 189, 226
coconut 29
coconut oil 29
coleus, native 105
colocynth 23
common thornapple 17, 50–51
comfrey 218
convolvulus, beach 17, 20, 29, 64, 65, 66, 170, 179, 227
coolibah 73, 84, 179
coondoo, blush 152
cooper clover 144
coral phlegm 121
corkwood 17, 40, 41, 70, 187, 192, 204, 205, 209, 211–212, 214
cotton tree 167
creeper, caustic 17, 229
    milk 166
cress 145
    wood 148
    yellow 148
crested crumbweed 109
crimson fuchsia bush 11
crinum lily 78, 171, 229
crowned sandpaper fig 166
crumbweed, green 106, 108
    crested 109
cunjevoi 36, 37, 179, 185, 229
curled dock 32, 52
currant bush 145–146
cuttlefish 122
cycad 163, 167
cypress pine, 93, 95–96
    black 185
    northern 95, 96
    white 12, 25, 95, 96, 165

daisy, European 59
dandelion 48, 49, 59

Darwin stringybark 85, 87, 226
Darwin woollybutt 84, 89, 137, 141
dead finish 14, 15, 17, 167
deadly nightshade 196, 205
death adder 171, 175, 179
death cap mushroom 59
denhamia 159, 227
desert fuchsia bush 10
desert hakea 78, 159
desert lime 144
desert mustard 147
desert poplar 13, 154
desert walnut 78
dingo 121
dock 32, 36, 49, 59, 176
    curled 32, 52
    yellow 52
dog's nuts 138
dogwood 75
doubah 159, 160
dugong oil 29, 40, 63-64, 117, 127, 132-134, 209
dysentery bush 17, 34, 63, 78, 138, 166, 226

echidna 128
eel 122
emu apple 227
emu bush 97-99
    weeping 97
emu oil 63-64, 117, 127, 129-132
English marshmallow 44
eucalypt 5, 9, 10, 12, 14, 32, 34, 39, 66, 78, 83-91, 93, 129, 140, 204, 224, 226, 228, 229
    lemon-scented 10
    peppermint 12, 84, 86
eucalyptus oil 36, 40, 41, 86-88, 90-91, 129, 152, 154, 209, 213
European centaury 45-46
European daisy 59
European raspberry 44

false sandalwood 101
false sarsaparilla 143, 147
fan flower, sea 78, 167, 169, 171, 184, 185, 227
fan palm 152, 153
fennel 32, 218
fern 94, 176
fig, cluster 140, 141, 226
    crowned sandpaper 166
    sandpaper 24, 63, 66, 167
finch, zebra 120
fish oil 127, 128
flax 45
    native 45
flaxleaf riceflower 183
flower, chaff 39, 40
    golden guinea 166, 167
    rice 182
    sea fan 78, 167, 169, 171, 184, 185, 227
fly agaric 205, 207
foam-bark 201, 202
forest mint 111
forest red gum 213
frangipanni 29
fringe myrtle 9, 10, 100, 165
    pink 9
frog, gastric brooding 125
    green tree 125
    platypus 125
fruit-salad plant 23, 109, 152
fuchsia bush 10, 13, 17, 26, 72, 75, 97-99, 152
    crimson 11
    desert 10
    harlequin 99
    narrow-leaf 97, 99
    rock 34, 97-99, 226, 227
    purple 99
    spotted 98, 99
    terpentine 99
fumitory 52
fungus, plate 81
    scarlet bracket 80, 81, 159

gardenia 78, 185
garlpu-nganing 227

gastric brooding frog 125
geebung 32
    pine-leaf 12, 17
gentian 47
    mountain 45, 47
geranium, native 3, 5, 78
ghost gum 84, 85, 164
ginseng 39
goanna 176, 178
    perenty 71
    sand 129, 131
    spiny-tailed 129
goanna oil 36, 64, 117, 127, 128-129, 134, 158
gold tops 206
golden beard grass 78, 182
golden guinea flower 166, 167
golden orchid 228
goldtop mushroom 188
goodenia, hop 141, 185, 229
gooseberry, Cape 202
gorse 59
grape, native 147, 152, 179, 228
grass, golden beard 78, 182
    lemon 10, 26, 63, 105-106, 152, 158, 172, 184, 185, 226, 227
    lemon-scented 28, 29, 106, 165
    scurvy 145, 147
    woollybutt 78
grasshopper, red leichhardt 100
grasstree 202, 203
great morinda 17, 66, 78, 166, 167, 172, 226
green amaranth 59
green cestrum 32, 59
green crumbweed 106, 108
green plum 17, 26, 167, 183, 227
green tree ant 117-118, 170
green tree frog 125
grey mangrove 170
grub 128
    witchetty 118, 121
guinea flower, golden 166, 167
gum tree 78, 83-84
    blue 85, 86, 87, 89-91, 154, 185, 228, 229
    brittle 89
    cider 189, 201
    forest red 213
    ghost 84, 85, 164
    manna 89
    narrow-leaved 138
    river red 13, 83, 84, 85, 164, 226
    rusty 14

hairy goodenia 191, 193
hakea, desert 78, 159
harlequin fuchsia bush 99
hawthorn 58, 59, 218, 225
headache vine 151, 226, 229
heartease 44
hemlock 17, 39, 59, 218
henbane 205
hollyhock, native 5, 6, 44, 221, 229
honey 75
honey-myrtle, broom 94
    narrow-leaved 95
honeypot ant 189
honeysuckle 32, 34
honeysuckle oak 78
hoop pine 185
hop bush 33, 185, 202, 228
    broad-leaf 171, 202
hop goodenia 141, 185, 229
hop, native 32, 179, 202, 203, 227
hops 204
horehound 32, 46, 47, 49, 52, 109, 152, 217-218, 224, 228
horseradish tree 154, 159
human oil 134
huon pine 96
hyoscine 17, 192, 204, 209, 212
hyoscyamine 204
hyptis 115

iguana 127, 128, 158
inkweed 52, 53
ipecacuanha 176, 219

ironwood 157, 158, 165, 172, 184, 227
    southern 14, 15
isotome, rock 17, 28, 192

jalap 89, 219
Jerusalem thorn 23
jirrpirinypa 26, 103, 106, 165, 226
jockeys' caps 99

Kakadu plum 149
kangaroo 26, 40, 122, 138
kangaroo apple 17, 209, 210-211, 214
Kangaroo Island mallee 86
kangaroo oil 134
kapok tree 80, 81, 154, 155, 226
karkalla 5
kava 192
kino 5, 14, 40, 78, 84, 164, 192, 203, 209
kite leaf poison bush 192
konkerberry 185, 227
kurrajong 32

lace monitor 129, 130
lady apple 20, 78, 167, 185, 227
laurel, camphor 59, 218
lawrencia, salt 147
leech 26, 122-125, 138
Leichhardt tree 154, 178, 179, 202, 228
lemon grass 10, 26, 63, 105-106, 152, 158, 172, 184, 185, 226, 227
lemon-scented eucalypt 10
lemon-scented grass 28, 29, 106, 165
lemon-scented tea tree 10, 95
lettuce 206
lignum, climbing 147
lilly pilly 190, 202
lily 166
    Cape York 159, 161
    crinum 78, 171, 229
    onion 17, 227
lime, desert 144
liniment tree 12, 95, 152, 226
lobeline 17
long yam 167
loose-flowered rattle-pod 182
loosestrife 48
lotus, sacred 39
LSD 205, 206
lucerne, paddy's 139

Madagascar periwinkle 17
Magic Ophthalmia Cure 36, 184
mallee, blue 86
    Kangaroo Island 86
mangrove monitor 19
mangrove worm 121-122
mangrove, grey 170
    river 185, 227
manna 78, 89
manna gum 89
mapoon shrub 159, 160
marijuana 204, 205
markura 165
marshmallow 6
    English 44
    native 221, 223
matchbox bean 159
mat-rush 185
meadow saffron 219
menthol 11, 13, 115
Mexican tea 59
milk creeper 166
milkwood tree 17, 159, 165, 227
milky plum 14, 63
mimosa 218
mint 5, 31
    apple 115
    forest 111
    native 9, 103
    river 71, 72, 111, 114, 224, 226, 228
    slender 113
mint-bush 5
    round-leaved 4, 100
    striped 17, 99-100, 101, 134
mistletoe 166, 167

molucca bramble 44
monitor, lace 129, 130
    mangrove 19
Moreton Bay chestnut 17, 214, 215
morinda, great 17, 66, 78, 166, 167, 172, 226
moss 94
mountain gentian 45, 47
mud 138, 167
mulga 191
mulga snake 125
mullein 52, 106
mushroom, death cap 59
    goldtop 188
    magic 205
musk basil 104, 105
mussel shells 158
mustard, desert 147
mustard oil 13
muttonbird oil 5
myrtle, pink fringe 9

nardoo 199
narrow-leaf fuchsia bush 97, 99
narrow-leaved gum tree 138
narrow-leaved honey-myrtle 95
native asparagus 141
native bauhinia 165
native bee 118, 119
native caper 165
native centaury 45
native cherry 33, 172
    broad-leaved 158, 161
native coleus 105
native flax 45
native geranium 3, 5, 78
native grape 147, 152, 179, 228
native hollyhock 5, 6, 44, 221, 229
native hop 32, 179, 202, 203, 227
native marshmallow 221, 223
native mint 9, 103
native nutmeg 167
native pea 147
native pennyroyal 112-113, 223, 228
native pepper 214
native peppermint 111
native raspberry 223, 226, 228
    pink-flowered 43, 44
native sarsaparilla 146, 147, 209, 210, 228
native thornapple 187, 188
native tobacco 17, 70-71, 188, 190-191, 204
nettle 32, 36, 46, 47, 52, 185, 227
    small 46
    scrub 46, 47
New Zealand spinach 145
ngarrik 62, 63
nicotine 16, 190, 196, 197, 198, 212
nightshade, deadly 196, 205
nipan 80, 179, 227
northern black wattle 16, 17, 185, 227
northern cypress pine 95-96, 161, 172
northern death adder 179
nutgrass 59
nutmeg, native 167

oak, honeysuckle 78
oil, aromatic 9, 129, 165
    beef 134
    cajuput 95
    cassowary 134
    coconut 29
    dugong 29, 40, 63-64, 117, 127, 132-134, 209
    emu 63-64, 117, 127, 129-132
    eucalyptus 36, 40, 41, 86-88, 90-91, 129, 152, 154, 209, 213
    fish 127, 128
    goanna 36, 64, 117, 127, 128-129, 134, 158
    human 134
    kangaroo 134
    mustard 13
    muttonbird 5
    sandalwood 40, 209
    snake 127, 128
    tea tree 41, 94-95, 209, 213-214

oil of wintergreen 129
oilgrass, silky 106, 157
old man saltbush 148, 158
onion lily 17, 227
opium 199, 201, 206
opossum 127
orange, wild 141
orchid 47, 139-140, 207
    black 140
    golden 228
    tree 78, 140, 159, 166, 226, 228
owl, powerful 6, 128

Pacific baza 88
paddy's lucerne 139
pale turpentine bush 2
palm, cabbage 63
    fan 152, 153
    walking stick 185
pandanus 78, 141, 152, 157, 158, 161, 167, 226
paperbark, tea tree 95, 226
    swamp 95
    wattle-flowered 172
passionfruit, stinking 23, 78
    wild 63
pea, native 147
pea vine 78, 141
peanut tree 171, 172, 227
pear, prickly 58, 202
pellitory 52
pemphis 183, 184, 227
pennyroyal 72, 112, 113, 223
    Brisbane 114, 229
    native 112-113, 223, 228
pennyweed 56, 228
pepper, native 214
pepper bush 192
peppermint 10, 115
    native 111
peppermint eucalyptus tree 12, 84, 86
    broad-leaved 84
perenty goanna 71, 117
periwinkle, Madagascar 17
petty spurge 50, 225
pigface 5, 172, 173, 229
pigweed 71, 147
pimpernel, scarlet 50, 51, 229
pine tree 32, 34, 95
    black cypress 185
    cypress 93, 95-96
    hoop 185
    huon 96
    northern cypress 95, 96, 172
    white cypress 12, 25, 95, 96, 165
pine-leaf geebung 12, 17
pink fringe myrtle 9
pink-flowered native raspberry 43, 44
pitcheri 198
pittosporum, sweet 185, 229
    weeping 78
pituri 17, 28, 40, 69-74, 187, 188, 190, 191, 192, 195-199, 204
plant, asthma 23, 58, 224, 228
    fruit-salad 23, 109, 152
    wild bean 229
plantain 51
plate fungus 81
platypus frog 125
plum, billygoat 148-149
    green 17, 26, 167, 183, 227
    Kakadu 149
    milky 14, 63
poison bush, kite leaf 192
pokeweed 52
poplar, desert 13, 154
poppy, bristle 201
porcupine 127, 128
powerful owl 6, 128
prickly pear 58, 202
processionary caterpillar 118
psilocybin 206
puffball 81
purple fuchsia bush 99
purslane, pink 145
python, carpet 127

quandong 80, 182
    bitter 188
quinine berry 152, 154, 161, 227, 228
quinine tree 17, 36, 40, 153, 155, 228

rabbit 4, 23, 121
ragwort 152, 191
    toothed 26, 27, 109, 165, 226
raspberry 44
    native 223, 226, 228
    pink-flowered native 43, 44
    roseleaf 44, 141
    wild 31, 141
rat 120
rattle-pod, loose-flowered 182
red ash 226
red gum, forest 213
    river 13, 83, 84, 85, 164, 226
red leichhardt grasshopper 100
red stringybark 88
regal birdflower 184, 227
reserpine 17
rice flower 182
    flaxleaf 183
ringworm shrub 23, 66
river mangrove 185, 227
river mint 71, 72, 111, 114, 224, 226, 228
river red gum 13, 83, 84, 85, 164, 226
river tea tree 95
rock fuchsia bush 34, 97-99, 226, 227
rock isotome 17, 28, 192
roseleaf raspberry 44, 141
round yam 77, 78, 158, 159
round-leaved mint-bush 4, 100
ruby saltbush 71, 72
rush, mat- 185
    spike 157, 226
rusty gum 14

sacred basil 11, 34, 104, 224, 226
sacred lotus 39
saffron, meadow 219
St John's wort 48, 217, 225
salep 47
sally wattle, black 197
salt lawrencia 147
saltbush, old man 148, 158
    ruby 71, 72
salvinia 218
sand goanna 129, 131
sandalwood 40, 80, 100, 101, 172, 207
    bastard 101
    false 101
    western 80, 101
sandalwood oil 209
sandarac 96
sandhill wattle 69, 73, 196, 197
sandpaper fig 24, 63, 66, 167
    crowned 166
sarsaparilla 32, 147
    false 143, 147
    native 146, 147, 209, 210, 228
sassafras, southern 228
    yellow 228
scarlet bracket fungus 80, 81, 159
scarlet pimpernel 50, 51, 229
scopalomine 204
scorpion 173
scrub cherry 202, 203
scrub nettle 46, 47
scrub yellowwood 215
scurvy grass 145, 147
sea almond 29
sea anemone 121
sea box 34, 36, 228
sea celery 145
sea fan flower 78, 167, 169, 171, 184, 185, 227
sea purslane 145
selfheal 48, 49
she-oak 202
    coast 183, 227
shepherd's purse 52
shrub, mapoon 159, 160
    ringworm 23, 66
sida-retusa 139, 228

silky oilgrass 106, 157
silky-heads 106, 107
slender mint 113
small nettle 46
snake, mulga 125
    whip 176
snake oil 127, 128
snakevine 17, 171, 179, 226, 227
sneezeweed 36, 109, 191
    spreading 184, 185, 229
soap bush 185
soap tree 185
solasodine 17, 210
sorrel, yellow wood 147
southern ironwood 14, 15
southern sassafras 228
sowthistle 5, 6, 26, 122, 138, 158, 206
spearmint 115
speedwell 45, 106
    trailing 45
    tropical 197, 198
spike centaury 45
spike rush 157, 226
spinach, bower 147
    New Zealand 145
spinifex 120
    coast 167
spiny-tailed goanna 129
spotted fuchsia bush 99
spreading sneezeweed 184, 185, 229
spurge 17, 73, 159
    petty 50, 225
squill 219
stalked brooklime 44
sting-ray 122
stinging tree 7, 36, 185, 188, 189, 227
stink-wood 5
stinking arum 189
stinking passionfruit 23, 78
stinkwort 51
strap wattle 185
stringybark 85, 228
    Darwin 85, 87, 226
    red 88
    youman's 88
striped mint-bush 17, 99–100, 101, 134
strychnine berry 16, 17
strychnine tree 182
sundew 52
supplejack 183, 227
swamp bloodwood 91, 166, 167
swamp paperbark 95
sweet bursaria 212
sweet pittosporum 185, 229

tamarind 23, 78, 141, 226
tannin 9, 13–16, 20, 78, 164
tansy 109
tea 141
    Mexican 59
tea tree 5, 9, 10, 12, 13, 40, 41, 93, 94–95, 152, 185, 213–214, 226

bottlebrush 94
broad-leaved 95
lemon-scented 10, 95
paperbark 95, 226
river 95
wattle-flowered paperbark 172
weeping 21, 61, 66, 94, 95
tea tree oil 41, 94–95, 209, 213–214
termite 28, 119, 120
terpentine fuchsia bush 99
thistle, variegated 52, 53, 59, 218, 224
thorn, Jerusalem 23
thornapple 196, 204, 205, 212, 218, 225, 228
    common 17, 50–51
    native 187, 188
Ti Ta 36, 94
tick-weed 13, 152, 185, 226, 227
toad, cane 124, 125
tobacco 187, 196, 199, 205
    bush 190
    coast 188, 189
    native 17, 70, 71, 188, 191, 204
    tree 23, 191
    wild 28, 198
tomato, wild 191
toothed ragwort 26, 27, 109, 165, 226
trailing speedwell 45
tree, boobialla 181
    bottlebrush tea 94
    broad-leaved tea 95
    cajuput tea 95
    cinchona 152
    coca 201
    cotton 167
    gum 78, 83–84
    horseradish 154, 159
    kapok 80, 81, 154, 155, 226
    Leichhardt 154, 178, 179, 202, 228
    lemon-scented tea 10, 95
    liniment 12, 95, 152, 226
    milkwood 17, 159, 165, 227
    narrow-leaved gum 138
    paperbark tea 95, 226
    peanut 171, 172, 227
    peppermint 84
    pine 95
    quinine 17, 36, 40, 153, 155, 228
    river tea 95
    soap 185
    stinging 7, 36, 185, 188, 189, 227
    strychnine 182
    tea 5, 9, 10, 12, 13, 40, 41, 93, 94–95, 152, 185, 213–214, 226
    wattle-flowered paperbark tea 172
    weeping tea 21, 61, 66, 94, 95
tree frog, green 125
tree orchid 78, 140, 159, 166, 226, 228
tree tobacco 23, 191
tropical speedwell 197, 198
trumpet, angel's 205
turkey-bush 97
turpentine bush 226

pale 2
tylophora, coast 214

variegated thistle 52, 53, 59, 218, 224
vervain 47, 228
vine, headache 151, 226, 229
    pea 78, 141
    snake 17, 171, 179, 226, 227

walking stick palm 185
walnut, desert 78
wattle 5, 32, 34, 40, 83, 140, 157, 158, 228
    black sally 197
    northern black 16, 17, 185, 227
    sandhill 69, 73, 196, 197
    strap 185
wattle-flowered paperbark 172
weed, billygoat 23, 55, 56, 166, 226
    caustic 70, 179, 228
weeping emu bush 97
weeping pittosporum 78
weeping tea tree 21, 61, 66, 94, 95
western bloodwood 34, 84, 86, 164, 184
western sandalwood 101
whip snake 176
white apple 179
white berry 170, 171
white cheesewood 17
white cypress pine 12, 25, 95, 96, 165
white root 177
wild bean 185
wild bean plant 229
wild orange 141
wild passionfruit 63
wild raspberry 31, 141
wild tobacco 28, 198
wild tomato 191
witchetty bush 182
witchetty grub 118, 121
wood cress 148
wood sorrel, yellow 147
woollybutt, Darwin 84, 89, 137, 141
woollybutt grass 78
worm, mangrove 121–122
wort, St John's 48, 217, 225
wurrumba 100, 165

yam 63, 210
    long 167
    round 77, 78, 158, 159
yarrow 109
yellow boxwood 167
yellow buttons 185
yellow centaury 45
yellow cress 148
yellow dock 52
yellow sassafras 228
yellow wood sorrel 147
yellowwood, scrub 215
youman's stringybark 88

zebra finch 120

Photographic Credits

(Numbers refer to pages)
All photos were taken by the author, except for the following:
Jeanette Covacevich 4 (below)
Murray Fagg 87 (below)
Glen Fergus 52–53 (sundew), 90
Owen Foley 120, 122–123, 132
Tom & Pam Gardner 135
Trevor Hawkeswood and Tim Low 3, 42–43, 178, 222–223
Anthony Preen 133
Queensland Herbarium 35 (left), 39
Queensland Museum 116–117, 119 below, 124, 125
Peter Slater 88
Steve Wilson 18–19, 130–131, 131
Anon. 7, 35 (below), 130
Art Gallery of South Australia 144 (bottom)

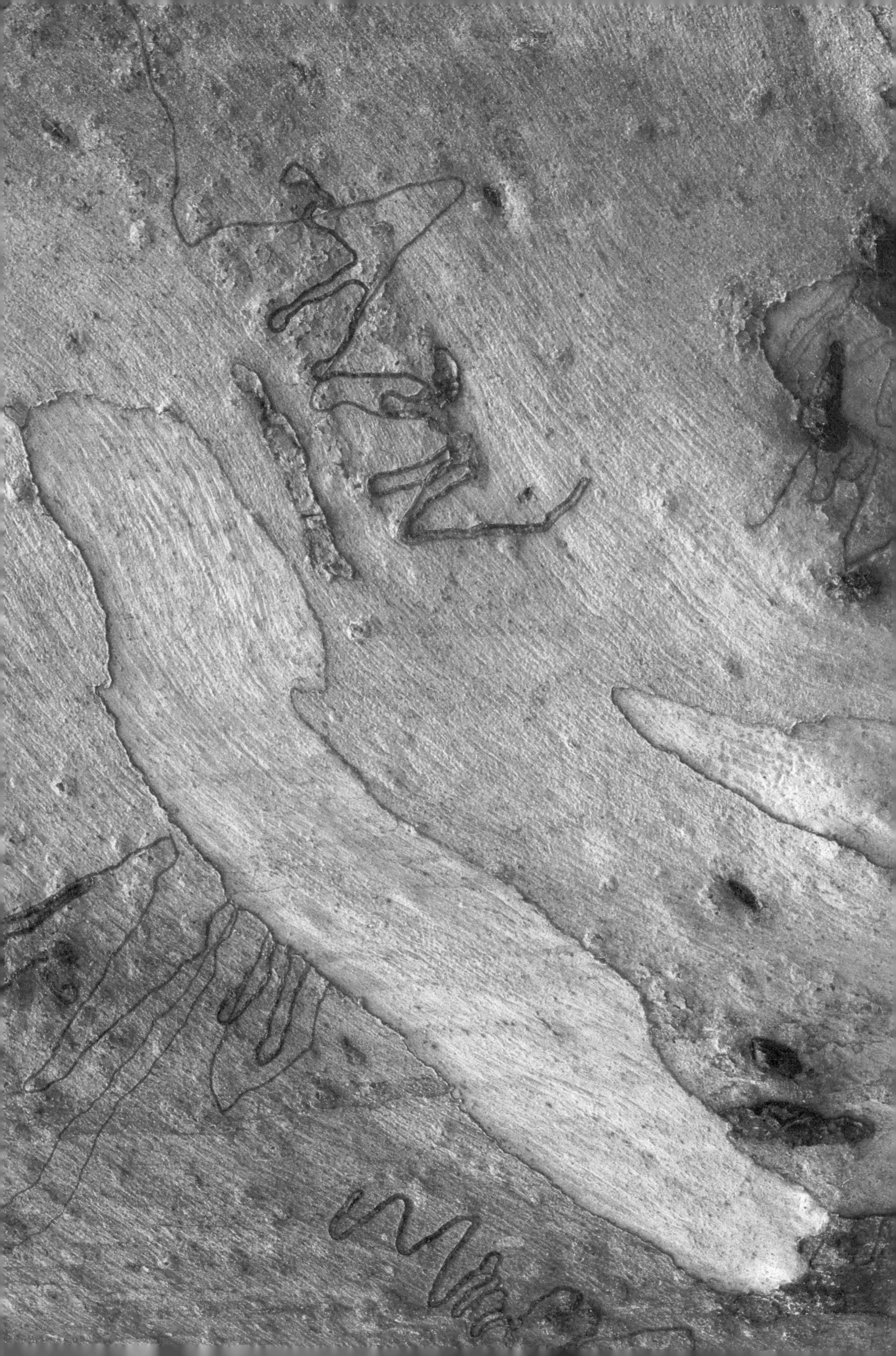

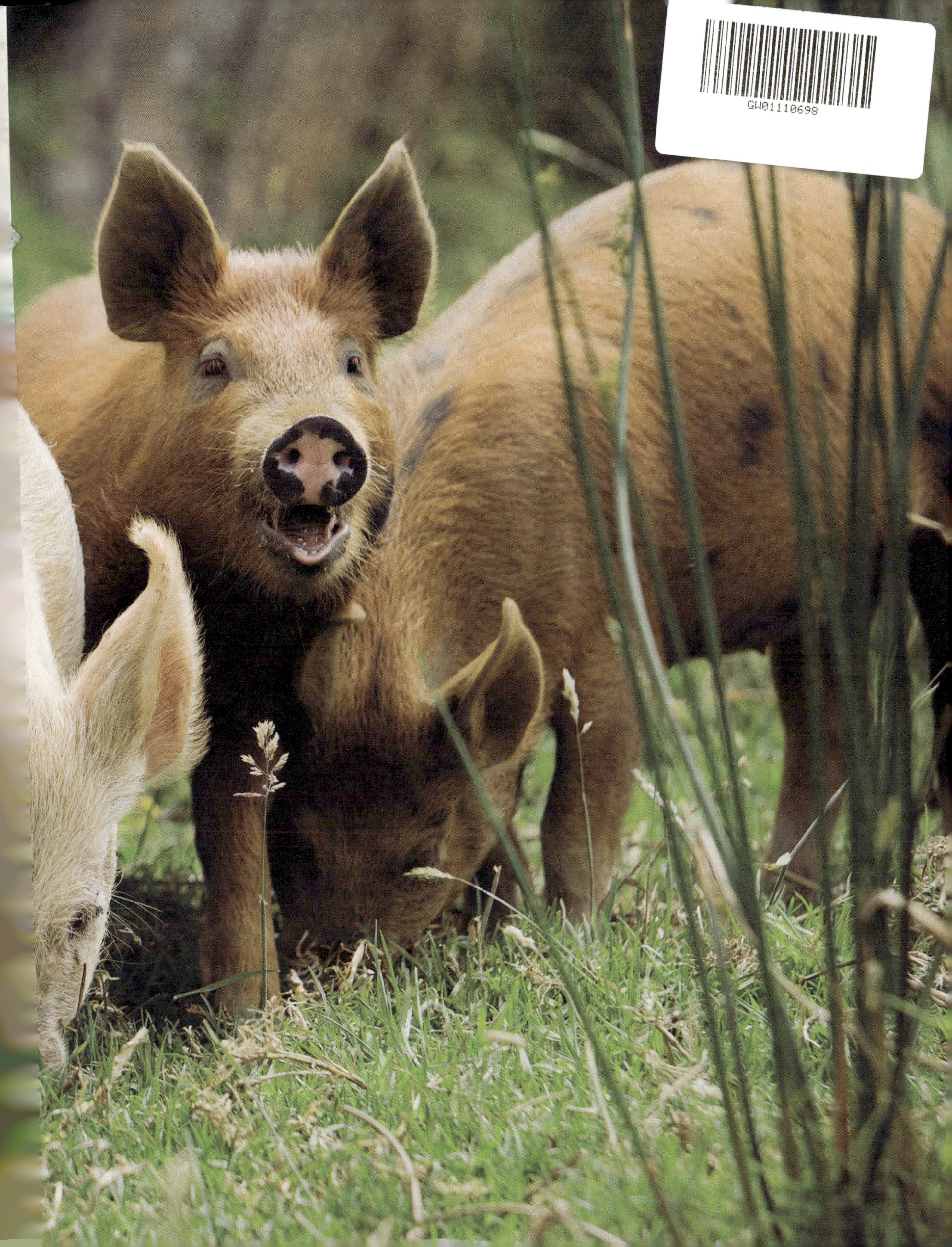

To you Didee on your birthday 2010
With loving wishes
Sasha

must eat

Russell Blaikie
with wine notes by Paul McArdle

photography by Craig Kinder

UWAP
UWA PUBLISHING

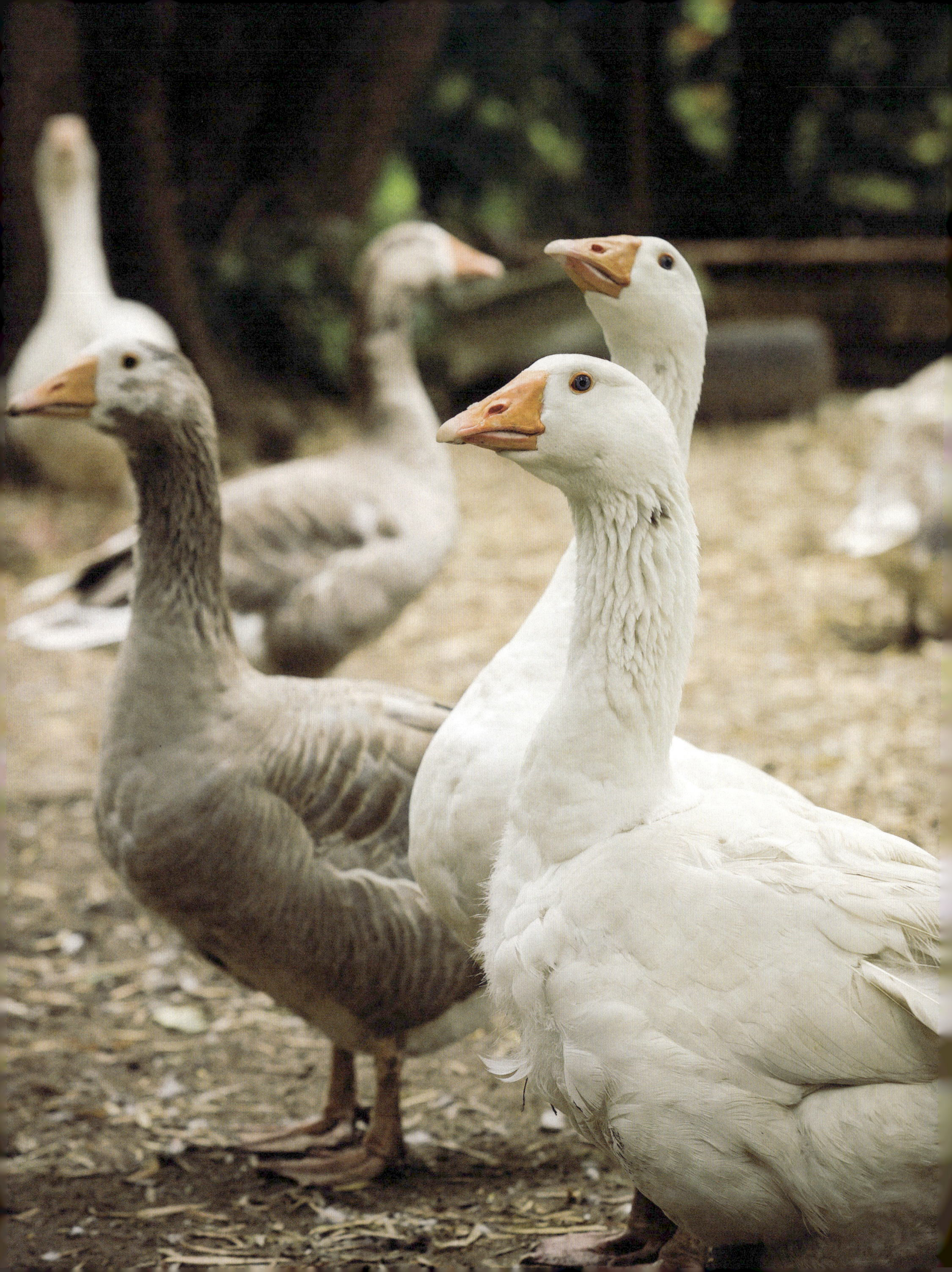

introduction	001
bar snacks	003
oysters	035
starters	051
charcuterie	113
mains	137
sides	193
desserts	215
cocktails	255
basics	263
select bibliography	293
thanks	294
index	295

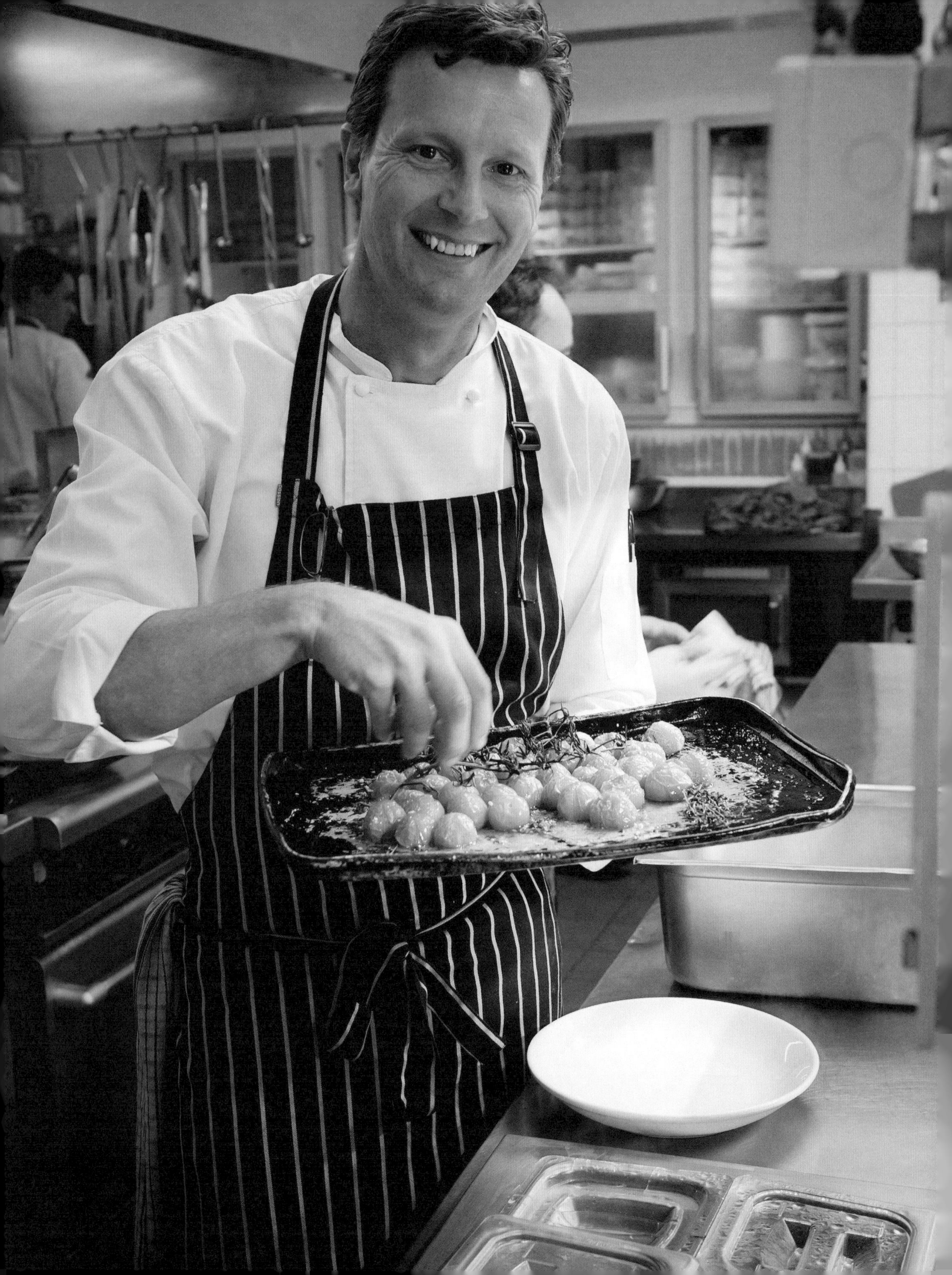

Russell Blaikie

I had a fantastic childhood. I was brought up on my family dairy and sheep property at Cowaramup, a few kilometres from Margaret River. Before breakfast every morning I'd bump down the track on my bike to the dairy where Dad would be finishing off the milking. Occasionally I'd catch a lump of gravel on a trainer wheel and spin out in a cloud of dust, brush myself off and then limp into the noisy milking shed looking for sympathy.

Dad knew what to do. My tears quickly vanished when I was given a bucket to feed the calves, their eager appetite and soft mouths sucking in the warm milk, and the odd hand if offered up!

The orchard around our house had trees heavy with apricots, pears, figs, grapes and peaches. We'd run nets over the stone-fruit trees before summer to keep the silvereyes off the fruit. A fig tree formed a massive umbrella over the outside toilet. Any farm kid knows that the fig makes the best climbing tree. In summer my brother Ray and I would take turns to sneak up the tree and throw the mushy fruits at our sisters; the giggling from the tree usually met with a muffled expletive from the victim, and the occasional volley of return fire.

For a special occasion, Mum or Dad would head to the chicken pen to select a young bird, taking it squawking and flapping to the woodheap to finish it off with the axe. We'd dunk it into the copper full of boiling water and then pluck the feathers. A lavish feast would follow, usually with roast potatoes from Uncle Frank's farm at Carbunup, maybe some freshly caught herring from Cowaramup Bay, a fruit pie made with apricots or peaches, and always, always, lashings of rich thick cream scooped off the top of the milk pails before they were sent to the dairy factory.

My early years on the farm gave me an education in provenance and seasonality that has been invaluable in my career as a chef. I have an understanding, respect and admiration for my suppliers and growers, all of whom are an integral part of my success as a restaurateur. Many of my suppliers are friends – friends that cope through hardship and flourish in the good times – and friends who will be receptive to suggestions and comments and feel free to do the same to me. This book is as much their story as it is about the recipes that use their products.

All the recipes in *Must Eat* have been served in the restaurant since we first opened in 2001. Some of these have sprung from an original idea, some from trying to re-create dishes I've eaten in Parisian bistros, while others hail from an idea or technique gleaned from another chef, or another cookbook.

These are restaurant dishes that ideally serve to inspire your own cooking at home. Cooking an entire dish, its sauce and accompanying garnishes can be time-consuming, so if you're short on time choose something you like the look of – perhaps a simple main, or a side dish, sauce or salad – to share the plate with some of your own recipes.

Our Must wine consultant of six years, Paul McArdle, has matched each recipe with a specific wine. If you are unable to find that particular wine then use Paul's notes as a guide, or ask a knowledgeable cellar person at your local bottle shop to recommend a wine with similar characteristics. Rather than listing vintages, Paul has chosen wines from producers that have, over years, shown consistency for that variety.

And if you are in a party mood, sommelier Aaron Commins has shaken and stirred a shortlist of classic and contemporary cocktail recipes to share with your friends.

True bistro food is cooked from the heart to feed the soul. It is not always rich, but it is rich in tradition, and made to be enjoyed in the company of good wine and a crusty baguette. Above all, bistro food is to be shared with friends and family, so please join my table...

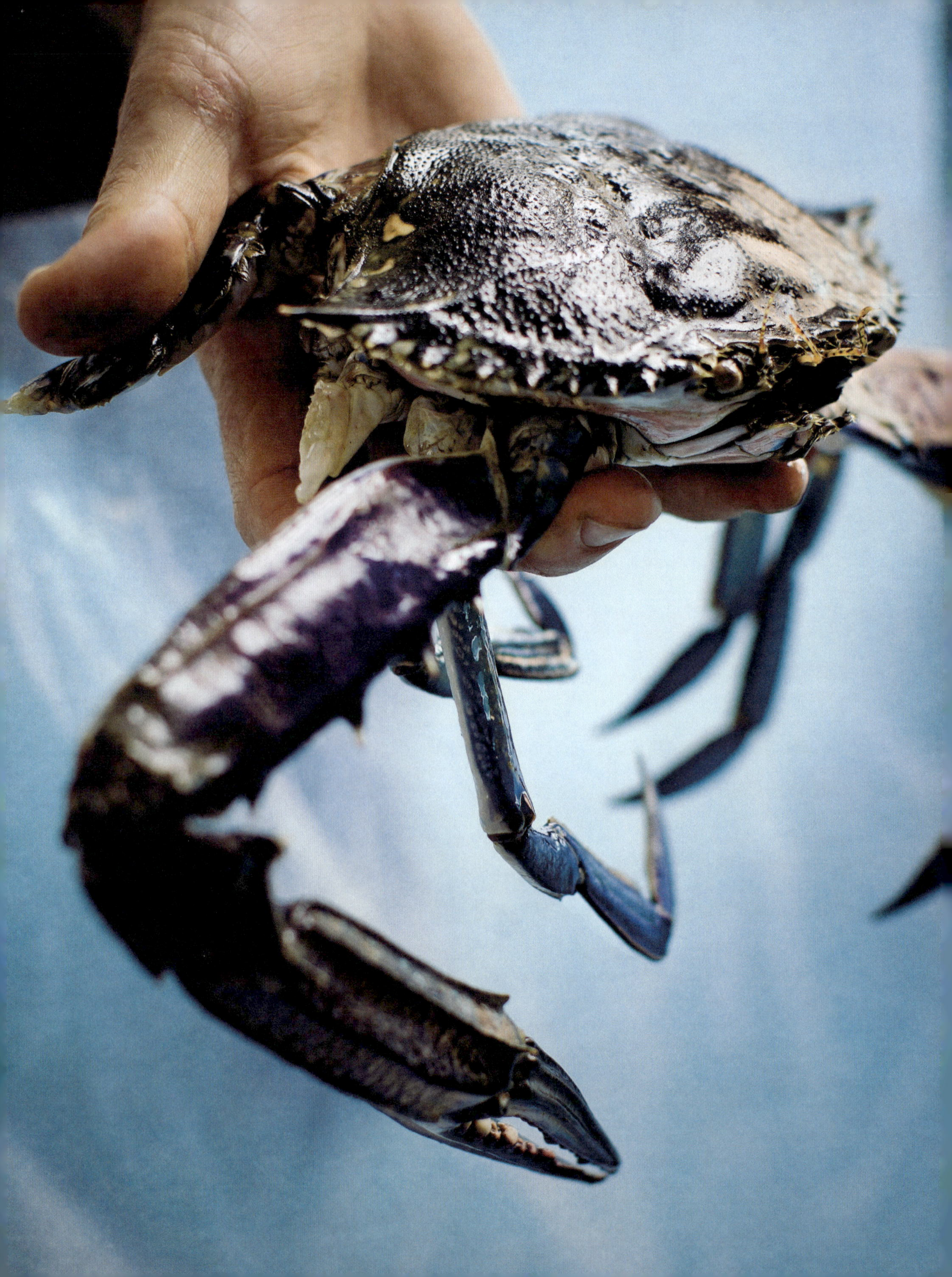

bar snacks

Jan and Kim Harwood embarked on a tree change twenty years ago. They ended up at Rosa Brook, setting up an egg farm, Margaret River Free-range Eggs, in the mistaken belief that farming cattle and sheep would be too hard. Like many food pioneers, they had a rude shock learning about their fickle birds, almost giving up the farm after five years (and two small kids). But they stuck it out and now produce fantastic quality free-range eggs from happy chooks.

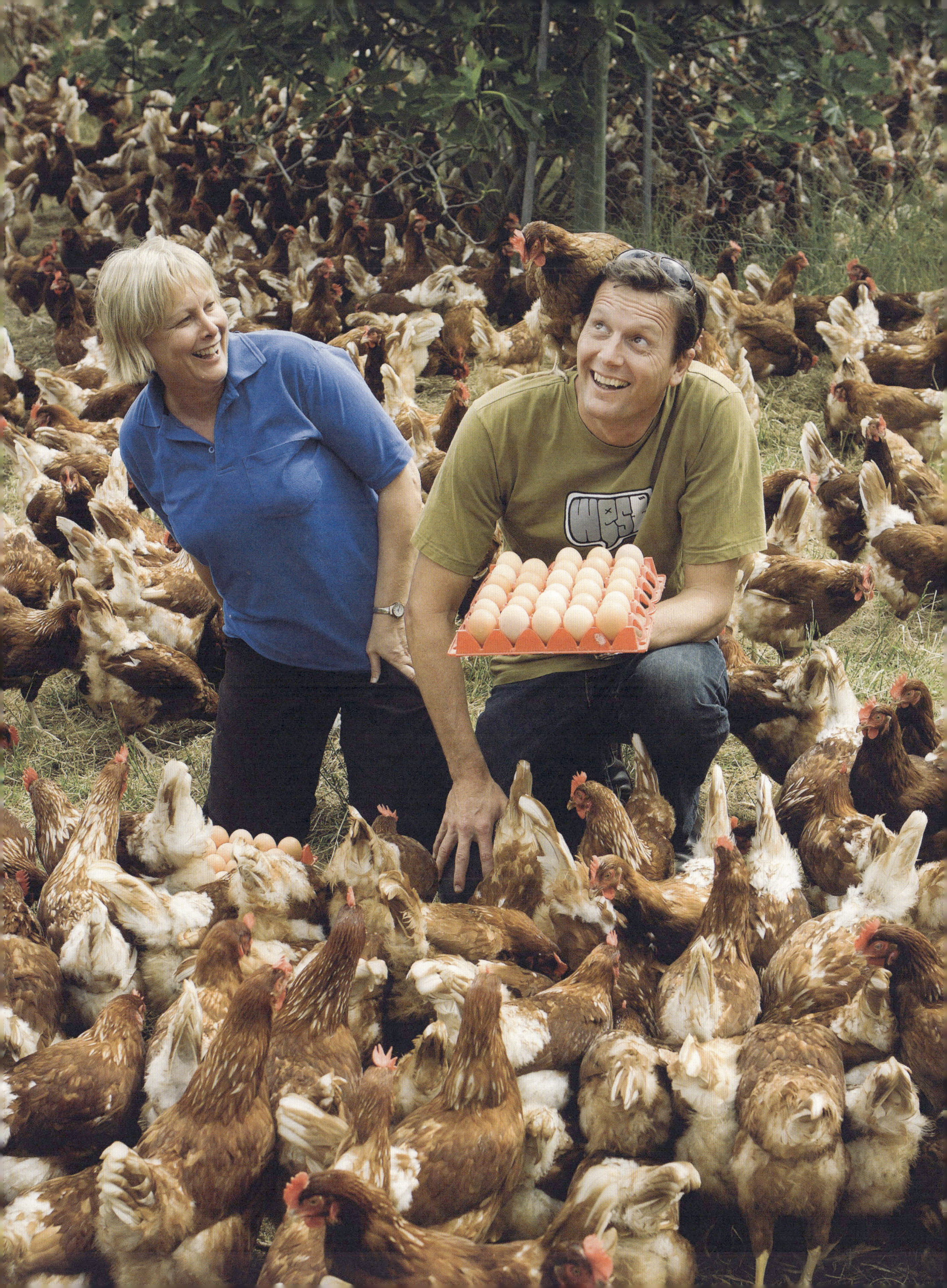

**Cantina Tolio
Montepulciano
d'Abruzzo**
Abruzzo, Italy

The juicy, generous flavours of the Montepulciano grape softens the flavours of these toasties. There is a rustic mouth-feel with gentle grainy tannins, which combines with the earthiness of the truffle and at the same time copes with the creaminess of the cheese. Not too pretentious, this little Italian wine fits like a glove.

Manjimup Black Truffle & Fontina Cheese Toasties

Eating fresh truffles is one of life's true pleasures. At Manjimup, the gods of soil and climate have joined together to ensure perfect growing conditions for this sensually aromatic fungi, and Manji' is the heart of truffle production in Australia. Al Blakers mid-season truffles – those he picks in July – seem to be the ripest, funkiest-smelling truffles, so hold out if you can.

Preheat the oven to 200°C.

Slice the baguette into 24 1 cm thick pieces.

Slice the Fontina and lay some onto one slice of bread. Shave the black truffles tissue-paper thin and add a few slices to the bread then sandwich together with another slice of bread; keep going until all of the bread is filled.

Lay the sandwiches onto a baking tray and drizzle generously with the olive oil, turn over and repeat. Place into the oven and cook for approximately 6 minutes until the toasties are crispy, golden brown and the cheese filling has melted.

1 baguette
90 g Fontina cheese
1 fresh black truffle
(a 15-20 g truffle would be plenty)
Extra virgin olive oil

MAKES 12 TOASTIES

bar snacks

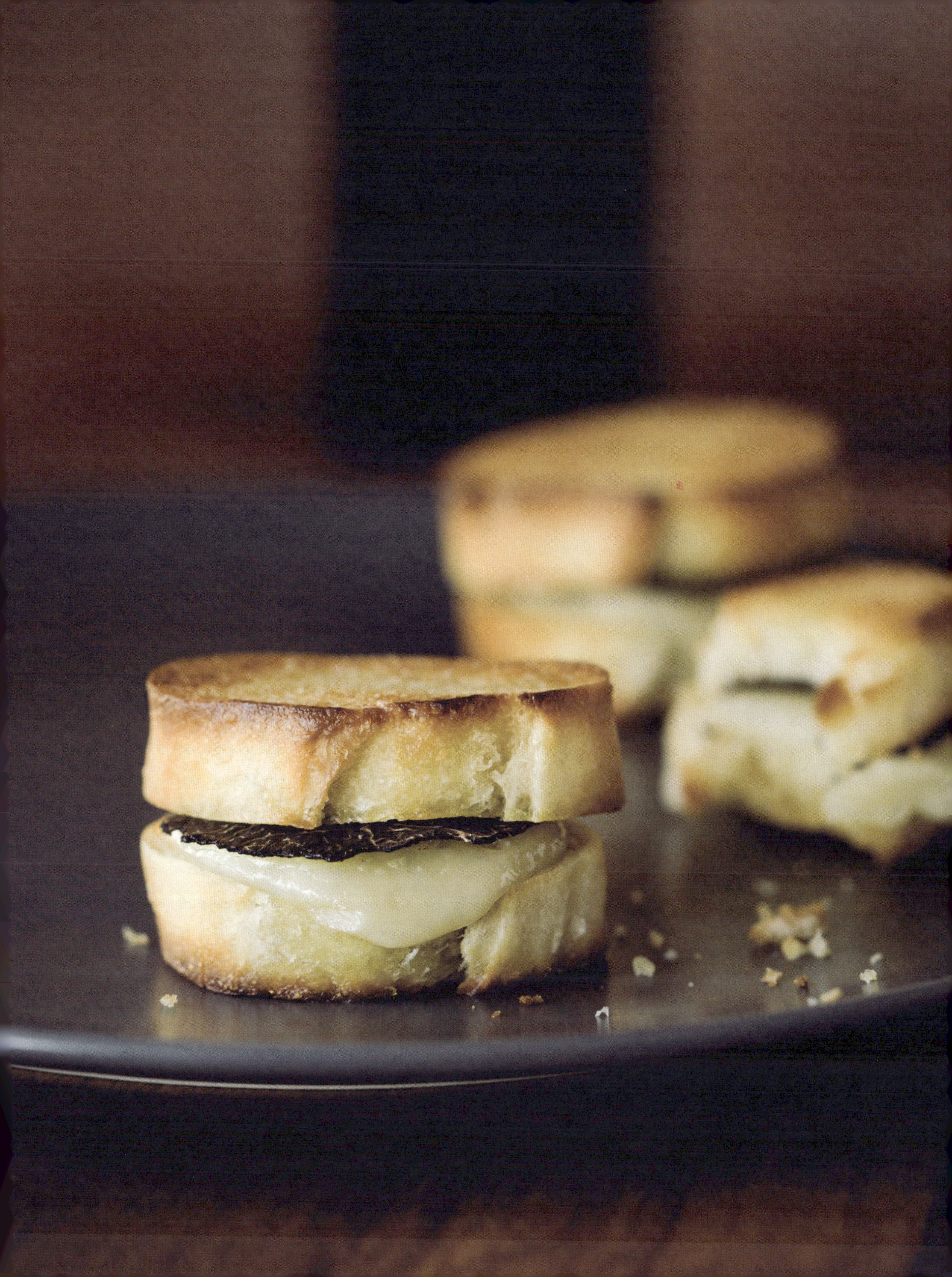

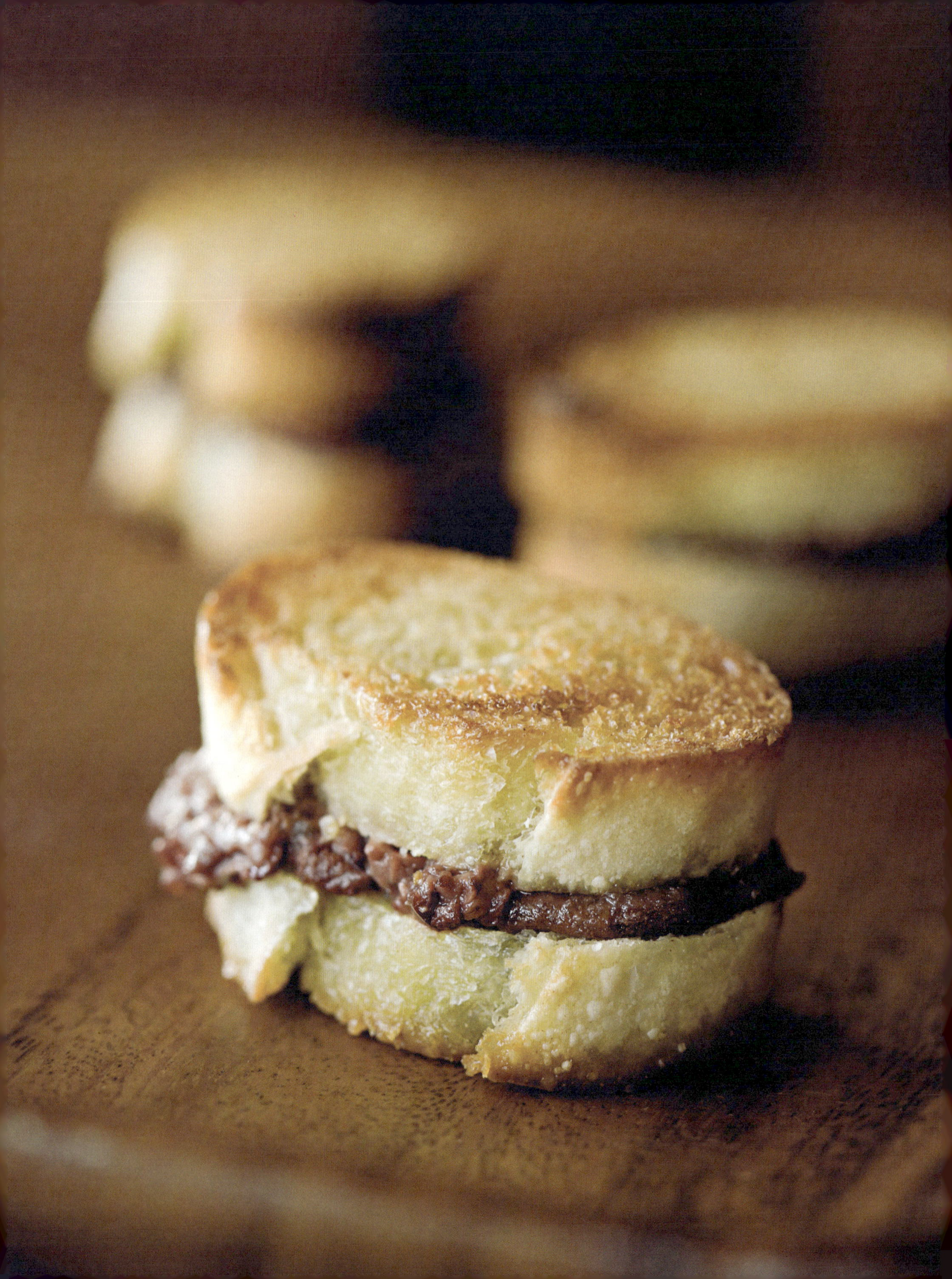

Black Olive & Gruyère Toasties

Everyone loves a toasted sandwich. This 'deluxe' version combines the richness of gruyère with the savoury-saltiness of freshly made tapenade.

....

1 baguette

40 g Comté (or gruyère) cheese, grated

60 g tapenade (p. 10)

Extra virgin olive oil

MAKES 12 TOASTIES

Preheat the oven to 200°C.

Slice the baguette into 24 x 1 cm thick pieces.

Combine the cheese and tapenade together in a bowl and stir well to combine.

Spread a generous quantity of the mix onto one slice of bread, then sandwich together with another; keep going until all of the bread is filled.

Lay the sandwiches onto a baking tray and drizzle generously with the olive oil, turn over and repeat. Place into the oven and cook for approximately 6 minutes until the toasties are crispy, golden brown and the cheese filling has melted.

Meáo 'Meandro'
Douro, Portugal

These deliciously crunchy toasties with their salty olive earthiness make an ideal match with this generous Portuguese red. It is calm, soft and opulent with layers of dark, brooding fruit, a sense of floral juiciness and a line of lingering acidity. The wine slices through the olive texture, providing a refreshing, clean finish.

Moric Blaufränkisch
Burgenland, Austria

This Blaufränkisch (red) is minerally and svelte and complements the flavour and texture of the salty olive tapenade. The taut, finely knit texture and red-crackly fruit is not showy or profound but juicy and quenching with this uncomplicated olive baguette.

Tapenade

This Provençale condiment is such a tasty invention. At home I often chop the ingredients with a sharp knife, which gives a rough, rustic texture to the earthy flavour. If you're using a food processor, don't blend the ingredients to a fine purée – keep it coarse.

••

Blitz the ingredients in a food processor until a coarse paste is achieved.

Scoop into serving dish, drizzle with extra virgin olive oil and serve with fresh crusty baguette.

120 g best quality black olives, pitted
1 teaspoon capers
1 anchovy
1 teaspoon thyme leaves
1 small garlic clove
2 tablespoons extra virgin olive oil, plus extra to drizzle over the top
1 teaspoon lemon juice
Twist of freshly ground black pepper

MAKES ABOUT 6 SERVES

bar snacks

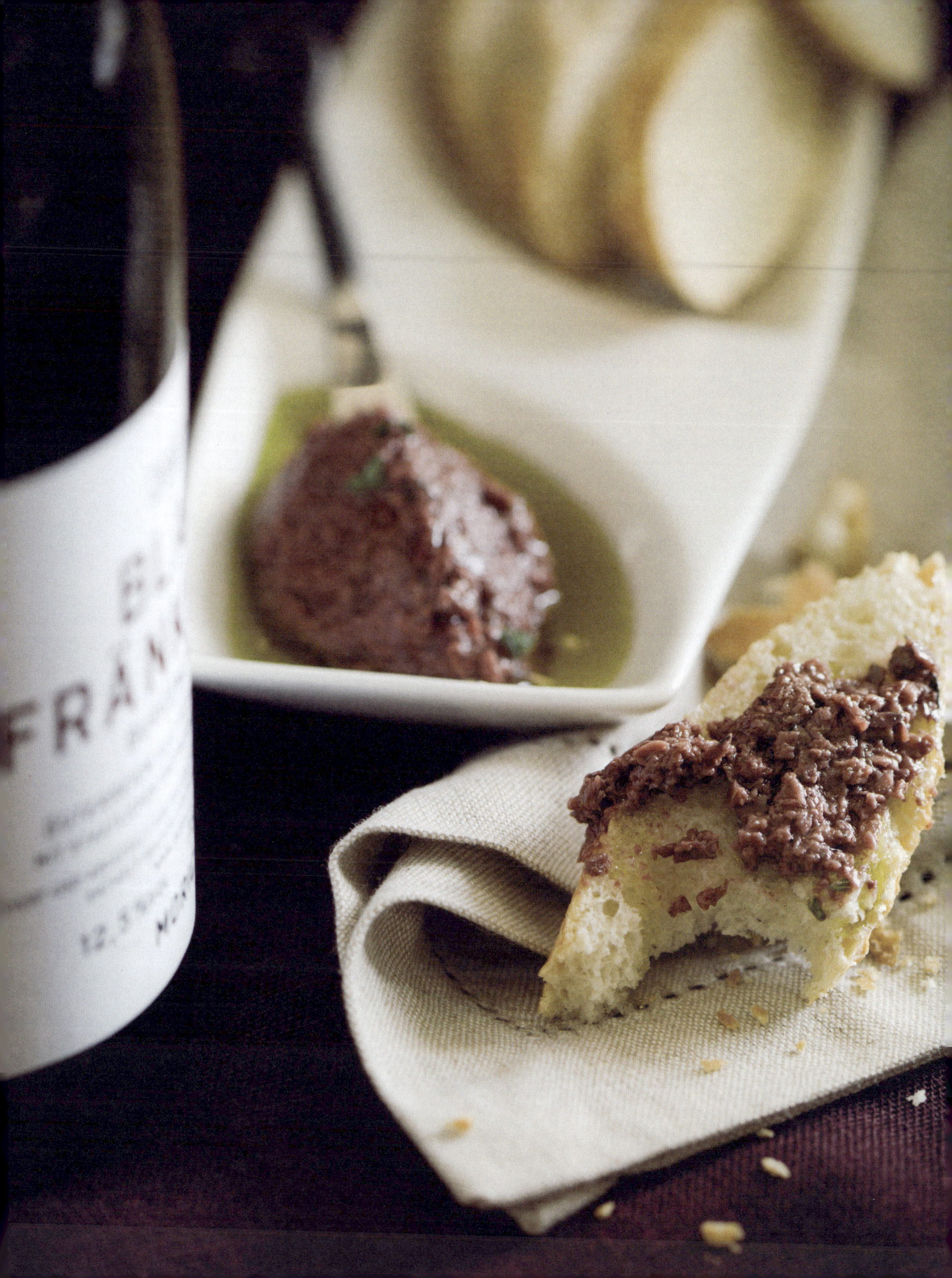

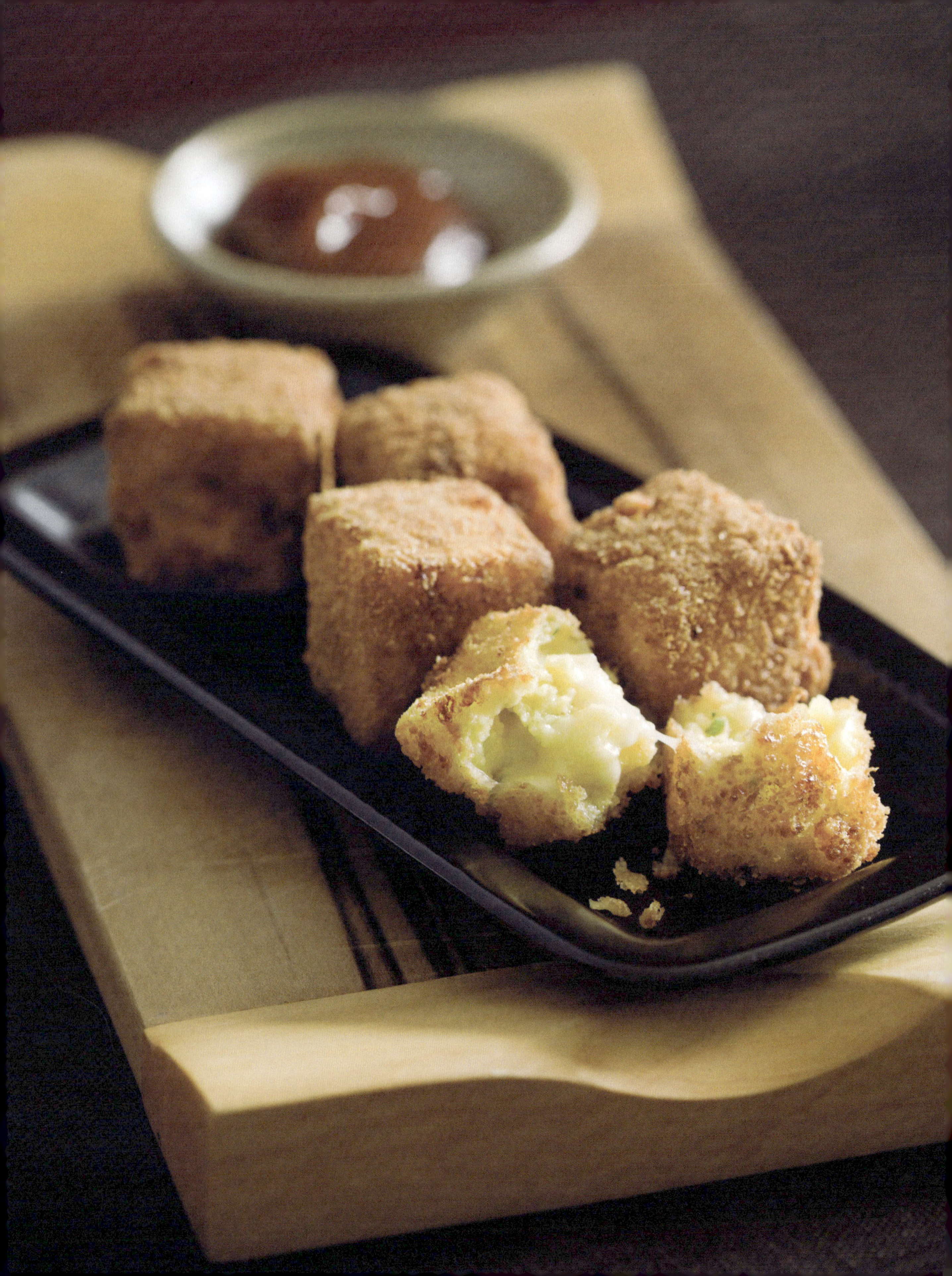

Polenta Bites & Chilli Tomato Jam

In the weeks before opening Must, I accidentally dropped extra cornmeal into a batch of boiling polenta. I couldn't bear to waste the resulting stodge so I flavoured it with several tasty cheeses and herbs, set it in a tray then sliced the mixture into cubes, which I crumbed and deep-fried. They had the most amazing texture and flavour and have been a hit on our bar snack menu ever since.

125 g feta cheese
125 g gruyère cheese
40 g parmesan cheese, grated
625 ml milk
150 ml water
1 teaspoon each sea salt & freshly ground black pepper
190 g fine instant polenta
110 g butter
$\frac{1}{4}$ cup chopped parsley

Crumb
Plain flour
1 egg and 50 ml milk for egg wash
Breadcrumbs

Oil, for deep-frying
Chilli tomato jam (p. 266)

MAKES ABOUT 30 BITES

Dice the feta and gruyère into $\frac{1}{2}$-cm cubes mix with the grated parmesan. Set aside.

Combine the milk, water, salt and pepper in a large saucepan and bring to the boil. Reduce the heat to low and drizzle the polenta into the liquid, stirring constantly to ensure no lumps form. Keep stirring the polenta over heat for 5 minutes; it will become very thick. Take off the heat and add the butter, stirring so that it is well combined.

Allow the mixture to cool for 3 to 4 minutes then add the three cheeses, followed by the chopped parsley, stirring to combine.

Line a 22 x 32 cm slice tray with baking paper and scoop the mixture onto the tray. Use a spatula to flatten the mixture evenly into the tray to about 2 cm deep, then refrigerate for 2 to 3 hours until cool and firm.

Carefully tip the slab of polenta onto a chopping board and cut into 2 cm cubes.

Crumb the bites by tossing through the flour, then dipping in egg wash then rolling in breadcrumbs. Cool in the refrigerator.

Heat the oil in a deep-fryer or deep saucepan to approximately 180°C. To test the temperature drop in a single polenta bite; it should take about 1 minute to turn golden brown; if it colours quicker the oil is too hot. Fry the polenta bites in batches, then drain on kitchen paper, keeping them warm in the oven.

Serve warm with a side dish of chilli tomato jam.

Pala Crabilis Vermentino di Sardegna
Sardinia, Italy

The Italian acidity and slightly bitter finish work in unison with the creaminess of these cheesy polenta bites. There is a suggestion of weight, which copes with the solidness of the polenta and also has a textural feel to balance the crunchy fried coating.

Billecart-Salmon Brut NV
Champagne, France

It is important to balance the level of butter and cheese in these gougères with a wine that has crisp acidity and is subtle in flavour, to complement rather than dominate the lightness of the pastry. This Champagne has the weight and acid freshness, with some toasty gentle flavours to partner these tasty bites.

Gougères

These tasty puffs are traditionally made with Comté, the French version of Swiss gruyère. We bake them every evening for our Champagne Lounge guests. These can be tricky to make if the flour mix is not properly cooked out, so follow the recipe closely.

....

- Preheat the oven to 200°C.
- Combine the water, milk, butter, salt, sugar and cayenne in a saucepan and place over moderate heat until the butter has melted.
- Pour in the flour and stir with a wooden spoon over heat until smooth and thick. This should take approximately 2 minutes. Remove the mixture from the heat and tip into a mixing bowl.
- Add the eggs to the hot mixture one at a time, beating constantly with a wooden spoon until they are completely incorporated.
- Add the grated cheese and stir through the mixture.
- Place into a piping bag with a plain nozzle. Line an oven tray with baking paper.
- Pipe round dobs around 2 cm in diameter onto the tray, keeping them well spaced.
- Bake for approximately 20 minutes until the puffs are golden brown on the outside and, when broken open, cooked (not sticky) inside.
- Remove and cool on wire racks.
- Gougères can be served at room temperature or, if you wish, warmed through in the oven for 2 to 3 minutes.

125 ml water
125 ml milk
100 g butter
1 teaspoon salt
1 teaspoon sugar
A pinch of cayenne pepper
150 g plain flour
5 x 50 g eggs
100 g Comté cheese, grated
(Marcel Comté is a great example; you can use gruyère as a substitute)

MAKES 30 GOUGÈRES

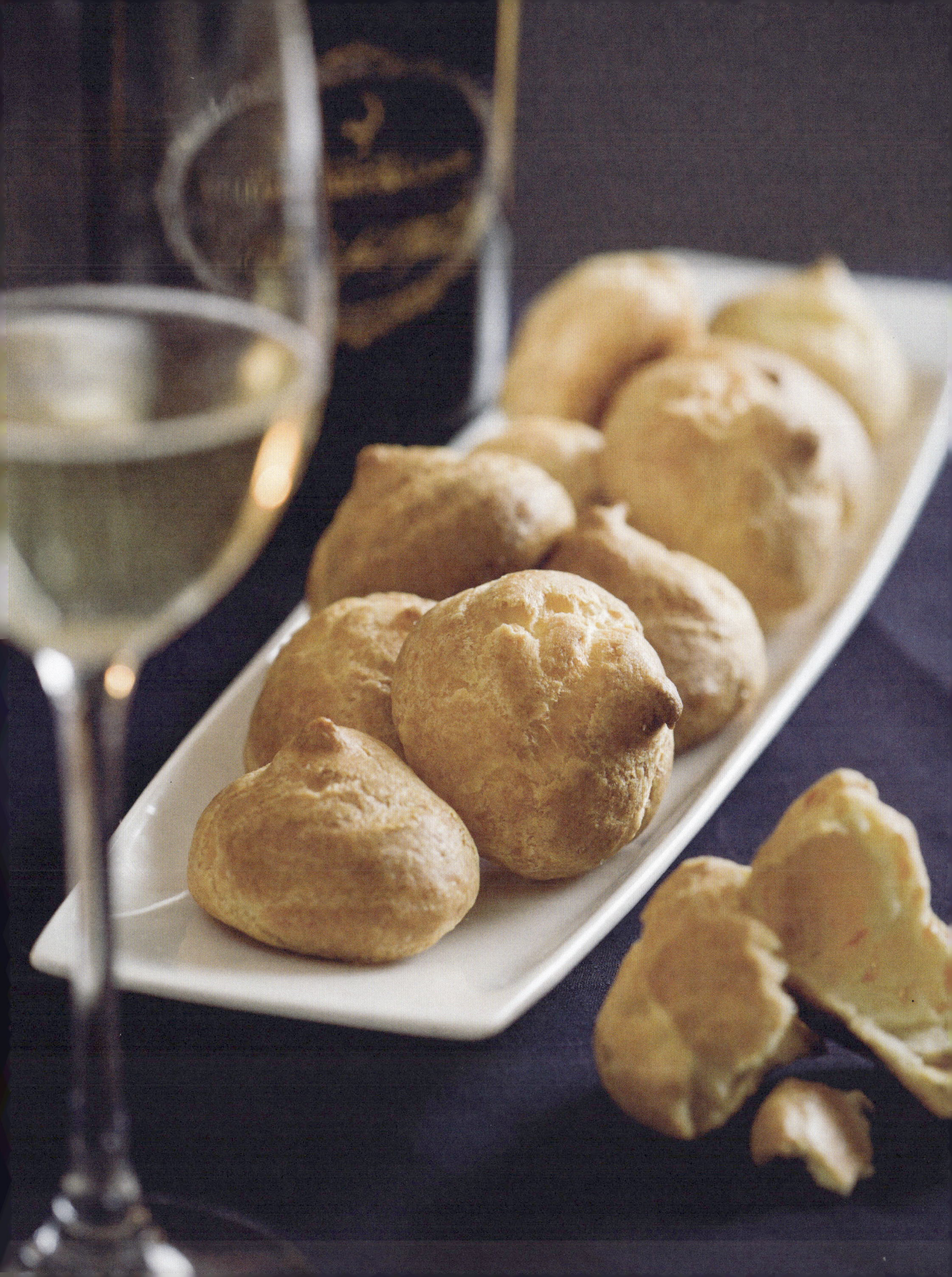

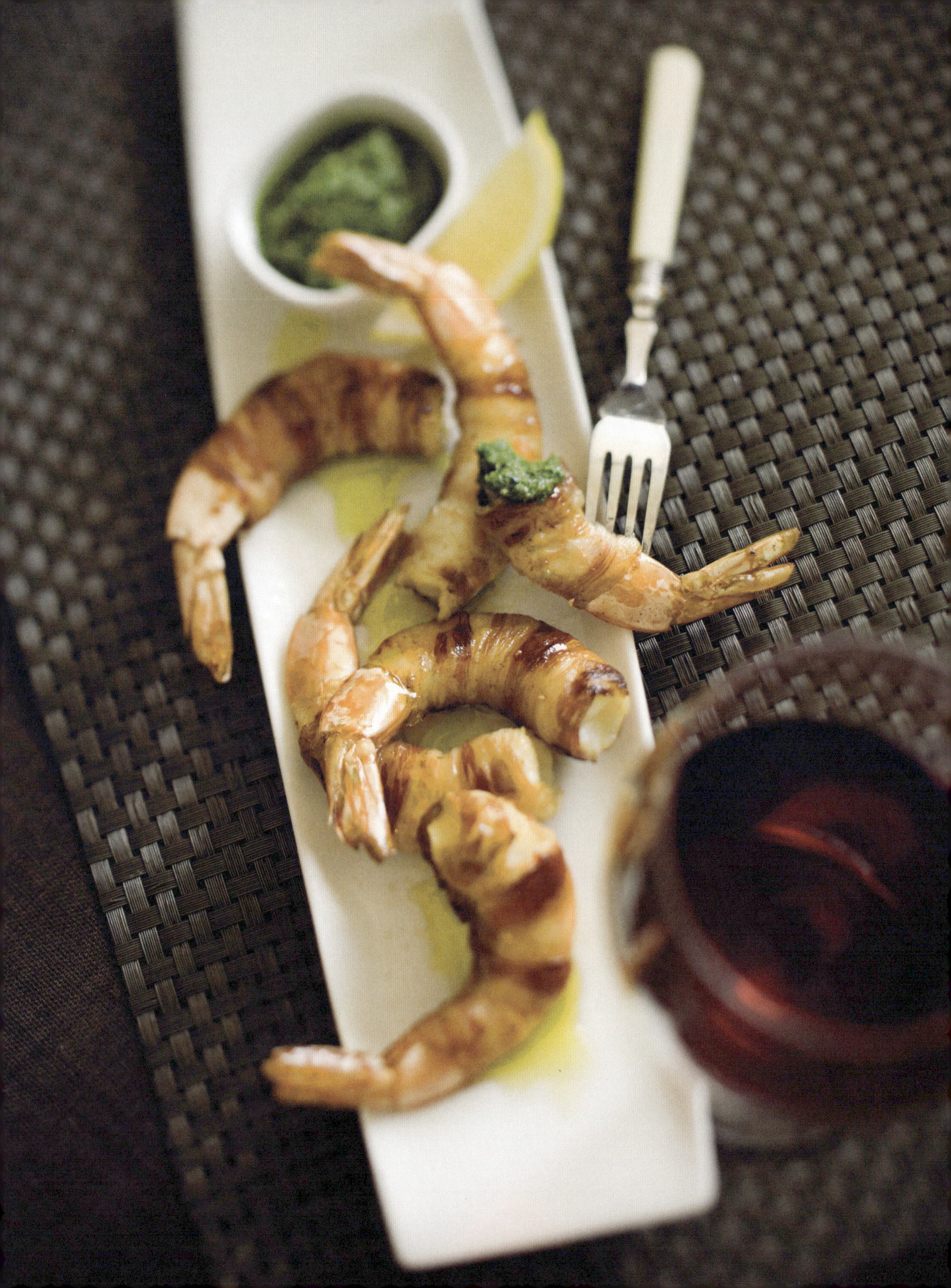

Pancetta-Wrapped Tiger Prawns & Sauce Verte

The World Heritage Listed Shark Bay is an abundant fishery where we source our scallops, pink snapper and tiger prawns. Trawlers work the bay, catching the prawns in nets then quickly sorting and snap-freezing them on the boat; the fast handling guarantees a 'crunch' to the bite and preserves their sea-sweet flavour. Sauce verte means 'green sauce' and works as a piquant foil to the richness of the seafood.

Vacheron Rouge Sancerre
Loire Valley, France

The dish is about contrast – salty pancetta and sweet, delicate prawns. It demands a wine that complements this contrast, and this cool-climate red is perfect. There is gentle fruit restraint and brisk acid with an undertone of sappy pinot flesh; it doesn't overpower the prawns, but highlights their flavour and succulence.

Sauce Verte

50 ml extra virgin olive oil
6 cornichons
1 clove garlic, crushed
20 g salted capers, rinsed well
2 anchovies
1 tablespoon lemon juice
½ cup flat-leaf parsley, tightly packed
Sea salt & freshly ground black pepper

12 large whole raw tiger prawns
Sea salt & freshly ground black pepper
12 thin strips pancetta
Extra virgin olive oil

SERVES 6

To make the sauce verte (about 180 ml), place all the ingredients together into the bowl of a food processor and blitz to a coarse consistency. Check seasoning and reserve in refrigerator.

Peel and devein the prawns, leaving the tail intact.

Season the prawns then wrap each one with a slice of pancetta.

Heat a large frying pan with a little oil over high heat. Place the prawns into the pan, cook for 2 minutes then flip over and cook a further 2 minutes.

Place onto a serving platter, drizzle with a little extra virgin olive oil and serve with a dish of sauce verte.

Lis Neris Pinot Grigio
Friuli, Italy

These croquettes are crunchy and crisp outside and soft, textural and creamy inside, so the ideal wine match provides fleshy viscous texture with mouth-watering acid brightness. This pinot grigio from Lis Neris has both weight and richness, which balances the fried edge and shows typical savoury bitterness that finishes clean, sharp and fresh.

Scallop & Jamón Croquettes

Croquettes are European fast food: deep-fried dumplings or fritters sometimes bound together with potato. I prefer the Spanish version, which has a thick béchamel as the base that turns to molten velvet when cooked.

Preheat the oven to 200°C.

Lay the jamón onto an oven tray lined with baking paper, cook for 5 minutes in the oven, remove, cool a little then dice into 1 cm pieces.

Melt the butter in a large heavy saucepan, add the flour and cook over low heat, stirring for 2 minutes to make a roux. Remove from the heat.

Place the milk in a separate saucepan, season with salt and pepper then bring to the boil. Add the chopped scallops and immediately remove from the heat. Allow the scallops to steep in the hot milk for another 30 seconds, then strain the milk into the roux mix (keeping the scallops aside to cool), stirring the mixture with a whisk to ensure no lumps form.

Place the mixture back over medium heat, and cook for a further 5 minutes, stirring constantly with a wooden spoon as it becomes very thick. Remove from the stove and allow to cool for 2 minutes.

Add the cooked scallops, jamón and chives to the thickened mixture, stirring together with a wooden spoon. Spoon out onto a tray, cover with cling film, cool on the bench for another 10 minutes then refrigerate for 2 hours until chilled and set.

Set up for crumbing and scoop balls of the chilled mix (about one tablespoon) into the flour. Roll the mix into round balls and drop into the egg wash, then roll in breadcrumbs. Keep refrigerated until you wish to cook the croquettes.

Heat the oil in a deep-fryer or deep saucepan to approximately 180°C. To test the temperature, drop in a single croquette; it should take about 1 minute to turn golden brown. If it colours quicker the oil is too hot. Fry the croquettes in batches, then drain on kitchen paper. Serve immediately.

130 g finely sliced jamón serrano (can be substituted with prosciutto)
100 g butter
125 g flour
500 ml milk
1 teaspoon sea salt
Pinch of white pepper
300 g fresh scallop flesh, chopped into 1 cm dice
3 tablespoons chopped chives

Crumb
Plain flour
1 egg whisked together with 150 ml milk for egg wash
Japanese breadcrumbs

Vegetable or cottonseed oil, for deep frying

MAKES ABOUT 30 CROQUETTES

18 bar snacks

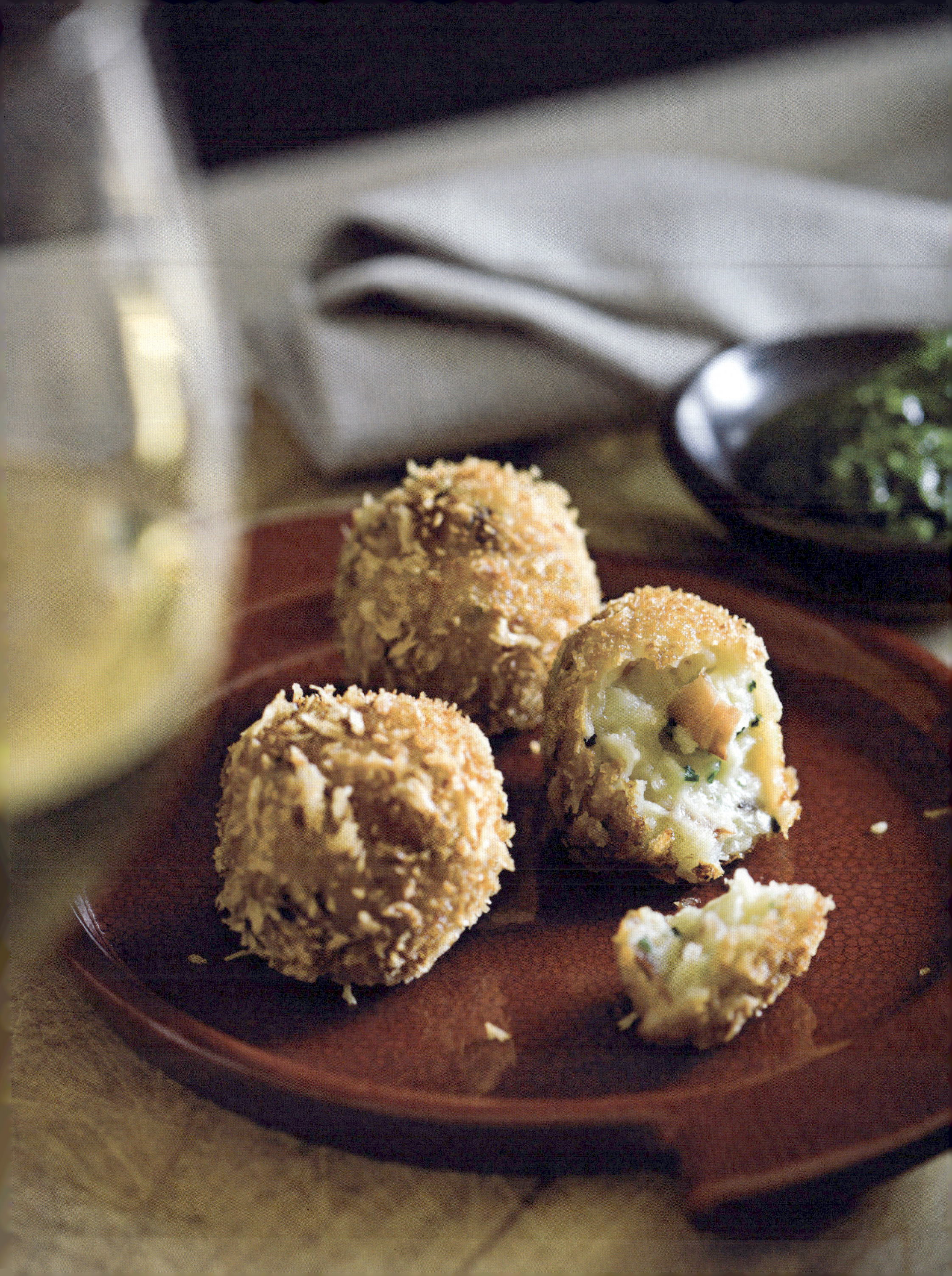

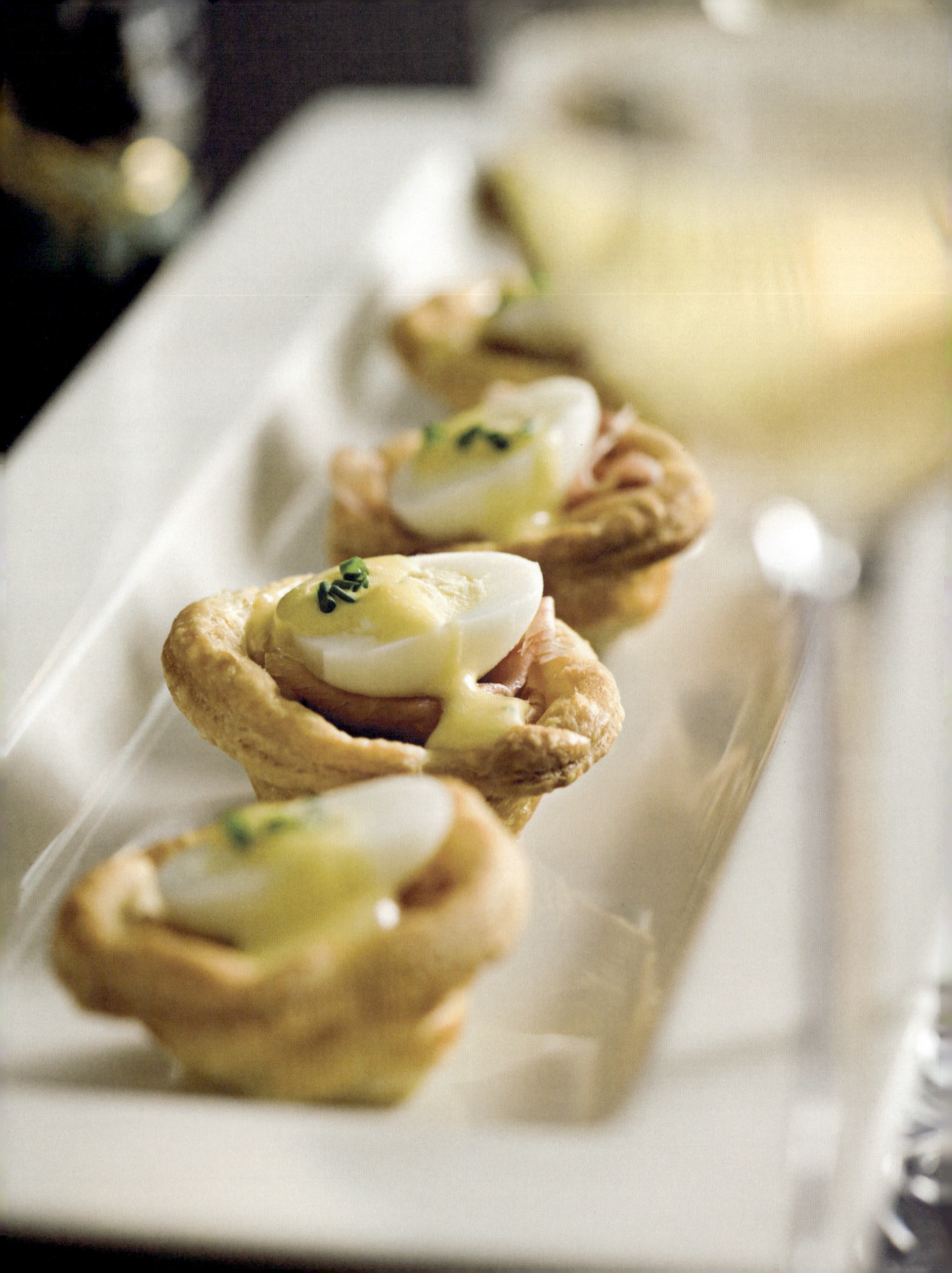

Quail Egg Benedict Tartlets

We serve these little gems at special events in our Champagne Lounge. If you'd rather not pay $400 a kg for jamón ibérico, substitute with a good-quality prosciutto or jamón serrano.

350 g pâté brisée (p. 285)

Butter and flour, to coat the tartlet moulds

12 quail eggs

24 fine slices jamón ibérico (I use 5J jamón ibérico de Bellota)

2 tablespoon finely chopped chives

150 ml hollandaise sauce (p. 280)

MAKES 24 TARTLETS

Preheat the oven to 200°C.

Brush a 24-hole mini tartlet tray with butter and dust with flour.

Roll the pastry out to 3 mm thick. Cut circles out of the pastry large enough to fit the tartlet holes. Lightly spike with a fork and press into the tartlet tray.

Line each tartlet with baking paper, drop in some dried pulses (lentils work well) and rest in the refrigerator for 20 minutes.

Place the tartlets into the oven and bake for 15 minutes, until golden brown. Remove the paper and pulses and reserve.

Fill a deep saucepan with water, add a pinch of salt and bring to a shimmering simmer. Carefully drop the quail eggs into the water and cook for approximately 2 minutes (the yolk should be just set).

Lift out with a slotted spoon and drain on kitchen towel for 1 minute until they have cooled a little.

Peel the eggs and cut in half with a small sharp knife.

Warm the tartlets.

Fold a small piece of jamón ibérico into each tartlet. Place a quail egg on top and carefully spoon a little hollandaise onto each egg. Sprinkle with chives and serve.

Nigl 'Kremser Freihart' Grüner Veltliner
Kremstal, Austria

These little tartlets are loaded with richness and flavour and require a wine such as this Gruner Veltliner, which has a slight celery-citrus character and is densely packed, juicy and taut.

bar snacks

TWR (Te Whare Ra) Sauvignon Blanc
Marlborough, New Zealand

The salty snapper, garlic and herbal influence requires a wine with substance, brightness and expression. This sauvignon blanc has a lively acidity and concentrated lemon/honey fruit character, which complements the cured fish and appropriately combines with the lemon infused olive oil … lip-smacking with the saltiness!

Snapper Brandade Tartlets

Brandade is a rich, garlicky indulgence traditionally made with salt cod. My version is lighter and instead uses fresh snapper that I salt for 12 hours.

Sprinkle both sides of the fish with rock salt, wrap in cling film and leave in the refrigerator overnight (about 12 hours).

Preheat the oven to 200°C.

Brush a 24-hole mini tartlet tray with butter and dust with flour. Roll the pastry out to 3 mm thick. Cut circles out of the pastry large enough to fit the tartlet holes. Lightly spike with a fork and press into the tartlet tray. Line each tartlet with baking paper, drop in some dried pulses (lentils work well) and rest in the refrigerator for 20 minutes. Place the tartlets into the oven and bake for 15 minutes, until golden brown. Remove the paper and pulses and reserve.

Wash the cured fish in lots of cold water, chop into 2 cm cubes and reserve.

Combine the cooking liquor ingredients in a saucepan and bring to the boil. Reduce the heat to a slow simmer, cooking for 5 minutes so that the aromatics release their flavour into the milk. Drop the fish into the milk and cook for 2 to 3 minutes, until the flesh begins to flake. Strain the fish out of the milk and put to the side, reserving all the cooking liquid and garlic cloves, discarding the herbs.

Return the milk and garlic to the stove. Drop in the diced potato and simmer for 10 to 15 minutes until tender. Strain the potato and garlic from the milk, again reserving the cooking liquid. Mash the potato and garlic, or press through a potato ricer.

Flake the fish into the potato purée, adding some of the cooking liquor to make a soft (not wet) consistency. Add the chopped herbs and check for seasoning, adding a little more sea salt if required.

Warm the tartlet cases if required, spoon in the brandade mixture, drizzle with a little lemon-pressed extra virgin olive oil and serve immediately.

250 g snapper (pink or gold band snapper are fine)

A handful of rock salt

Butter and flour, to coat the tartlet moulds

350 g pâté brisée (p. 285)

350 g royal blue potatoes, peeled and diced

1 heaped tablespoon chopped flat-leaf parsley

1 heaped tablespoon chopped chives

Sea salt, to taste

Cooking Liquor

500 ml milk

1 fresh bay leaf

2 garlic cloves, peeled

A few sprigs of thyme

3 white peppercorns

Lemon-pressed extra virgin olive oil

MAKES 24 TARTLETS

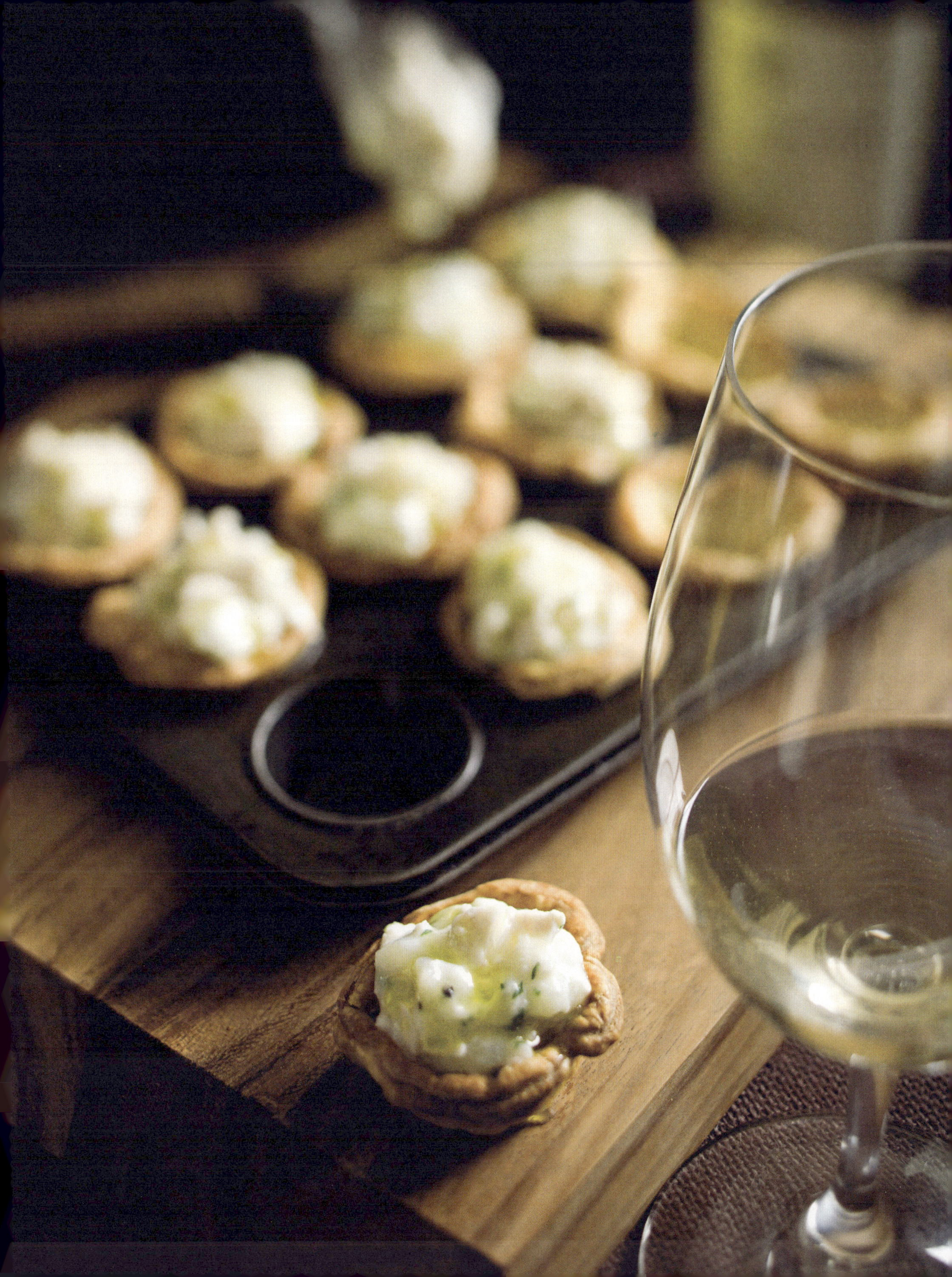

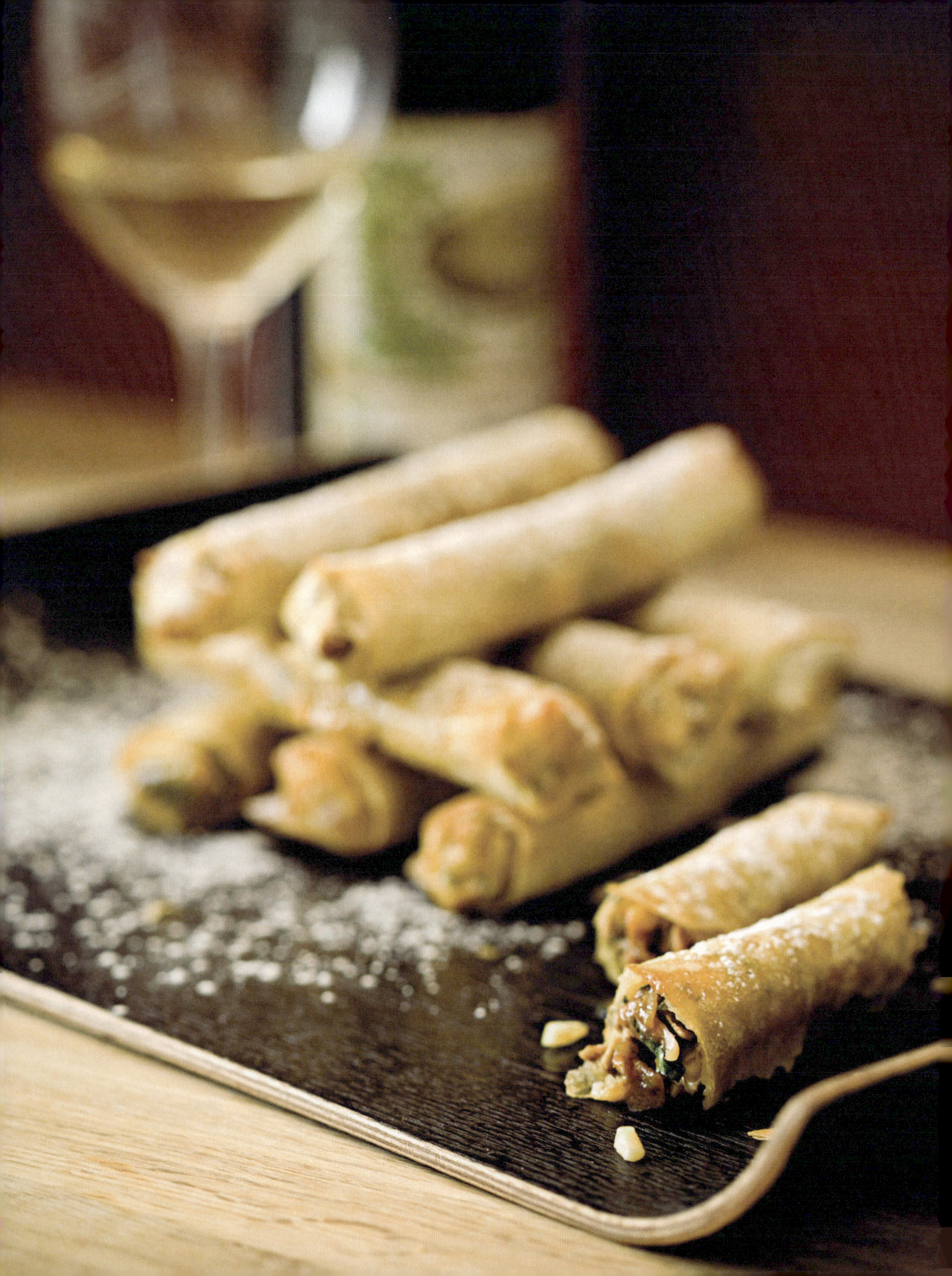

Pastilla of Duck

A traditional Moroccan 'bisteeya' (the French call it 'pastilla') is a pie made from squab pigeon, with an egg 'custard' and crisp, fine pastry crust. I've retained the spicing of the traditional dish but removed the egg, wrapping the filling into a spring roll shape with Tunisian brik pastry. The recipe specifies duck but it works equally well using free-range chicken thighs or lamb.

Andre Kientzler Riesling
Alsace, France

This riesling has the expression and extract to offset and pair with the gamey character of duck. It has wonderful purity and piercing acidity, with a suggestion of sweetness to balance the touch of cinnamon and dusting of icing sugar.

---

2 tablespoons extra virgin olive oil
4 large duck legs
200 g onion, finely chopped
2 garlic cloves, chopped
1 heaped teaspoon cumin powder
½ teaspoon coriander powder
A pinch of saffron strands
¼ teaspoon cayenne pepper
Juice of ½ a lemon
3 cm piece cinnamon stick
500 ml chicken stock
300 ml tomato purée
½ teaspoon ras el hanout (p. 272)
⅓ cup fresh coriander leaves
⅓ cup fresh parsley leaves
⅓ cup fresh mint leaves
100 g blanched almonds, chopped coarsely
6 large sheets Tunisian brik pastry
A little egg wash
Mixture of 1 tablespoon cinnamon powder to 2 tablespoons icing sugar

MAKES PASTILLA FOR 6

Preheat the oven to 210°C.

Heat the oil in a large heavy saucepan over high heat. Season the duck with salt and colour in the pan. Remove the duck, reduce the heat, add the onion and garlic and cook until translucent.

Combine the cumin, coriander and saffron on a small tray and roast in the oven for 2 minutes.

Add the roasted spices, cayenne, lemon juice, cinnamon stick, stock and tomato purée to the pan. Bring to the boil and simmer until the flavours combine; about 10 minutes. Add the duck legs and cook until very tender (approximately 1 hour).

Add the ras el hanout powder and cook a further 5 minutes. Cool.

When the duck has cooled to room temperature, remove the pieces of leg from the pan and tear the flesh off the bone, discarding the skin and bone, reserving the flesh in a large bowl.

Place about ½ cup of the braising liquid in a small saucepan over medium–high heat until reduced by half. Add to the bowl, then allow to cool to room temperature.

Add the herbs and almonds and toss well.

Lay out the sheets of the brik pastry, brush around the edges with egg wash and place some of the filling in the centre of each sheet.

Tightly roll each into a large 'spring roll' shape and reserve on a baking tray. Place the pastillas into oven and bake for approximately 12 minutes, until golden brown and crisp.

Place onto a serving platter, dust with the icing sugar-cinnamon mix and serve immediately.

Ravensworth Shiraz Viognier *Heathcote, Victoria*

This wine is rich with flavours of roasted black fruits with exotic spices and floral notes. It has a seamless expression with lush, succulent juicy fruit and wonderful textural opulence that combines with the baharat spices and meaty, earthy flavours of the garlic lamb.

Baharat-Spiced Lamb Meatballs

Baharat is a Middle Eastern spice mix and the perfect partner for lamb. Use lean lamb, and moisten the breadcrumbs with wine before combining them into the mix – this 'lightens' the meatballs.

Preheat the oven to 210°C.

Place the chopped onion and oil in a saucepan and cook over low heat for 20 to 25 minutes, stirring regularly until the onion softens and becomes a rich golden brown. If the onion becomes dry or catches on the bottom of the pan, moisten with a little water.

Add the garlic, cayenne and baharat to the onion, and cook for a further 3 minutes, then add the tomato paste, cook for 2 minutes and cool.

Tip the breadcrumbs into a bowl and moisten with the white wine. Add the caramelised onion mix, lamb mince, egg and salt, then mix well to combine.

Scoop a tablespoon of the mix into your hand, roll into a round ball and place onto an oven tray lined with baking paper; repeat until you have rolled all of the mix.

Place the meatballs into the oven and cook for 8 to 10 minutes (break one in half to check that they are just cooked through). Once cooked, remove from the oven and place in a large casserole dish. Reduce the oven temperature to 180°C.

To make the sauce, place the onion, garlic and oil into a large saucepan over low-medium heat to sweat for 5 minutes. Add the tomato paste and cook while stirring for a further 2 minutes.

Add all the other ingredients except for salt and pepper, increase heat to medium-high and cook for 30 minutes. Season and pour over the cooked meatballs.

Bake for approximately 15 minutes until heated through.

Meatballs

250 g finely chopped onion
50 ml extra virgin olive oil
3 garlic cloves, peeled and chopped
$\frac{1}{2}$ teaspoon cayenne pepper
10 g baharat (p. 271)
120 g tomato paste
60 g breadcrumbs
100 ml white wine
1 kg lamb mince
1 egg
20 g sea salt

Sauce

100 g finely chopped onion
2 garlic cloves
50 ml extra virgin olive oil
60 g tomato paste
A few sprigs of fresh thyme
1 bay leaf
1.25 litres tomato purée
$\frac{1}{4}$ teaspoon cayenne pepper
1 teaspoon sugar
Sea salt & freshly ground black pepper

MAKES 30 TO 40 MEATBALLS

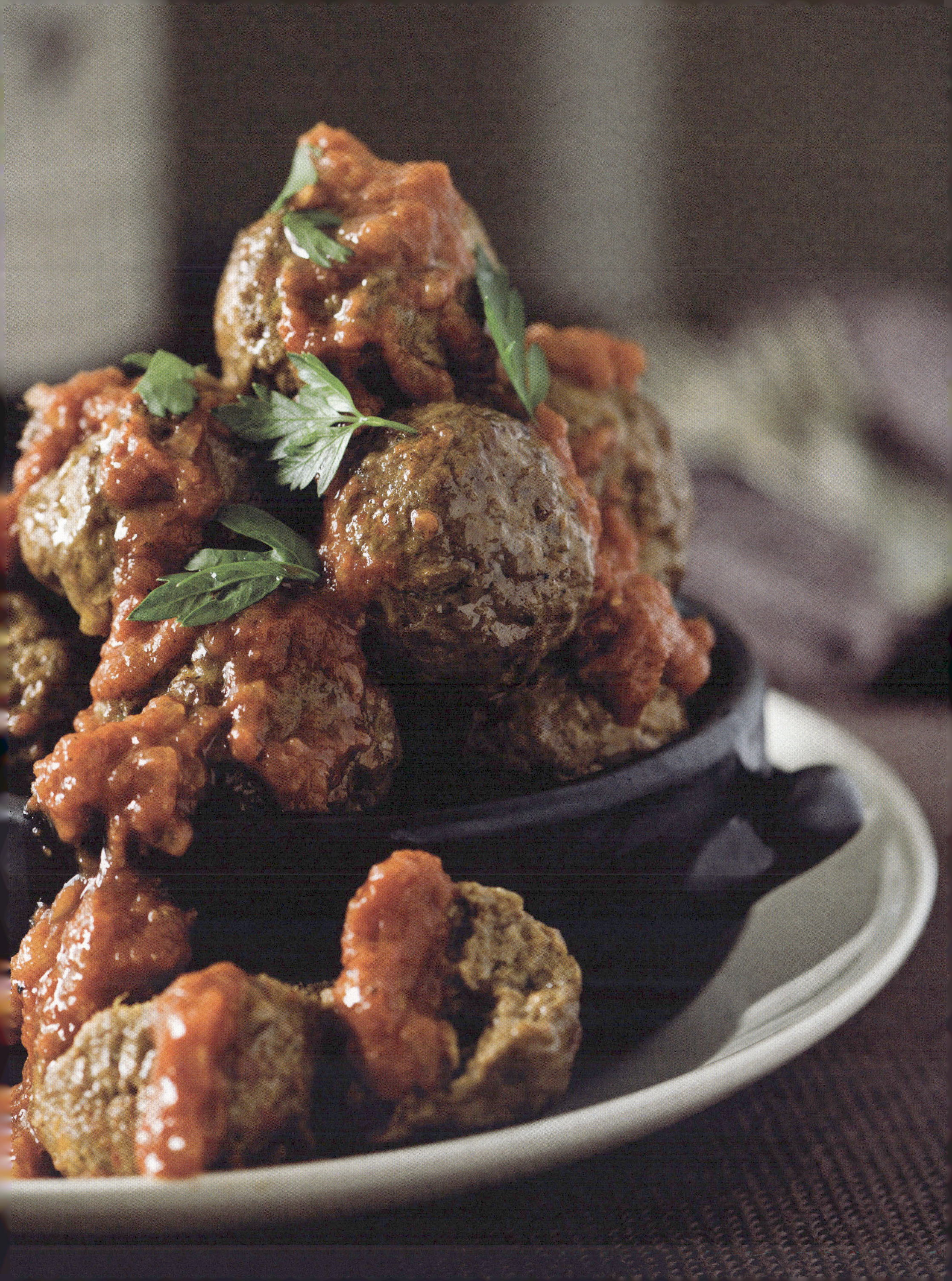

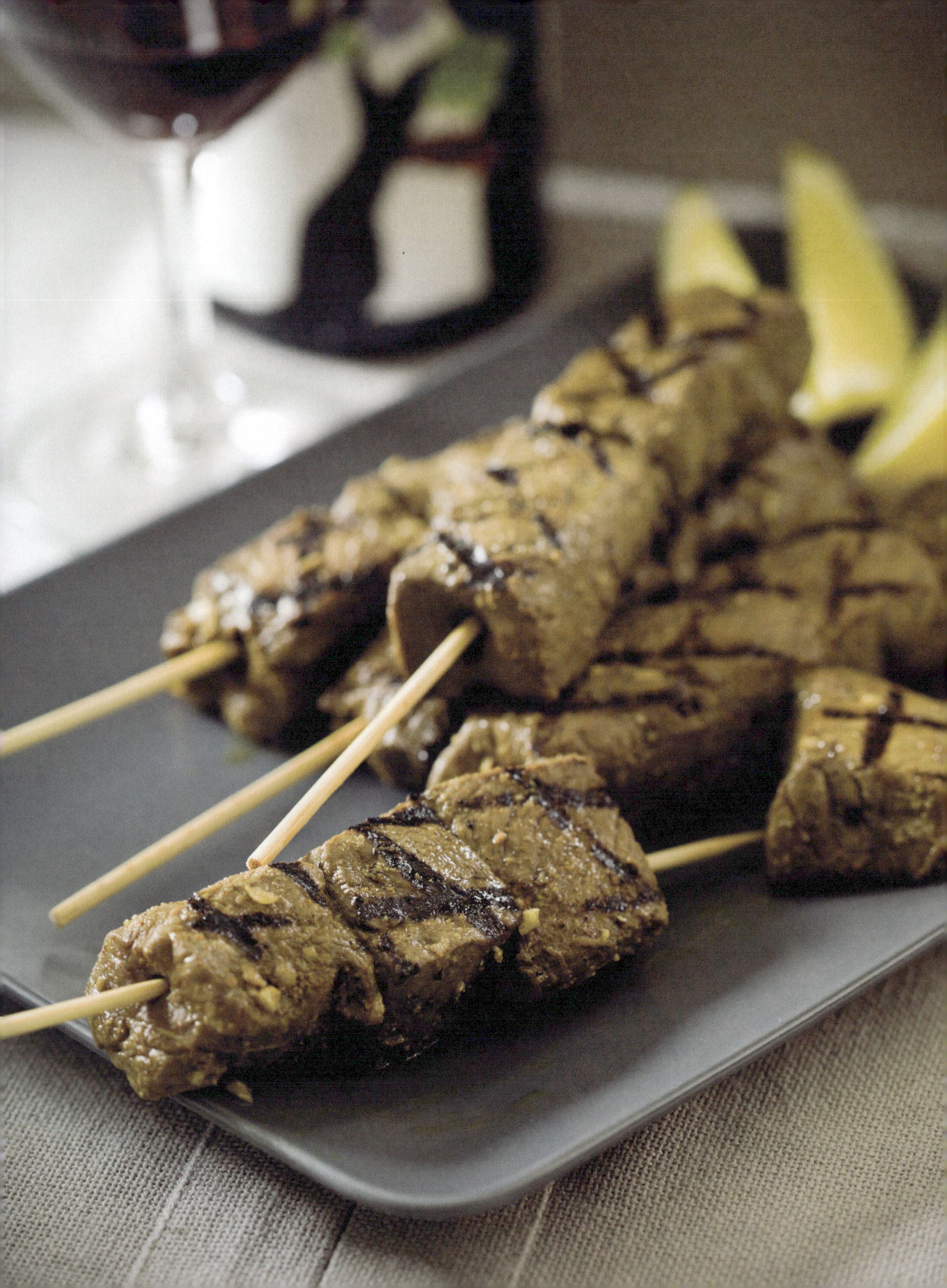

Moorish Spiced Beef Skewers

At Must Margaret River our head chef Chris serves three types of steaks once the meat has been hung for 28 days. When he trims down the rump, he uses a tender 'cap' of muscle (we believe it's the best part of the rump) for these skewers. His spicing then adds another dimension of flavour, and this marinade works equally well on lamb.

1.2 kg rump cap, cut into 1.5 cm dice

Marinade

2 tablespoons cumin powder
1 tablespoon sweet paprika powder
½ teaspoon smoked paprika powder
1 teaspoon nutmeg powder
1 teaspoon turmeric powder
2 teaspoons sea salt
½ teaspoon cayenne pepper
½ cup flat-leaf parsley leaves, finely chopped
2 garlic cloves, minced
125 ml Pedro Ximénez sherry
80 ml extra virgin olive oil
1 lemon, finely sliced

Extra virgin olive oil, for drizzling
Lemon wedges

MAKES 12 SKEWERS

- Mix the marinade ingredients together in a large non-reactive bowl, add the diced rump cap, mix well, cover and refrigerate for at least 4 hours (preferably overnight).
- Preheat a char-grill or barbecue, thread the rump cap pieces onto bamboo skewers that have been soaked in water (or use metal skewers if you have them) and cook over medium–high heat for 2 minutes each side.
- Remove from the barbecue and rest in a warm spot for 3 to 4 minutes.
- Place skewers onto serving platter, drizzle with oil and serve with lemon wedges.

DJP 'Petalos' Mencia
Bierzo, Spain

Mencia (red) is almost rustic in style and is multi-layered, packed with coffee, spice, earth and roasted black fruits. It has the strength and power to sit alongside the meaty richness and spice of these skewers. The protein in the meat has the added benefit of softening the sturdy tannins and calming the warming alcohol.

bar snacks

Cricketers Arms (beer)
Victoria, Australia

This beer has all the flavour and weight to cope with the juicy succulence of the beef and is crisp and quenching, which is a perfect pairing for the crunchy, salty frites.

Must Steak Sandwich

Warm grilled baguette, aged beef cooked to pink perfection, crisp fried onions and rich mayonnaise...our steak sandwich is always a popular fixture on the bar menus at Perth and Margaret River.

....

Preheat the oil in a deep-fryer or deep saucepan to 190°C.

Cut each baguette into three pieces, then split lengthways and reserve.

Season each steak generously, brush with oil and place on a barbecue or char-grill pan, cooking for about 1 minute on each side. Remove and keep warm.

Slice the onion into thick rings, dust with flour and drop into the chickpea batter.

To test if the oil is hot enough, drop in a cube of bread; if it sizzles immediately the oil is ready to use.

Carefully drop the onion rings into the hot oil, cook for 2 minutes until golden brown then scoop from oil and place onto kitchen paper.

Brush the cut side of the baguettes with oil then drop on to a hot char-grill plate to warm through.

Arrange one piece of baguette on each serving plate, place rested steak on top, then pile with sliced tomato, rocket and onion rings. Drizzle with mayonnaise, place the 'lid' of baguette on top and serve with pommes frites (p. 203).

2 baguettes
6 x 150 g sirloin steaks
Sea salt and freshly ground black pepper
Olive oil
1 brown onion
Plain flour
Chickpea batter (p. 44)
Oil, for deep frying
2 ripe Roma tomatoes, sliced
A handful of rocket
Mayonnaise (p. 281)

MAKES 6

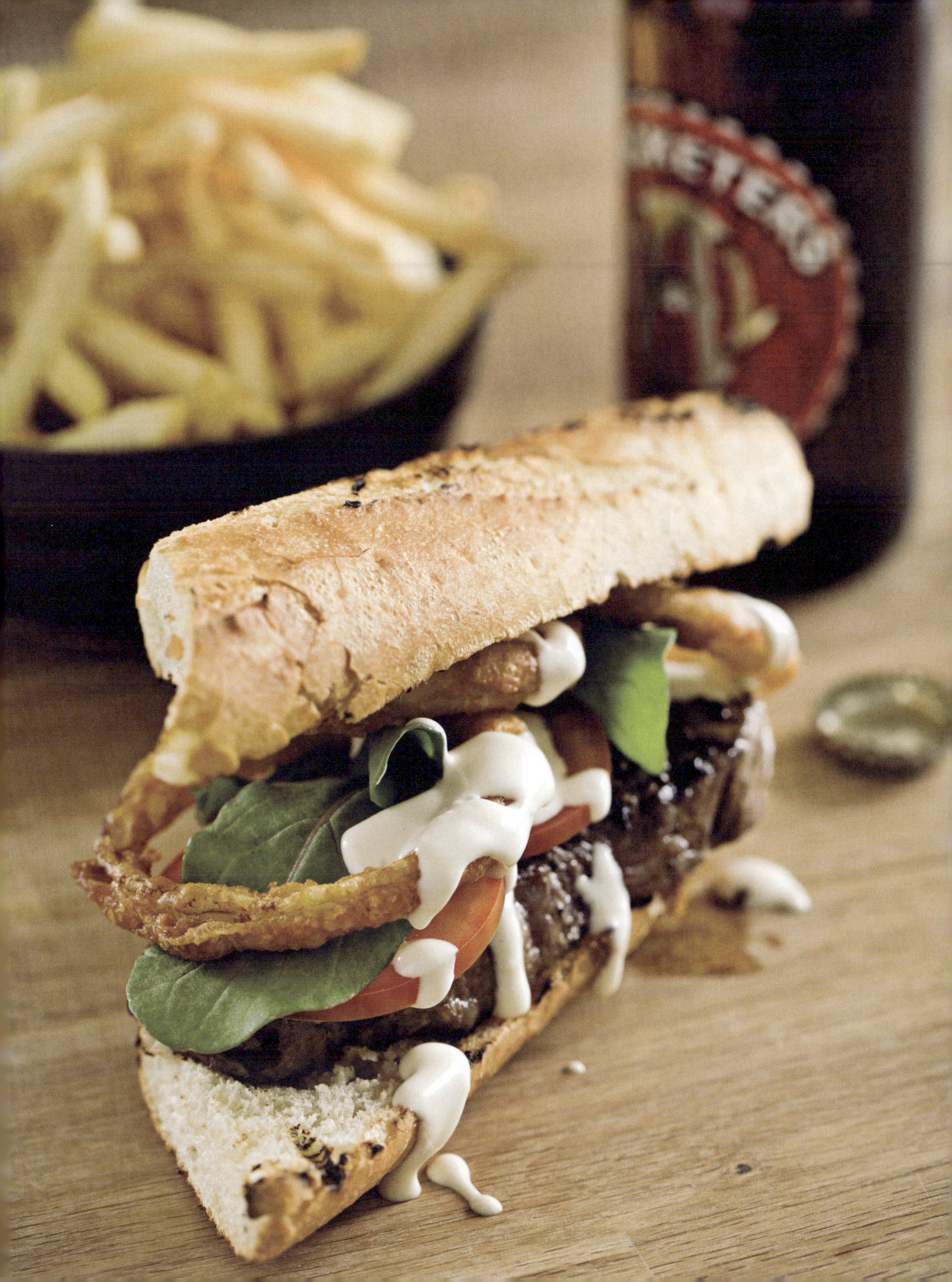

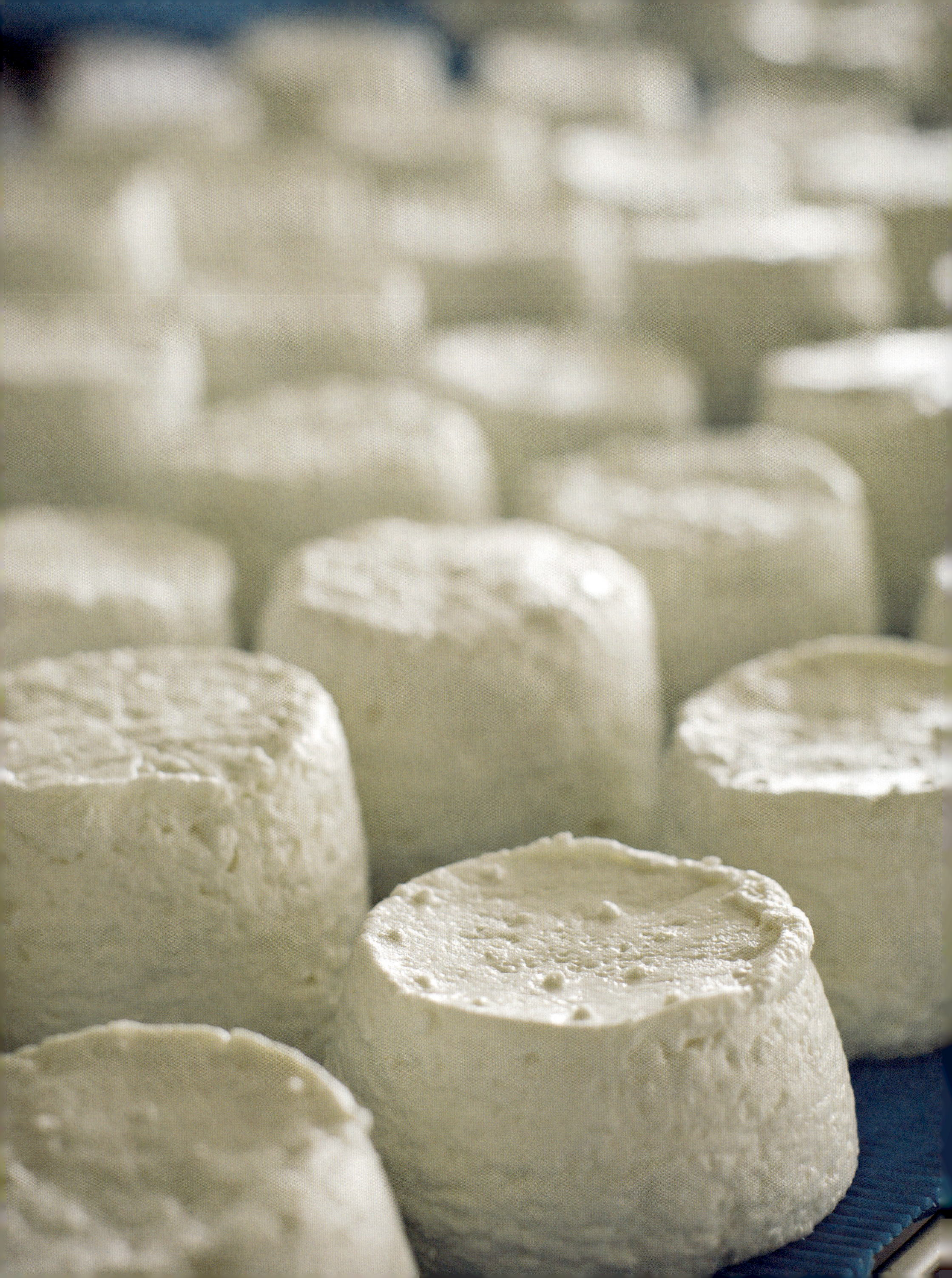

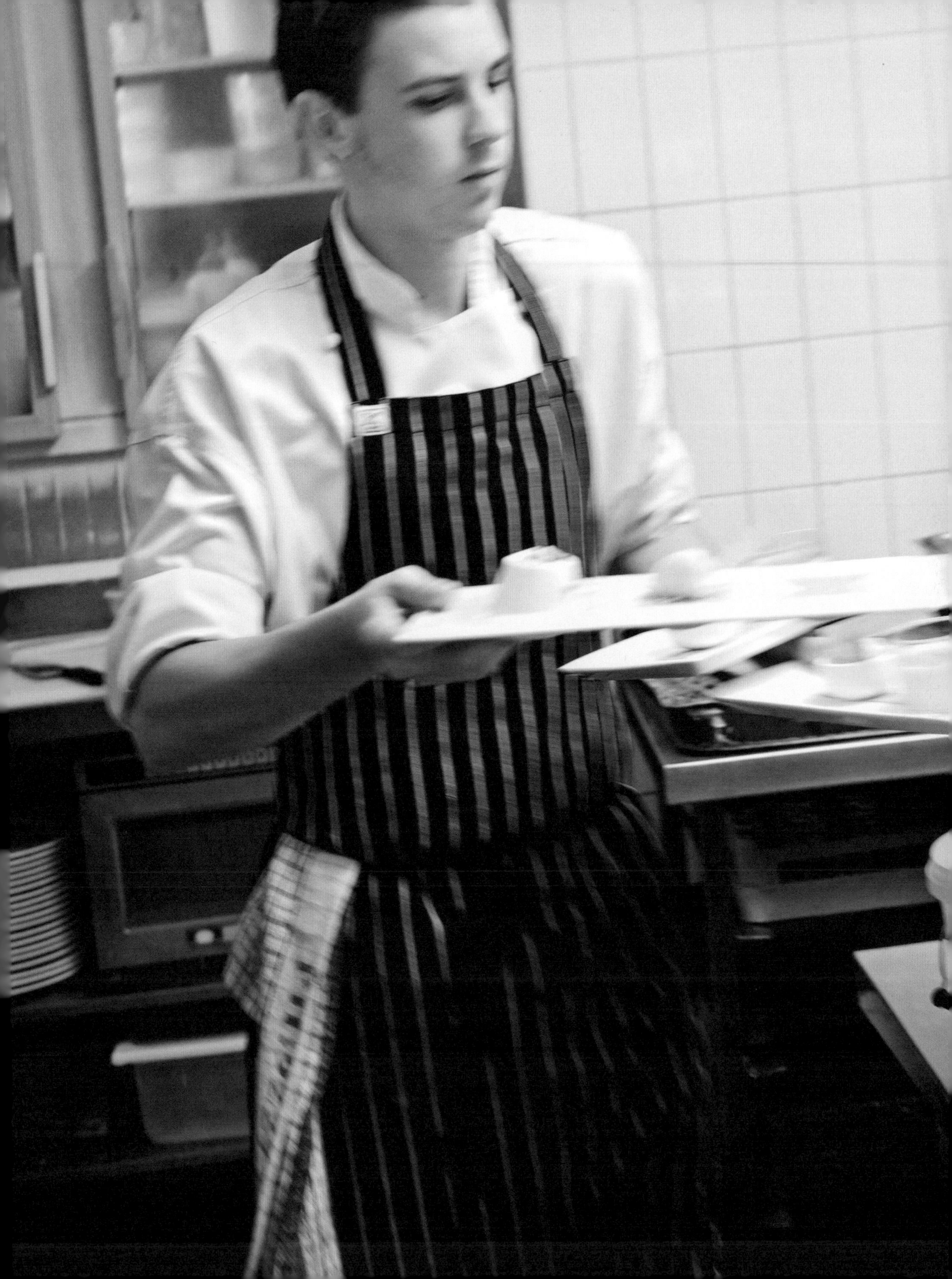

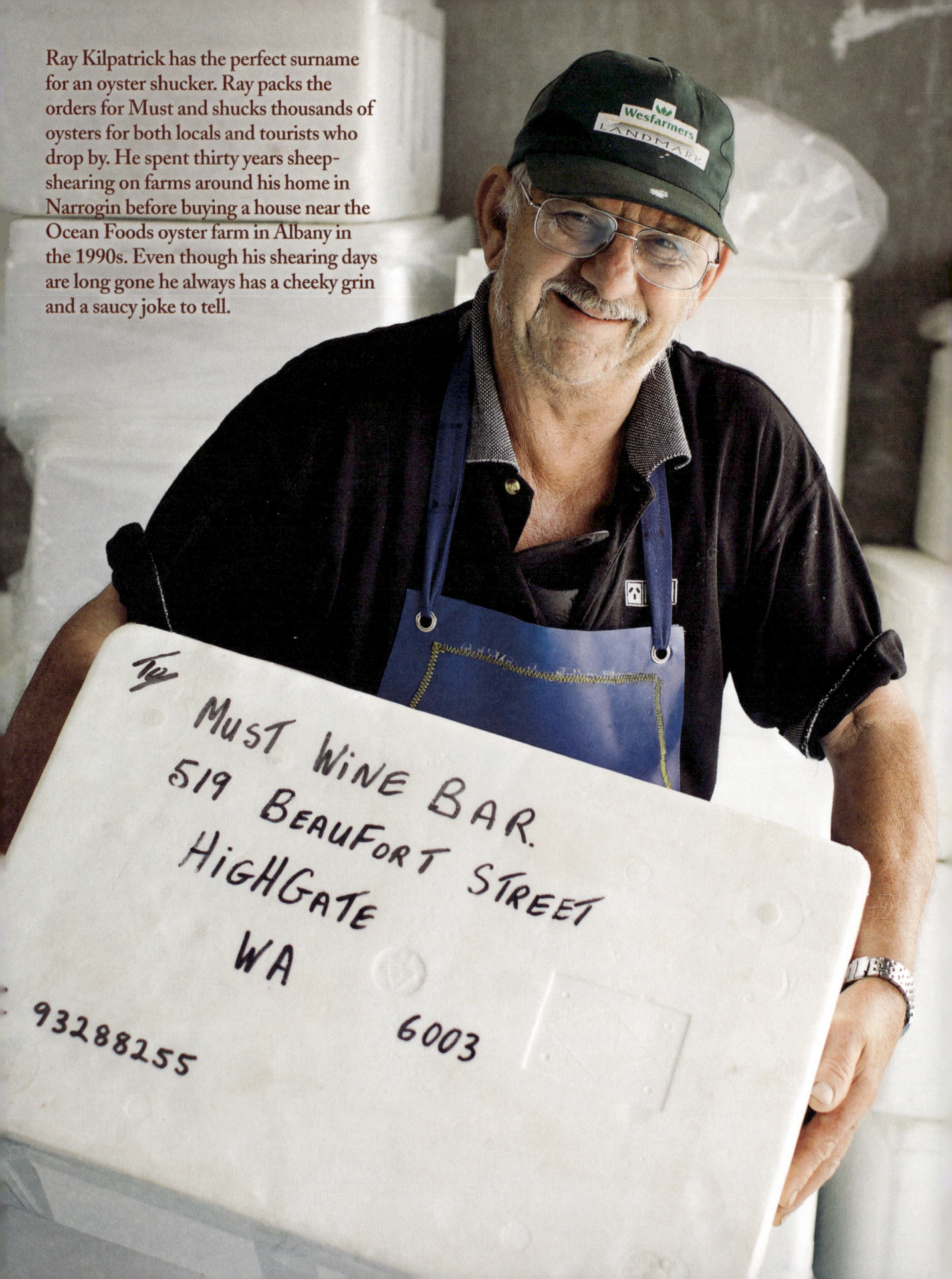

Ray Kilpatrick has the perfect surname for an oyster shucker. Ray packs the orders for Must and shucks thousands of oysters for both locals and tourists who drop by. He spent thirty years sheep-shearing on farms around his home in Narrogin before buying a house near the Ocean Foods oyster farm in Albany in the 1990s. Even though his shearing days are long gone he always has a cheeky grin and a saucy joke to tell.

oysters

Jerry Fraser is the champion of oysters in Western Australia. He shucks regularly at Must, and travels internationally with the very adult 'shuck me, suck me, eat me raw' message wherever he works. Jerry describes the Albany rock oyster as 'the most delicate of all oysters in Australia'. He also reckons they are the best variety for 'oyster virgins' as they are delicately flavoured and not too big.

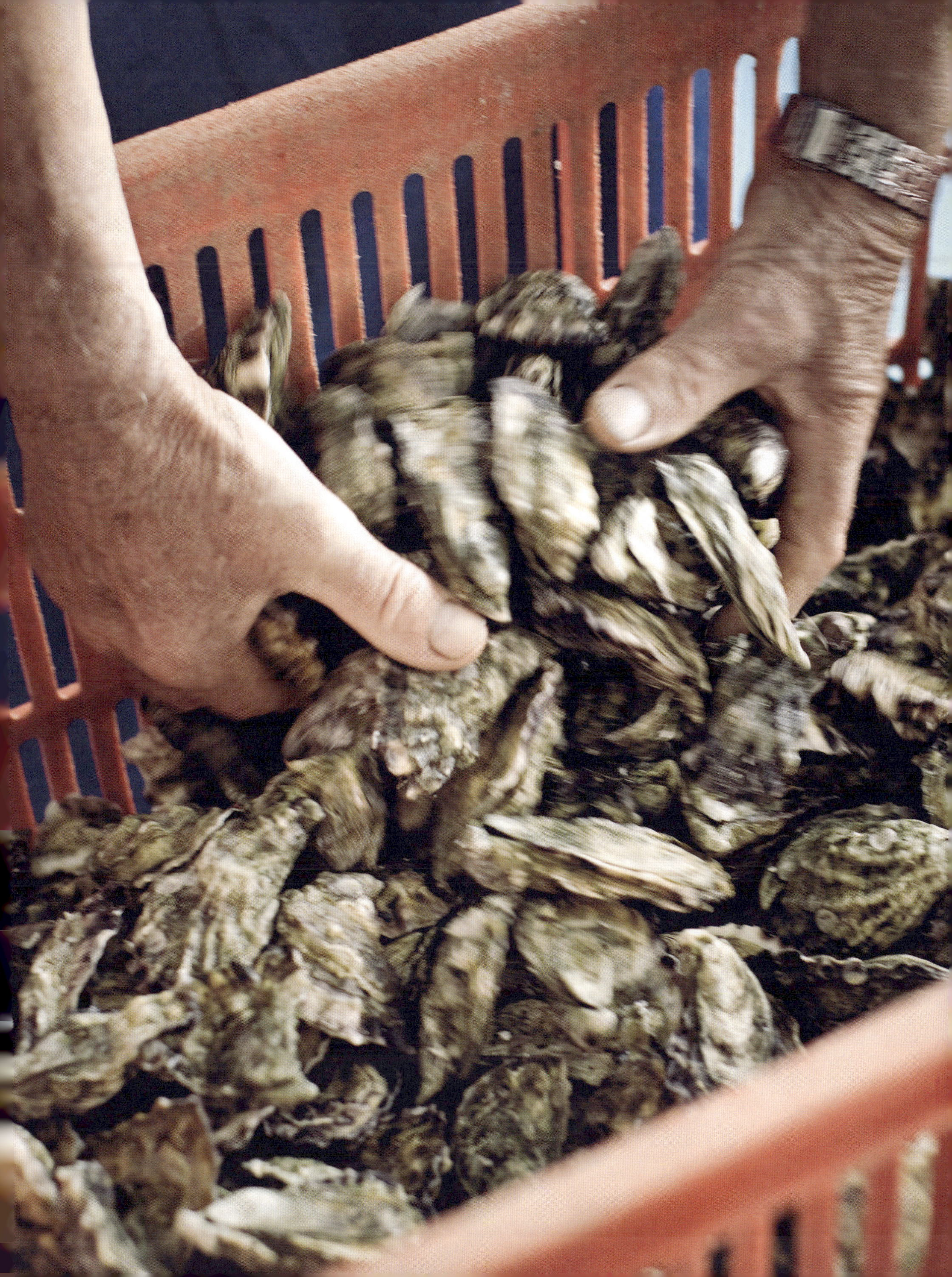

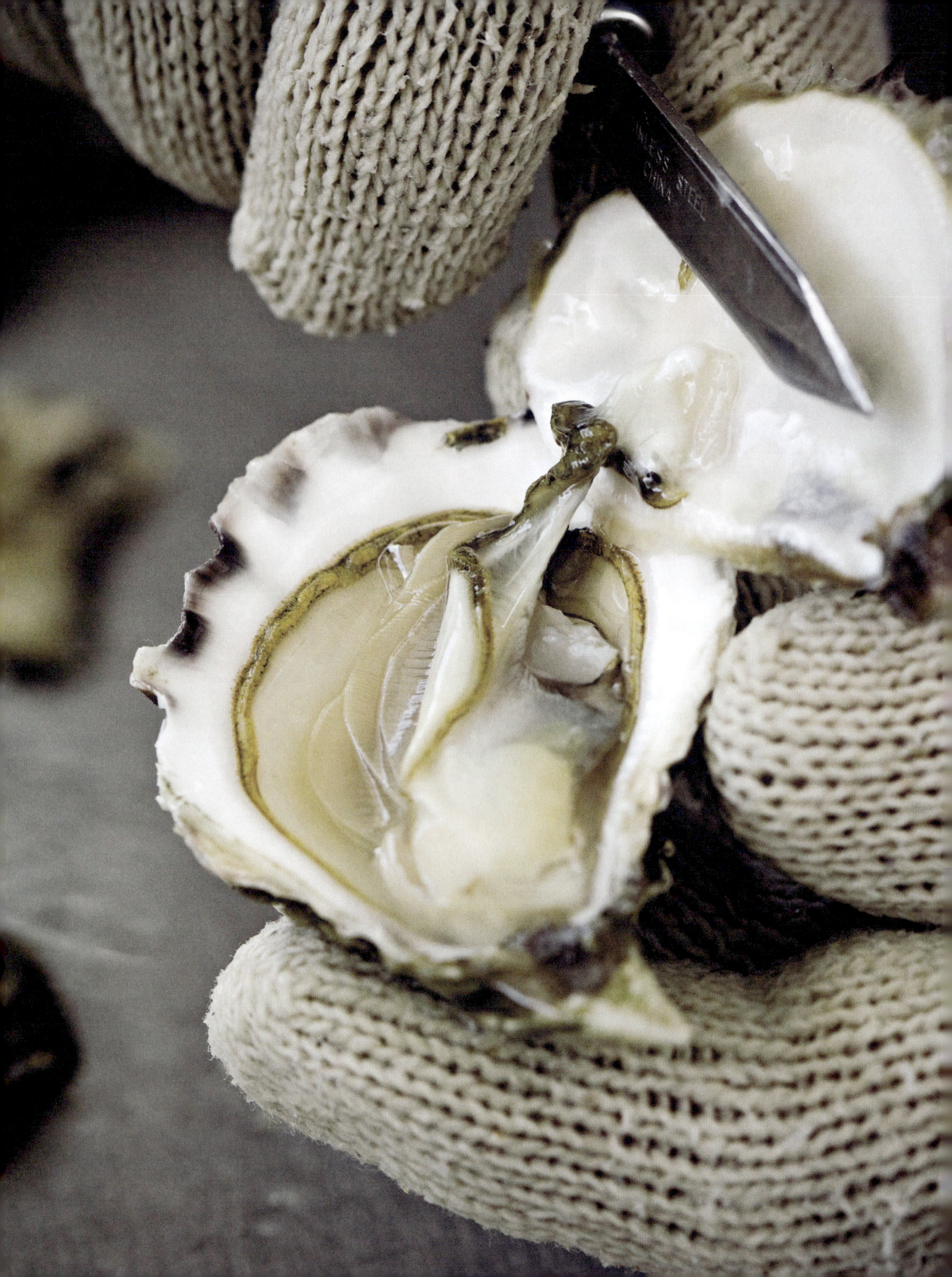

Oysters

Oysters are the quintessential French bistro food enjoyed by people from all walks of life.

I have fond Parisian memories of watching oysters being enjoyed by people spilling into the streets from dedicated huitreries and bars: rouged old ladies rubbing shoulders with dusty tradesmen, suave businessmen and students. Parisians know that the icy cold winds of winter bring the finest oysters, and all follow the common etiquette for eating them – carefully forming a ramp with the shell to allow the flesh to slip into the mouth.

Closer to home, Oyster Harbour in Albany is home to the only commercial oyster farm in Western Australia. Ray Kilpatrick packs whole oysters into eskies for delivery to Perth, in between shucking them for locals and 'in the know' tourists. Albany rock oysters are an amazing food secret, as many locals are unaware that the farm on their doorstep exports to the world. Plucking an oyster here, quickly flicking off its shell then slurping it down is one of the most delicious food experiences in Australia.

Albany rock oysters have an initial salty impact, the flesh has a golden hue with a buttery texture, and Jerry Fraser says they have a unique nutty flavour, a light tannic 'tang' and a lingering mineral finish. Each of these oyster recipes uses a dozen oysters – whether that serves one person or two, or more, is up to you.

Champagne remains the quintessential wine match for oysters. The elegance, vibrant acidity and earthy mineraliness offset the creamy, salty 'meaty' flavours of the oysters. It is best to match more delicate blanc de blanc Champagnes such as Egly-Ouriet, with oysters that have gentle, neutral-based 'toppings' and the fuller, richer Champagnes, such as Gosset Grand Reserve, with oysters Rockefeller.

Egly-Ouriet Brut Tradition Grand Cru
Champagne, France

This wine is individual and iconic. It complements the creamy, earthy raw oyster, rather than the oyster/vinegar combination. It is minerally with chalky-dryness and provides the perfect background to the salty flavours of the oysters.

Oysters Natural & Shallot Vinegar

The classic bistro accompaniment for natural oysters (apart from eating them 'nude') is to splash them with a little shallot vinegar. For an alternative flavour that is a great match with Champagne, substitute the red wine vinegar with French Champagne vinegar as we do in The Champagne Lounge.

··

- Combine the chopped shallots with the vinegar; allow to steep in the refrigerator until chilled.
- Pour the rock salt or ice onto a serving platter, arrange the oysters on top and serve with a small dish of the shallot vinegar.

2 or 3 shallots, peeled and finely chopped
100 ml good-quality red wine vinegar
Rock salt or ice
1 dozen freshly shucked oysters

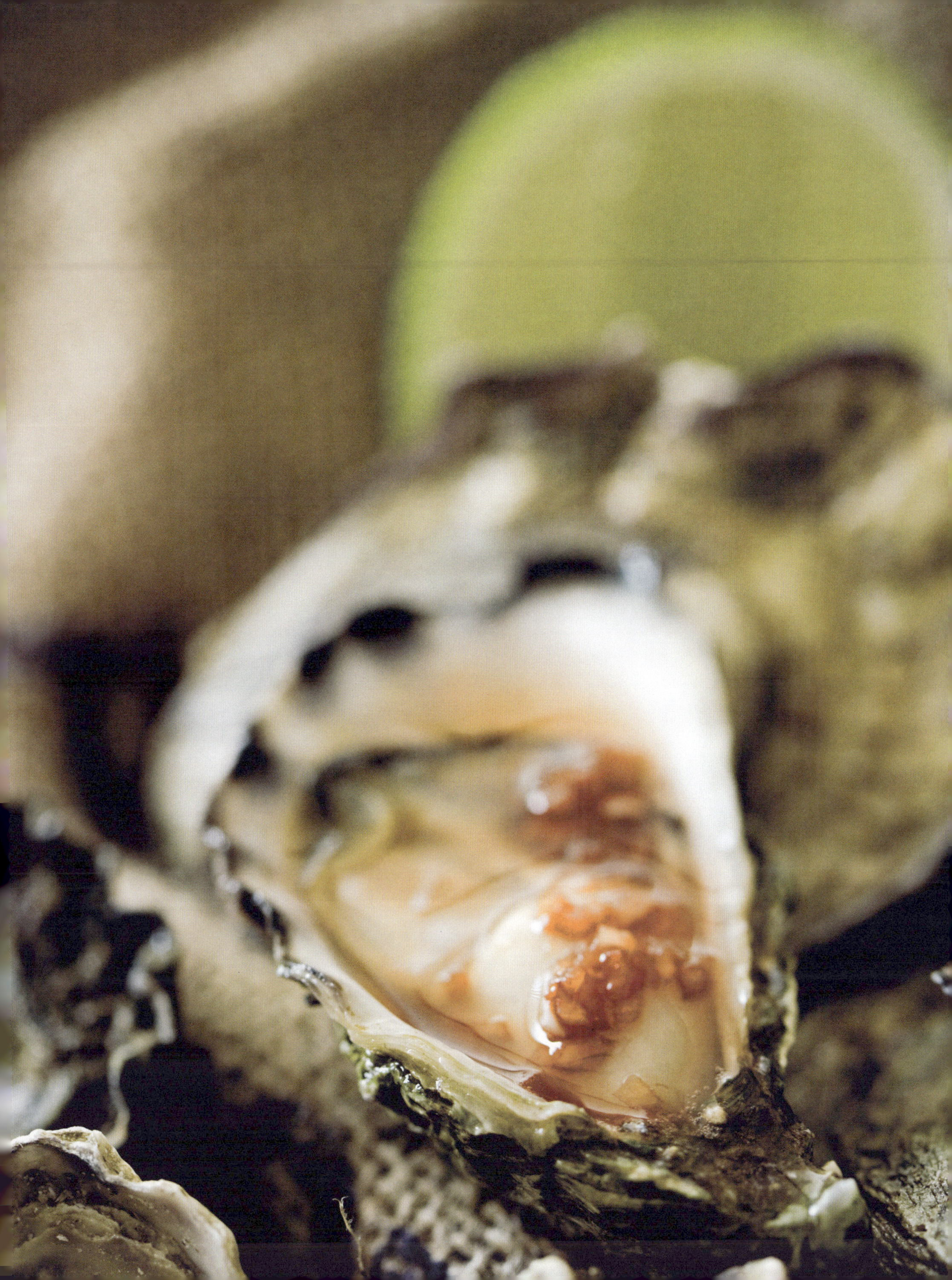

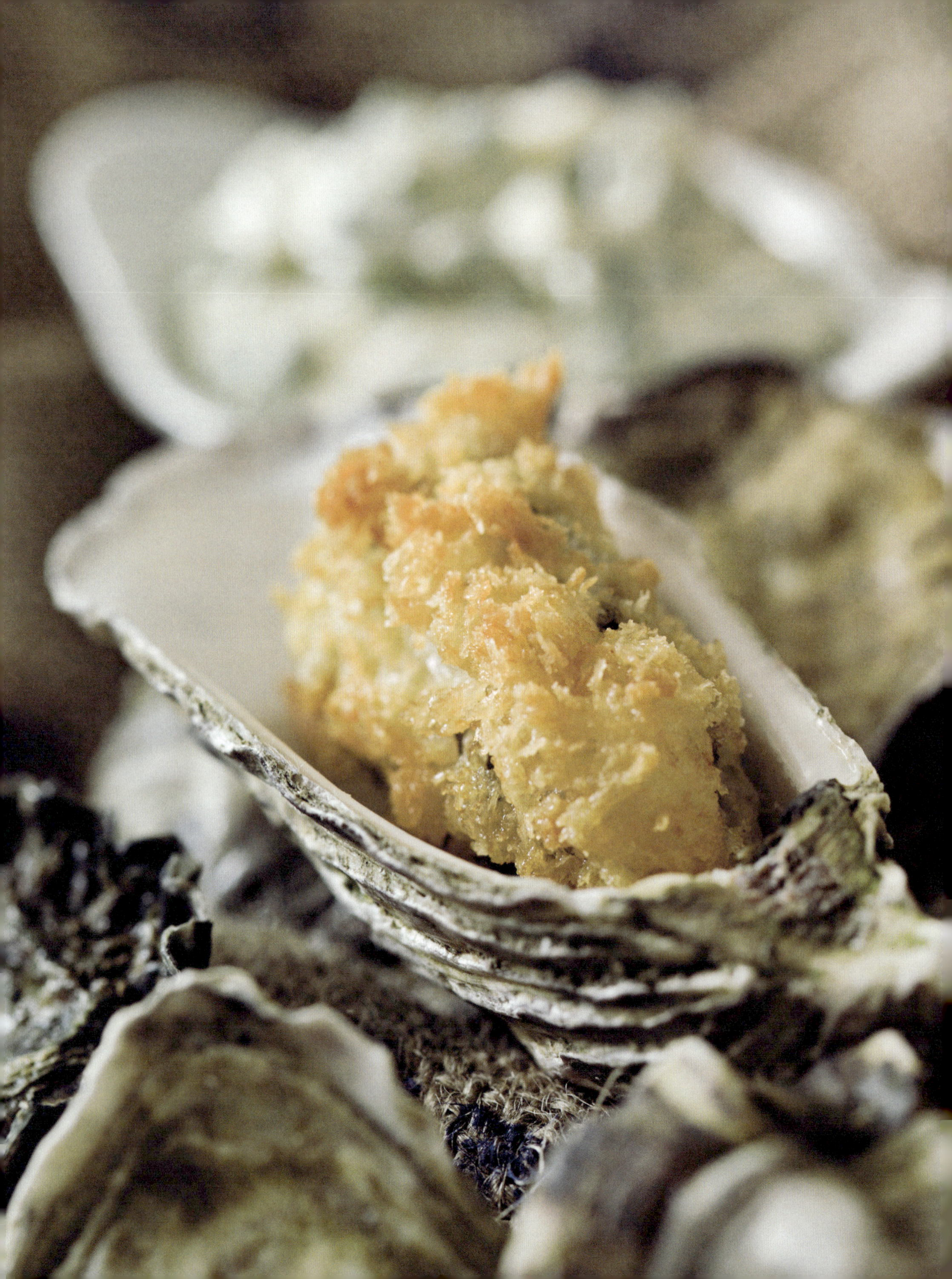

Crumbed Oysters & Tartare Sauce

Japanese breadcrumbs give a delicate golden crust to crumbed oysters – a perfect textural foil to their rich creamy flesh.

1 dozen freshly shucked oysters
Oil, for deep-frying

Crumb

50 g plain flour
1 egg whisked together with 125 ml milk for egg wash
100 g Japanese breadcrumbs

Tartare sauce (p. 282)
Rock salt
Lemon wedges

Remove the oysters from their shells, reserving the shells.

Crumb the oysters by tossing through the flour, then dipping in eggwash, then rolling in breadcrumbs.

Place the oyster shells in a warm oven to dry.

Heat the oil in a deep-fryer or deep saucepan to 180°C. Drop the oysters in the oil for no more than 30 seconds until golden, then scoop from the oil and place onto kitchen paper.

Pour rock salt onto a serving platter, arrange the oyster shells on top and drop an oyster into each shell.

Place a dish of the sauce in the middle of platter, and finish with a wedge of lemon.

Champagne Barons De Rothschild Brut
Champagne, France

This Champagne is accessible rather than austere, with forward fruit expression. This contrasts with the crunchy mouth-feel and balances the creamy tartare sauce. It also has a suggestion of sweetness, but with perfect brisk acidity.

Perrier-Jouët Blason Rosé NV
Champagne, France

The red fruit influence in this rosé Champagne is important to this match. It is pure and delicate with energetic, fresh berry flavours that complement the salsa, yet it avoids overwhelming the charm and expression of the oysters.

Chickpea-Battered Oysters & Avocado Salsa

Chickpea flour is a secret ingredient that provides a nutty flavour and crisp texture to this batter, which also makes a great coating for fish fillets and shellfish.

Combine all the batter ingredients together in a mixing bowl and whisk until combined. Rest the batter in the refrigerator or at least 2 hours.

Combine the salsa ingredients together in a mixing bowl and gently combine. Keep refrigerated until required.

Remove the oysters from their shells, reserving the shells.

Heat the oil in a deep-fryer or deep saucepan to 180°C.

Drop the oyster flesh into the batter; place the shells into a warm oven to dry.

Drop the oysters into the oil and cook for 30 seconds until golden. Scoop from oil and place onto kitchen paper.

Pour rock salt onto a serving platter, arrange the warm oyster shells on top, spoon salsa into each shell, place an oyster on top and finish with lime wedges.

1 dozen freshly shucked oysters
Oil, for deep-frying

Batter
80 g chickpea flour (besan)
80 g plain flour
100 ml beer
90 ml water
A pinch of sea salt

Salsa
1 ripe avocado, finely diced
1 ripe Roma tomato, finely diced
2 tablespoons red capsicum, finely diced
1 tablespoon chopped chives
1 teaspoon finely chopped red chilli
Juice of 1 lime
Sea salt & freshly ground black pepper

Rock salt
Lime wedges

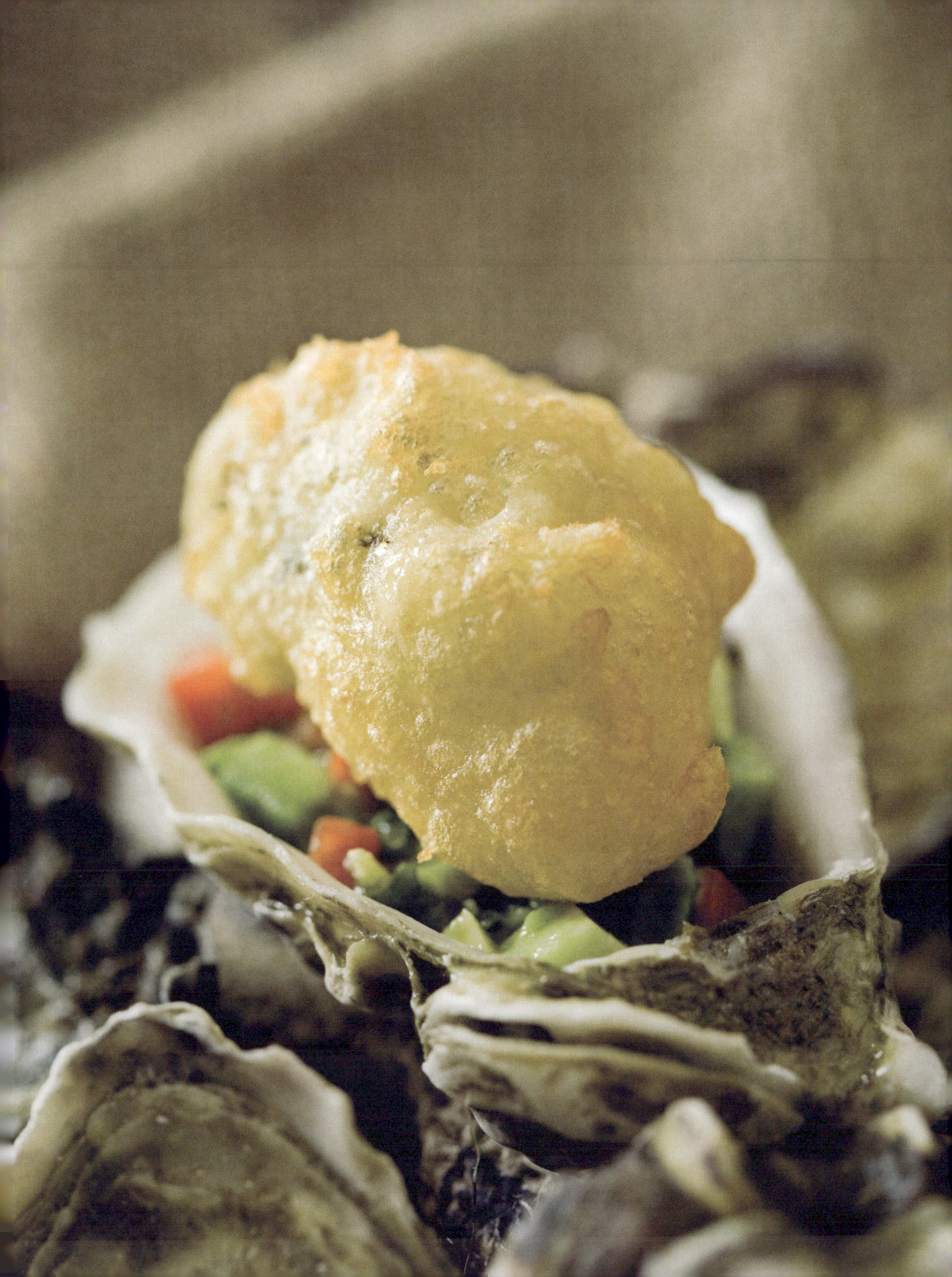

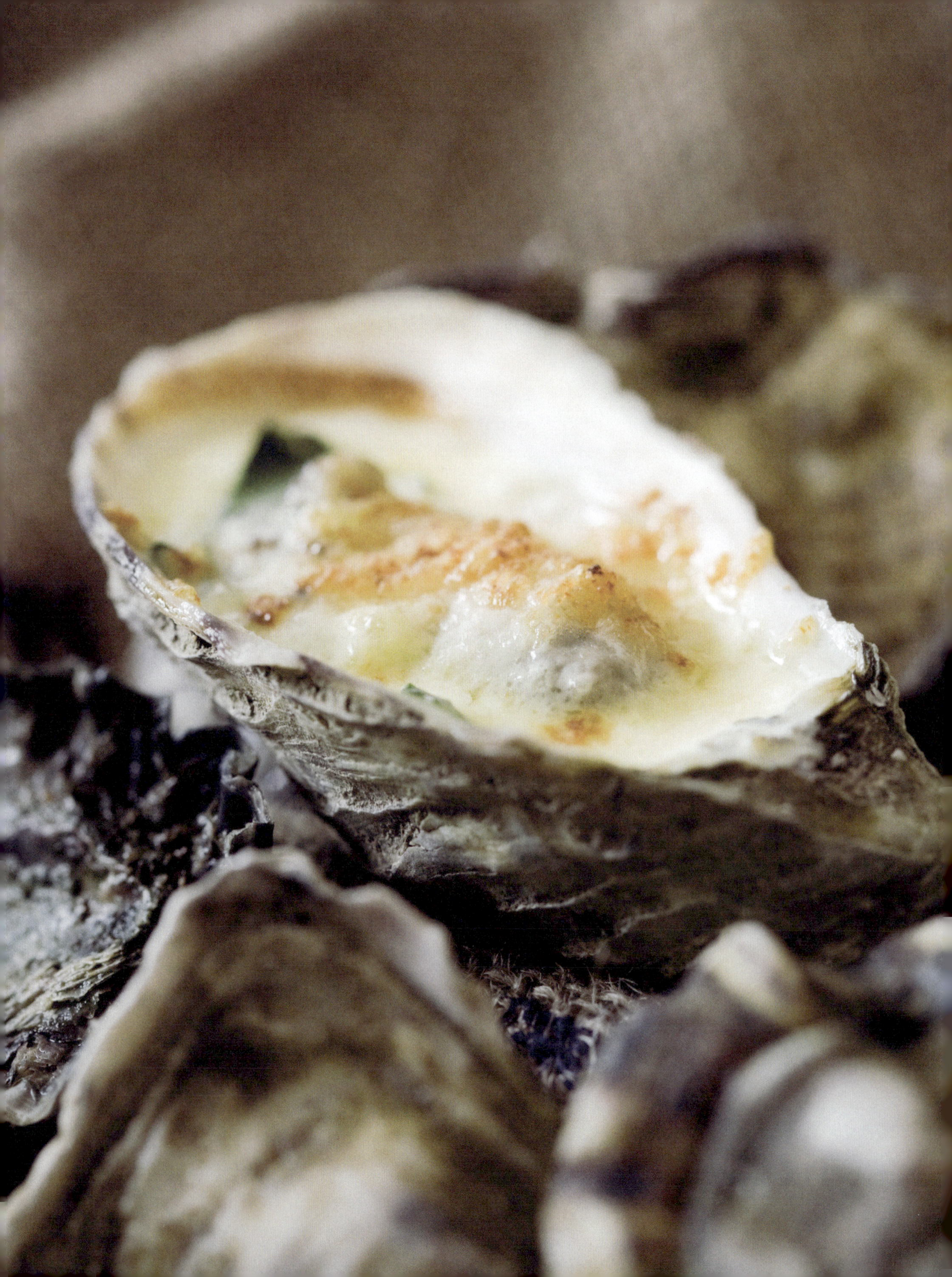

Oysters Rockefeller

Rich creamy oysters baked on a bed of spinach with a crust of cheese and waft of aniseed from Pernod...it sounds a little eighties but this is one of the most popular hot oyster dishes on our menus.

---

1 dozen freshly shucked oysters
1 bunch English spinach
10 g butter
1 garlic clove
200 ml cream
30 ml Pernod
Sea salt & freshly ground white pepper
50 g finely grated Cantal cheese (gruyère can be used instead)
50 g breadcrumbs

Rock salt
Lemon wedges

Trim the stalk ends of the spinach and wash well to rid it of any grains of sand. Bring a large pot of salted water to the boil, drop in the spinach, allow the spinach to wilt for about 30 seconds, stirring, and then tip the spinach into a colander. Once the hot water has drained, throw a handful of ice into the colander and run cold water over the spinach to quickly cool. Remove the spinach and squeeze to drain the excess water out.

Remove the oysters from their shells and line each shell with a few blanched spinach leaves, placing the oyster flesh back on top. Fill an ovenproof dish with a 1 cm layer of rock salt, sit the oyster shells on top so that they are flat; refrigerate until required.

Melt the butter in a medium sized saucepan over low heat, mince the garlic and add, sweating it until fragrant; about 1 minute.

Add the cream and increase the heat to medium-high. Allow the cream to boil and reduce a little, being careful not to let it boil over. This should take about 3 to 5 minutes. If the cream separates, add a little more to the saucepan and whisk well to bring the sauce back together. Remove the cream from the heat, add the Pernod, season and reserve.

Preheat your oven on grill setting to medium-high. Warm the oysters under the grill for about 1 minute, then pour a little of the sauce into each oyster shell, being careful that they are level so the sauce does not seep out.

Mix the grated cheese and breadcrumbs together and sprinkle onto the oysters. Place under the grill in about the centre of the oven to brown for about 3 minutes (not too close to the element as the oysters should cook slowly). The oysters will be ready when golden brown and bubbling.

Gosset Grande Reserve Brut NV
Champagne, France

These grilled oysters require Champagne with muscle and fullness. Gosset has the weight and power to balance the richness and flavour profile, while the tartness and piercing acidity cut through the soft dairy influence.

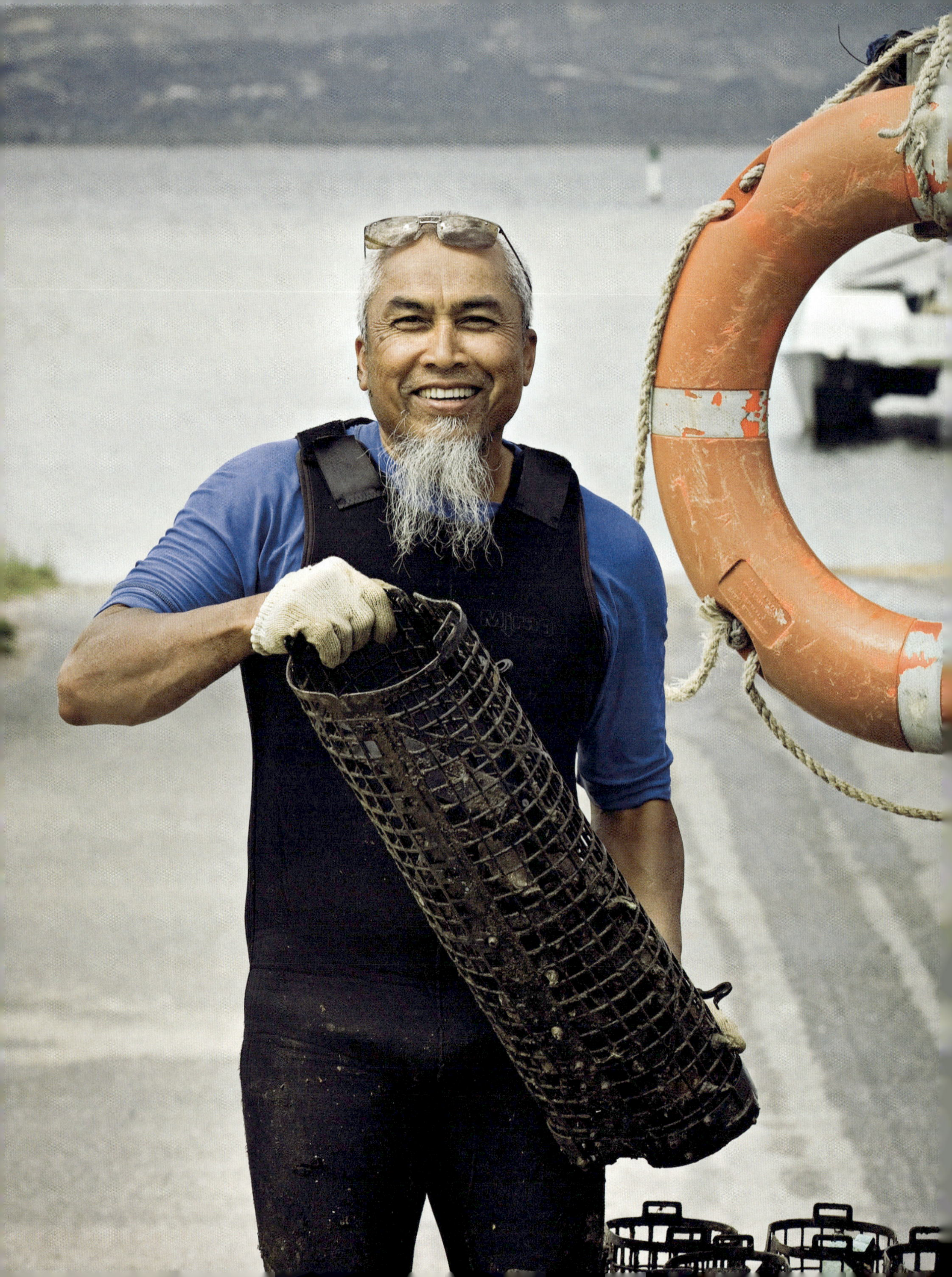

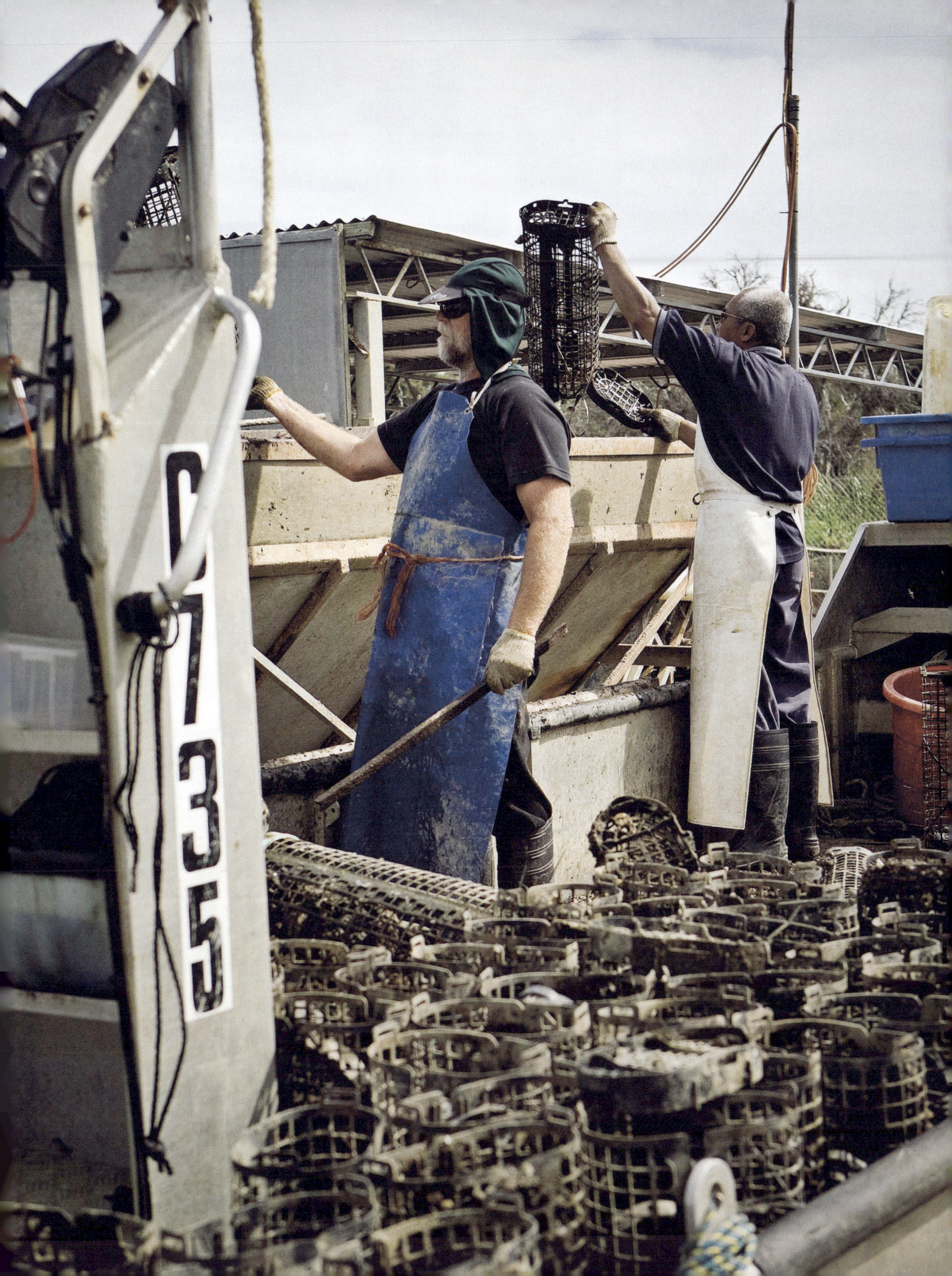

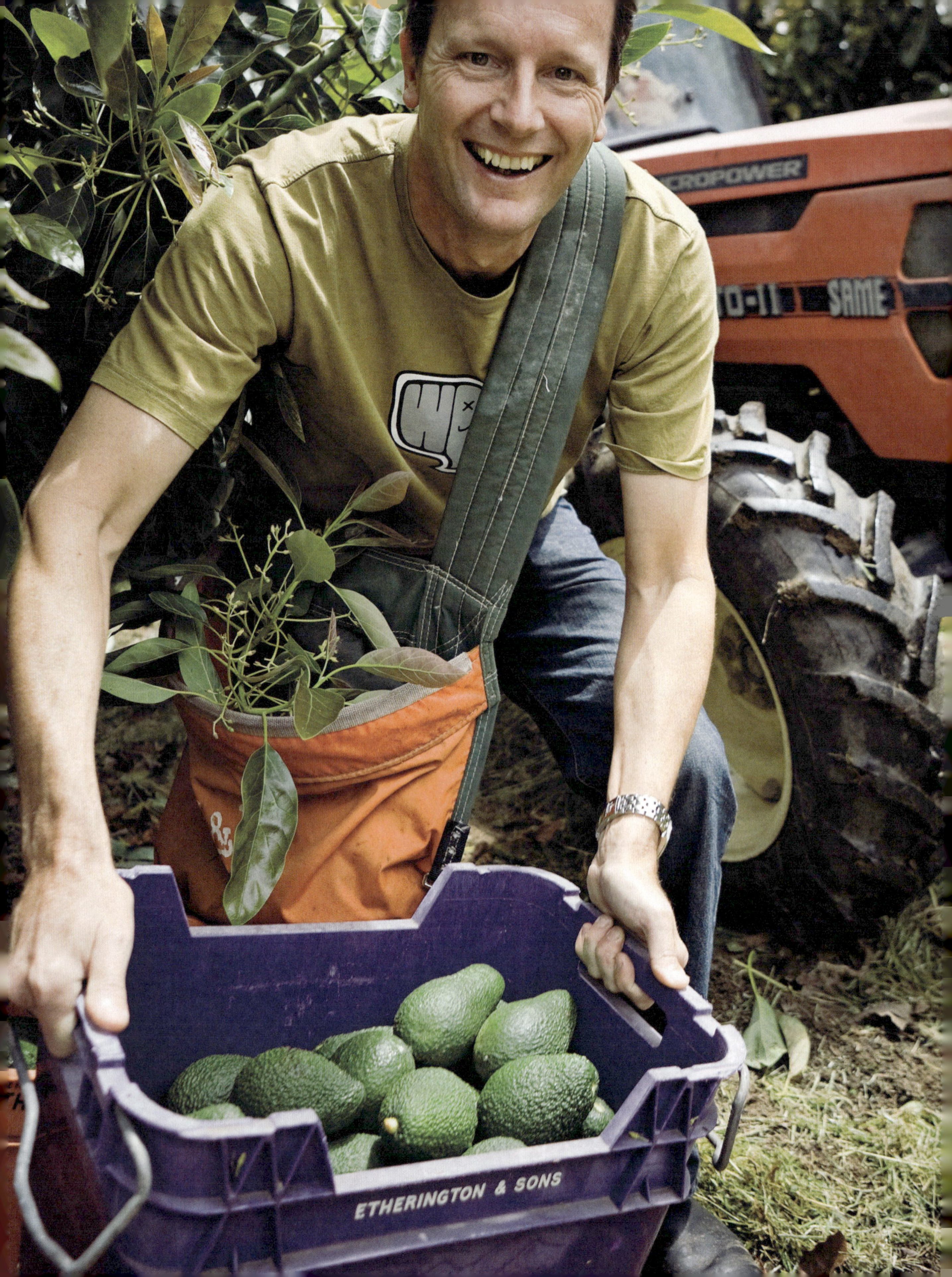
SAME
ETHERINGTON & SONS

starters

FRESH W.A. ASPARAGUS
Store At 1°C
BEST FOR QUALITY
UNITED CRATE CO-OP

In 1970, Phillip and Sheelagh Marshall left Surrey and, as 'ten pound Poms', bought a cattle farm at Torbay. The beef market promptly collapsed, so they planted WA's first commercial crop of asparagus, harvesting the first spears in 1974; it took the rest of the decade to persuade consumers to ditch the tinned variety for their exceptional product.

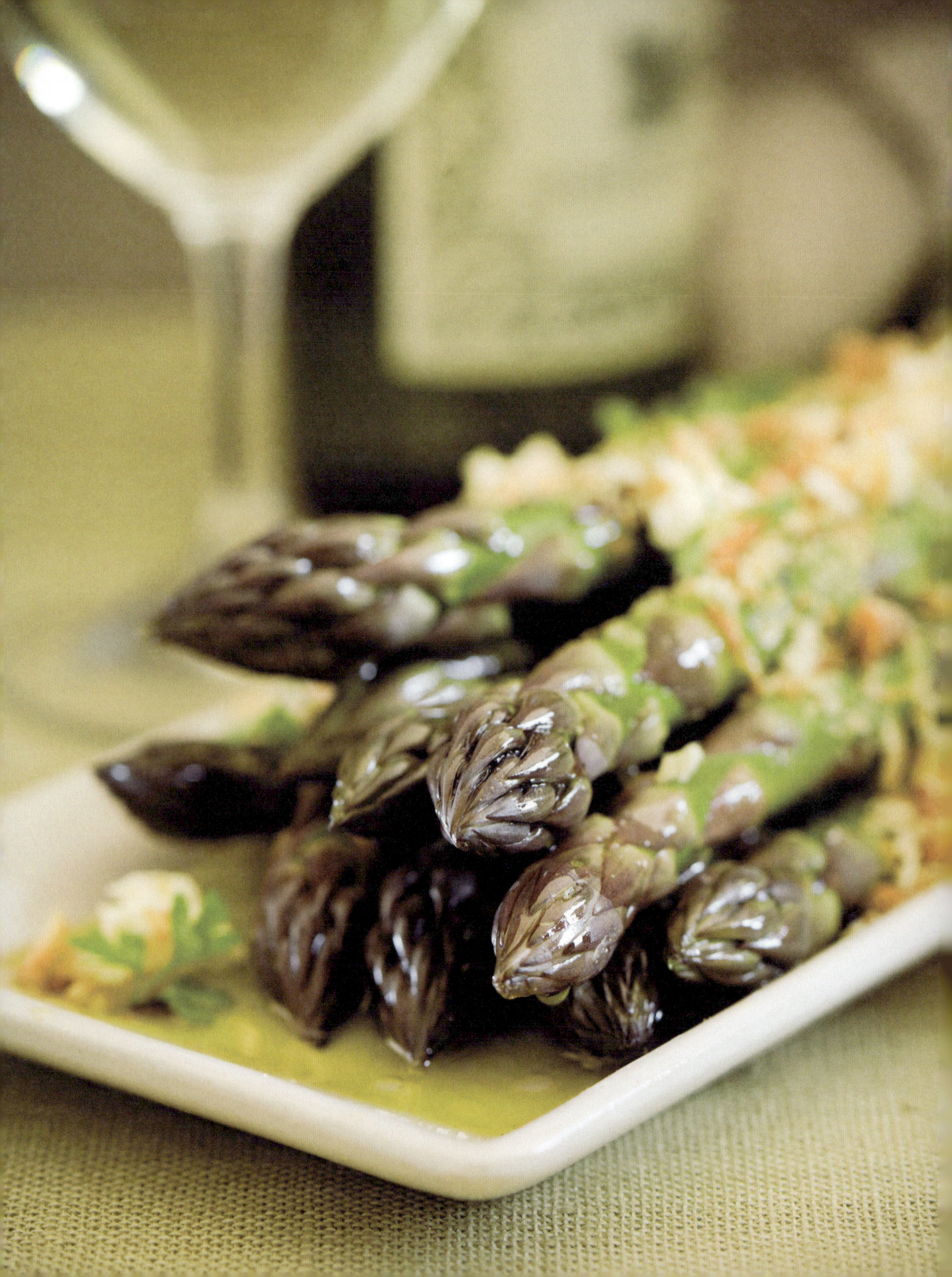

Asparagus, Over the Moon Organic Feta, Macadamia Crumble & Verjuice Dressing

At Must I only use the largest asparagus spears, they are the most tender and impressive on the plate. I designed this dish for a Great Southern Slow Food lunch and it's been on our menus (during asparagus season) ever since.

....

24 large asparagus spears

150 g Over the Moon organic feta (most good growers markets sell this, but it can be substituted with a soft sheep's or cow's milk feta cheese)

Dressing

75 ml verjuice (unfermented grape juice, available from good gourmet shops)

150 ml good extra virgin olive oil

Sea salt & freshly ground black pepper

Macadamia Crumble

150 ml good extra virgin olive oil

2 garlic cloves

100 g Japanese breadcrumbs.

150 g unsalted macadamia nuts, coarsely crushed

A few sprigs of fresh thyme

SERVES 6

First make the dressing. Whisk the verjuice and the olive oil together and season.

Crumble the feta into small pieces and keep in a bowl until you are ready to put the dish together.

Heat the olive oil in frying pan over medium heat flame, smash the garlic cloves with the back of a knife and add to the oil; cook for 30 seconds.

Toss in the Japanese breadcrumbs and macadamia nuts, stirring gently over heat until golden brown.

Tip the breadcrumb mix onto a tray to cool, remove the smashed garlic cloves and discard. Strip the thyme leaves off their stem and sprinkle over the breadcrumb mix. Season with salt and freshly ground black pepper.

Check the base of the asparagus when you buy them; if the cut face is dry or the base shrivelled it is not fresh. Hold the spear at the tip (be careful, fresh asparagus will be brittle) and peel the thick bottom section of the spear, to about 5 cm from the base.

Bring a large pot of salted water to boil, drop in the asparagus spears, cook for 2 to 3 minutes, remove and drain, then place onto a warm serving platter.

Sprinkle the warm asparagus with crumbled feta and generously spoon the dressing over the top. Sprinkle with the crumble and serve immediately.

Palacio de Fefiñanes Albariño
Rias Baixas, Spain

Albarino is bright and energetic and has the opulence and fruit profile (grapefruit, green mango, fresh herbs) to work alongside the pronounced character of this dish. The texture and substance balances the tangy creaminess and green edge feel.

Château Caillou 2006
Barsac
Bordeaux, France

There are ingredients here that can make a wine match difficult. The sweetness and saltiness of this combination requires a wine with the weight and residual sugar to work in unison – not to overpower or be dominated. The classic pairing is Sauterne or Barsac.

Caramelised Fig, Jamón Ibérico & Roquefort Salad & Walnut Vinaigrette

This is a simple salad for sunnier months when figs are abundant. If jamón ibérico is too pricey, opt for a good jamón serrano (made from white pigs) or Dorsogna prosciutto.

Preheat the oven to 160°C.

Heat a non-stick pan over a medium-high heat.

Slice the figs in half lengthways and dredge the cut side with caster sugar.

Carefully place into the hot pan, cut side down – the sugar will caramelise and smoke.

Use tongs to remove onto a tray lined with baking paper before the caramel burns.

Place the tray into the oven for 5 minutes while the remainder of the salad is prepared.

In a large bowl crumble the Roquefort cheese, together with the frisée lettuce and walnuts. Drizzle some of the vinaigrette over the salad and toss lightly.

Remove the figs from the oven, and lay two halves on the base of each plate.

Lay two slices of jamón on top of the figs, top with the salad and drizzle with a little extra dressing.

6 ripe figs

100 g caster sugar

12 paper thin slices jamón ibérico de Bellota (I use 5J)

150 g Roquefort cheese (I use Papillon Roquefort)

200 g frisée lettuce, washed and picked

60 g walnuts, roasted and roughly crushed

Walnut vinaigrette (p. 268)

SERVES 6

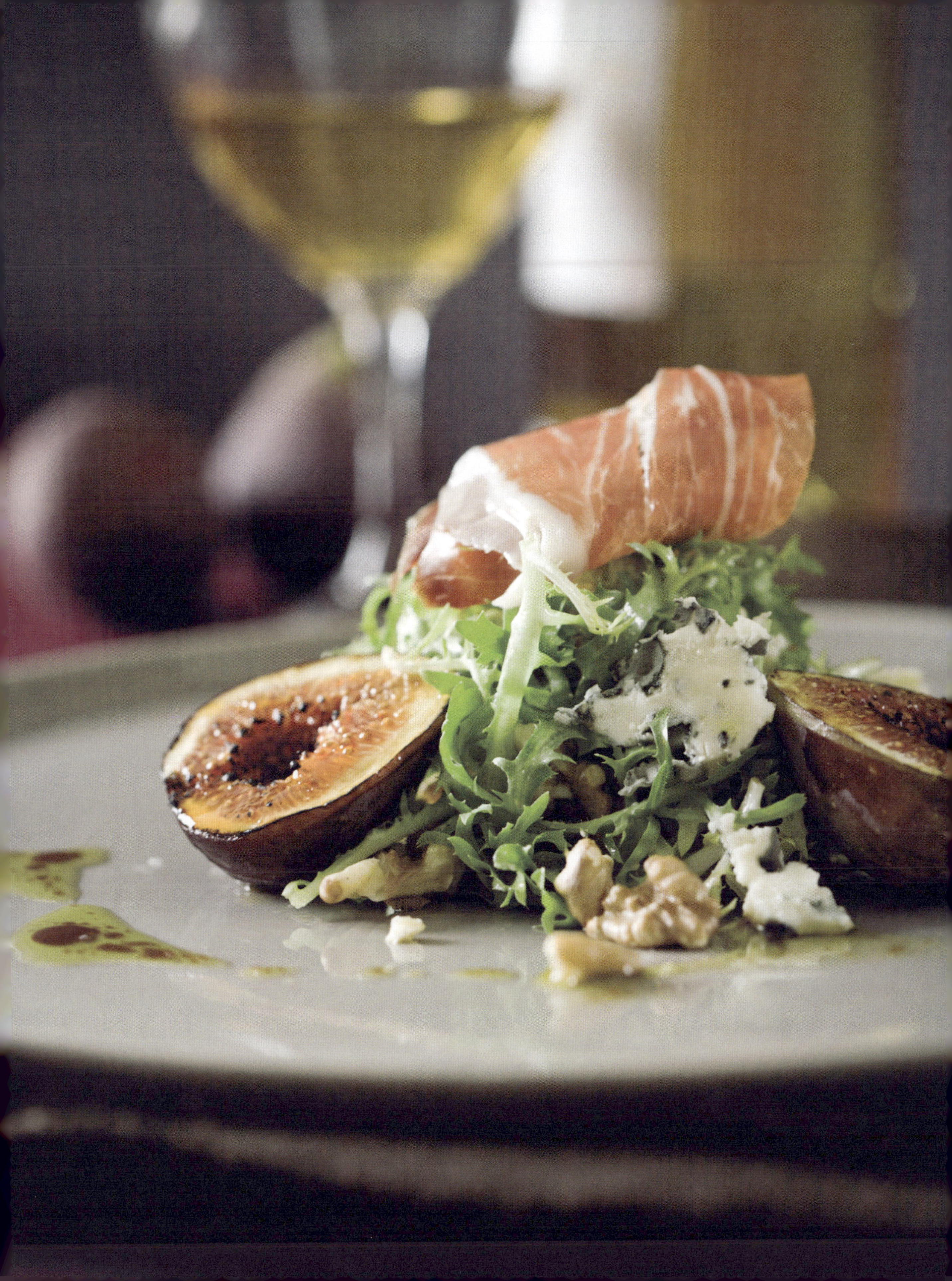

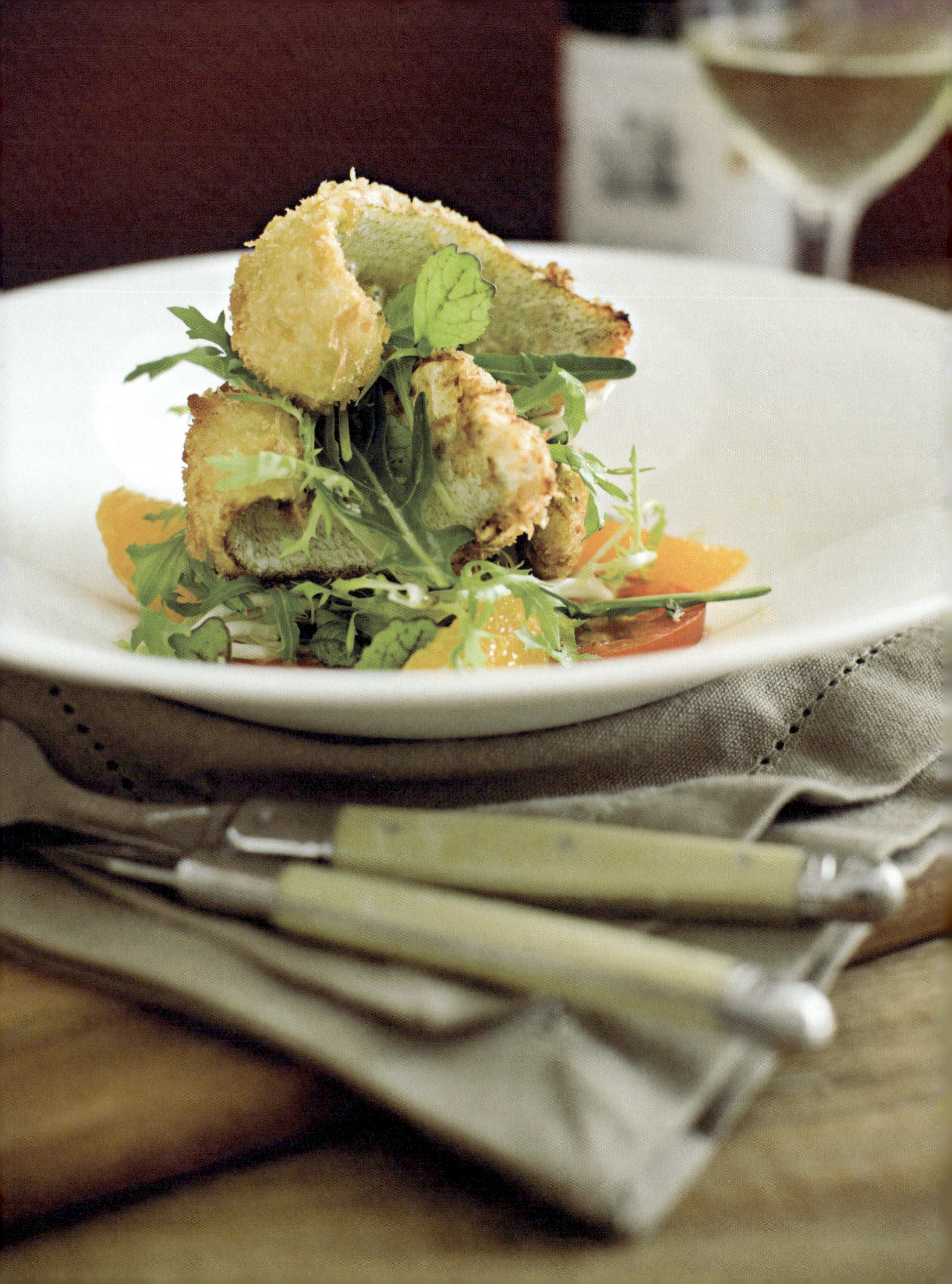

Half-Crumbed Whiting Fillets, Vine-Ripened Tomato & Citrus Salad

I like to leave the skin on the fillets so it becomes crisp when fried, crumbing only the flesh side, thus 'half crumbed'.

18 silver or yellow fin whiting fillets

250 ml oil, for pan-frying

Sea salt & freshly ground black pepper

Crumb

Plain flour

1 egg whisked together with 150 ml milk for egg wash

Japanese breadcrumbs

Salad

200 g mixture of wild rocket, frisée and chard lettuce

2 oranges, peeled and segmented

2 large vine-ripened tomatoes

Citrus vinaigrette (p. 268)

Mayonnaise (p. 281), thinned with a little lemon juice

SERVES 6

Dip both sides of the whiting fillets in flour. Dip the flesh side of the whiting fillets into the egg wash, then coat just that side with breadcrumbs.

Pan-fry the fillets in hot oil, crumbed side first, then turn over and increase heat to crisp the skin side. Remove, drain on paper towels and season.

Slice the tomato and lay onto the base of serving plates.

Toss the salad leaves together with the oranges and dressing, laying the salad together with the whiting fillets on top of the tomato slices.

Drizzle with the thinned mayonnaise and serve.

Robert Weil Trocken Dry Riesling
Rheingau, Germany

This is racy, taut and pure with medium alcohol (11.5 per cent) – we don't want too much volume here, it's all about delicacy and subtle balance, allowing the fish to integrate with the wine's gentle lemony/apple flavours.

starters

Trevor Price is a third-generation fisherman of Augusta's Blackwood River. His great-grandfather started netting the lower reaches for whiting, bream and mullet in 1944, and he is now the only remaining commercial fisherman operating on the river. In winter he often sees dolphins fifty kilometres or more upriver, chasing the mullet that have come in from the ocean.

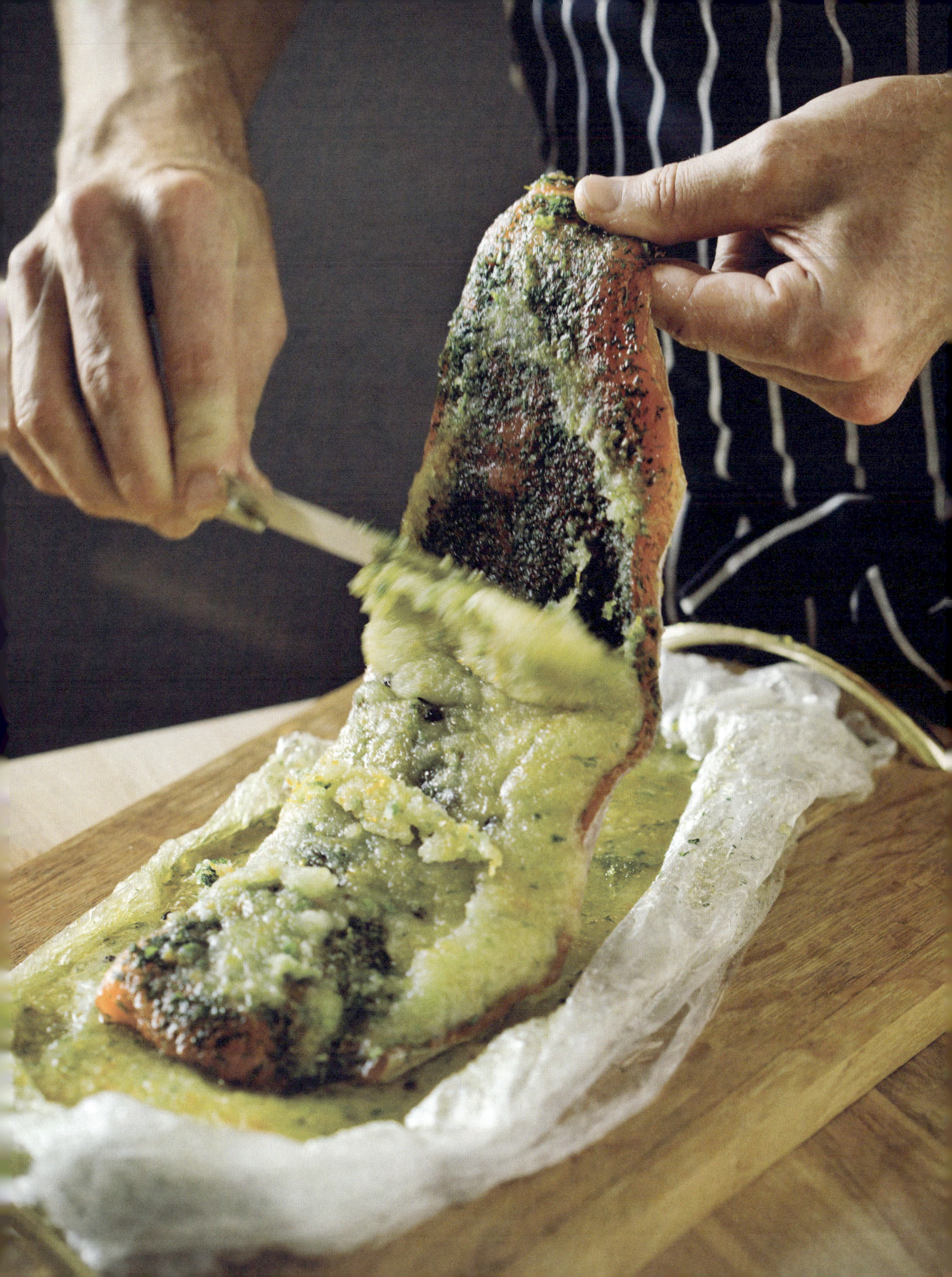

Domaine de Rimauresq 'R' Rosé
Provence, France

The cured ocean trout requires a slightly more robust rosé – one with texture and full mouth-feel. Rimauresq 'R', from Provence, has wonderful weight, and although it appears light there is concentration and seamless textural balance, which is just perfect with the flavours and underlying richness of the fish.

Cured Ocean Trout, Dill & Grain Mustard Dressing

When I worked at the Dorchester in London, iced cases of wild-caught salmon and ocean trout would arrive from Scotland. The menus called for the fish to be smoked, poached or cured. Curing the fillets with salt, sugar and herbs resulted in silky, sublime-tasting fish. When lifting the cured fillet, make sure you cradle the fillet with both hands to lift so the connective tissue between the flakes of flesh is not torn.

- Remove the pin-bones (if they haven't already been taken out), they run in a lateral line along the thickest part of the fillet. Use a pair of tweezers to grip the top of each bone, pulling it upwards and toward the head end of the fillet – the bone should release easily. Scale the fillet, wipe down the skin with kitchen paper and slash the skin ½ cm deep five times across the thickest part of the fillet.
- Mix the cure ingredients (except the Cognac) together in a large bowl.
- Lay several layers of cling film onto your bench top, place half of the cure in a line onto the cling film, lay the fillet, skin side down, then place the rest of the cure evenly over the top. Sprinkle the Cognac over the top. Wrap the cling film up and around the fillet so that is creates a tightly sealed bundle, then place it onto a deep tray. Place another tray on top of the fillet, and place a heavy weight (about 3 kg) on top. Place in the refrigerator for 24 hours.
- Once the fish is cured, unwrap it, scrape off the excess cure, then give the fillet a light rinse under cold water. Pat dry with kitchen paper and keep refrigerated until you wish to use – it will keep (wrapped) in the refrigerator for 4 to 5 days.
- Lay the fillet on a board, skin side down. Moisten a sharp thin-bladed knife (a filleting knife is good) with a little warm water, lay it flat on the flesh of the fish about three-quarters of the way down the fillet, and shave the flesh in thin slices; using a sawing motion down towards the tail. Repeat this slicing action, slowly moving the starting point of the slice back towards the thick end of the fillet.
- Lay the slices onto serving platters and drizzle with the dill and grain mustard dressing. Add a drizzle of thin mayonnaise, some capers and herbs to finish.

1 fillet of ocean trout (about 800 g), skin on

Cure
500 g salt
500 g caster sugar
Zest of 1 large orange
Zest of 2 lemons, finely grated
Zest of 1 lime, finely grated
1 cup fresh dill leaves
1 tablespoon white peppercorns
30 ml Cognac
(don't add this to the cure mix)

Dill and grain mustard dressing (p. 269)
Some mayonnaise (p. 281), thinned with a little water
Tiny capers
A few chopped chives
Some small herb leaves, parsley or rocket

SERVES 6-8

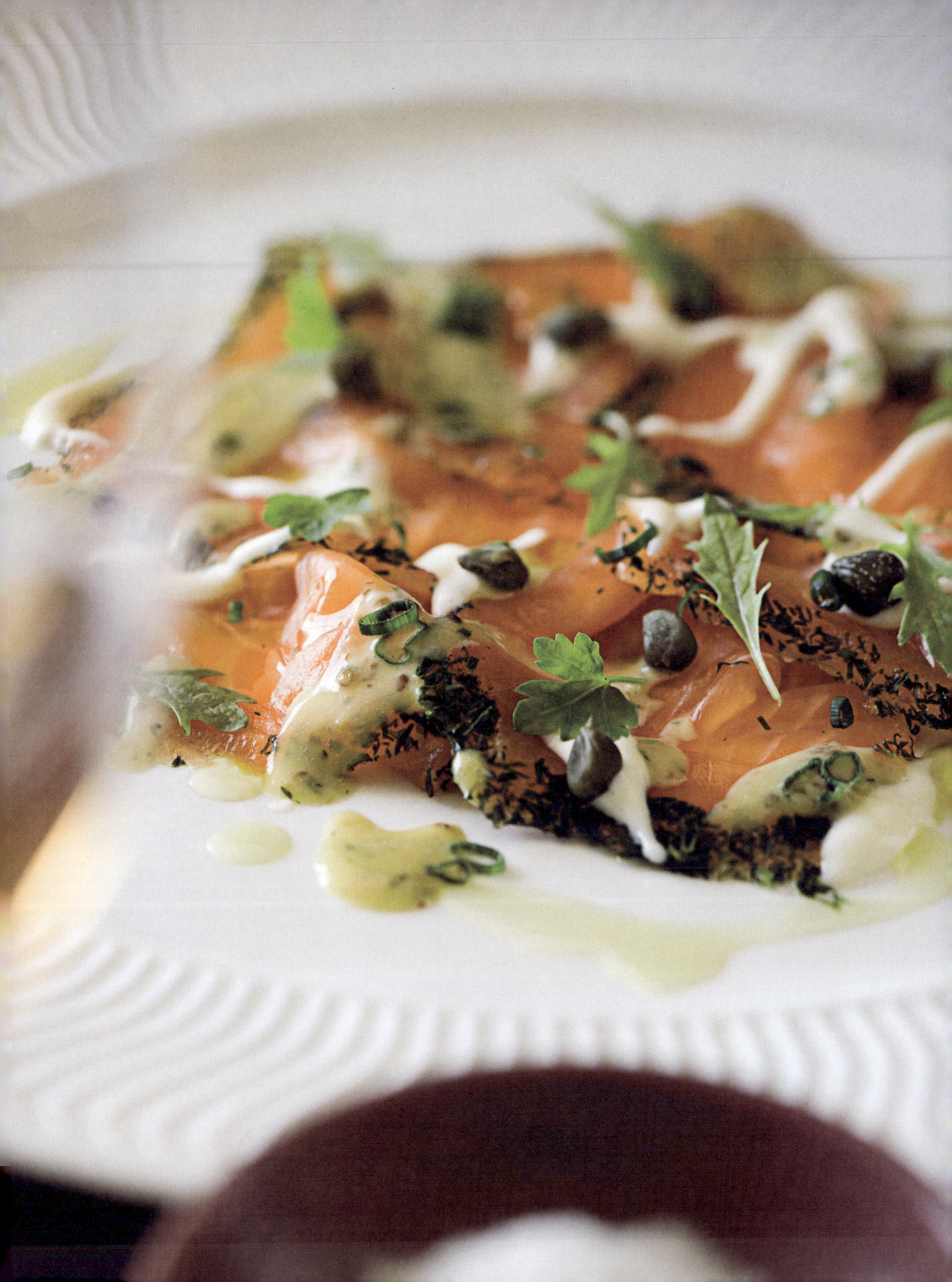

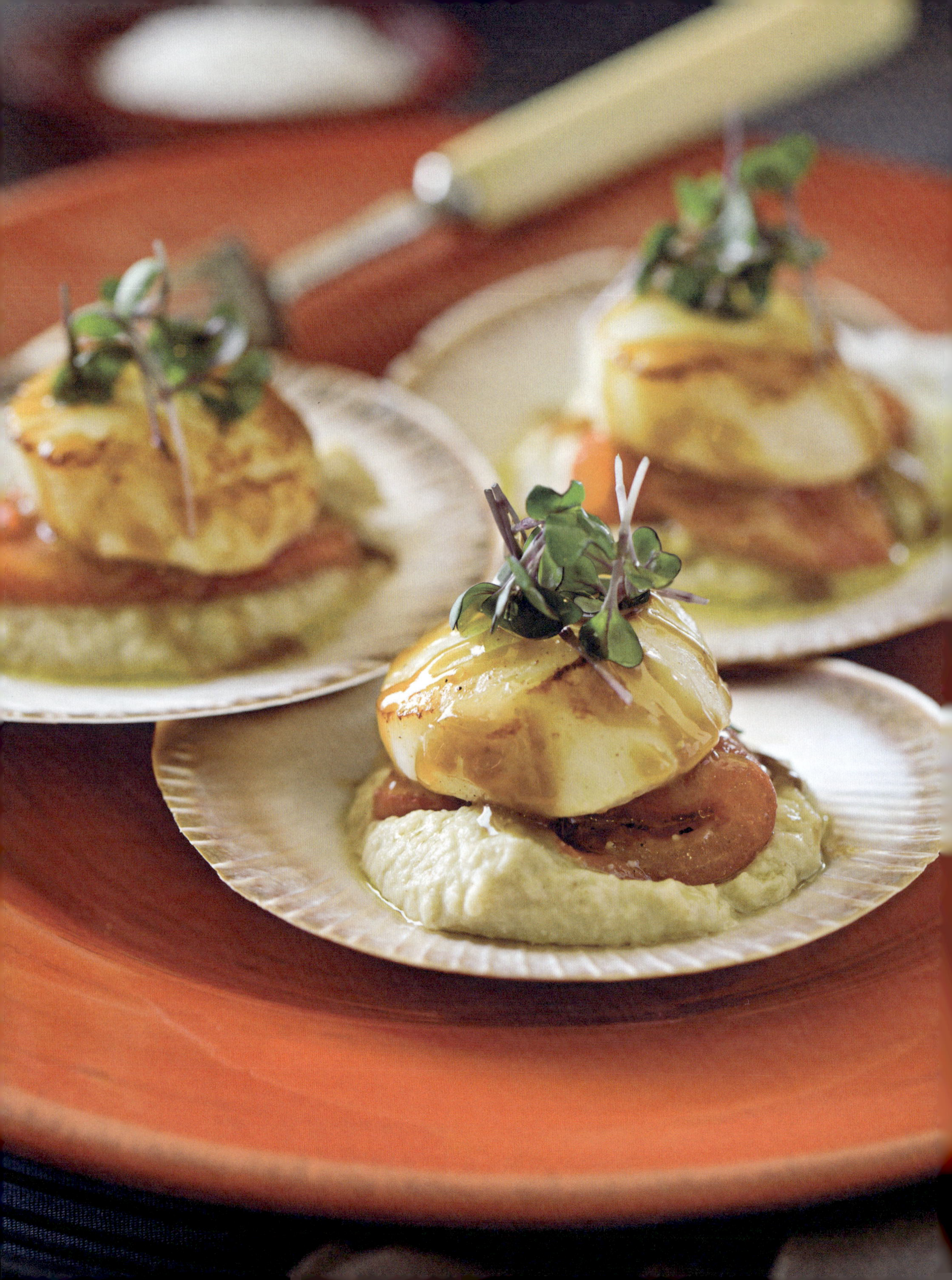

Seared Fremantle Scallops, Hummus, Confit Tomato & Pomegranate Dressing

West coast scallops have a fine, shallow shell filled with rich sweet flesh that matches beautifully with the smooth lemony hummus. Err on the side of undercooking the scallops as they are best eaten 'medium rare'.

Domaine Belle Croze Hermitage (Blanc)
Rhone, France

This dish needs some weight, but still requires the wine to have a sense of fruit restraint. The Blanc Crozes-Hermitage is sturdy and shapely and flatters the scallops, while coping with the lively flavours of the supporting condiments.

Hummus

100 g dried chickpeas
1 garlic clove, finely minced
1 teaspoon tahini
A pinch of salt
A pinch of cumin powder
1 tablespoon extra virgin olive oil
Juice of 1 lemon
Sea salt & freshly ground black pepper

12 large scallops on the half shell
Sea salt & freshly ground black pepper
Olive oil
Confit tomatoes (p. 267)
Pomegranate dressing (p. 268)
Chopped chives

SERVES 6

To make the hummus, soak the chickpeas in water overnight.

Rinse the chickpeas and place into a large saucepan, cover with plenty of water and simmer over medium heat until tender (do not add salt; it hardens the skin of the chickpeas, making them coarse and unpalatable).

Drain the chickpeas (keep a little of the cooking liquid), pick out any loose skins if you have the time and process with the rest of the hummus ingredients until smooth, adding a few tablespoons of cooking liquid to make a wet consistency.

Season and reserve at room temperature if using straight away, or reserve in the refrigerator for up to two days.

Remove the scallops from their shells by running a flexible knife between the flesh and the shell.

Wash the shells and pat them dry with kitchen paper.

To keep the shells in place, place two small dobs of hummus on 6 serving plates, then lay two scallop shells onto each plate, with the hummus keeping them in place.

Place a large dob of hummus onto each shell, then a piece of the confit tomato.

Heat a non-stick frying pan over high heat, season the scallops and brush with a little olive oil.

Place the scallops into the pan and sear until light brown on one side (about 30 seconds), then turn over and briefly cook on the other side.

Remove the scallops from the pan and place on top of the confit tomato. Drizzle with pomegranate dressing and sprinkle with chives, then serve immediately.

Bindi 'Quartz' Chardonnay
Gisborne, Victoria

A wonderful, modern Australian chardonnay that avoids being obvious or flashy, yet accentuates the pure delight of the prawns. It has everything – balance, elegance and acidity – which provides crisp freshness and soothes the richness of the sauce.

Warm Prawn Cocktail

This dish is inspired by memories of my apprentice days at the Sheraton Hotel in Perth. We served thousands of cocktails the old-fashioned way – lettuce, prawns and 'Marie rose' sauce. This revitalised prawn cocktail has freshly cooked prawns on buttered spinach topped with warm choron sauce (a tomato hollandaise). Great bistro food!

Prepare the sauce choron and reserve in a warm place.

Combine the court bouillon ingredients in a large saucepan with 1.5 litres water and bring to the boil.

Drop in the prawns, cook for only 1½ to 2 minutes. (They will keep on cooking after you take them out!) Use a slotted spoon to remove the prawns, drain and keep warm.

While the prawns are cooking, wilt spinach: melt the butter in a large saucepan, drop in the spinach and cook over high heat with the lid on for one minute, giving the saucepan an occasional flip to toss the contents.

Season and drain the spinach then lay into the bottom of your serving 'coupe' (large wine or cocktail glasses work well).

Arrange the prawns on top, zest some of the lemon over the prawns, pour over the wonderful sauce and serve immediately.

1 quantity choron sauce (p. 280)

36 large tiger prawns, peeled and de-veined

2 bunches English spinach, washed, picked and chopped

30 g butter

Sea salt & freshly ground black pepper

1 lemon

Court Bouillon

¼ cup each of chopped celery, leek, carrot and onion

1 bay leaf

4 white peppercorns

½ cup white wine

1 tablespoon salt

Sprig of thyme

SERVES 6

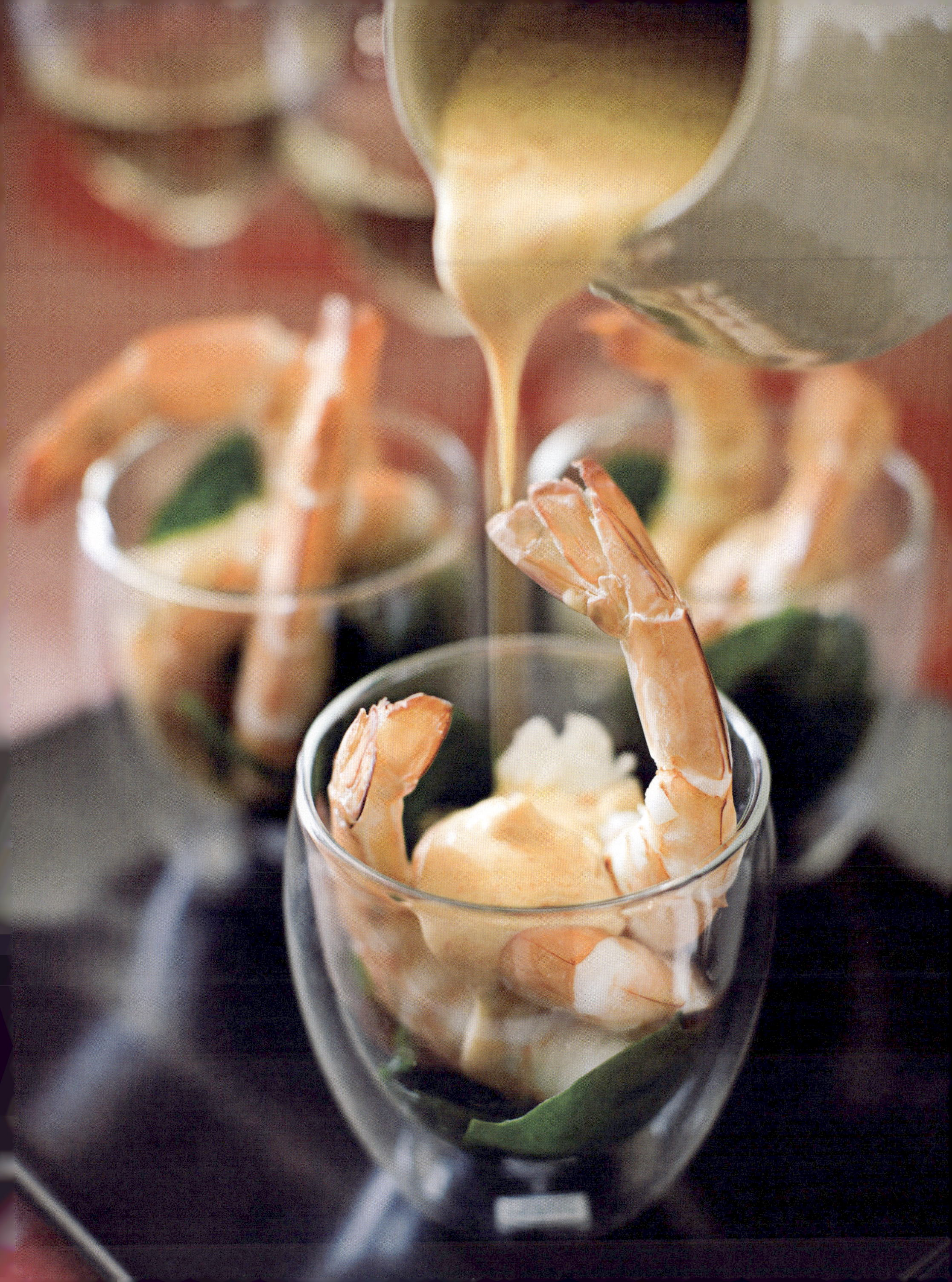

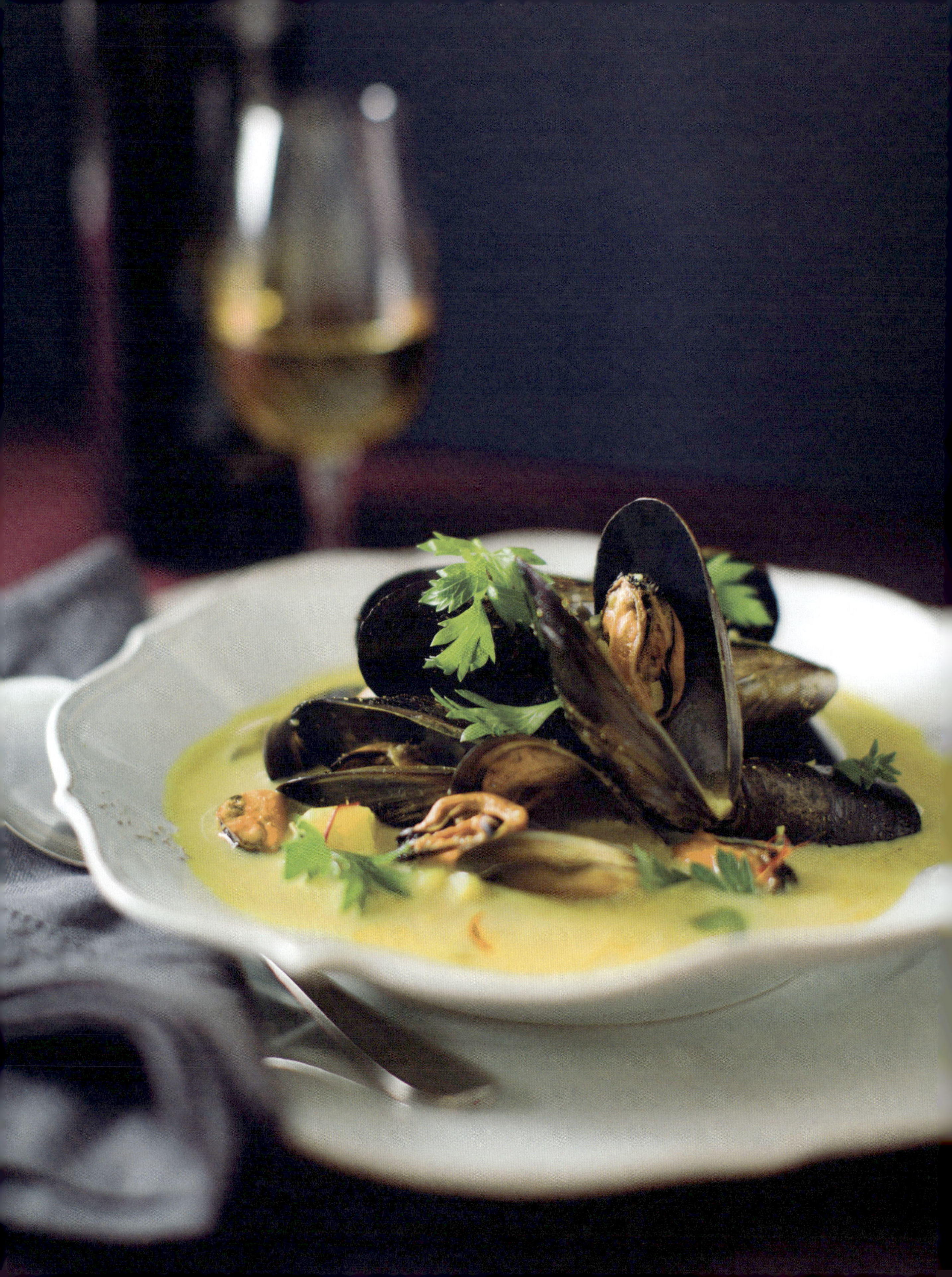

Mouclade

Mouclade is a classic dish from the Bay of Biscay region in southwestern France. Sea-fresh mussels are steamed with aromatic spices, leaving a tasty sauce to mop up with fresh bread. The flavours lend themselves best to a soup, so I've added stock and crème fraîche to extend the warmly spiced broth once the mussels are cooked.

90 g butter

3 large shallots

2 garlic cloves, crushed

A good pinch of saffron threads

375 ml dry white wine

A good sprig of thyme

1.5 kg fresh mussels, de-bearded and scrubbed

3 potatoes, peeled and diced

1 leek white, finely chopped

1.5 litres fish or white chicken stock (p. 275 or p. 277)

300 ml crème fraîche

2 heaped teaspoons curry powder, roasted until fragrant

Sea salt & freshly ground black pepper

2 tablespoons flat-leaf parsley, chopped

SERVES 6

Melt half of the butter in a large-lidded saucepan, add the shallots and garlic and sweat until soft.

Soak the saffron threads in the white wine.

Add the thyme and wine (with saffron) to the shallots, bring to a fast boil, add the mussels, fit the lid tightly and steam until the mussels open (approximately 1 minute).

Strain through a colander, reserving the liquid.

Remove the mussel flesh from the shells, reserving a few in their shells to garnish the soup.

Heat the remaining butter in saucepan, add the potatoes and leek and sweat slowly for 2 minutes.

Strain the mussel cooking liquid through a fine strainer, then add to the vegetables. Add the stock and bring to the boil. Simmer until the vegetables are tender.

Pour the mixture through a colander, reserving both the vegetables and cooking liquid. Return the cooking liquid to a clean saucepan. Purée the vegetables in a food processor, then return to the pan with the cooking liquid.

Add the crème fraîche and roasted curry powder, season and gently warm.

Sprinkle the reserved mussel flesh and whole mussels into the soup and serve immediately with lashings of chopped parsley.

Romate 'NPU' Amontillado
Jerez, Spain

The nutty tones, weight and texture of this wine, along with a suggestion of briny sharpness, provide the flavour combination and important mouth-feel to balance the potato and curry spice of this soup.

starters

Blue Manna Crab Bisque

Arlewood Chardonnay
Margaret River, Western Australia

This bisque recipe works with slender, cool-climate white wine. It is also important to balance the richness and sweetness of the crab with some strength of fruit. Arlewood craft stylish chardonnays that have grace, warmth and concentration; there is vibrancy and flavour that complements this seafood-based soup.

The French bistro chef never wastes a thing: trimmings of meat, fish and fowl together with herb stalks are used for stock; egg whites are saved for meringue; and even orange rinds gain a second life as part of the marinade for boeuf bourguignon. And it would be such a waste to throw away crustacean shells when they could help flavour a scrumptious soup.

If using whole crabs, flip up the triangular flap on the underside of the crab and pull off the carapace.

Remove the dead man's fingers, take the flesh from the claws (to finish the soup) and reserve in the refrigerator.

In a large saucepan sweat the onion, carrot and celery in olive oil over medium heat, add the herbs, paprika and crab shells, cooking for 5 to 6 minutes until the shells become orange in colour.

Crush the shells with a heavy wooden spoon then add the tomato purée and half of the Cognac.

Tilt the saucepan to ignite the Cognac. When the flame has died, add the white wine and cook until almost fully reduced (about 5 minutes), then add the stock.

Bring to the boil and simmer for 30 minutes until richly flavoured.

Season to taste, strain through a fine sieve into a clean saucepan and bring to the boil cooking over high heat, skimming regularly, until reduced by one-third (about 30 minutes).

Whisk 90 ml of the cream to soft peaks and reserve.

Add the remaining cream to the stock and simmer over medium heat for 5 minutes. Add the whipped cream and whisk over heat – the soup will become frothy as the air in the cream expands. Drop in the raw crab meat and stir through soup.

Pour a little Cognac into each serving bowl and ladle the soup over, then serve immediately.

1.5 kg blue manna crab shells and legs (or use whole crabs if you wish)
1 onion, chopped
1 carrot, chopped
2 celery sticks, chopped
2 tablespoons olive oil
A sprig of thyme
1 bay leaf
A sprig of flat-leaf parsley
1 heaped teaspoon sweet paprika
2 tablespoons tomato purée
90 ml Cognac
200 ml white wine
2 litres fish stock (p. 275) or water
100 ml cream plus 90 ml cream, lightly whipped
Sea salt & freshly ground black pepper

SERVES 6

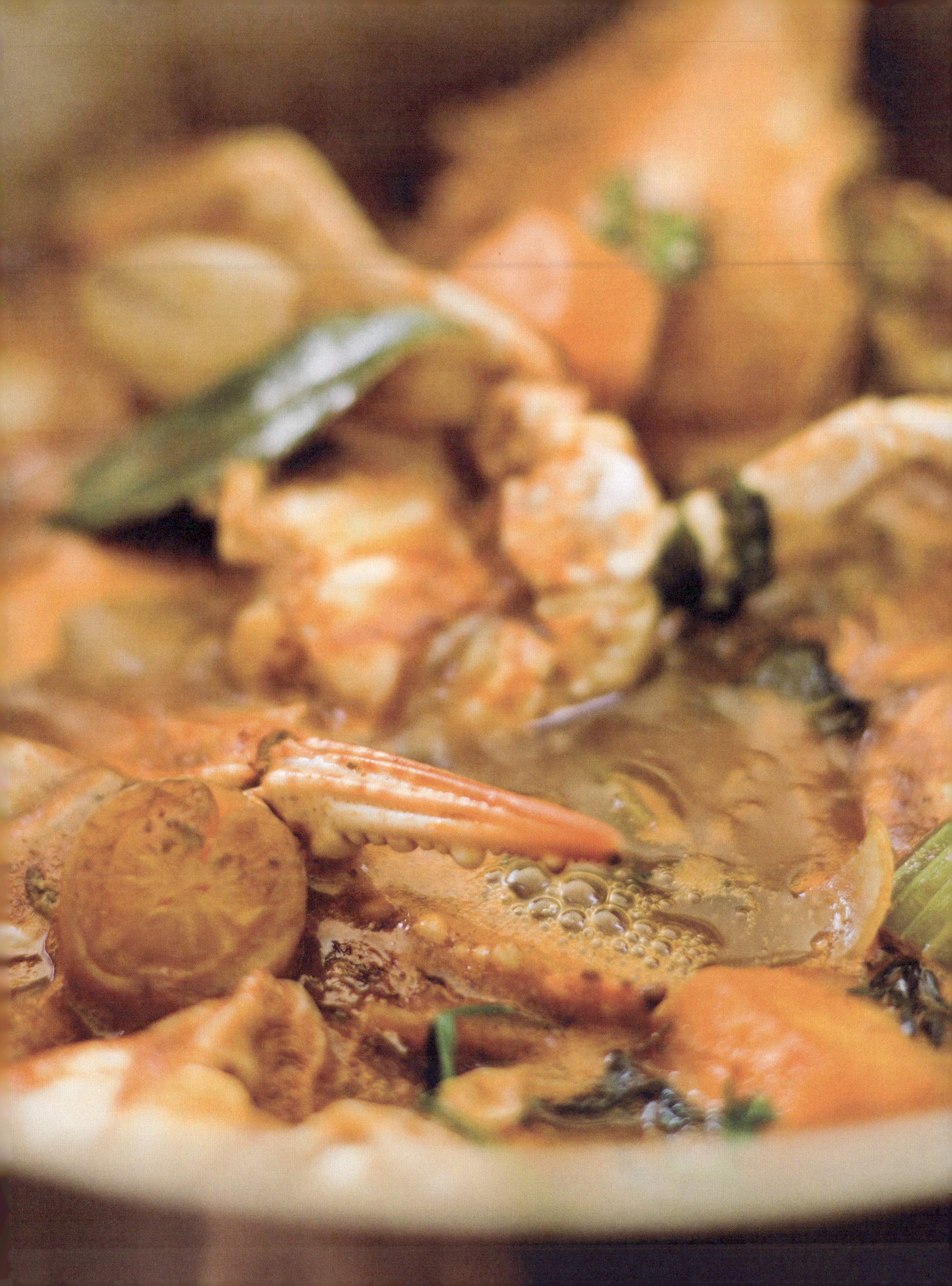

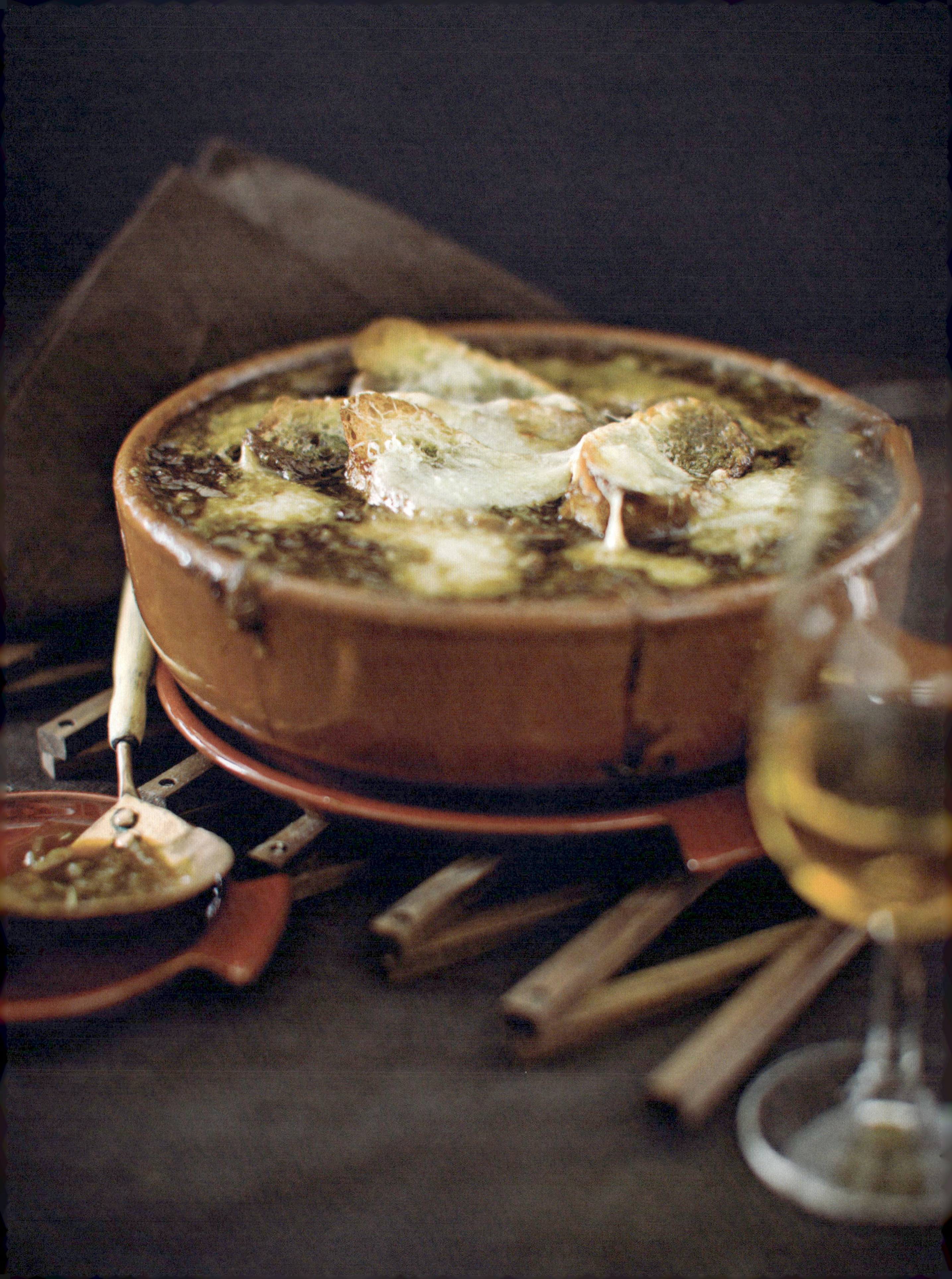

French Onion Soup Gratinée

At Margaret River our head chef Chris makes a fabulous French onion soup – the key is waiting patiently for the onions to caramelise.

Henriques and Henriques Madeira 'Sercial'
Madeira, Portugal

The dry, nutty and quite high-toned character of this wine has the power to harmonise the savoury, salty, cheese profile of this soup.

---

60 g unsalted butter

750 g brown onions, peeled and sliced

1 teaspoon sugar

3 garlic cloves

150 ml amontillado or manzanilla sherry (or dry white wine)

1.7 litres good beef or brown chicken stock (p. 274)

Salt & freshly ground black pepper

12 slices good-quality baguette

150 g Comté cheese (or gruyère), finely grated

SERVES 6

Melt the butter in a large heavy saucepan, add the sliced onion and sugar. Cook over medium heat for about 20 minutes to give the onion a golden brown colour.

Reduce the heat, add the garlic and cook for 45 minutes to an hour until the onion is meltingly soft.

Add the sherry, increase the heat and cook to reduce the liquid until it has almost vanished. Add the stock and bring to the boil, simmer for 5 minutes, season and keep warm.

Pour the soup into serving dishes, top with a slice or two of baguette and sprinkle generously with cheese. Place under grill to melt cheese, this may take 2 to 3 minutes. Serve piping hot.

Pieropan Soave 'Classico'
Veneto, Italy

This soave has some flavour and fullness, but still retains a minerality and austerity to pair with the sturdiness of potato and the light herbal character of leeks and chives.

Leek & Potato Soup

The key to this simple soup is to cook the leeks and potatoes quickly so they don't discolour.

Preheat the oven to 210°C.

Cut the sourdough bread into 1 cm cubes, place onto a baking tray, drizzle with extra virgin olive and bake until golden, reserve.

Melt the butter in a large heavy saucepan, add the leeks and cook over low heat to sweat for 5 minutes.

Add the potato cubes and stock, bring to a rapid boil and cook until the potatoes are tender.

Add the milk and bring back to the boil, then season.

Remove from the heat and purée soup with a stick blender or in a food processor

Return the soup to the saucepan, add the cream and bring to a gentle simmer.

Serve in bowls with lashings of croutons and chopped chives.

Croutons

A few slices of sourdough bread

Extra virgin olive oil

3 tablespoons butter

3 leeks, white part only, washed and chopped into rough 1 cm cubes

1 kg desirée or royal blue potatoes, peeled and cut into rough 2 cm cubes

1.5 litres white chicken stock (p. 277)

375 ml milk

Sea salt & freshly ground white pepper

200 ml cream

2 tablespoons chopped chives

SERVES 6

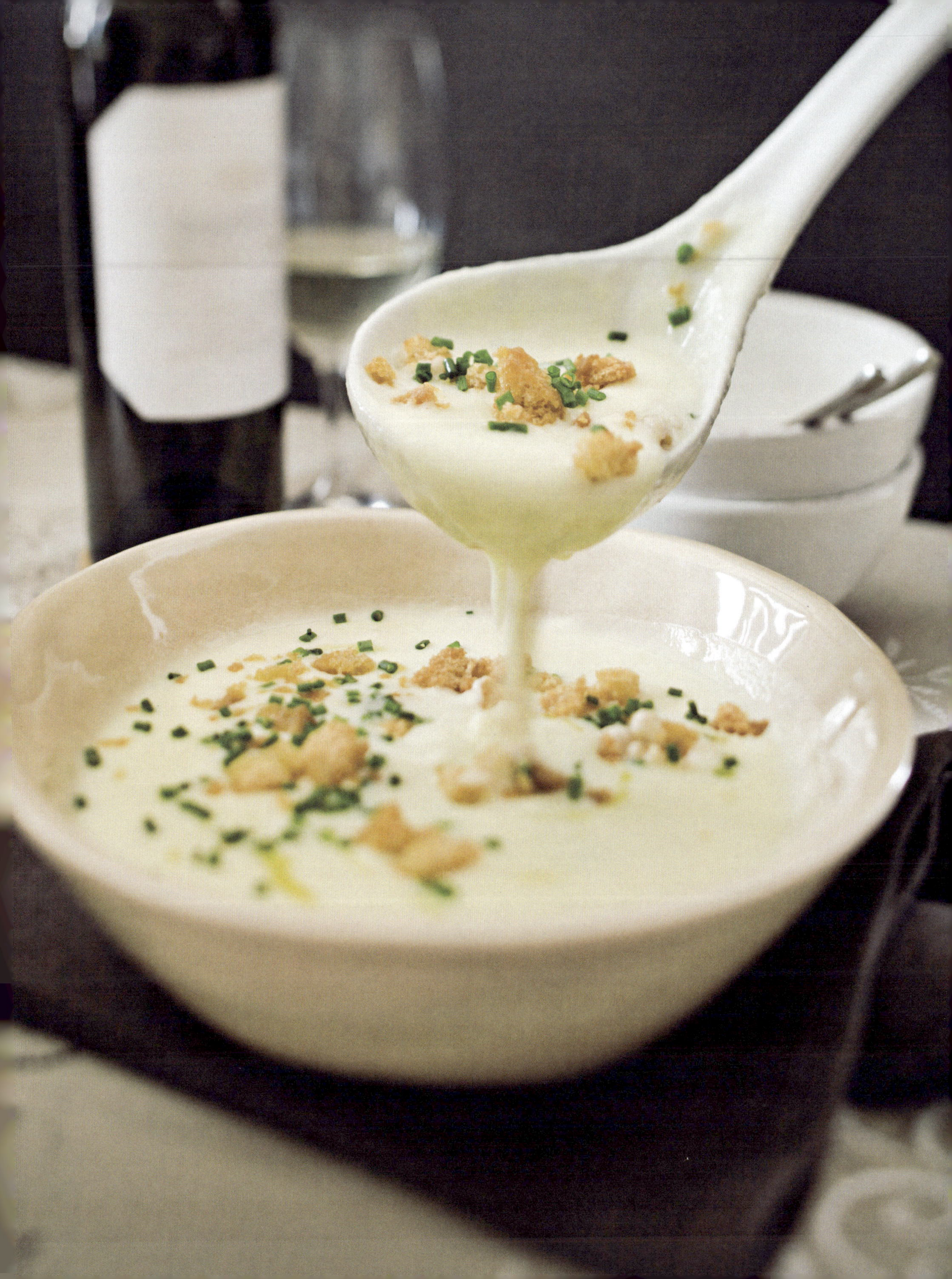

Gnocchi Tips

The keys to light gnocchi are light handling and dry potatoes. We peel and boil the potatoes (rather than roast them in the oven), but we do 'dry' them in the oven for a few minutes after draining.

* Look for a floury potato to make gnocchi – royal blue is the best option as the creamy yellow flesh becomes light and fluffy when mashed.
* Carefully 'dry' the potatoes after cooking. Pop them on a tray in the oven for a couple of minutes to get rid of excess moisture. Not too long though or the potatoes will end up as lumpy mash.
* Never use a machine to mash or mix. Use your hands, with a light touch, or a wooden spoon if you must. Too much mixing will result in the potatoes turning into sticky 'glue'.
* Some recipes suggest that the gnocchi is cooked once it has floated to the surface of the pot; this is not the case for the recipe opposite.

Gnocchi

We make gnocchi every day at Must. This recipe originated years ago when I ran the kitchens at 44 King Street, Perth. King Street's brilliantly talented head chef Bernard McCarthy made a similar recipe; with his soft hands and deft touch the result was sensational. It has remained in my repertoire ever since.

....

600 g (peeled weight) royal blue potatoes

125 to 150 g bread-making flour, plus extra flour for rolling gnocchi

A pinch of freshly grated nutmeg

1 teaspoon table salt

A pinch of ground white pepper

1 egg, lightly whisked

SERVES 6

Preheat the oven to 150°C.

Chop the potatoes into 4 cm chunks and place into a large saucepan of salted water, then simmer until tender (a knife will pierce them easily).

Pour the potatoes into a strainer placed over the sink, drain for 4 to 5 minutes. Place onto an oven tray into the oven for another 4 to 5 minutes to remove any excess moisture from the potatoes.

Crush the potatoes through a potato ricer, or push through a coarse sieve, into a large bowl. Add 125 g flour, the nutmeg, salt, pepper and egg.

Mix lightly for no more than 30 seconds with your hands or a wooden spoon to a smooth but soft dough; if it is very sticky add the extra flour and gently work through.

Flour a large board or the back of an oven tray, and gently roll the mix in batches into large sausages, about 2 cm thick. Cut into 2 cm lengths with a knife, dipping the knife in flour to prevent it sticking.

Fill your largest saucepan with salted water and bring it to the boil. Carefully roll the gnocchi off the tray or board and into the pot (you may need to cook the gnocchi in several batches) and cook at a very slow 'shimmering' simmer for 7 to 8 minutes.

The gnocchi will float to the top of the pot as they are cooking; firmly tapping the side of the pot with a wooden spoon will help release any gnocchi that may be stuck to the bottom.

Leave the gnocchi to cook in the shimmering water for 5 minutes after it bobs up to the surface.

Remove gnocchi from the pot with a slotted spoon and drop into a sinkful of cold (preferably iced) water until cool, then drain and refrigerate until required.

The gnocchi can be used directly from the pot, but we pan-fry ours at Must.

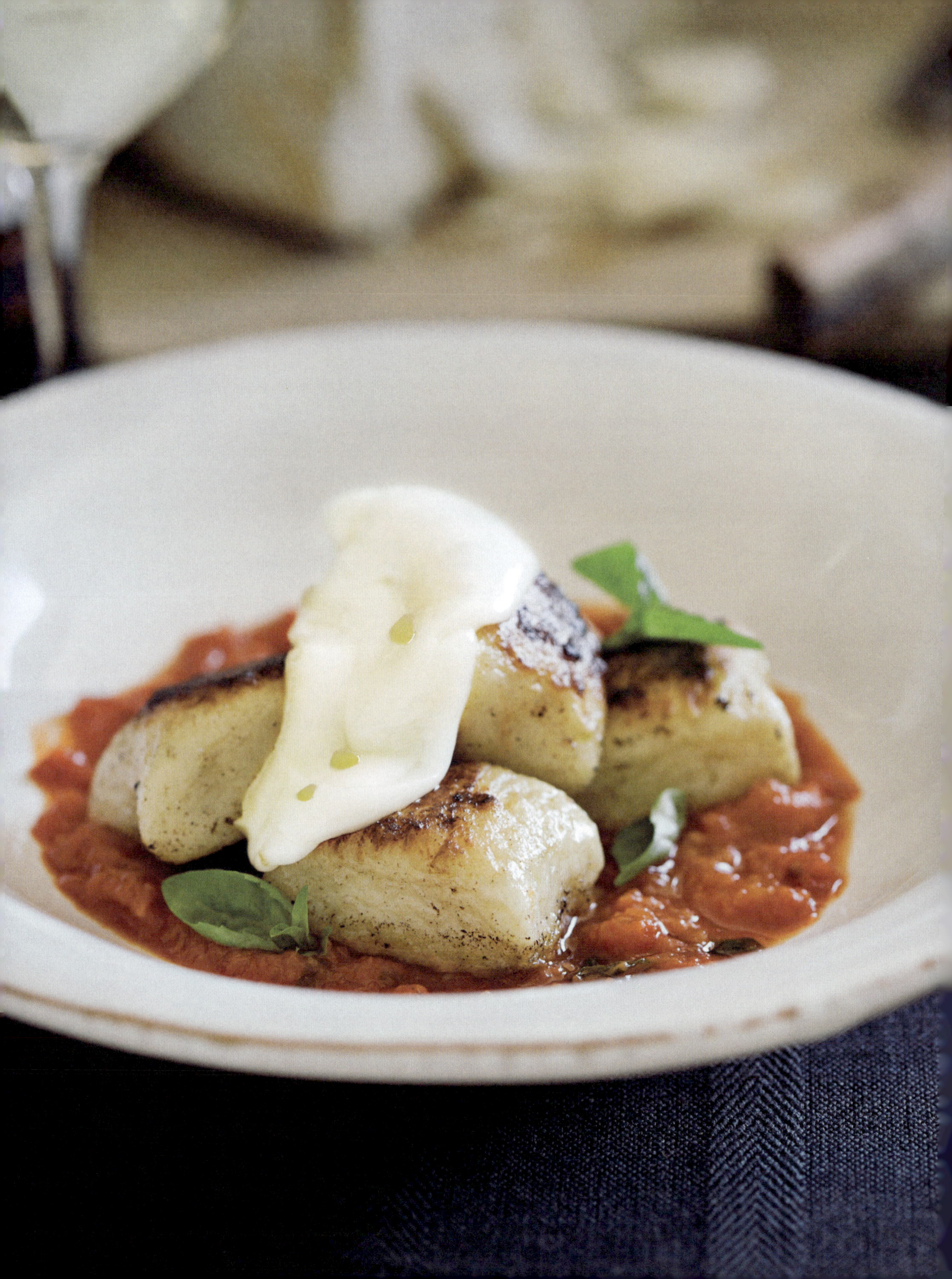

Pan-Fried Gnocchi, Tomme de Chèvre & Organic Tomato Coulis

I make this tomato coulis with organic sugo from Sona and Harry Toutikian in Gidgegannup. Bottled Italian sugo will work well as a substitute, but check that the label only lists tomatoes and salt to be certain of a pure flavour.

Craggy Range 'Te Muna' Sauvignon Blanc
Martinborough, New Zealand

This wine is punchy and vibrant and has the fullness to balance the flavour of the gnocchi. It also has the typical herbal fruit profile that goes 'hand-in-hand' with the chèvre.

Tomato Coulis

1 litre organic tomato sugo
4 tablespoons extra virgin olive oil
1 medium-sized onion, chopped finely
2 cloves garlic, finely sliced
Sea salt
½ teaspoon caster sugar (optional)

A handful of fresh basil leaves
60 g unsalted butter
1 quantity potato gnocchi (p. 77)
180 g Tomme de Chèvre cheese
Extra virgin olive oil

SERVES 6

To make the tomato coulis, place the onion, garlic and olive oil in a deep saucepan and sweat over medium heat.

Add the tomato sugo, reduce to a simmer and cook for at least 40 minutes. You may need to add some water towards the end of the cooking time to prevent the sauce from becoming too thick.

Add the basil leaves and cook a further 5 minutes.

Add salt to taste; if the sauce tastes a little acidic, add the caster sugar.

Preheat oven on high grill setting.

Stir ½ of the basil leaves through the tomato coulis and reserve, keeping warm.

Heat a large non-stick pan over medium-high heat, add butter and once it has started to turn brown add the gnocchi to the pan.

Leave the gnocchi to colour on one side for about a minute, then toss the pan to flip them over, using tongs to turn over the stragglers.

Cook another minute and a half, turning the gnocchi to give a golden brown crust.

Place a spoonful of the coulis into each of 6 serving dishes, top with a pile of gnocchi and a generous slice of the cheese.

Place under the grill to gently melt the cheese, finish with a sprinkle of basil leaves and extra virgin olive oil before serving.

starters

Stag's Leap Cabernet Sauvignon *Napa Valley California, USA*

The meatiness of this gnocchi and overall weight requires the strength of a full-bodied red wine. The blackfruit, tannins and slight herbal austerity of this cabernet will complement the Wagyu and also work with the leafy parsley.

Pan-Fried Gnocchi, Slow-Braised Wagyu Beef Shin & Parsley Purée

If Wagyu shin is difficult to find, beef cheek or brisket makes a great substitute.

To make the purée, place all of the ingredients into the bowl of a food processor and blitz until finely ground, reserve in the refrigerator.

Place all of the marinade ingredients together in a large bowl with the Wagyu shin. Cover and allow to marinate in the refrigerator for 24 hours.

Strain the beef and vegetables from the marinade and reserve the liquid. Separate the meat from the vegetable/herb ingredients.

Pat the beef dry, and sauté in oil in a large deep saucepan until well browned on all sides. Reserve.

Sauté the vegetables in the saucepan (again in some more oil) until they are lightly browned. Add the marinade to the saucepan, bring to the boil over high heat, simmer for 1 minute then add the stock and bring to the boil.

Add the beef and reduce the heat to barely a simmer. Cook the beef a further 3 hours until it is tender and beginning to fall apart. The cooking liquor should have reduced by about half by this stage.

Strain off the cooking liquid, reserving the beef, stripping it into smaller chunks and discarding the vegetables. Pour the cooking liquid back into the pot and bring back to simmer. Reduce to a glossy thick consistency (about one-fifth of the original quantity) and reserve.

Heat a large non-stick pan over medium-high heat, add the butter and, once it has started to turn brown, add the gnocchi. Leave the gnocchi to colour on one side for about 1 minute, then toss the pan to flip them over, using tongs to turn over the stragglers. Cook for another $1\frac{1}{2}$ minutes, turning the gnocchi to give a golden brown crust. Add the beef, caramelised onion and cooking liquor. Increase to high heat for another minute to combine ingredients.

Check for seasoning, adding a little more salt and freshly ground black pepper if required, then serve into warm pasta bowls. Top with parsley purée, scatter with parsley leaves and serve.

One quantity potato gnocchi (p. 77)
60 g unsalted butter
$\frac{1}{2}$ cup caramelised onion (p. 266)

1.2 kg Wagyu beef shin (off the bone) cut into 5 cm dice
Olive oil
2 litres beef stock (p. 274)

Marinade
1 bottle red wine
1 large carrot, chopped
2 celery sticks, chopped
1 onion, chopped
$\frac{1}{2}$ leek, chopped
2 garlic cloves, chopped
1 tablespoon fresh thyme
1 fresh bay leaf
A few parsley stalks
1 cinnamon stick
Rind of $\frac{1}{4}$ orange
4 black peppercorns

Parsley Purée
1 cup (firmly packed) flat-leaf parsley leaves, picked and washed
100 ml extra virgin olive oil
Juice of 1 lemon
Sea salt & freshly cracked black pepper

Flat-leaf parsley leaves, for garnish

SERVES 6

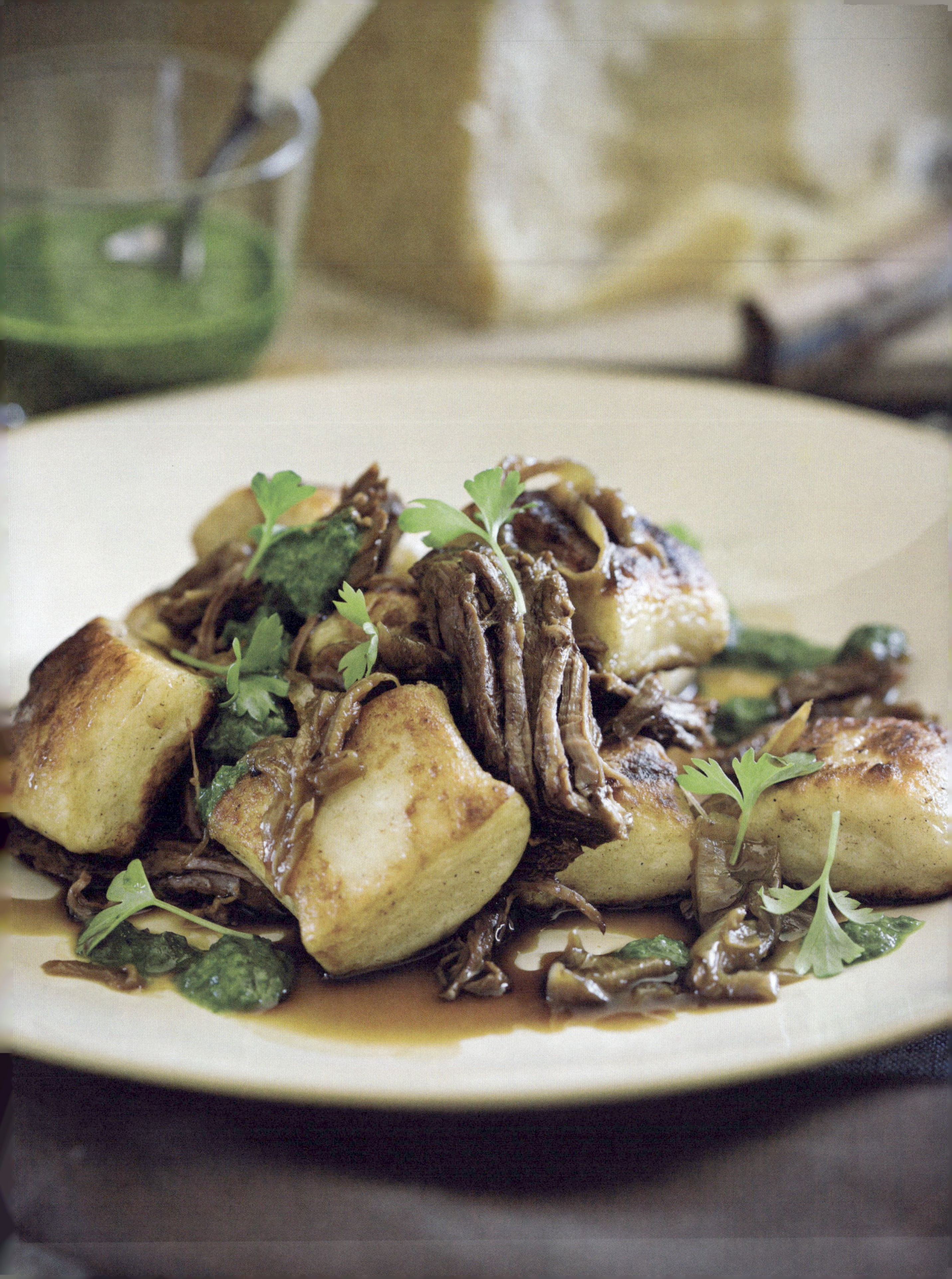

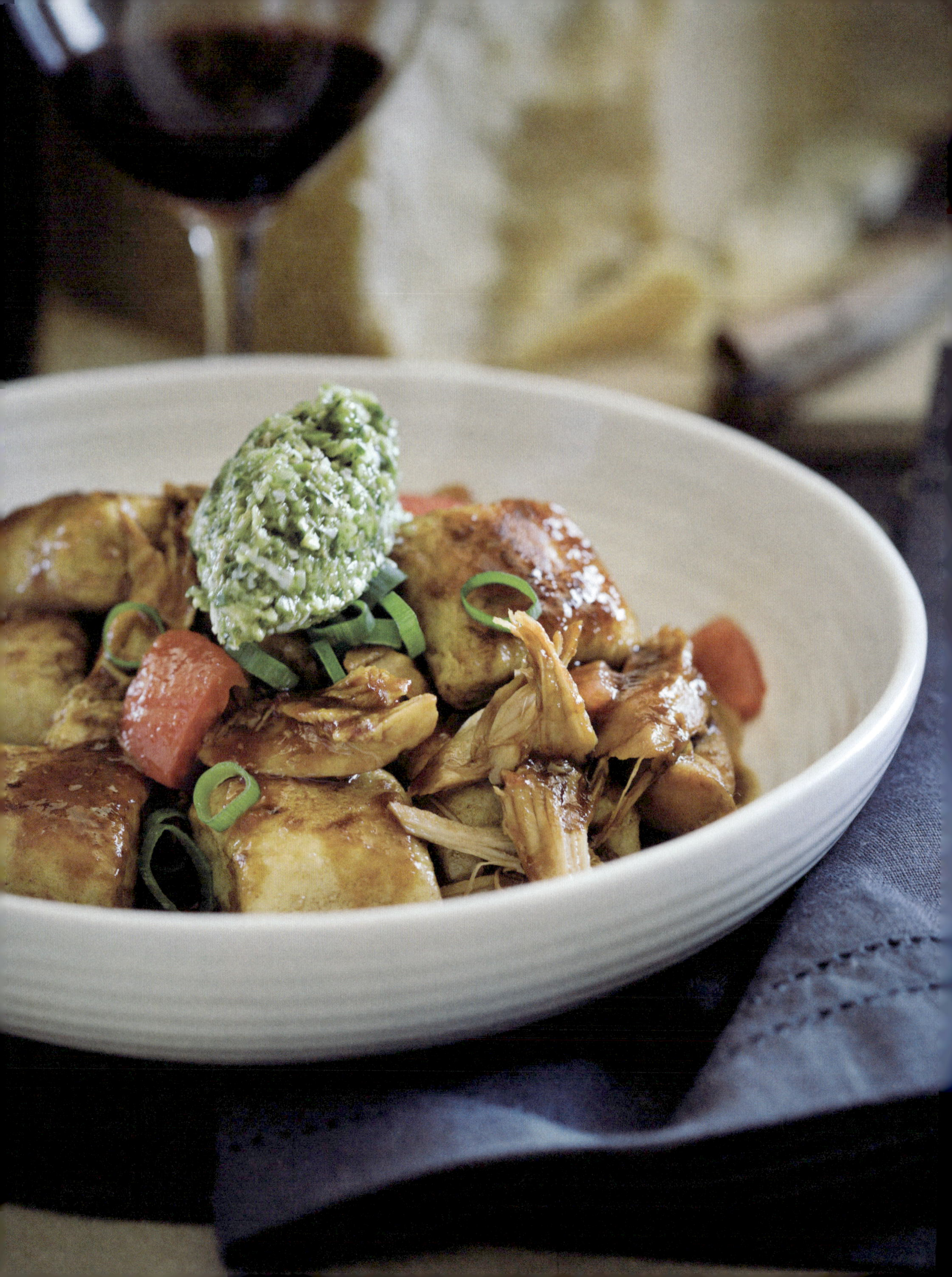

Pan-Fried Gnocchi, Rabbit Ragoût & Green Olive Tapenade

Choose a large white rabbit for the ragoût. If you ask your butcher to joint the rabbit, he will separate the legs and slice through the saddle in seconds!

Feudi di San Gregorio 'Rubrato' Aglianico
Campania, Italy

The rabbit and tapenade almost demand red Italian wine. Rabbit flesh and plump gnocchi suit this powerful Aglianico, which has generosity and warmth, with layers of rich flavours, typical Italian acidity and grainy finishing tannins – perfect with this rustic dish.

Ragoût

1 whole white rabbit (about 1 to 1.2 kg)
90 ml extra virgin olive oil
1 onion, chopped
2 garlic cloves, minced
1 celery stick, chopped
2 large carrots, chopped
A few sprigs of thyme
1 small sprig of rosemary
1 bay leaf
8 juniper berries
6 white peppercorns
400 ml white wine
400 g ripe tomatoes, chopped
Juice of 1 lemon
2 litres brown chicken stock (p. 274)
Sea salt

60 g unsalted butter
1 quantity gnocchi (p. 77)
4 large Roma tomatoes, de-seeded and chopped into 1 cm dice
Sea salt & freshly ground black pepper
3 spring onions, finely sliced

Green olive tapenade (p. 271)

SERVES 6

To make the ragoût, cut the rabbit into joints.

Heat the olive oil over medium heat in a large heavy saucepan, drop in pieces of rabbit and brown lightly. Remove.

Drop in the onion, garlic, celery, carrot, herbs, juniper and peppercorns, sauté for 5 minutes then add the wine and increase the heat, cooking for 5 minutes.

Add the tomato and cook for a further 5 minutes. Add the chicken stock and bring to a simmer.

Add the rabbit, reduce the heat to a slow simmer and cook for approximately 1 hour until the rabbit begins to fall off the bone.

Strain the cooking liquid, separating the rabbit from the vegetables. Pick out the juniper, peppercorns and herbs and place the vegetables in the bowl of a food processor; blitz to a fine purée.

Pick the rabbit flesh from the bone, being careful to separate all of the fine rib bones out from the flesh; reserve.

Return the strained cooking liquor to the stove to reduce over high heat for approximately 20 minutes so that about 800 ml of stock remains. Return the vegetable purée to the reduced stock, season with salt and reserve.

Heat a large non-stick pan over medium-high heat, add the butter and, once it has started to turn brown, add the gnocchi to the pan.

Leave the gnocchi to colour on one side for about 1 minute, then toss the pan to flip them over, using tongs to turn over the stragglers.

Cook for another $1\frac{1}{2}$ minutes, turning the gnocchi to give a golden brown crust.

Add the cooked rabbit, the tomato dice and 1 cup of the reduced cooking liquid, tossing the gnocchi over high heat for another minute to combine the ingredients.

Check for seasoning, adding a little more salt and freshly ground black pepper if required, then serve into warm pasta bowls. Top with green olive tapenade and spring onion and serve.

Pisa Range Pinot Gris
Central Otago, New Zealand

Blue manna crab is best served with a bright white wine, but the chilli, tomato and depth of flavour require the fullness of this pinot gris, which has a suggestion of sweetness and underlying acidity that cuts through the richness and finishes clean. If you increase the chilli, I suggest a pinot noir, which will help neutralise the heat.

Angel Hair Pasta, Blue Manna Crab, Chilli, Tomato, Basil & Extra Virgin Olive Oil

This dish combines the glorious sweetness of our local blue manna crab with organic olive oil from trees grown south of Margaret River, freshly picked herbs and ripe chilli. A perfect regional dish from the west coast of Australia, this is one of the most popular dishes on the menu at Must.

....

Place a large saucepan of salted water on to boil.

Pour the olive oil into a large heavy frying pan. Add the garlic, chilli and tomato, then cook over medium heat for 1 minute.

Increase the heat, add the cream and bring to the boil, then add the crab meat, tossing the meat so that it cooks evenly. Cook for a further 1 minute.

Drop the angel hair pasta into boiling water, cook until al dente, strain and place into pan with crab. Sprinkle in the basil, toss and season.

Serve immediately in deep pasta plates.

90 ml extra virgin olive oil (I use 34 Degrees South)

3 garlic cloves, finely minced

2 red chillies, de-seeded and finely diced

250 g tomato, finely diced

150 ml pouring cream

350 g raw blue manna crab meat

420–450 g fresh angel hair pasta (p. 270)

3 tablespoons freshly picked basil, finely shredded

Sea salt & freshly ground black pepper

SERVES 6

starters

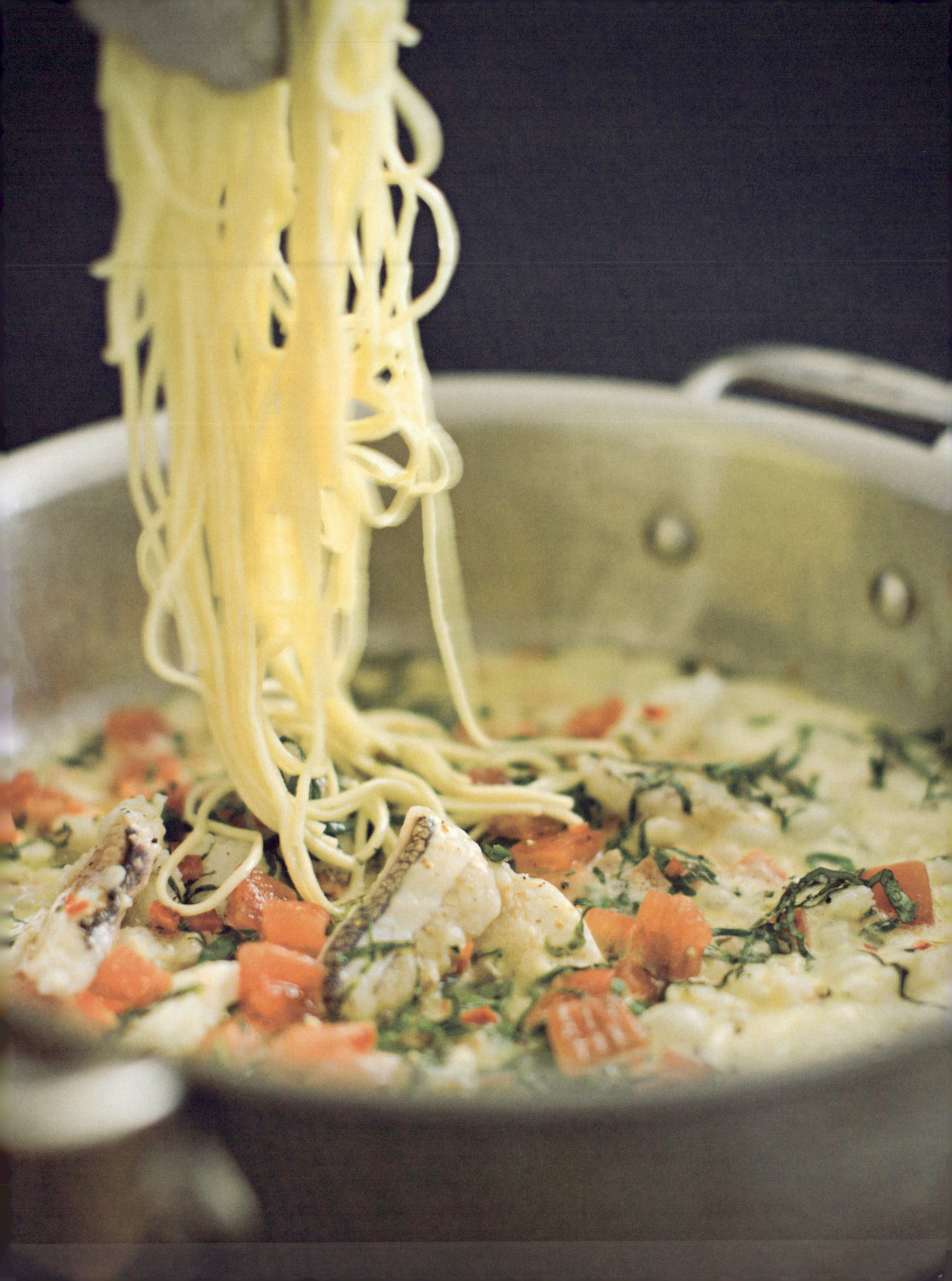

Lillian
Marsanne
Rousanne
Pemberton,
Western
Australia

This impressive interpretation of a Rhône white has volume and layers of unfolding flavours. It has just the right balance to work with the sweetness of pumpkin and meld with the dairy texture of goat's cheese and butter.

Pumpkin Ravioli, Ringwould Blanc Goat's Cheese, Red Capsicum Coulis & Sage Butter

For 25 years Gabrielle Kervella made the finest goat's cheese in the country from her herd at Gidgegannup. When Gabrielle and partner Alan Cockman decided to sell the farm, John Saunders and his wife Toni Louise bought the animals, dairy and cheese-making rooms and relocated everything to Redmond, near Albany. Today they produce delightful cheese under their Ringwould brand.

....

Start the filling a day before you wish to serve the ravioli.

Preheat the oven to 200°C.

Place the whole pumpkin onto a roasting tray and bake for approximately 1 hour, until the pumpkin is soft when prodded with the handle of a spoon. Remove and allow to cool.

Line a large strainer with muslin or a clean tea towel and sit over a bowl to catch the drained juice.

Cut the pumpkin in half lengthways, scoop out the seeds then carefully scoop the flesh away from the skin with a spoon, placing it into the lined strainer.

Once all the flesh has been scooped out, gather the top of the cloth and tie it firmly with kitchen twine. Place a heavy weight on top and allow to drain in the fridge for about 24 hours.

To make the filling, simply turn the drained pumpkin into a large bowl, add the remaining ingredients and mix well, ensuring the filling is well seasoned.

To make the ravioli, cut the pasta into 10 cm circles and lay onto a lightly floured bench. (Only cut 3 or 4 at a time so they do not dry out.)

Beat the egg with a little cold water and brush onto the surface of each pasta circle.

Place a heaped teaspoon of the filling in the centre of each circle, then fold over, tightly pressing at the edge to ensure a firm seal, checking that there are no air bubbles left inside the ravioli, or air will expand when heated and explode the delicate parcel.

Repeat until all of the pasta is used, counting on about 3 ravioli per person as a starter.

Carefully store the ravioli between sheets of kitchen paper, so that they do not touch, in the refrigerator until required.

Filling

1 small butternut pumpkin (about 1 kg)

30 g almond meal

30 g breadcrumbs

100 g parmesan cheese, grated

A good pinch of sea salt & freshly ground black pepper

A pinch freshly grated nutmeg

Egg pasta dough (p. 270)

An extra egg for sealing the ravioli

Red Capsicum Coulis

2 large red capsicums,
de-seeded and roughly diced

1 small brown onion,
peeled and roughly diced

2 garlic cloves, peeled and chopped

60 ml extra virgin olive oil

4 ripe Roma tomatoes, roughly diced

A sprig of fresh thyme

Sea salt & ground white pepper

1 tablespoon tomato paste

500 ml vegetable stock

18 freshly picked sage leaves

2 tablespoons unsalted butter

150 g Ringwould blanc goat's cheese

SERVES 6

To make the coulis, place all the ingredients except the tomato paste and vegetable stock in a large saucepan over low heat and sweat for 5 minutes until the ingredients begin to soften.

Add the tomato paste and stock, increase the heat to medium-high and cook for a further 10 to 15 minutes until the onion and capsicum are tender.

Pour the entire contents of the saucepan into a food processor and blitz to a fine purée. Pass the purée through a fine sieve; keep warm.

Bring a large pot of salted water to the boil, drop in the ravioli and cook for 6 minutes, until al dente.

Meanwhile, melt the butter in a large frying pan over medium heat until it starts to foam and turn light brown. Drop in the sage leaves and take the pan away from the heat.

Drain the cooked ravioli and drop into the frying pan, tossing through the sage butter, adding a little sea salt and freshly ground pepper.

Place a tablespoon of the red capsicum coulis into the base of 6 warmed pasta bowls, lay the ravioli on top and finish with a dob of goat cheese. Top with any stray sage leaves.

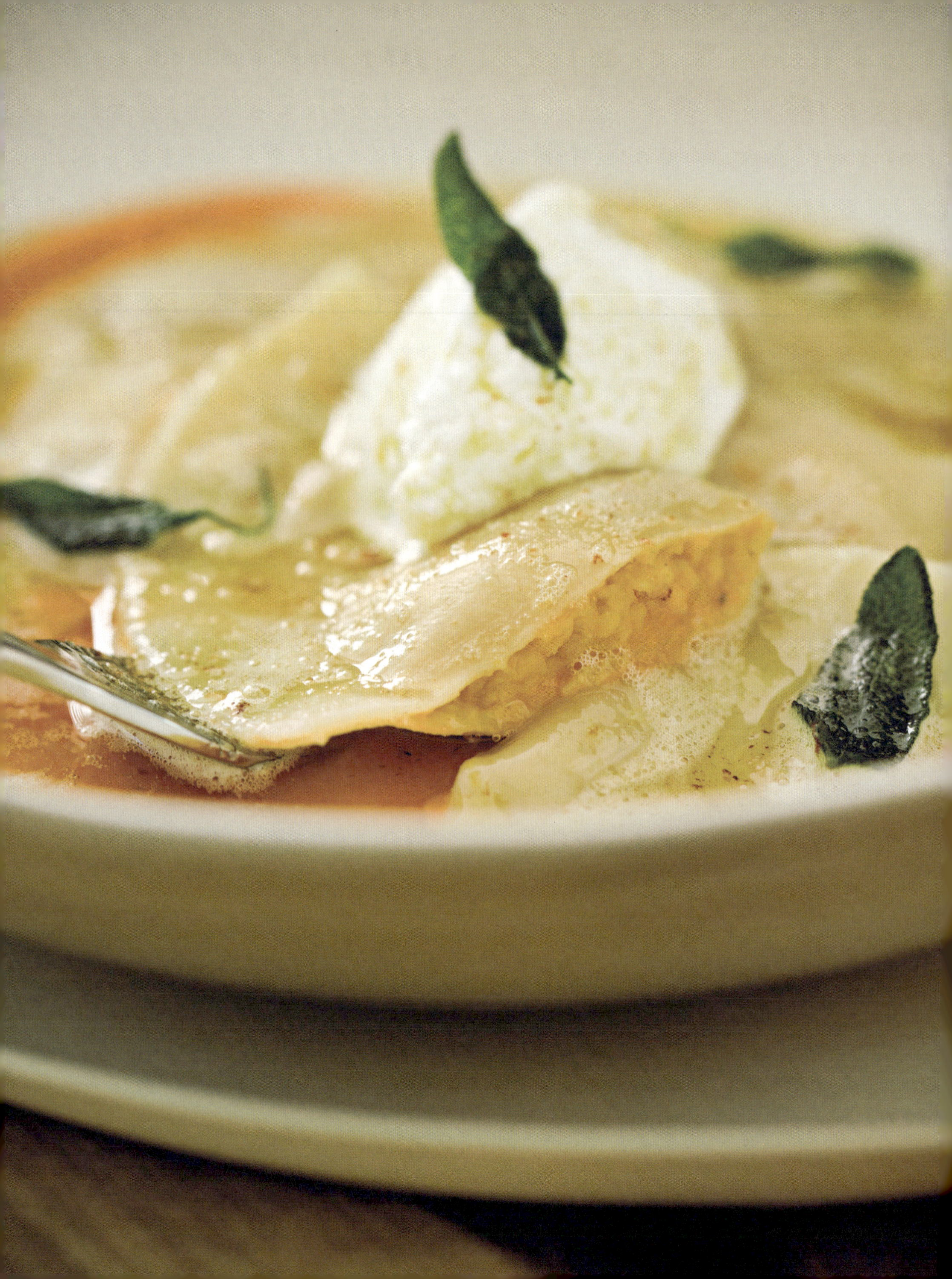

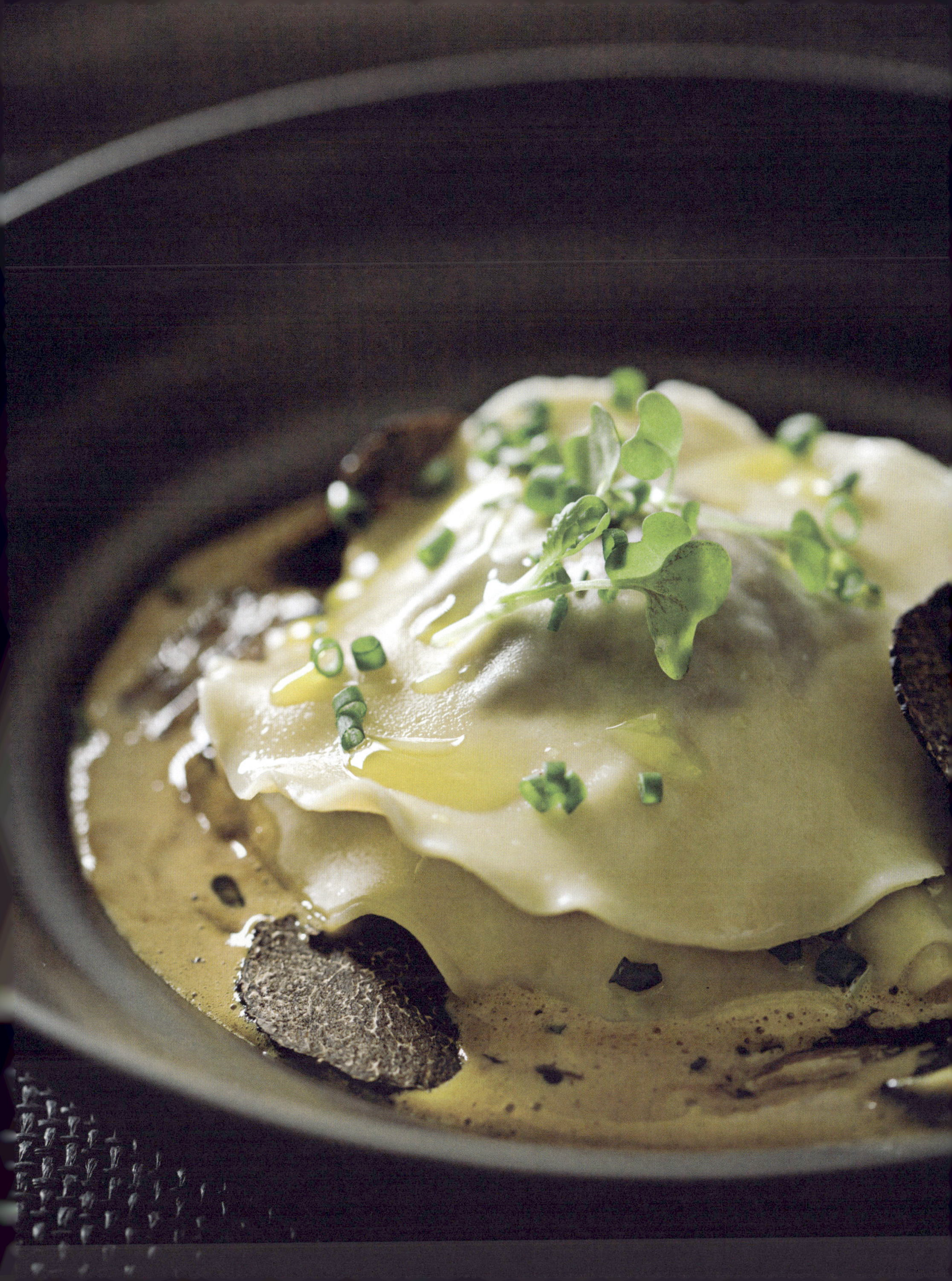

Tua Rita Rosso Dei Notri
Toscana, Italy

The richness and textural tenderness of the beef cheek demands a wine that contrasts this opulence and yet highlights the flavours and weight. This Italian blend is packed with briary black fruit that retains a slender, savoury expression, but is still dense, generous and forthright.

Butterfield Beef Cheek Ravioli, Mushroom Crème & Manjimup Black Truffles

I first developed this dish for a Shin, Shank and Cheek with Blockbuster Reds dinner in 2002, which we've since held annually. Perhaps it's the combination of the foamy smooth sauce with the rich gelatinous slow-cooked beef cheek that makes the recipe so popular. It's very versatile and also works as a main course.

....

To cook the beef cheeks, heat a little olive oil in a large casserole dish over medium-high heat. Add the beef cheeks, in batches if necessary, and cook until lightly browned. Remove and set aside.

Deglaze the pan with the red wine, add all the other ingredients and bring to the boil, drop in the cheeks and reduce barely to a simmer. Place a lid on the pot and cook for about 4 to 4½ hours until meltingly tender.

Alternatively, the cheeks can be cooked in a slow cooker on high for about 6 hours.

Allow the beef cheeks to cool in the cooking liquor. When cool, remove the cheeks with tongs and drain, reserving the cooking liquor. Tear the flesh apart with your fingers and add all the other filling ingredients. Mix well and refrigerate.

Strain the cooking liquor, skim off fat and reduce over high heat to a viscous syrup, there should be about three-quarters of a cup remaining; reserve.

To make the ravioli, cut out 36 x 8 cm circles of the pasta dough. Lay flat onto chopping board, and brush with egg wash. Place small piles of filling (the size of a 50-cent piece) onto the circle then lay another sheet on top. Carefully seal by pressing around the edge of each circle with your fingers. Cover and refrigerate until needed.

24 fresh egg pasta sheets (p. 270)
A little egg wash
(1 egg and 50 ml milk)

Beef Cheeks

Extra virgin olive oil
4 beef cheeks
200 ml red wine
500ml beef stock (p. 274)
A few sprigs of fresh thyme
1 bay leaf
3 white peppercorns
1 garlic clove

Filling

1 tablespoon chives, chopped
1 teaspoon fresh oregano, chopped
20 g breadcrumbs
1 egg yolk
30 g butter, melted
30 g parmesan, freshly grated
Sea salt & freshly ground black pepper

Mushroom Crème

2 shallots, chopped finely
1 garlic clove, crushed
A little olive oil
15 g dried porcini mushrooms, soaked in warm water, drained and chopped
2–3 Portobello mushrooms finely sliced
100 ml white wine
125 ml cream, whipped
A little truffle oil
20 g Manjimup truffle
A few chopped chives

SERVES 12 AS A STARTER OR 6 AS A MAIN

Sweat the shallots and garlic in olive oil.

Add the porcini and Portobello mushrooms, cook for 1 minute, then add white wine. Increase the heat and cook for 8 to 10 minutes until most of the escaped juices have evaporated.

Add the reduced beef cheek sauce and cook for a further minute; keep warm.

Meanwhile, cook the ravioli in boiling salted water for approximately 6 minutes.

Boil the sauce and add the whipped cream, whisking well to combine over the heat – the sauce will expand and bubble.

Immediately toss the drained ravioli through the sauce. Serve topped with shaved truffle, a sprinkle of chopped chives and a drizzle of truffle oil.

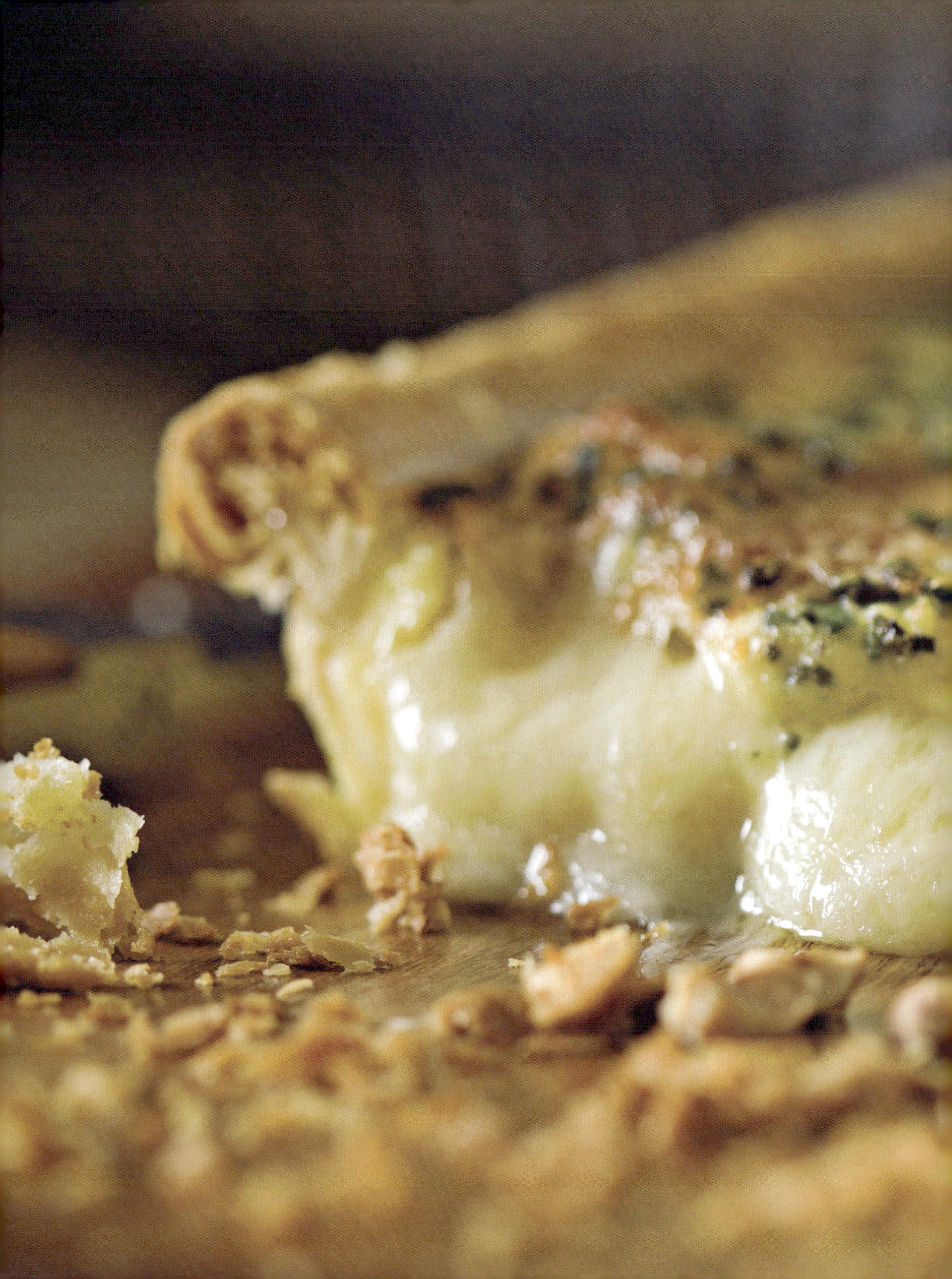

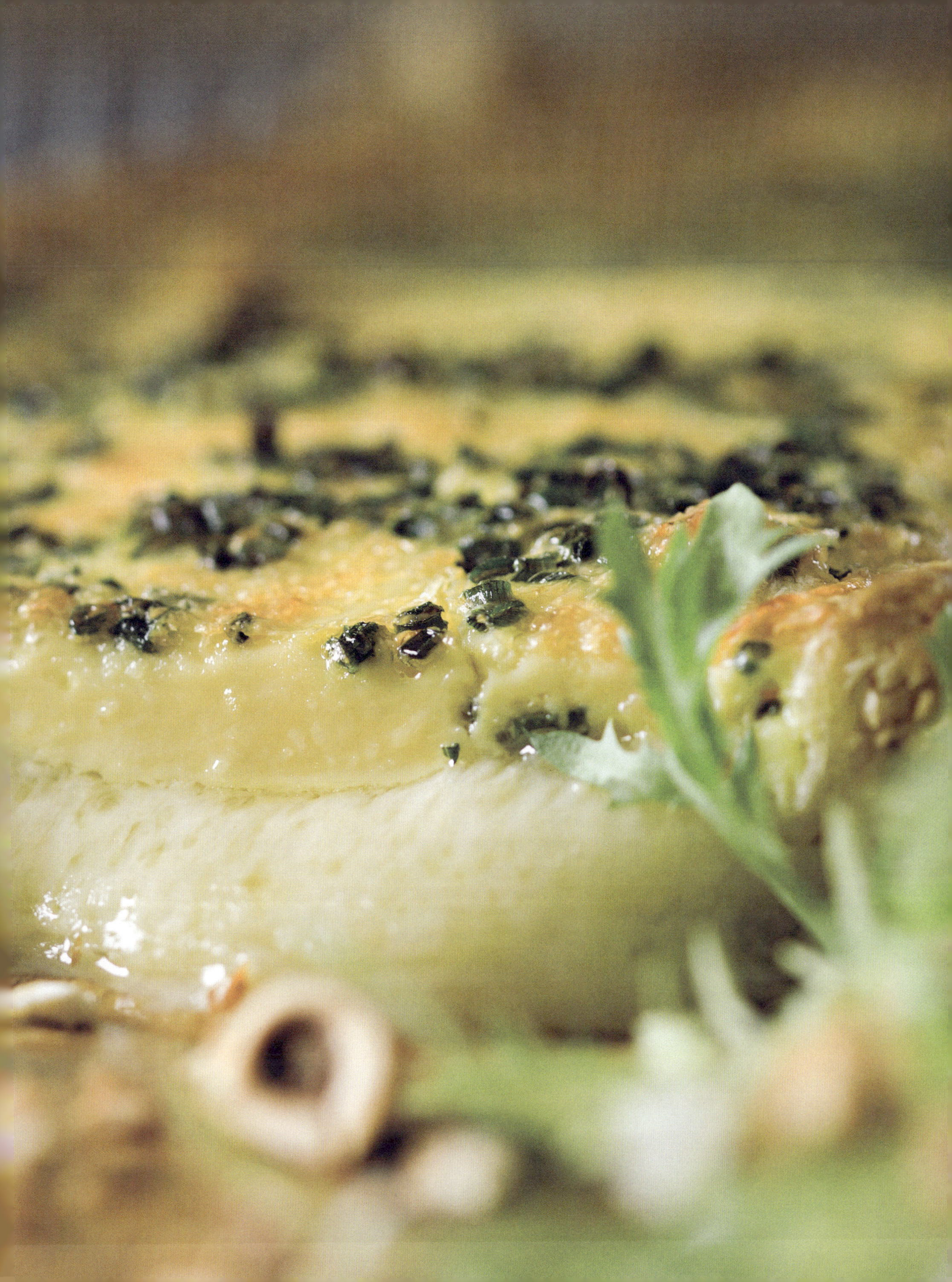

Maxine Magnon Rozeta
Corbieres, France

This biodynamic wine is pure of fruit and slightly herbal and sappy. It is not too heavy or rich, but provides sufficient weight for the pastry and the herbal flavour profile works with the frisée.

Comté Tart, Frisée & Hazelnut Salad, Cabernet Vinaigrette

A great tart is as much about the crisp crust that yields to the touch of a fork as it is about velvety texture and the flavours of the filling. My top tips for pastry making are to roll the pastry to an even thickness and ensure the edges are precisely formed. A low oven temperature and slow cooking means velvety textured tarts. Eggs will set at 70°C – heat the filling any hotter and you'll risk scrambling it!

Preheat the oven to 180°C.

Brush a 36 x 13 cm rectangular tart mould with butter and dust with flour.

Roll the pastry out to 3 mm thick. Line the tart mould with pastry, leaving an excess pastry edge 1 centimeter above the edge of the tin.

Brush the inside of the overlapping pastry with a little egg wash, then double it over, in towards the centre of the tart, pressing firmly between your fingers. Do this all the way around the edge of the tart; the result will be an edge that is about 5 to 10 mm raised above the top of the tart mould.

Line the tart with baking paper, drop in some dried pulses (lentils work well) and rest in the refrigerator for 20 minutes.

Place the tart into the oven and bake for 15 to 20 minutes, until golden brown. Remove the paper and pulses and allow to cool to room temperature. Reduce the oven temperature to 140°C.

Sweat the onion in butter with the thyme, white wine and seasoning until soft, about 20 minutes, then allow to cool.

Line the base of the tart with the onion mix, slice the gruyère into 1 cm thick slices and lay evenly over the top of the cooked onion base.

Whisk the eggs and cream together, season and carefully pour into the tart and sprinkle with chopped herbs.

Cook the tart for approximately 40 to 50 minutes. When cooked; the tart will be puffed in the centre and it will not wobble when shaken.

Place the frisée and hazelnuts together in a mixing bowl, toss through dressing and serve with slices of warm tart.

650 g pâté brisée (p. 285)
1 brown onion, finely chopped
50 g butter
1 tablespoon fresh thyme leaves
½ cup white wine
Sea salt & freshly ground black pepper
200 g Comté cheese (or gruyère)
3 x 50 g eggs
260 ml cream
2 tablespoons chopped herbs (parsley or chives)

Egg Wash
1 x 50 g egg
beaten with 50 ml milk

Salad
150 g frisée lettuce, washed and picked
60 g hazelnuts, roasted and peeled
Cabernet vinaigrette (p. 268)

SERVES 6

Fourme D'Ambert Tart, Caramelised Pear & Walnut Salad

Use the Comté tart recipe (p. 94) substituting that cheese with Fourme D'Ambert, a mild and creamy blue from the Rhône-Alpes region of France.

---

1 beurre bosc pear

¼ cup caster sugar

¼ cup Californian walnuts, lightly roasted

100 g baby chard or rocket lettuce

Sherry-walnut vinaigrette (p. 268)

SERVES 6

Prepare the tart using the same method as for the gruyère tart – simply substitute the cheese.

Heat a non-stick pan over a medium-high heat.

Core and slice the pear into thin slivers, dredge heavily with caster sugar and carefully place into the hot pan. The sugar will caramelise and smoke.

Use tongs to turn the pear over before the caramel burns, cook a further 30 seconds or so until the caramel is a rich dark brown. Remove onto a tray lined with baking paper.

Place the chard and roasted walnuts together in a mixing bowl, toss through the dressing and caramelised pear. Serve with slices of warm tart.

Francois Chidaine 'Clos Baudoin' Vouvray
Loire Valley, France

The slightly creamy blue cheese and suggestion of pear sweetness begs for a wine that has some depth, but without being too heavy. Francois Chidaine Vouvray is a serious chenin blanc with brisk acidity to balance the richness of the dish and it has sweetness, which offsets the cheese and caramelised pear. I would also recommend a Sauternes or a Loupiac.

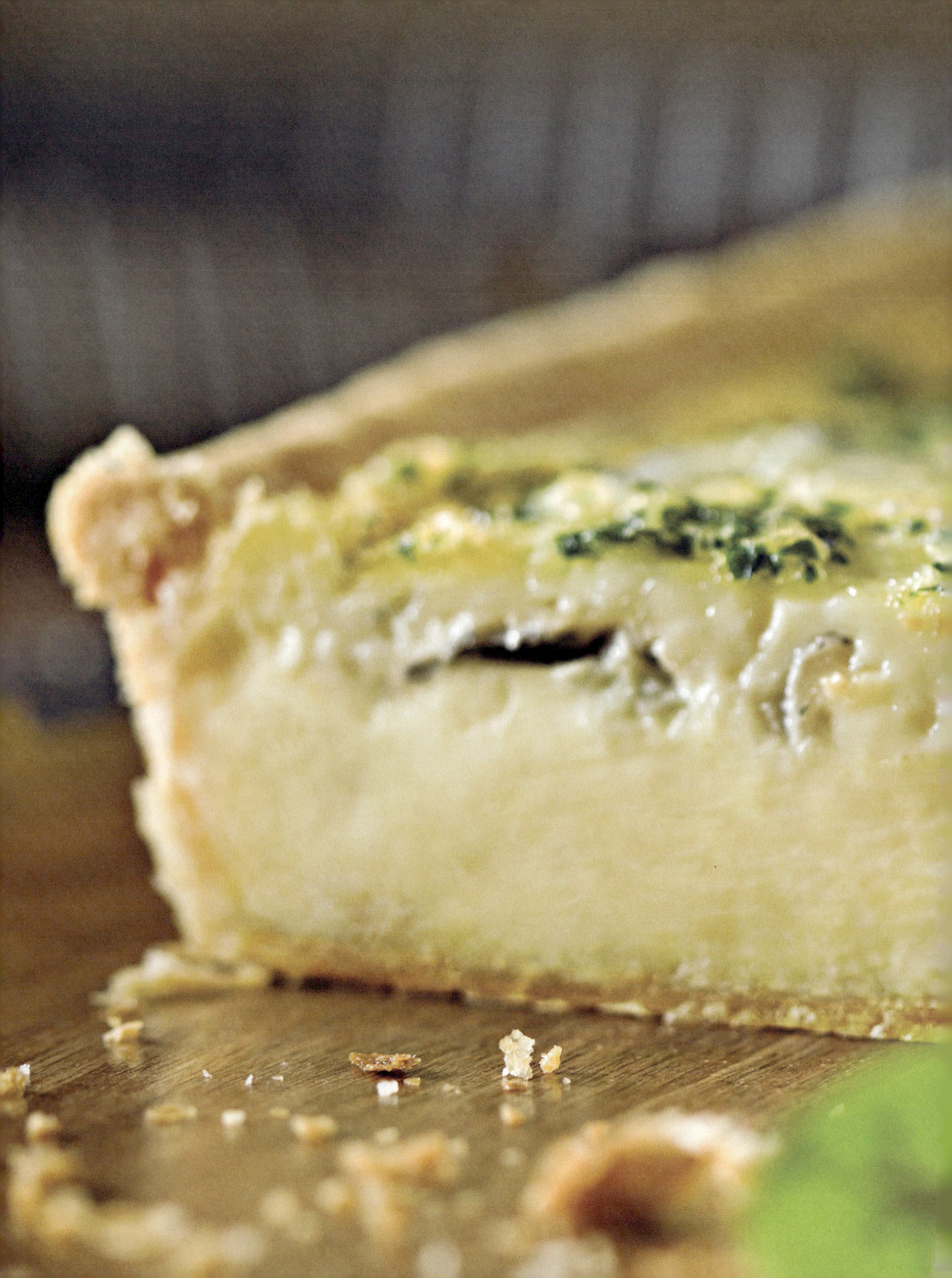

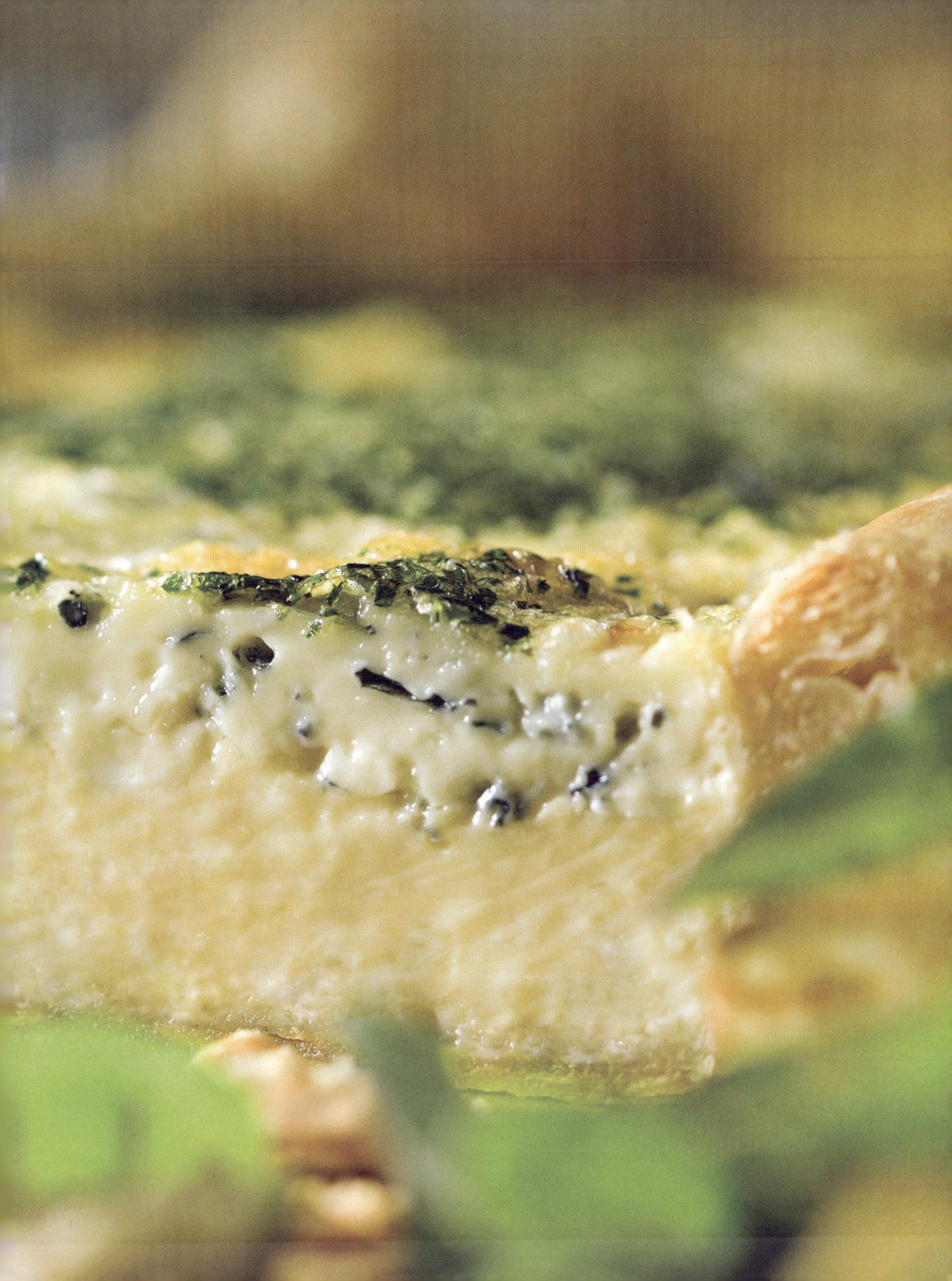

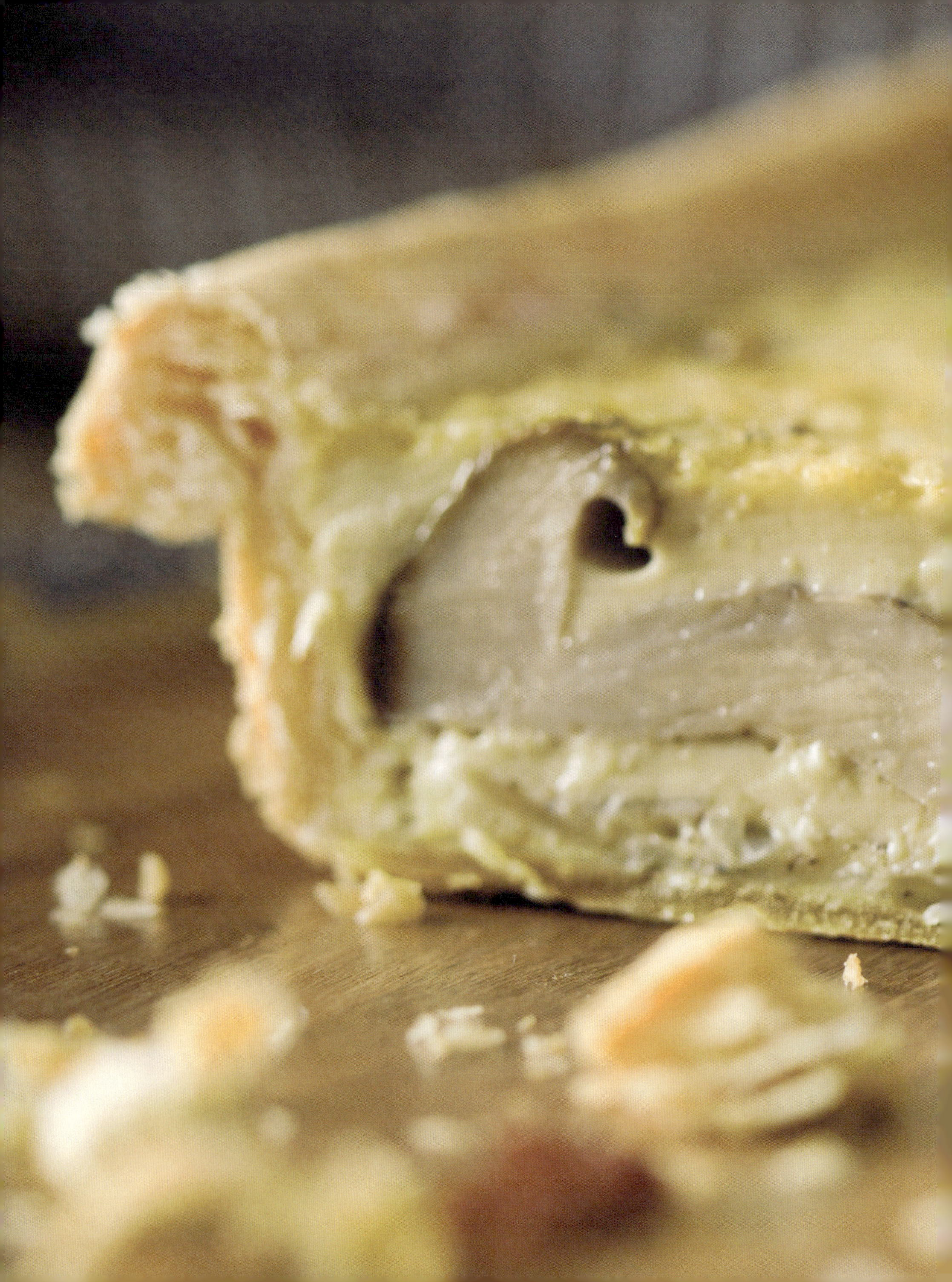

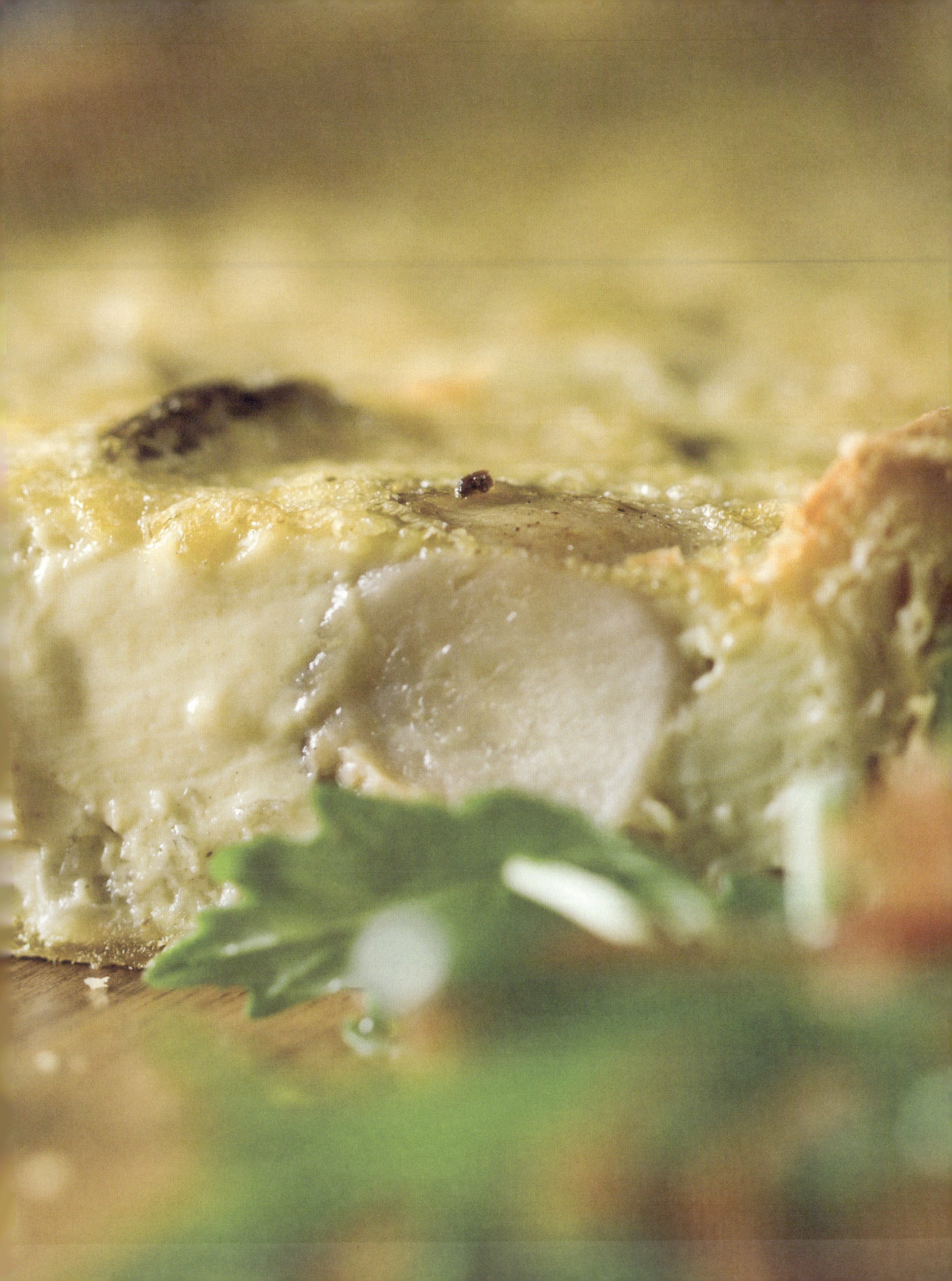

Thierry Puzelat 'La Tesniere' Pinot d'Anuis Touraine
Loire Valley, France

Pinot d'Anuis is an oddity from the Loire variety – not too heavy, more medium-bodied, lean and angular. This biodynamic wine walks on a knife's edge, but it's interesting and appropriate with the earthy, rustic nature of mushrooms and the lively salty pancetta.

Cepe Mushroom Tart, Crisp Pancetta & Rocket Salad

Use the Comté tart recipe (p. 94), substituting the cheese with frozen cepe mushrooms, also known as porcini mushrooms.

Prepare the tart using the same method as for the Comté tart, using cepe mushrooms instead of the Comté cheese. Thaw the cepe mushrooms at room temerature, then cut into 1.5 to 2 cm chunks. Lay evenly over the onion base in the tart shell. Sprinkle with grated parmesan cheese before flooding the tart with the egg mixture.

Preheat the oven to 200°C.

Lay a sheet of baking paper onto a flat oven tray, place the pancetta slices on top, and top with another sheet of paper and another flat baking tray. Place into the oven for 6 to 10 minutes, until the pancetta is golden brown and crisp.

Remove and allow the pancetta to cool to room temperature on the tray.

Place the rocket and crisp pancetta in a mixing bowl, carefully toss through dressing. Serve with slices of warm tart.

300 g frozen cepe mushrooms
60 g parmesan cheese, finely grated

Salad
12 fine slices rindless pancetta
100 g rocket
Red wine vinaigrette (p. 268)

SERVES 6

Yabby, Confit Tomato & Chive Tart

Use the Comté tart recipe (p. 94), substituting the cheese with yabbies, tomatoes and grated parmesan.

1.5 kg fresh yabbies
9 confit tomatoes (p. 267)
60 g parmesan cheese, finely grated
1 bunch chives

SERVES 6

Place the yabbies in the freezer to send to sleep.

Bring a large pot of salted water to the boil, drop the yabbies in, bring back to the boil and cook for 2 minutes.

Drain and refresh with cold water.

Remove flesh from yabby tails and reserve.

Spread the onion mix across the bottom of the tart, then lay half of the confit tomatoes across the onion mix.

Place the yabby tails on top then lay the rest of the confit tomatoes on top of the yabby tails.

Sprinkle with parmesan and flood with the custard mix.

Finely chop the chives and sprinkle over the top before cooking the tart.

Prager 'Hinter der Burg' Gruner Veltliner Federspiel
Wachau, Austria

Gruner Veltliner has the attributes of weight and acidity, which provides the perfect accompaniment to the yabby flesh in the light tart casing. It also has a distinct celery/herbal character, which adds interest to the tomato confit, parmesan and chives.

starters 101

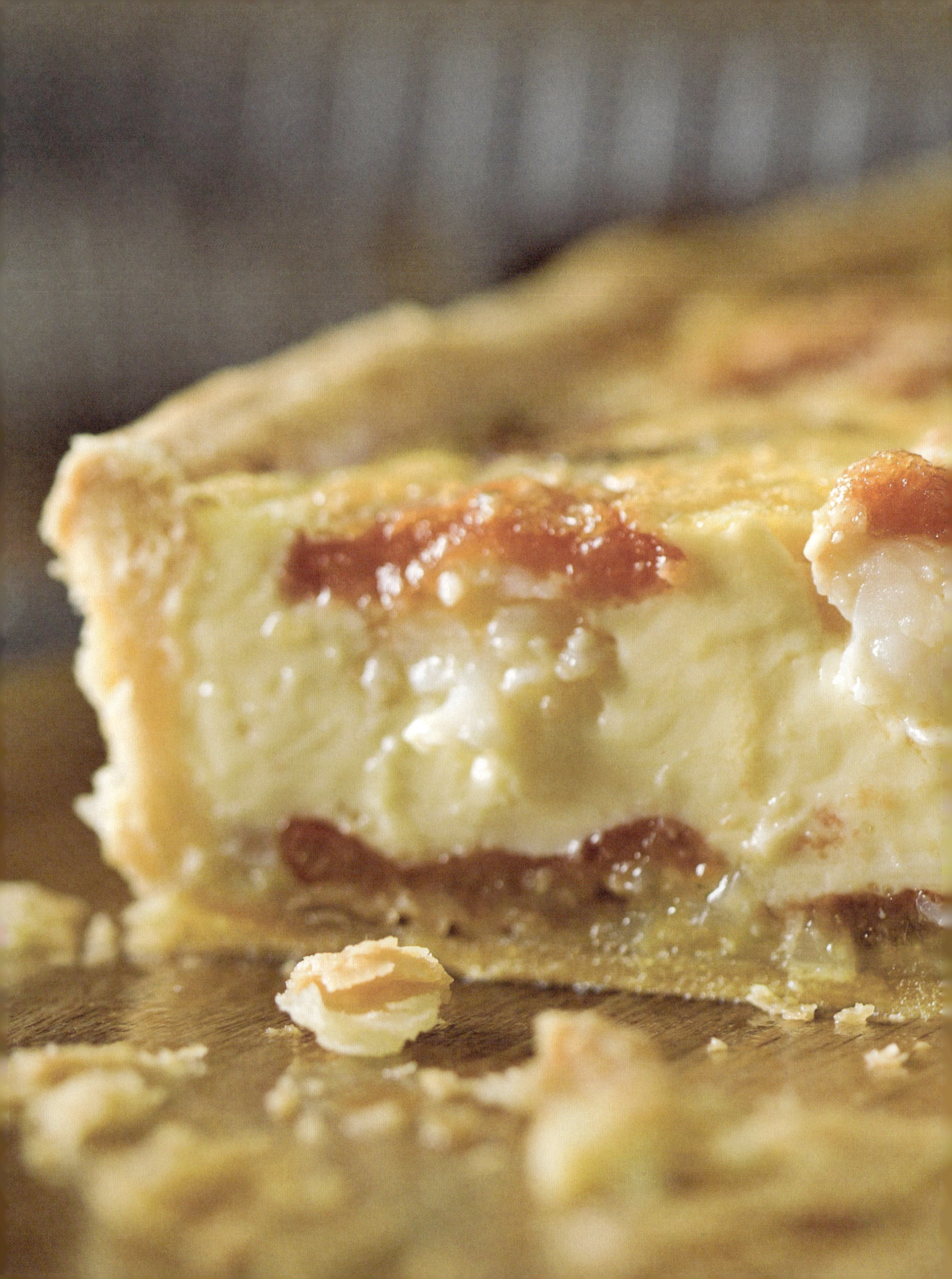

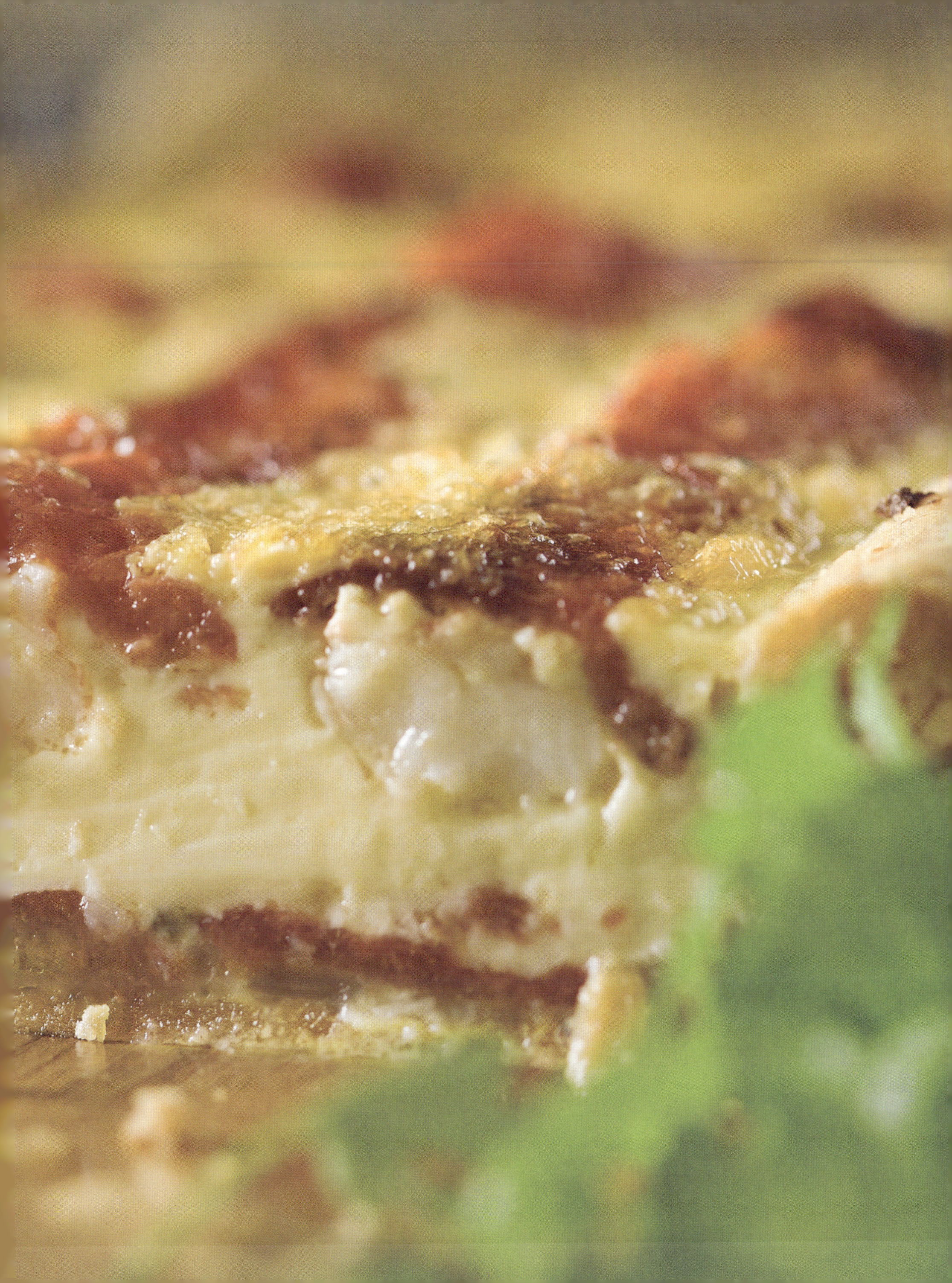

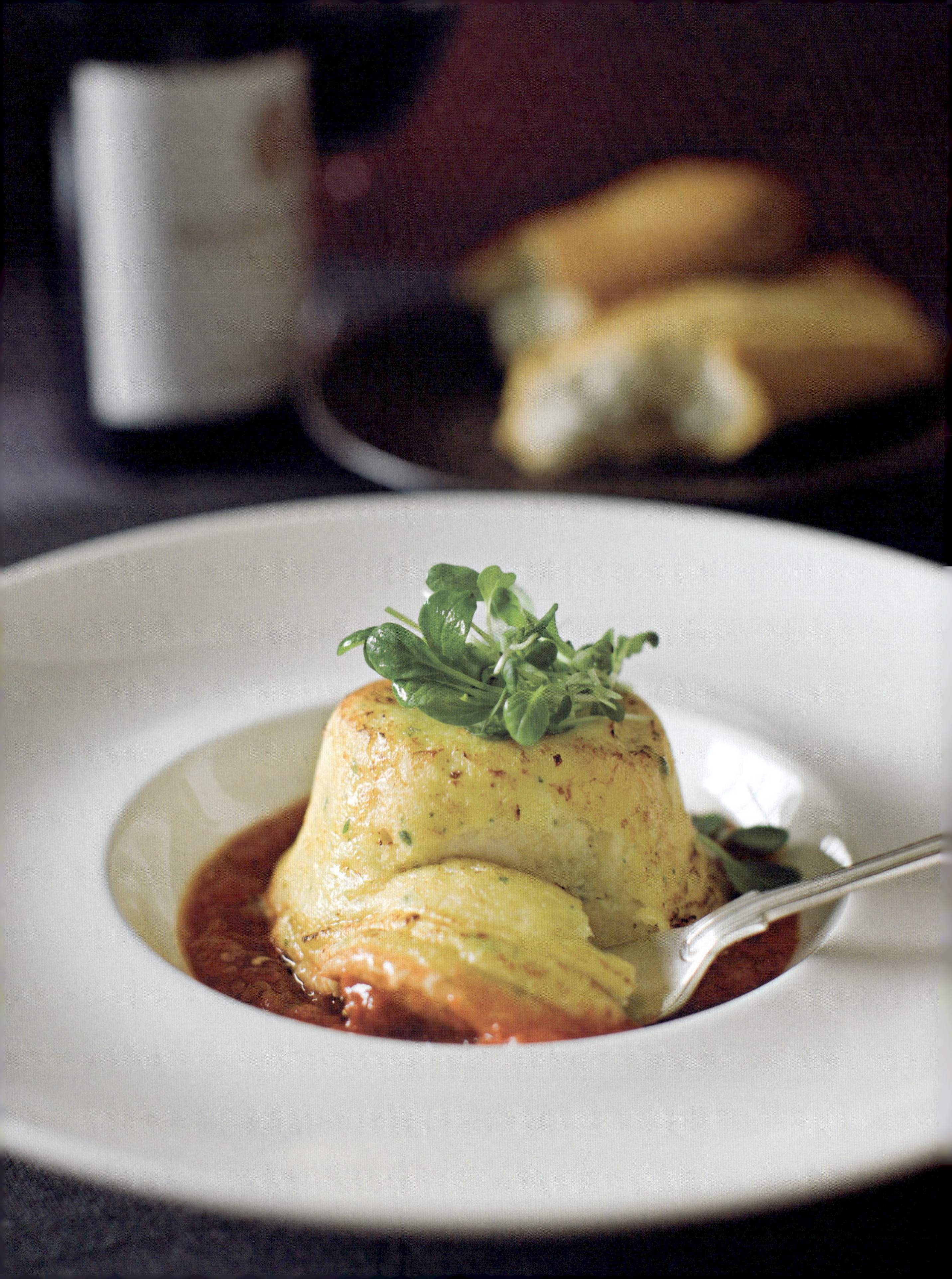

Twice-Baked Taleggio Soufflé & Organic Tomato Coulis

Soufflés are sensual, and this combination of melt-in-the-mouth lightness with the heady aroma and flavour of ripe taleggio cheese is habit-forming.

....

70 g unsalted butter, melted plus butter to brush into moulds

500 ml milk

100 g plain flour

1 teaspoon sea salt & a pinch of ground white pepper

90 g taleggio cheese

50 g parmesan cheese,

½ tablespoon flat-leaf parsley, chopped

4 large egg yolks

6 large egg whites

250 ml whipping cream

Tomato coulis (p. 79)

Baby lettuce or microherb leaves

SERVES 6

Preheat the oven to 200°C.

Select the moulds you wish to use; soufflé dishes approximately 7 cm wide by 5 cm deep are perfect.

Brush the inside of each dish or mould with melted butter. Place the moulds into a large deep casserole dish.

Pour the milk into a medium saucepan and bring to a simmer, remove from stove and keep warm.

Melt 70 g butter in a saucepan, add the flour and cook over medium heat for 2 minutes, stirring with a wooden spoon.

Add salt and pepper then gradually add the milk, stirring constantly to ensure no lumps form.

Reduce the heat to low and cook for a further 3 minutes, stirring the thick mixture constantly.

Remove from the heat, rub taleggio cheese through a coarse grater into the mix and add parmesan, stirring well to combine.

Allow to cool for about 2 minutes, then add the chopped parsley and yolks, stirring well.

Whisk the egg whites lightly (when the beater is removed from the whipped whites they should quickly fall off the whisk). Stir one spoonful of whites into the thickened milk/cheese mixture, then pour that back into the whites and fold together gently.

Divide the soufflé mixture evenly into the moulds, flood the bottom of the casserole dish with 1 cm of boiling water and place into the oven to cook for approximately 10 minutes. Reduce the oven temperature to 180°C and cook a further 8 to 10 minutes.

Check if the soufflés are cooked; they will be puffed and golden, and a skewer inserted into the centre will come out clean.

Remove from the oven and cool to room temperature. Don't be surprised to see the soufflés deflate – they will have a

---

Pittnauer Saint Laurent Klassik
Neusiedlersee, Austria

St Laurent is a red variety from Austria. It has a soft, calm feel and mirrors the gentleness of the soufflé but also has sufficient flavour to complement the soufflé. Not too many wines pair with tomato coulis, but this is shapely and succulent enough to cope.

starters 105

second rising when reheated. Once at room temperature, carefully slip the soufflés out of their moulds.

The soufflés can be refrigerated for up to 2 days before re-heating.

Preheat the oven to 180°C.

Place the soufflés onto a sheet of baking paper in a deep casserole.

Boil the cream and pour over the top of the soufflés, then put into the oven for approximately 8 minutes, basting the soufflés with the cream once or twice during the re-heating process.

The soufflés will be cooked when they have puffed to a size a little larger than when originally baked.

Ladle some coulis onto each plate, carefully slide a spatula under each soufflé and place on top. Finish with some baby lettuce leaves.

Twice-Baked Tiger Prawn Soufflé & Champagne Crème

Take care not to overwhip the egg whites when making your soufflé – they should be quite wet, not firm and peaked. Too much air in the whites from overbeating will cause a soufflé 'eruption' in the oven when they are baked!

Substitute taleggio cheese in the previous recipe (p. 105) with 300 g finely chopped tiger prawns.

Champagne Crème

1 shallot, chopped finely

150 ml sparkling wine or Champagne

2 white peppercorns

Juice of one lemon

200 ml cream

50 g unsalted butter, at room temperature

Sea salt

2 tablespoons chopped chives

SERVES 6

To make the champagne crème, place the shallot, wine, peppercorns and lemon juice into a small saucepan and boil for about 5 minutes until almost completely reduced.

Add cream and simmer for about 8 to 10 minutes until thickened. Season with a pinch of sea salt.

Remove from heat and slowly add knobs of the butter, whisking the sauce constantly. When the butter has been completely combined, strain the sauce and reserve in a warm place.

Add the prawns to the milk when it is at a simmer, remove from the heat, leave for 30 seconds then strain the prawns from the milk, reserving the warm milk and prawns separately.

Proceed with the previous recipe, adding the prawns back to the soufflé mix with the egg yolks.

To serve, place hot soufflés onto deep serving plates and spoon warm Champagne crème over the top. Finish with finely chopped chives.

Marchand & Burch Chardonnay
Great Southern, Western Australia

This chardonnay has power without being broad or muscle-bound, but it needs some strength to work alongside this soufflé. It is not encumbered with obvious oak and has a gentleness to allow this combination to work in harmony.

Dry River Pinot Noir
Martinborough, New Zealand

The delicacy of this beef is best combined with a wine that does not dominate the refined Wagyu and its gentle texture. Despite the Dry River pinot being a bigger style of pinot, it complements the beef in texture and flavour. The slightly higher acid lift in the pinot helps cut through the cream and horseradish.

Wagyu Beef Carpaccio, Pickled Beetroot, Horseradish Crème & Lime-Pressed Extra Virgin Olive Oil

Wagyu is delicious raw. The fat of this breed literally melts in your mouth. Ask your butcher for a marbling score of 5 or 6, this is the amount of fat that is peppered through the muscle.

- Sandwich each slice of beef between two slices of cling film, lay flat onto a bench and pat out with a meat hammer or rolling pin until paper thin.
- Place two slices of the flattened beef onto each serving plate.
- Sprinkle some of the pickled beetroot over each plate.
- To make the horseradish crème, finely grate the horseradish into a mixing bowl with the cream and whisk until the cream forms soft peaks. Drizzle a little of the cream over each plate.
- Sprinkle each plate with chopped chives, a generous drizzle of lime-pressed extra virgin olive oil, capers, sea salt flakes and a grind of pepper. Serve with freshly baked baguette.

12 fine slices Wagyu beef sirloin
Pickled beetroot (p. 202)

Horseradish Crème

2 cm knob fresh horseradish
100 ml whipping cream
1 tablespoon chopped chives
Lime-pressed extra virgin olive oil
Tiny capers
Sea salt flakes & freshly ground black pepper

SERVES 6

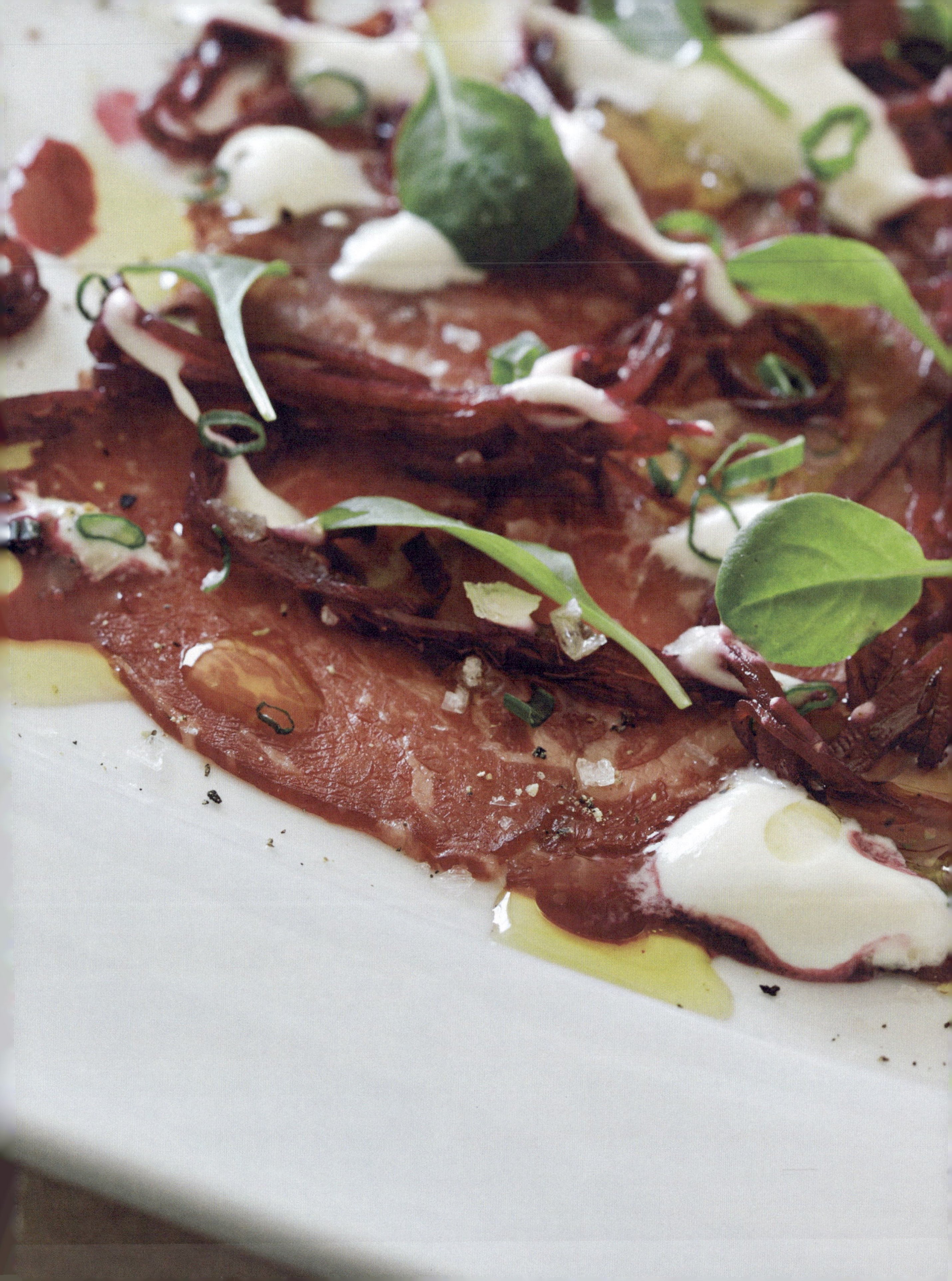

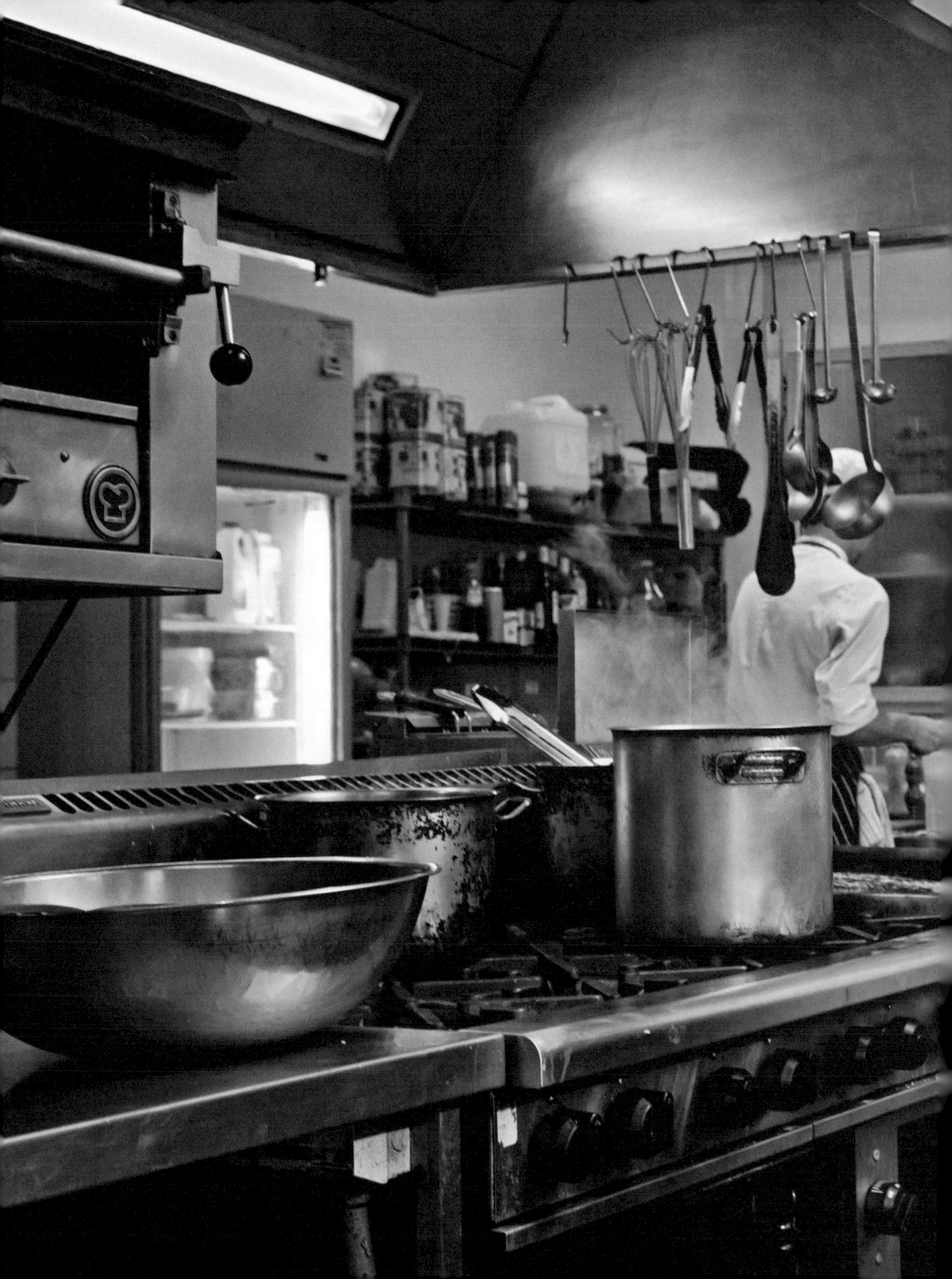

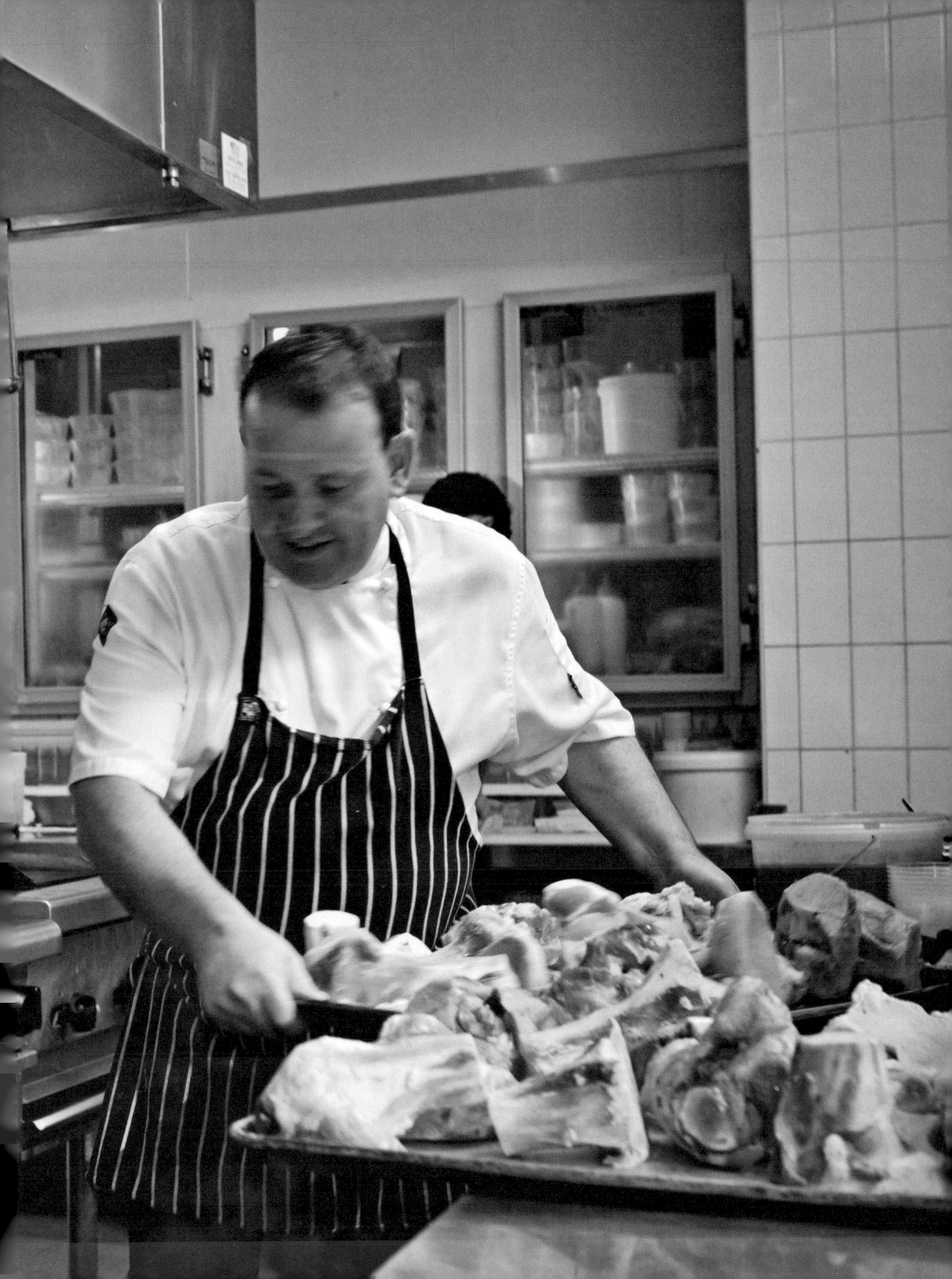

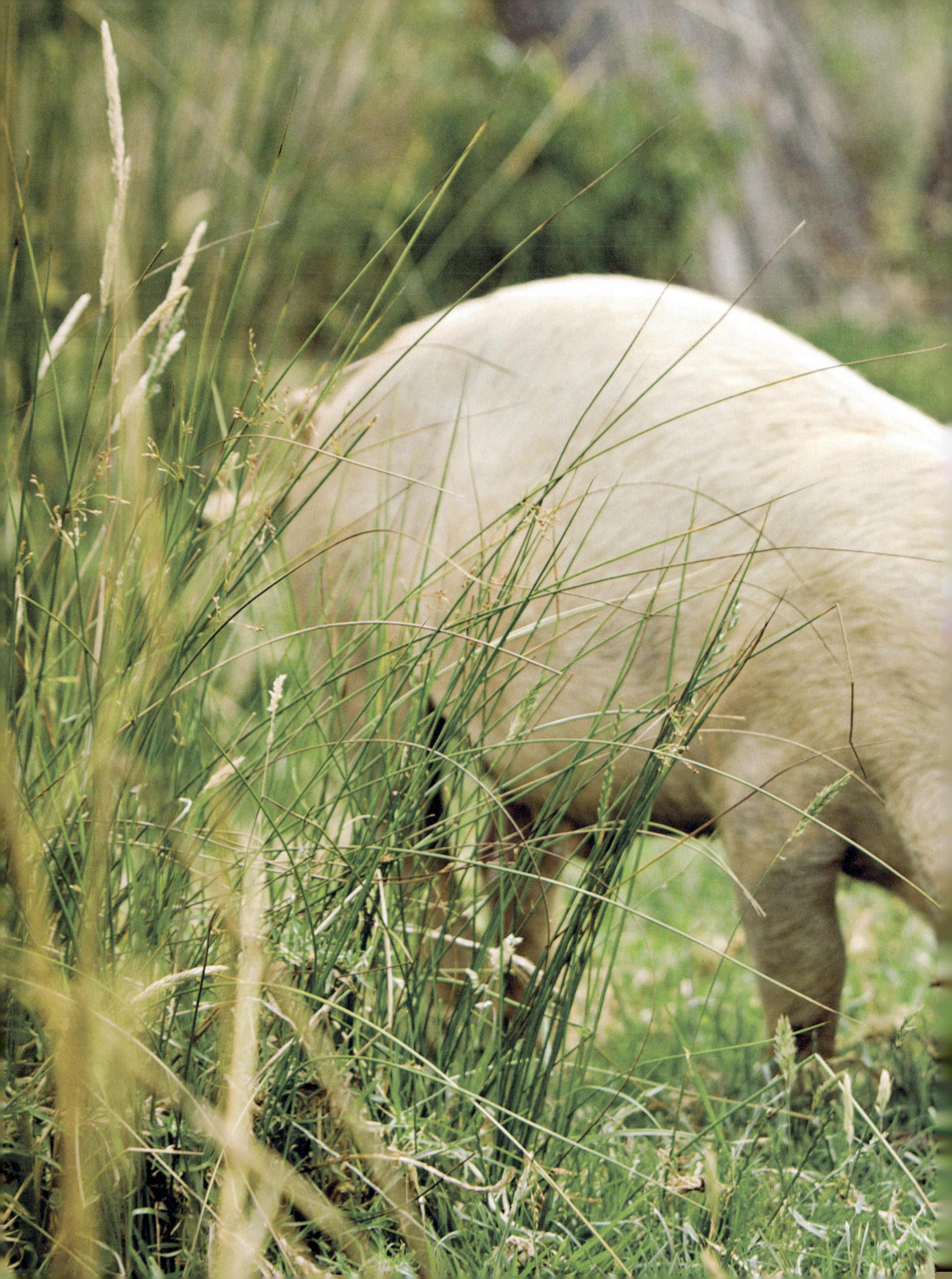

charcuterie

David Hohnen has the DNA of a hunter-gatherer. He was raised in Papua New Guinea, had his first 12-gauge shotgun when he was nine and hunted for food with the locals. They taught him both to respect his prey, and to relish the astonishing flavours and textures that wild-caught food has. He has taken this ethos to heart and his Arkady lamb and Jarrahdene pork are all free range.

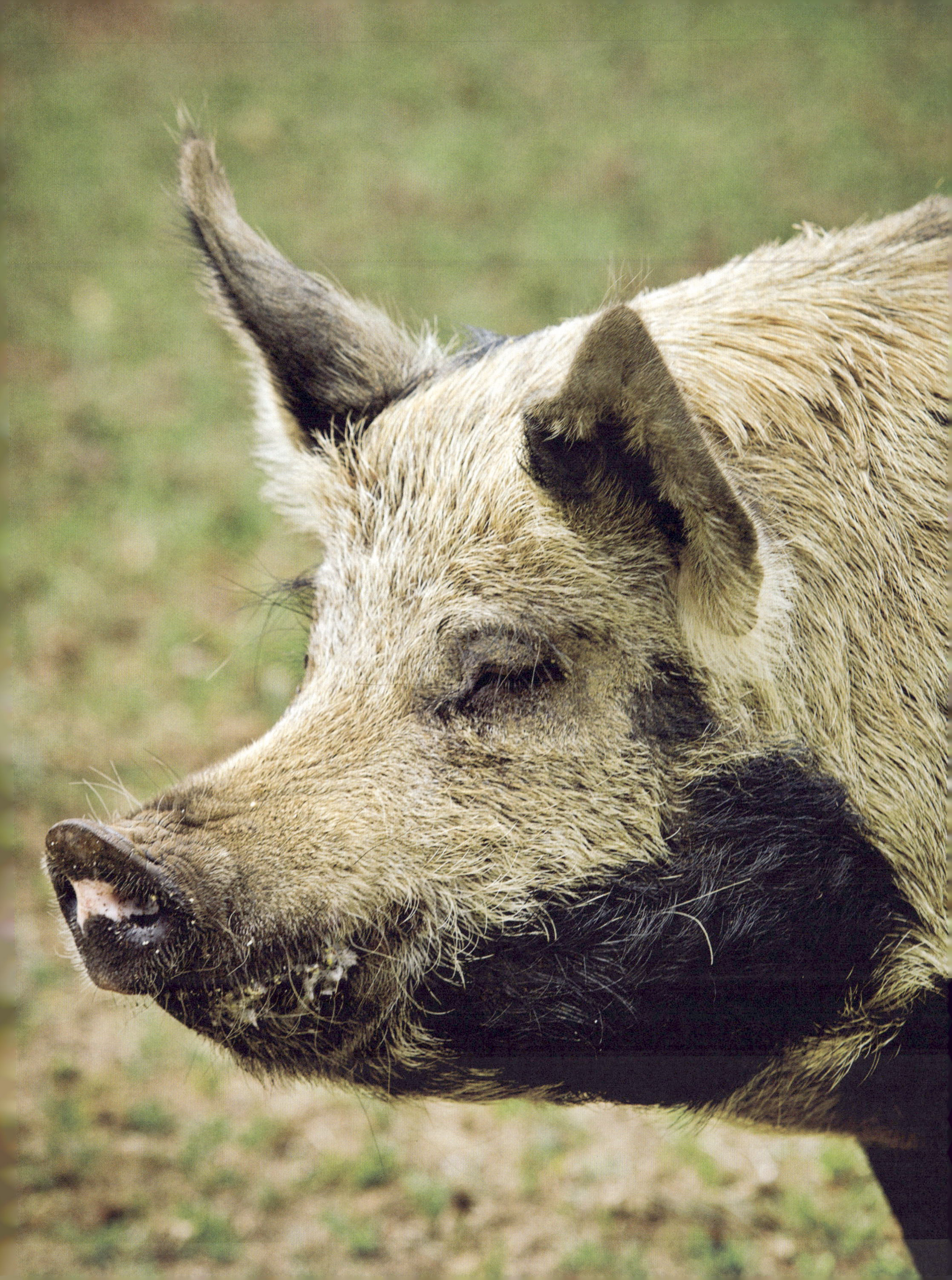

Charcuterie

Charcuterie means 'cooked meat', essentially that from a pig. Over centuries, French charcutiers have elevated their trade to an art form, converting every square centimetre of the animal to some delicacy – whether it is ears, intestines, trotters, shoulder, belly, head or tail.

Whenever I chance on a charcuterie in the main street of a little French village or town, I cannot help but go in and spend some time sampling and buying the product – a habit my family sometimes finds a little tedious.

Charcuterie is an important foundation cuisine of the French bistro menu. Many chefs, too busy to make the delicacies themselves, have a favourite charcutier that produces for them, often just around the corner. Sadly, I'd argue that the trade is slowly dying in France, as large marché and hypermarkets sell mass-produced terrines, patés, sausages and cured meats that are a far cry from their artisanal origins.

While we may eat a range of charcuterie dishes throughout the meal (such as the recipes for sausages and boudin blanc in the following pages) at Must charcuterie is prominent at the start of the menu: our shared charcuterie plate is wine-friendly and is served in the bar, as well as a starter in the bistro.

The key to making great charcuterie is to have plenty of time. Mis en place (everything in its place) is also critical – prepare, pre-weigh and pre-cut all the ingredients before starting the recipe. Super-sharp knives, good digital scales and a digital probe thermometer will also help.

Another essential aspect to charcuterie, or indeed successful cooking of any kind, is to have a good idea of what a dish tastes like before you make it. The only way to become an experienced cook is to remember the many different tastes of dishes – only then can you expect to re-create them successfully. My own 'palate memory' has been essential for so many recipes in this book.

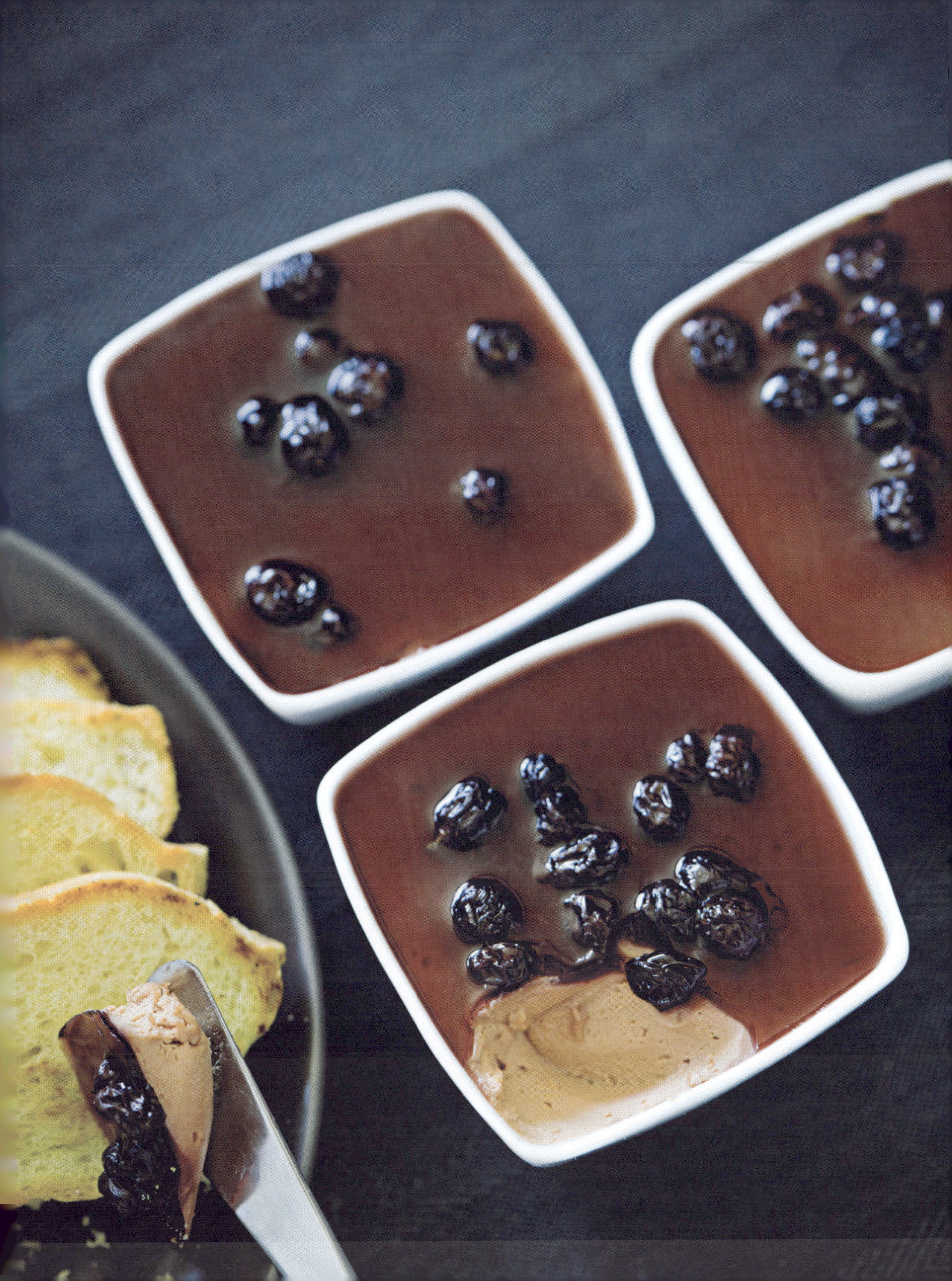

Jean Foillard 'Morgon' Gamay
Burgundy, France

This wine is crunchy, fresh, pure fruited and softly tannic. The slightly rustic flavour profile pairs well with the chicken liver opulence and the vibrant freshness of the wine balances the rich flavours.

Chicken Liver Parfait, Grenache Jelly & Melba Toast

This recipe has one critical moment – when you add the butter to the livers make sure the livers are not 'fridge cold' or the butter too hot. You'll need a digital probe thermometer to make this recipe. The parfait will keep for up to one week, covered, in the refrigerator.

Preheat the oven to 140°C.

To make the Melba toasts, lay the baguette slices onto a baking tray and bake for 20 to 30 minutes or until the toasts have become crisp.

Increase the oven temperature to 150°C.

I cook our parfaits in ceramic dishes that have a volume of about 120 ml. Arrange 12 ramekins into a deep oven tray before you start this recipe.

Place the madeira, port, shallots, garlic and thyme into a large saucepan over high heat and boil for about 25 minutes until reduced by 80 per cent.

Pour into the bowl of a food processor and blitz until the contents are completely smooth, then pass through a fine strainer and reserve.

Place the butter in a large saucepan over low heat to melt.

Trim the chicken livers of any blood vessels or fat, place into a colander, wash with cold water and drain well at room temperature (if they are too cold they will not combine with the butter).

Place the livers into the bowl of food processor and blitz for 1 to 1½ minutes until you have a a smooth purée. (At this stage I normally remove the livers and pass them through a fine strainer, then return them to the food processor, but you can skip this if you wish.)

Add the eggs, wine and shallot reduction, sea salt and pepper to the livers and blitz well to combine.

Heat the butter to 90°C and, with the food processor running on its highest speed, slowly drizzle the butter into the liver mix, this should take 30 seconds or so.

The mix will lighten in colour and thicken as you add the butter. (If the butter is too cold, or you add it too quickly the mix may 'split' and become grainy looking; if so, you'll need to start the entire recipe again.)

Quickly ladle the parfait mixture into the prepared dishes to within 5 to 8 mm of the rim (so the dishes can be topped with the jelly later).

Melba Toasts
1 day-old baguette, sliced as thinly as possible

Parfait
200 ml Madeira
130 ml port
40 g chopped shallots
1 garlic clove, minced
½ tablespoon fresh thyme leaves
650 g chicken livers
560 g unsalted butter
2 x 59 g eggs
15 g sea salt
A few twists of freshly ground black pepper

Jelly

4 gelatine leaves

375 ml grenache or grenache-shiraz wine

150 ml water

40 g sugar

30 ml lemon juice

4 juniper berries

1 clove

1 bay leaf

5 black peppercorns

1 teaspoon sea salt

40 g currants

SERVES 12

Flood the base of the deep oven tray with 2 cm of boiling water. Immediately seal the top of the tray with several layers of cling film, then cover that with a sheet of aluminium foil.

Carefully place the tray in the oven, immediately reducing the temperature to 110°C, and cook for 15 minutes.

To check the parfaits are cooked, insert a probe thermometer to the centre of a parfait – the temperature should be 70°C. Another way to test the parfaits is to press the cooked surface with your finger; if cooked they will be lightly springy to touch, and will look a little 'puffed'.

Remove from the oven, remove the foil and film, cool to room temperature then place in refrigerator for at least 2 hours hour to chill.

To make the jelly, soak the gelatine sheets in cold water.

Combine all the remaining ingredients except the currants in a large saucepan, place over moderate to high heat and bring to a simmer.

Squeeze water from the softened gelatine sheets. Add to the saucepan, stir to dissolve and remove from the stove.

Strain the hot jelly into a bowl and sprinkle in the currants. Allow to cool to room temperature.

Strain the currants out of the jelly, sprinkle a few on the top of each parfait dish, carefully flood with jelly and refrigerate until set. Serve with Melba toasts.

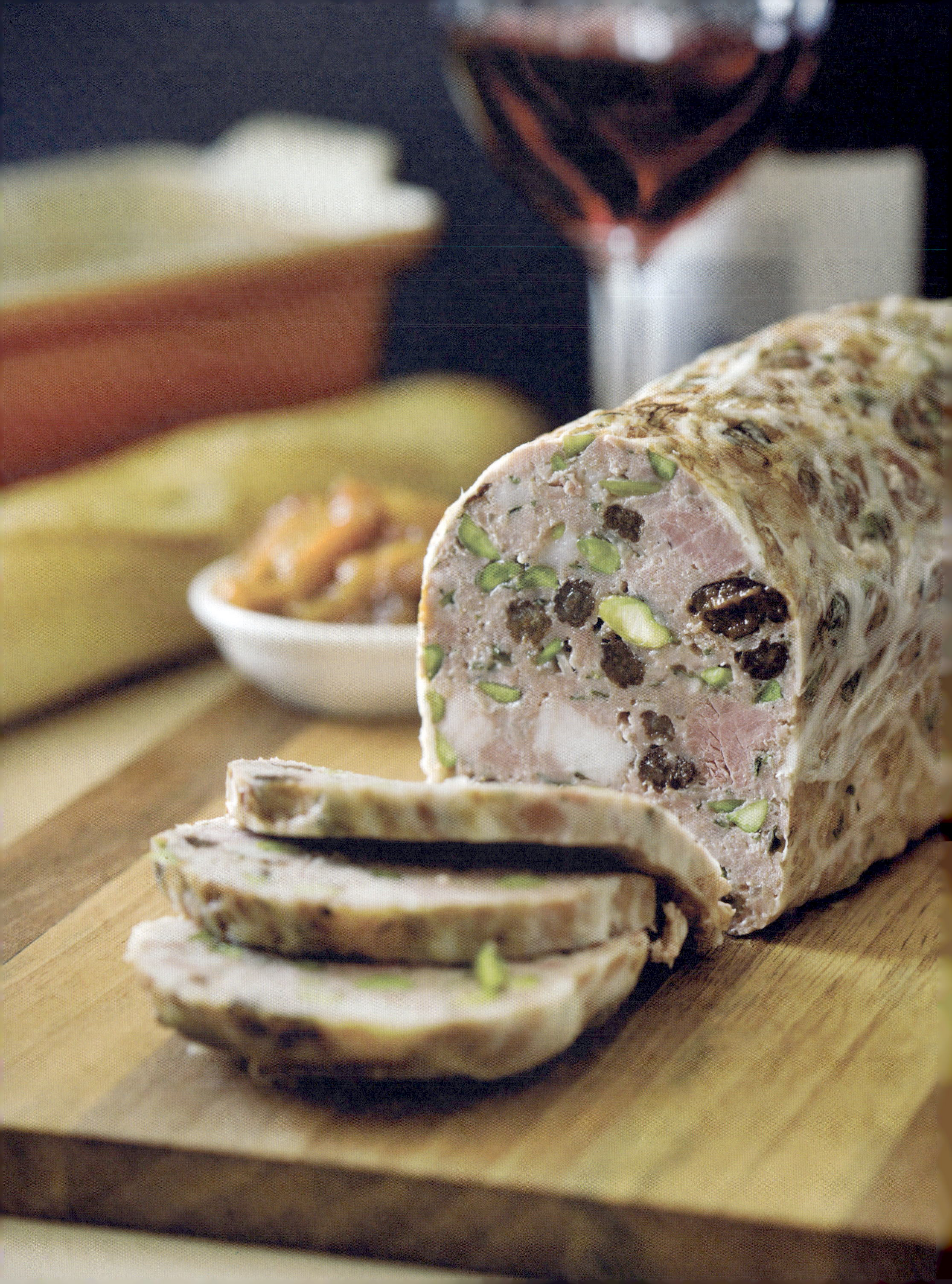

Country Terrine of Pork, Pistachios & Dates

This rustic pâté is much like the 'terrines de campagne' displayed in hundreds of charcuterie windows across France. It contains pork and pork fat, combined with either game or poultry. Each charcutier believes the recipe they have used and protected for decades 'est le meilleur' (is the best). If you can't source the fine caul fat to line the terrine, a good alternative is finely sliced pancetta.

---

1 large piece of 'crepinette' or pork caul fat, washed and soaked in salted water overnight. It is important to wash away all the blood so that there is no pink colour left.

150 g pork tenderloin, cut into 2 cm square dice

550 g minced pork shoulder (run through a 10 mm mincing plate)

150 g pork back fat cut into 1 cm dice

90 g onion, finely minced

8 g garlic (2 cloves), finely minced

14 g (1 tbsp) salt

1 pinch ground white pepper

12 g (2 tbsp) chopped parsley

1 x 59 g egg

75 ml white wine

20 ml cognac or brandy

3 g (2 tsp) thyme leaves, picked and chopped

65 ml madiera jus or red wine jus

80 g unsalted pistachio kernels

100 g pitted dates, chopped into 1 cm dice

60 g raisins

Peach chutney (p. 273)

SERVES 12

Preheat the oven to 220°C.

Line a 25 cm x 10 cm terrine mould with washed crepinette, allowing it to overhang the sides so that it can be tucked back in.

Mix all of the terrine ingredients together in a large bowl, working the mixture vigorously with a wooden spoon for 5 minutes to ensure all the ingredients are combined.

Fill the lined terrine with the mixture, ensuring there are no air bubbles in the mix.

Fold the crepinette back over and tuck it back in along the sides, so that the top of the terrine has no joins and is completely sealed. You can use a trimmed extra piece of crepinette to lay over the top and tuck down the sides to ensure there is no hole in the top of the terrine.

Place the terrine into a large tray of hot water, so that the water comes 2 cm up the sides of the terrine.

Place into the oven for 15 minutes. Reduce the heat to 140°C and cook approximately 1 hour more, until a thermometer inserted in the centre of the terrine reads 72°C.

Remove from the oven, cool to room temperature then refrigerate overnight.

Remove the cooked terrine from the mould. You may need to warm the mould by dipping it in a warm water bath. Clean any jelly or residual fat from the outside of the terrine with kitchen paper, slice into thin slivers and serve with fresh baguette and peach chutney.

Cantina Pra La Morandina Valpolicella Superiore
Veneto, Italy

The combination of savoury meat and a touch of sweetness (dates and currants) in this terrine, is perfect with a wine that has an earthy, savoury profile, noticeable acidity and a suggestion of sweet, rounded fruit. This valpolicella ticks all the boxes.

Jean Paul Schmitt 'Rittersberg' Riesling
Alsace, France

This Alsace wine is typically bracing, dry and textural. The flavours of the rillettes are quite delicate, yet it has textural depth and requires a wine that is both polite and focussed.

Duck Rillettes

Even though these are described as duck rillettes, we use pork shoulder in the mix to carry the flavour and enhance the texture. For a special indulgence I use goose instead of the duck and pork. Goose lends a generous flavour and silky texture to the dish, it is well worth the effort and expense in sourcing one.

Joint the duck, de-bone pork and cut the flesh into 4 to 5 cm diced pieces.

Place the flesh and bones into a large heavy saucepan. Add remaining ingredients, turn the heat to low until the duck fat has melted, then increase the heat to a lazy, slow simmer and cook for approximately 3 hours, until the pork flesh is tender and the duck flesh is falling off the bone.

Alternatively, if you have a slow cooker, it is ideal for cooking rillettes. Use the high setting and cook the ingredients for 4 to $4\frac{1}{2}$ hours until the meat falls off the bone.

Remove from the heat and cool for 45 minutes to 1 hour.

Pour the fat, flesh and bones into a strainer, reserving the fat and allowing it to cool to room temperature.

Strip the flesh from the bone, discarding the bones and any pieces of skin or fat. Tear the flesh into strips, tossing it into a large bowl.

Ladle $1\frac{1}{2}$ cups of the fat back into the flesh, stir well and refrigerate. As the rillettes cool, give an occasional stir to combine the fat with the flesh.

Once the rillettes are cold, place it into a tub or resealable jar and pour a little of the fat on top to give a final seal, refrigerate and use within 1 week.

Crunchy acidic cornichons are the perfect foil to the richness of rillettes, served, of course, with a fresh baguette.

1 small duck, or 4 duck legs
1.3 kg of pork neck (bone in)
150 g brown onion, finely sliced
15 g or 4 garlic cloves, minced
2 bay leaves
A large sprig of thyme
1.5 kg duck fat
25 g ($1\frac{1}{2}$ tbsp) salt
$\frac{1}{2}$ teaspoon ground white pepper
150 ml white wine

SERVES 12

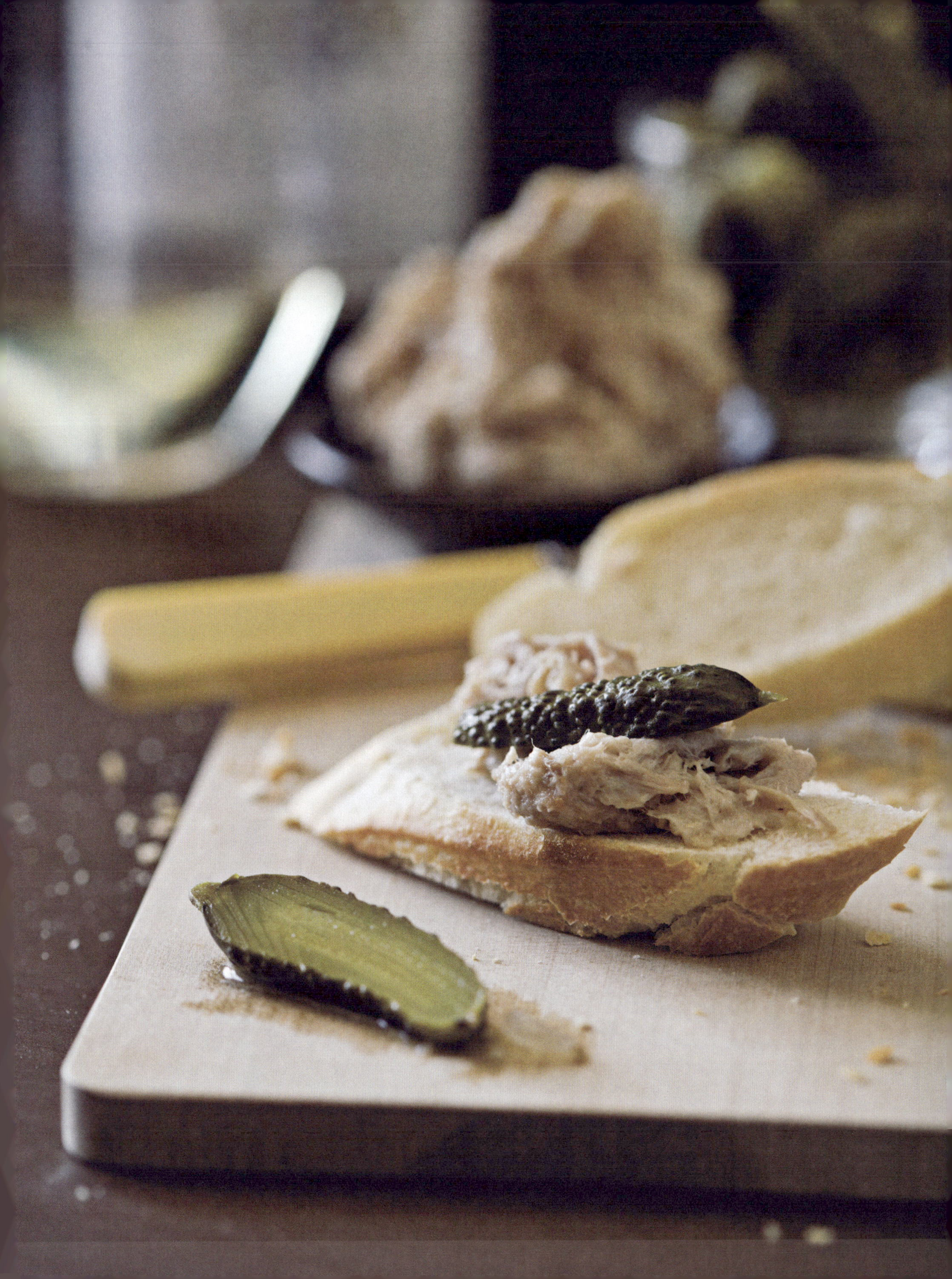

Andre Mahe is our head chef at Must in Perth. He comes from Moëlan sur Mer on the rugged north-west coast of France. At seventeen he began an apprenticeship with his father, who was a traiteur/charcutier. Andre then followed the attractions of an endless summer, so toured Australia working in many restaurants before settling in Perth. I believe he is the finest charcutier in Australia.

My Defining Moment

My first visit to France was in 1985 with my sister, Jenny. We stayed with the Wassermans, American wine negociants who lived in an ancient converted barn near Beaune.

One morning in the tasting room Mr Wasserman placed a dozen bottles of red wine onto a huge map of the Côte de Beaune. We were to taste wines from vineyards situated at different elevations along the Côte, from the highest vineyards to the valley floor: this was to be a lesson in Terroir.

The tasting demonstrated amazing differences in the wines – elegant subtle flavours from the higher vineyards contrasted with the chunkier, fuller wines from the valley floor.

After the tasting we visited each vineyard that produced the grapes made into those twelve wines. I was stunned by their proximity to each other – in some cases a few hundred metres separated premier cru vines from those that produced standard bourgogne.

We then motored for an hour north to Dijon, the historic capital of Burgundy. Once there, we sat at a bistro table on a cobblestone square. This was where I first tasted a ham and parsley terrine: Jambon Persillé. Eaten with a crackling fresh baguette, it was true comfort food: the gelée encasing the pink chunks of salty ham was rich and freshened with the addition of brilliant green parsley.

But then I washed the flavours down with a glass of village white burgundy, and it was then that the combination of food and wine absolutely hit me. This one extraordinary day was a defining moment in my life: there and then I decided to open a French bistro.

Domaine Leflaive Macon-Verze
Burgundy, France

This classily seductive chardonnay has the force and vibrancy that partners well with the meaty pork chunks and rustic terrine flavours of this Burgundy dish. It also has the flavour profile – minerals, warm nuts and citrus pith – and textural softness to balance the sharpness of the herbal mayonnaise that sits alongside the terrine.

Jambon Persillé (Ham & Parsley Terrine)

A classic terrine from the Burgundy region. A quintessential French snack is a slice of this terrine smothered with mayonnaise or tapenade, a crusty baguette and a glass of crisp white wine.

- Place the pork hocks in a large saucepan and cover with water. Bring it slowly to the boil to draw out excess fat and salt, drain off the water and refill the pot with cold water.
- Return to the stove over moderate heat, add the stock ingredients and simmer gently for 3 to 3½ hours until meat is tender and falling off the bone.
- Remove the pork hocks from the stock and place on a tray to cool down until warm to touch.
- Strain the stock through a fine strainer, reserving about 1 litre in a clean saucepan. Return to high heat and reduce to approximately 600 ml.
- Meanwhile, soak the gelatine leaves in cold water for 5 minutes, remove, squeeze and place into the stock stirring as it melts, then remove from the heat and allow to cool to body temperature.
- Trim the skin and fat off the hocks then cut the meat off in large (4 cm) chunks placing them into a bowl. Discard the bones and skin.
- Add the garlic, shallot, parsley, vinegar, and the warm jelly to the bowl. Stir all ingredients gently. Check for seasoning and add a little salt and pepper as required.
- Line the terrine with cling film and place onto a deep tray. Spoon the mixture into the terrine, cover with cling film and lay a tray on top with a kilogram or two of weight to maintain pressure as it sets.
- Place in the refrigerator overnight.
- Turn out, peel off the wrap and cut 1 cm cold slices with a freshly baked baguette, tapenade and lashings of ravigote mayonnaise.

2 large pickled pork hocks (approximately 3 kg)

Stock

1 brown onion, chopped
1 carrot, chopped
1 celery stick, chopped
5 white peppercorns
1 bay leaf
A few parsley stalks
A sprig of thyme
12 gelatine leaves
2 garlic cloves, chopped finely
50 g shallot, chopped finely
160 g flat-leaf parsley, chopped finely
10 ml white wine vinegar
Table salt & ground white pepper

Ravigote mayonnaise (p. 282)
Black olive tapenade (p. 10)
Fresh baguette

SERVES 10

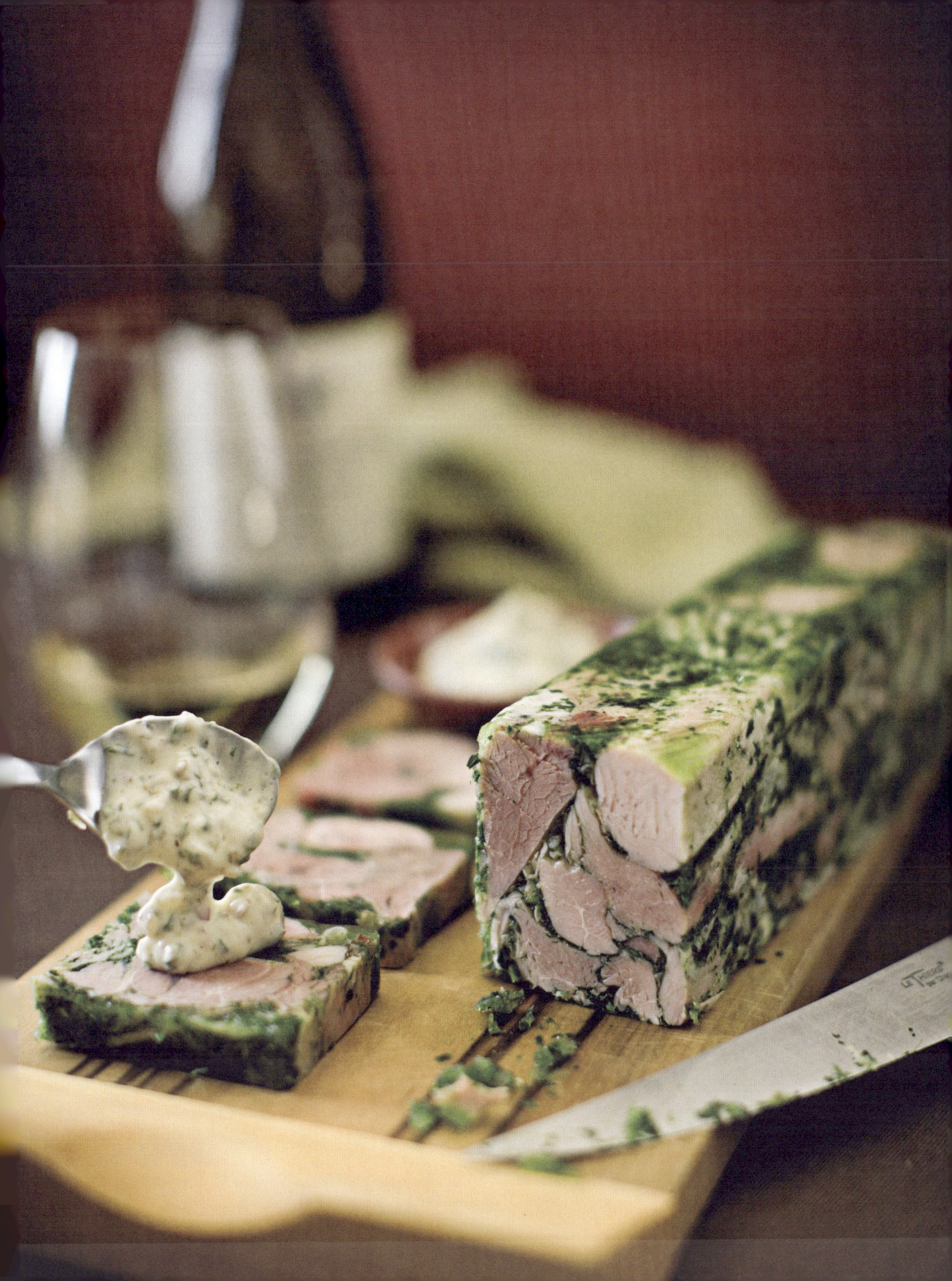

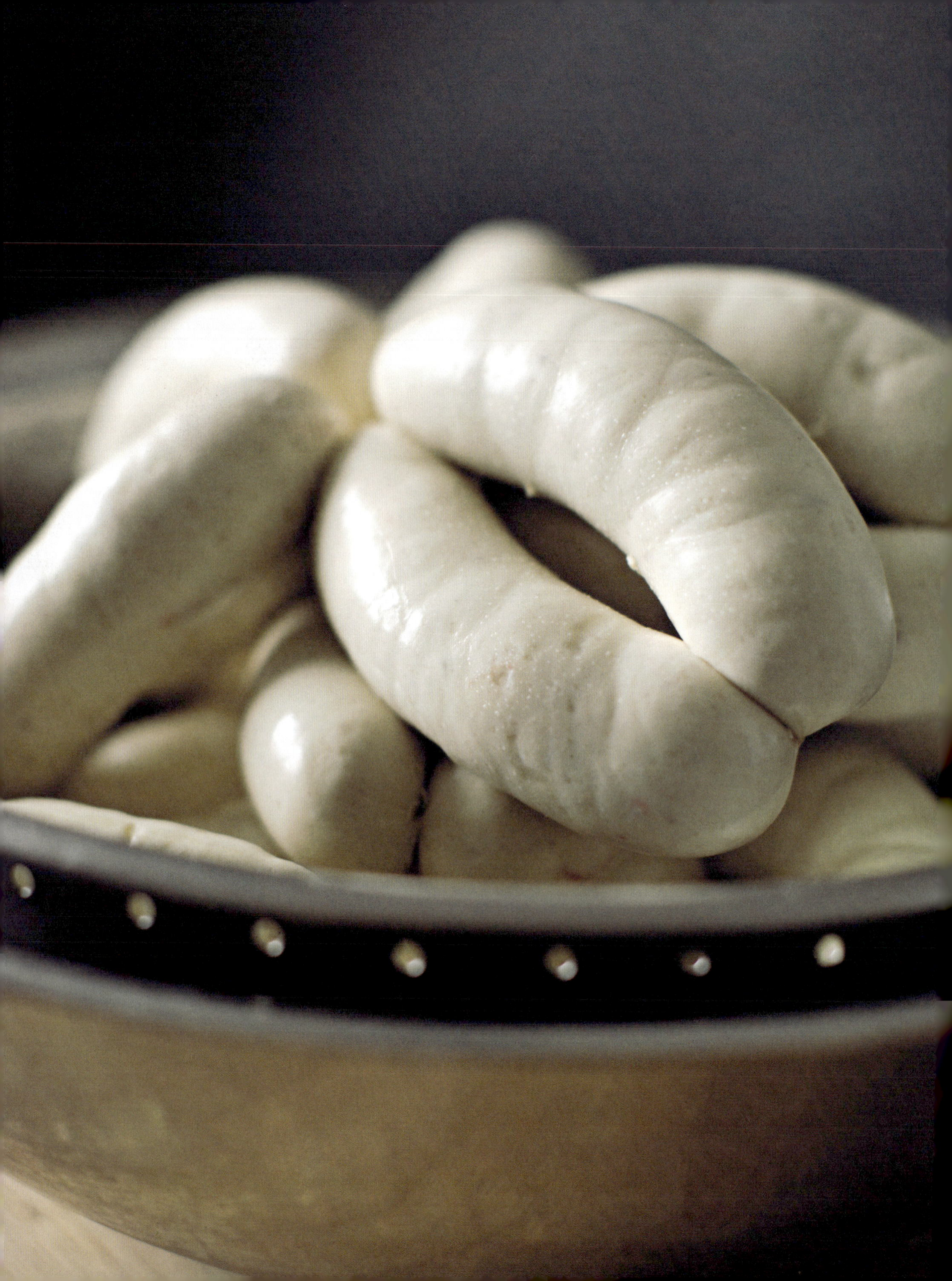

Mount Barker Chicken & Lobster Boudin Blanc

This recipe came about after a discussion on combining two classic partners, chicken and shellfish. Boudin blanc is a white sausage made with light meats, bound with milk, bread, eggs and cream. You'll need a sausage filler to make this recipe.

---

100 g brown onion, chopped

500 ml full-cream milk

1 bay leaf

A pinch of powdered nutmeg

5 slices soft white bread, crusts removed

400 g raw lobster flesh

1 kg chicken breast, finely minced

2 eggs

20 g (1½ tbsp) salt

2 g (1 tsp) ground white pepper

30 ml white vermouth

300 ml whipping cream

35–38 mm pork sausage casing

Butcher's twine

MAKES 10 TO 12 SAUSAGES

Combine the onion, milk, bay leaf and nutmeg in a large saucepan, place over moderate heat and simmer until reduced to barely 100 ml of liquid.

Strain the milk into a bowl, add the bread, mix well and cool in the refrigerator.

Heat a large saucepan of water to a simmer, add a generous handful of salt, dice the lobster into 1 cm cubes and drop into the water to cook for 1 minute. Strain and run a little cold water over the lobster to cool, drain and refrigerate.

Place the chilled soaked bread into the bowl of a food processor and blitz to a fine purée; pour this into a large mixing bowl. Add the chicken mince, eggs, salt, pepper and vermouth and mix well.

Place half of the mix back into the food processor, blitz for 1 minute then add half of the cream in a steady stream (this should take 20 seconds). Remove the first batch and repeat, then combine the two batches. Add the chilled cooked lobster to the mix and stir to combine.

Soak the sausage skins in cold water then rinse well.

Thread sausage skins onto the sausage filling tube, tie the end of the casing with string and fill entire casing firmly with filling, making sure there are no air bubbles. Tie off the end of the casing then tie off sausages to your preferred length (10 to 15 cm).

Heat a large saucepan of water to 80°C and drop sausages in to cook for approximately 20 minutes. Monitor the temperature constantly to ensure that the water does not go over 80°C.

Remove the blanched sausages and drop into cold water – ice on top helps to halt the cooking process.

Drain well and pat dry, these will keep in the refrigerator for 3 to 4 days until required. See p. 152 on how to cook and serve.

Gerhard Tischner has found 'the good life' on his property near Augusta, where he lives with his wife, Jocelyn. A pastry chef by trade, Gerhard worked in hotels in Germany, London, Sydney and Perth before becoming a partner in the Augusta bakery, which he has since sold. Now he invests his time and passion in raising the most glorious chickens, geese and ducks.

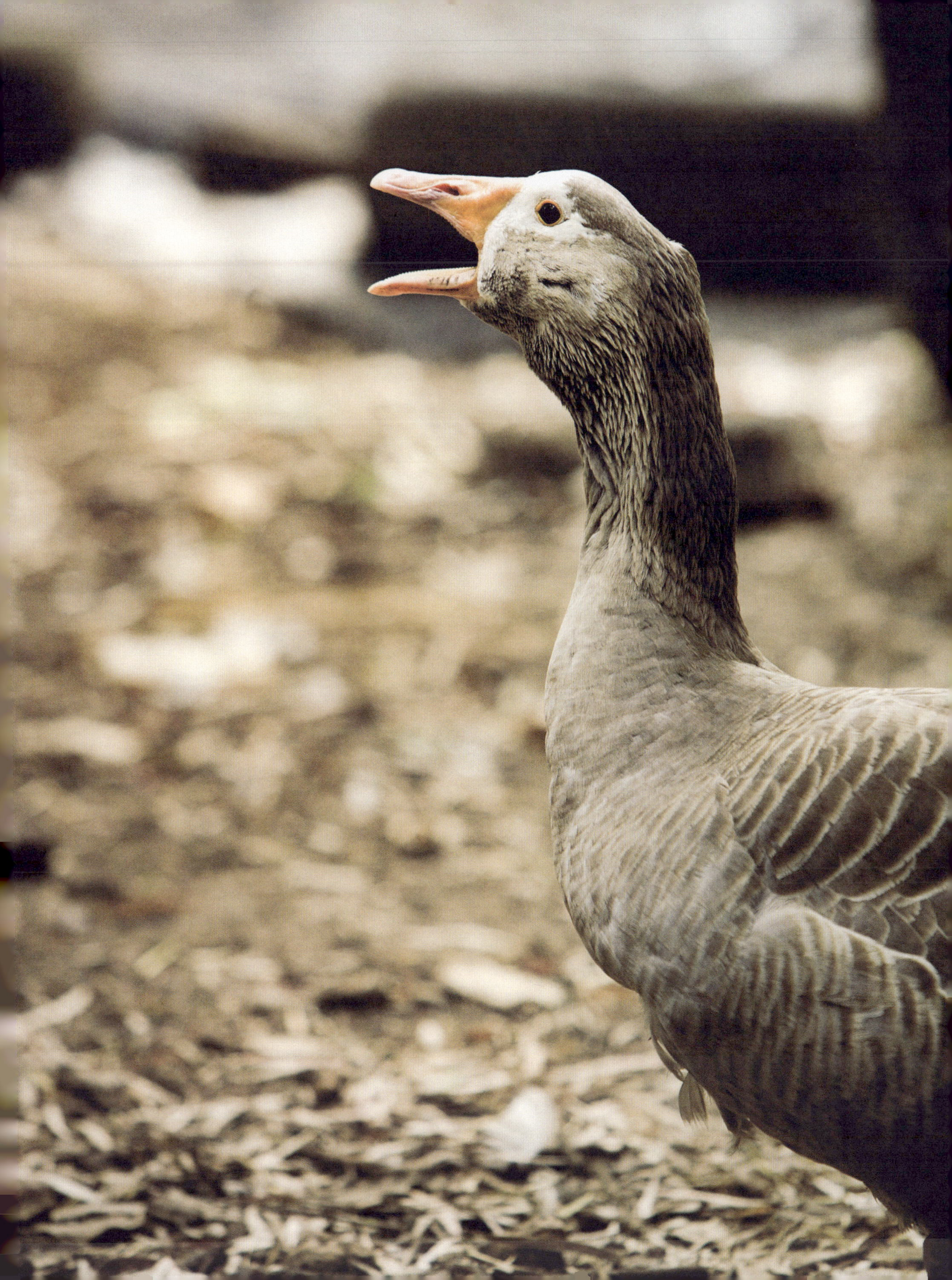

Must Pork Sausages

This is Must Perth's head chef Andre's recipe and it has been on our menu since 2001. Ask your butcher to mince the back fat and pork neck together through a coarse (10 mm) mincer plate. Make sure you work the mix well – the kneading process engages the proteins and helps the contents stick together.

- Combine all the filling ingredients in a large bowl and knead with your hands (use gloves if you must) for 5 minutes.
- Soak the sausage skins in cold water then rinse well.
- Thread sausage skins onto the sausage filling tube, tie the end of the casing with string and fill entire casing firmly with filling, making sure there are no air bubbles. Tie off the end of the casing then tie off sausages to your preferred length (10 to 15 cm).
- Heat a large saucepan of water to 80°C and drop sausages in to cook for approximately 20 minutes. Monitor the temperature constantly to ensure that the water does not go above 80°C.
- Remove the blanched sausages and drop into cold water – ice on top helps to halt the cooking process.
- Drain well and pat dry, these will keep in the refrigerator for 3 to 4 days until required. See p. 177 on how to cook and serve.

500 g pork back fat, minced
1.5 kg lean pork neck, minced
1 tablespoon thyme, chopped
1 tablespoon rosemary, chopped
3 tablespoons flat-leaf parsley, chopped finely
30 g (2 tbsp) salt
2 g (1 tsp) ground white pepper
5 g (1 tbsp) quatre épices (p. 272)
40 g (6 cloves) minced garlic
200 ml white wine
3 egg whites

35–38 mm pork sausage casing skins
Butcher's twine

MAKES 10–12 LARGE SAUSAGES

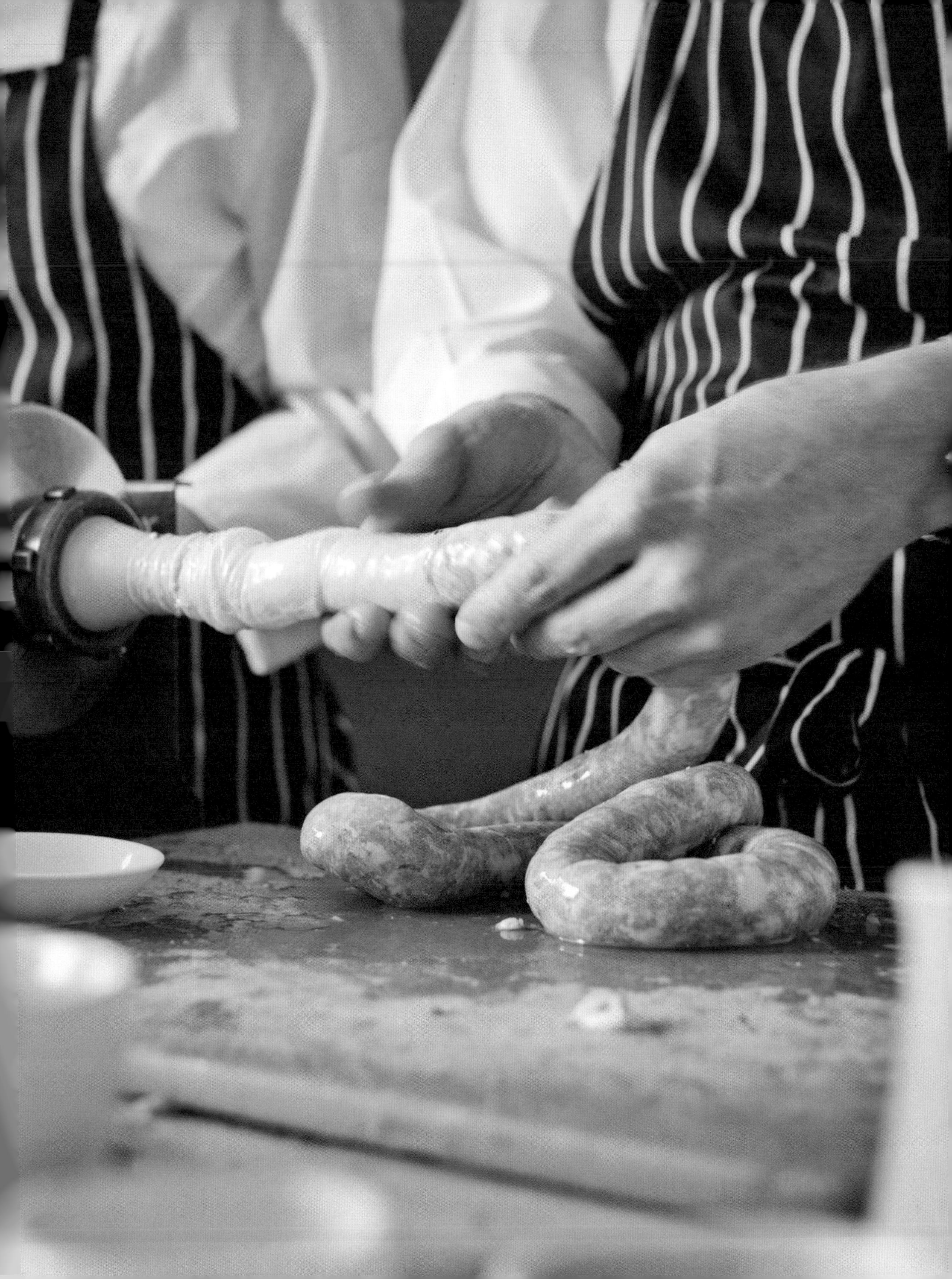

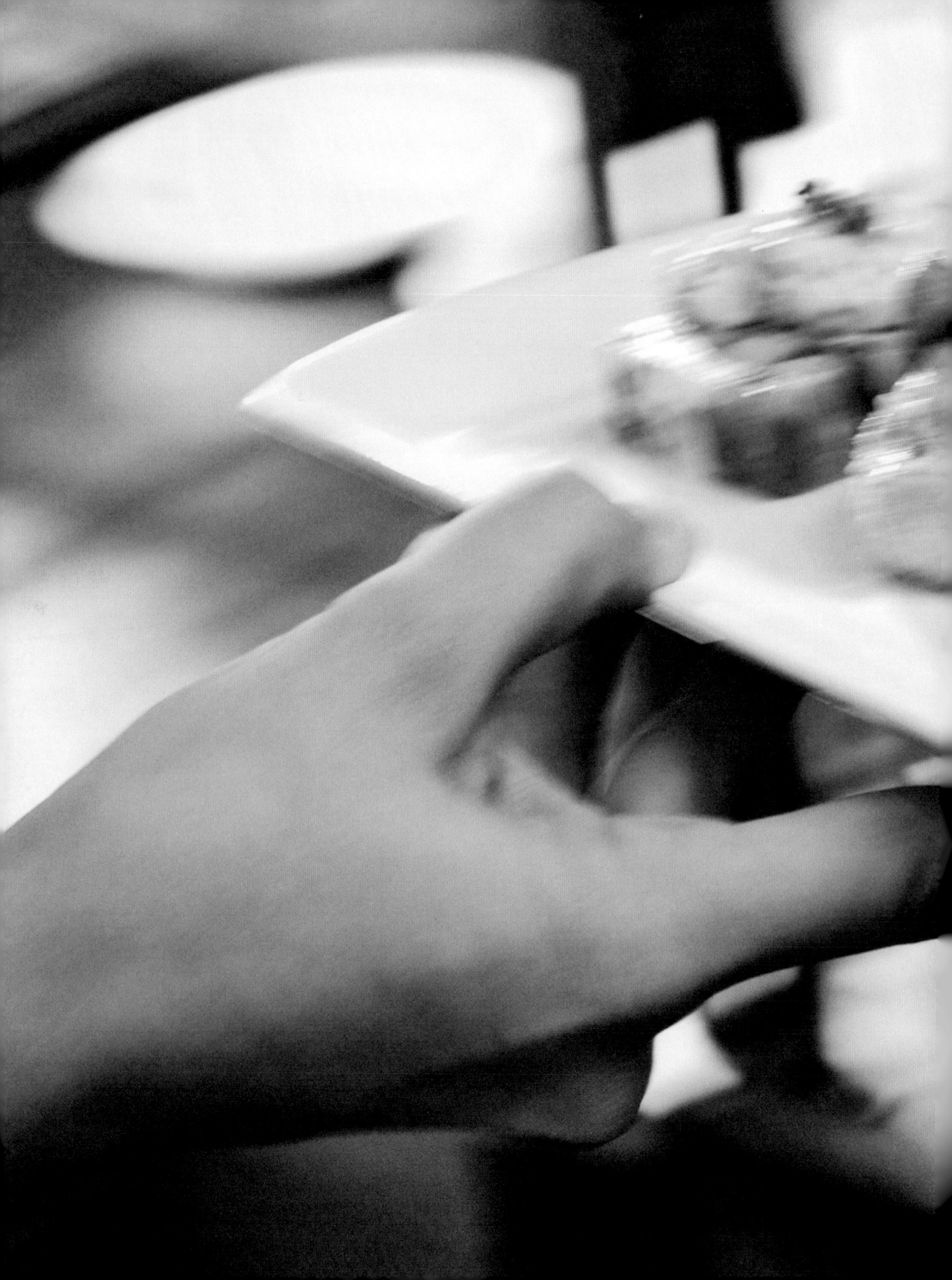

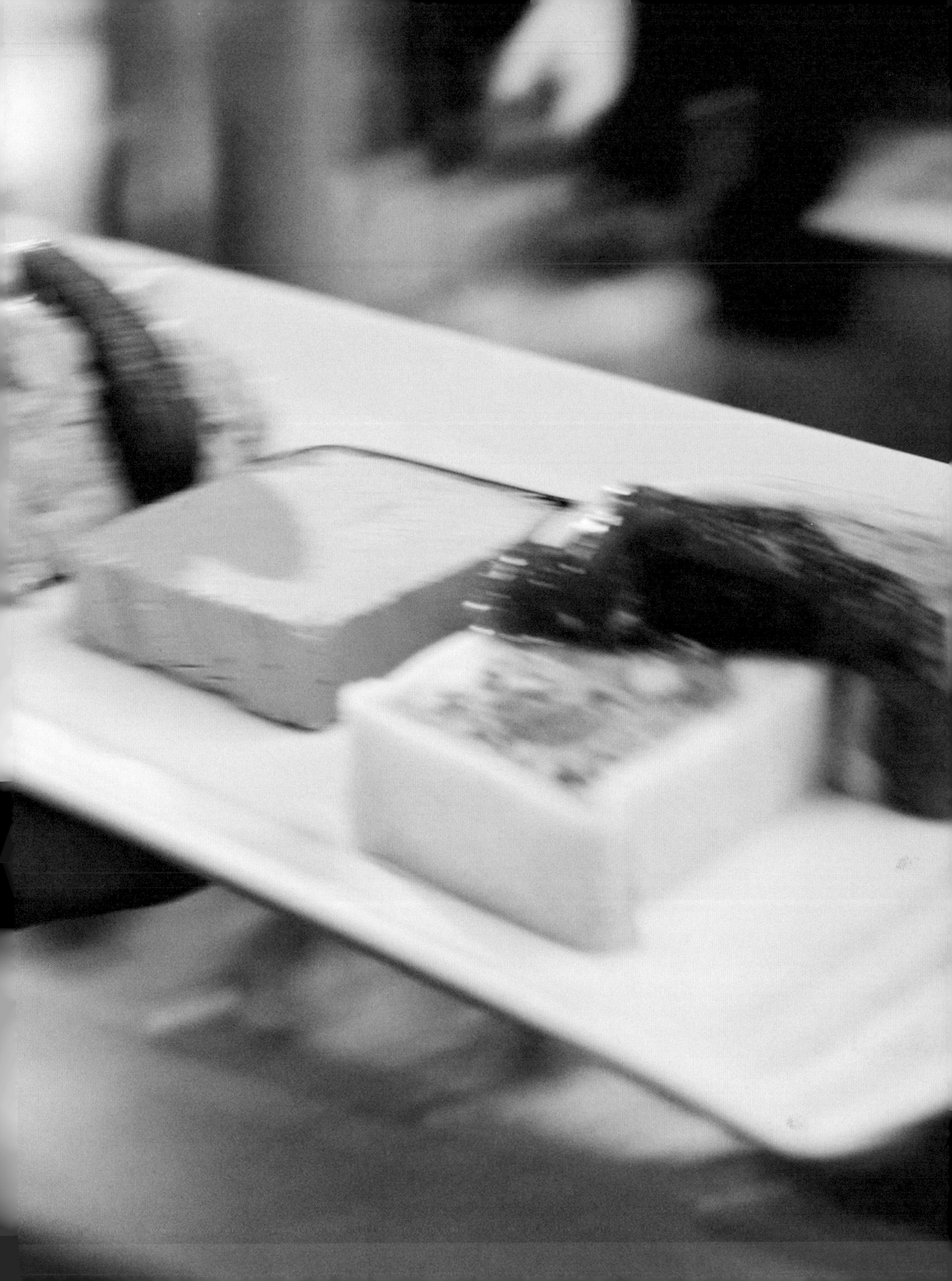

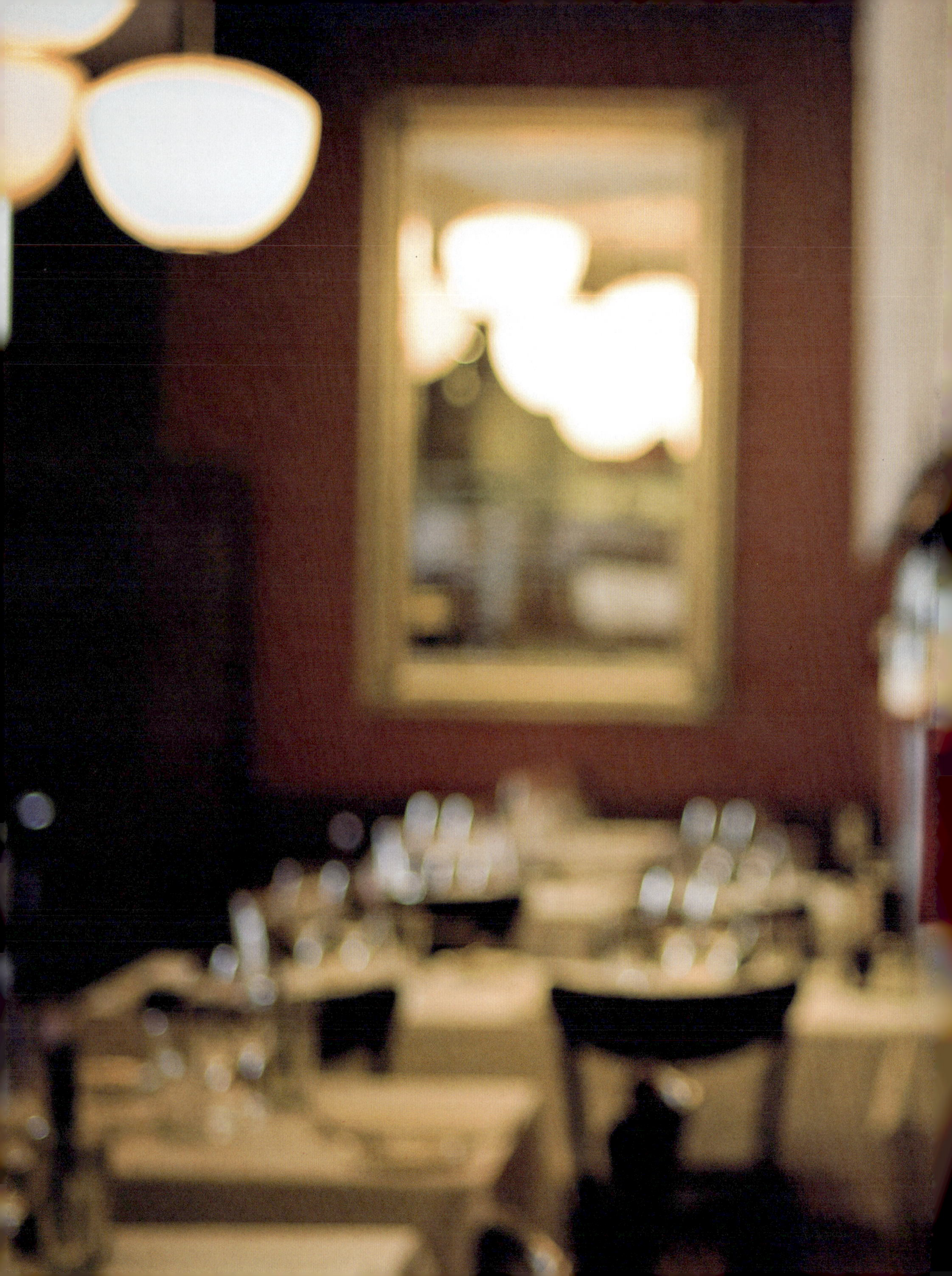

main

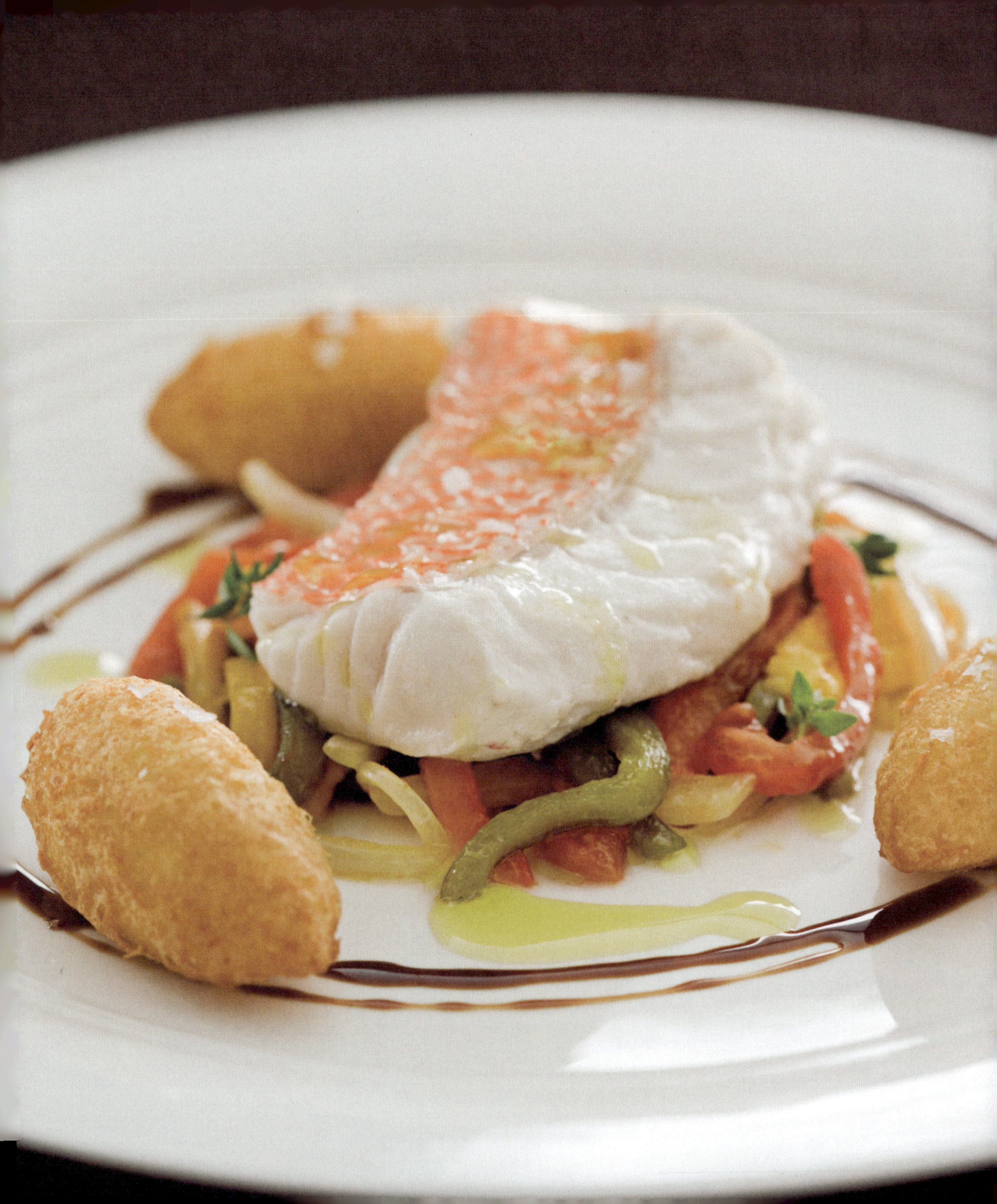

Steamed Augusta Nannygai Fillet, Piperade & Dauphine Potatoes

Brothers Brody and Jarrod Craven supply Must Margaret River with the most amazing seafood from fishermen in Augusta and Windy Harbour. One of the fish is nannygai, which has a succulent sweet white flesh that contrasts with its bright orange skin. Ask your fishmonger to scale the fillets for you as this fish has razor-sharp scales.

---

6 x 150–180 g nannygai fillets with skin (can substitute with pink snapper)

Sea salt & freshly ground pepper

Piperade

50 ml extra virgin olive oil

1 large brown onion, sliced

2 garlic cloves, crushed

A sprig of fresh thyme

1 red capsicum, de-seeded and sliced

1 green capsicum, de-seeded and sliced

2 large tomatoes, peeled, de-seeded and roughly chopped

¼ teaspoon paprika

Sea salt & freshly ground black pepper

Dauphine Potatoes

60 g butter

¼ cup water

¼ cup milk

1 teaspoon salt

½ cup sifted flour

2 x 50 g eggs, lightly beaten

A pinch of grated nutmeg

A pinch of white pepper

500 g royal blue potatoes

Canola or light olive oil, for frying

MAKES 6 MAIN COURSES

To make the piperade, heat the olive oil in a large saucepan over medium heat. Add the onion, garlic and capsicums and sauté for 1 minute, reduce heat, add the garlic, paprika and thyme, stirring occasionally, for 10 minutes, until the vegetables have softened. Add the tomatoes and simmer, uncovered, for 15 to 20 minutes, until most of the liquid has evaporated. Season and keep warm.

To make the potatoes, place the butter, water, milk and salt into a saucepan, bring to a simmer then add the flour, stirring well to ensure no lumps form. Stir over low heat for 2 minutes until the mixture leaves the sides of the saucepan and forms a ball when stirred. Remove from the heat.

Add the eggs to the hot mixture one at a time, beating constantly with a wooden spoon until incorporated. Place a piece of baking paper over the flour mixture and reserve.

Cook the potatoes in salted water until tender; drain, allow to steam off for a couple of minutes then pass through a potato ricer or fine strainer. Add the flour mixture, nutmeg and white pepper and stir through potatoes. Taste for seasoning. Refrigerate.

Heat the oil in a saucepan or deep fryer to 190°C, scoop 3 cm oval shapes with a spoon and drop into the hot oil. Cook until golden and crisp on the outside. Drain on kitchen paper; keep warm.

Season fish well and place on a plate or tray. Place this into a large steamer set over boiling water to cook for 6 minutes. Check to see if the fish is cooked by pressing the flesh with your finger – it will separate and flake when cooked. Place a dob of piperade onto each plate, topped with the fish and surrounded by 2 or 3 dauphine potatoes. You could drizzle the dish with a little Pedro Ximénez lacquer (p. 282) if you wish.

Ocean Eight 'Verve' Chardonnay
Mornington Peninsula, Victoria

This very contemporary chardonnay, with brightness, elegance and mineral complexity, pairs well with the fish's juicy flesh. It is slender and energetic and lifts the flavours, providing a clean, bright finish.

Viva Matilda Falanghina
Campania, Italy

Although Falanghina is obscure, it has typical Italian charm – quite rich in flavour, with a sense of opulence and textural softness. It complements the sweet flesh of the barramundi, and its savoury elegance contrasts with the soft potato and artichoke.

Crisp-Skinned Cone Bay Barramundi, Warm Potato, Artichoke, Olive & Truss Tomato Salad

Cone Bay barramundi is farmed in the pristine Buccaneer Archipelago north of Broome. The pristine sea water provides a clean flavour that land-locked ponds can't produce. The sorrel butter came about when one of our customers in Margaret River, Wendy Ellis, dropped into the kitchen with a bunch of homegrown sorrel. It gives a unique lemon-herbal flavour to the beurre blanc.

....

Preheat the oven to 220°C.

Wash the potatoes and slice lengthways. Season well and drizzle with olive oil. Place onto a baking tray and roast for about 20 minutes until golden brown and tender. Reserve in a warm place.

Drizzle the truss tomatoes with olive oil and season well. Place onto a baking tray and into the oven for 5 minutes until the skin begins to blister; reserve.

Heat a large (preferably non-stick) frypan over medium-high heat.

Season the barramundi on both sides, drizzle a little oil into the pan and place the fish skin side down into the pan. Cook for 1 to $1\frac{1}{2}$ minutes, then carefully turn over.

Reduce heat and cook a further 1 minute. If your frying pan is ovenproof, place into the oven; if not, place the fish on an oven tray and into the oven.

Cook for a further 3 minutes. Check to see if the fish is cooked by pressing on the fillet with your finger – the barramundi is cooked when this feels very soft, as though you could press your finger all the way through the flesh.

Divide the potatoes and salad ingredients among plates, top with the fish, mix the sorrel with the butter sauce and drizzle over the top.

6 Kipfler potatoes
Extra virgin olive oil
6 bunches of baby truss tomatoes
6 barramundi fillets, with skin
6 baby artichokes
24 kalamata olives
2 tablespoons of capers
Sea salt & freshly ground pepper
Lemon beurre blanc (p. 284)
About $\frac{1}{4}$ cup of fresh sorrel leaves, sliced finely

SERVES 6

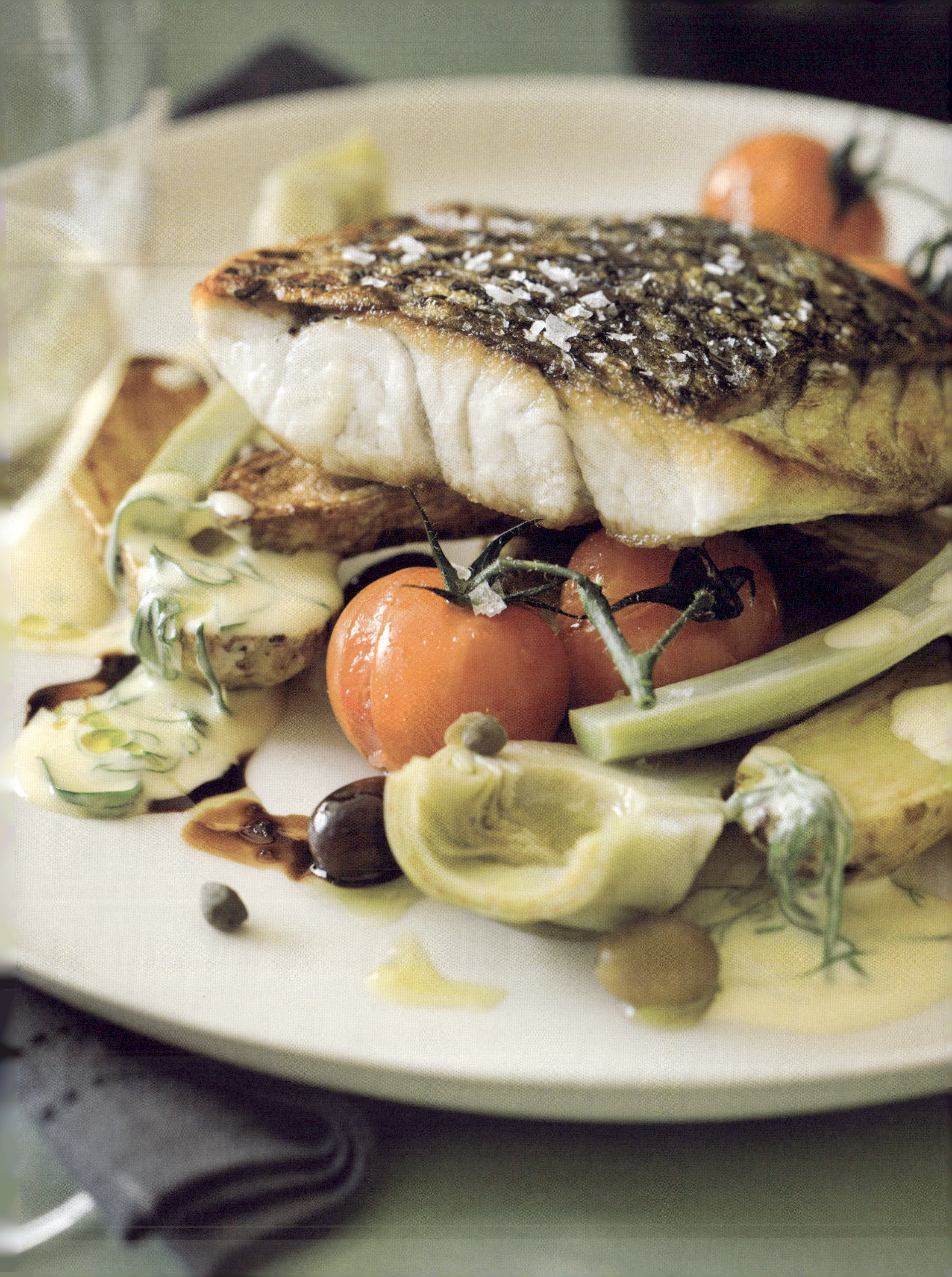

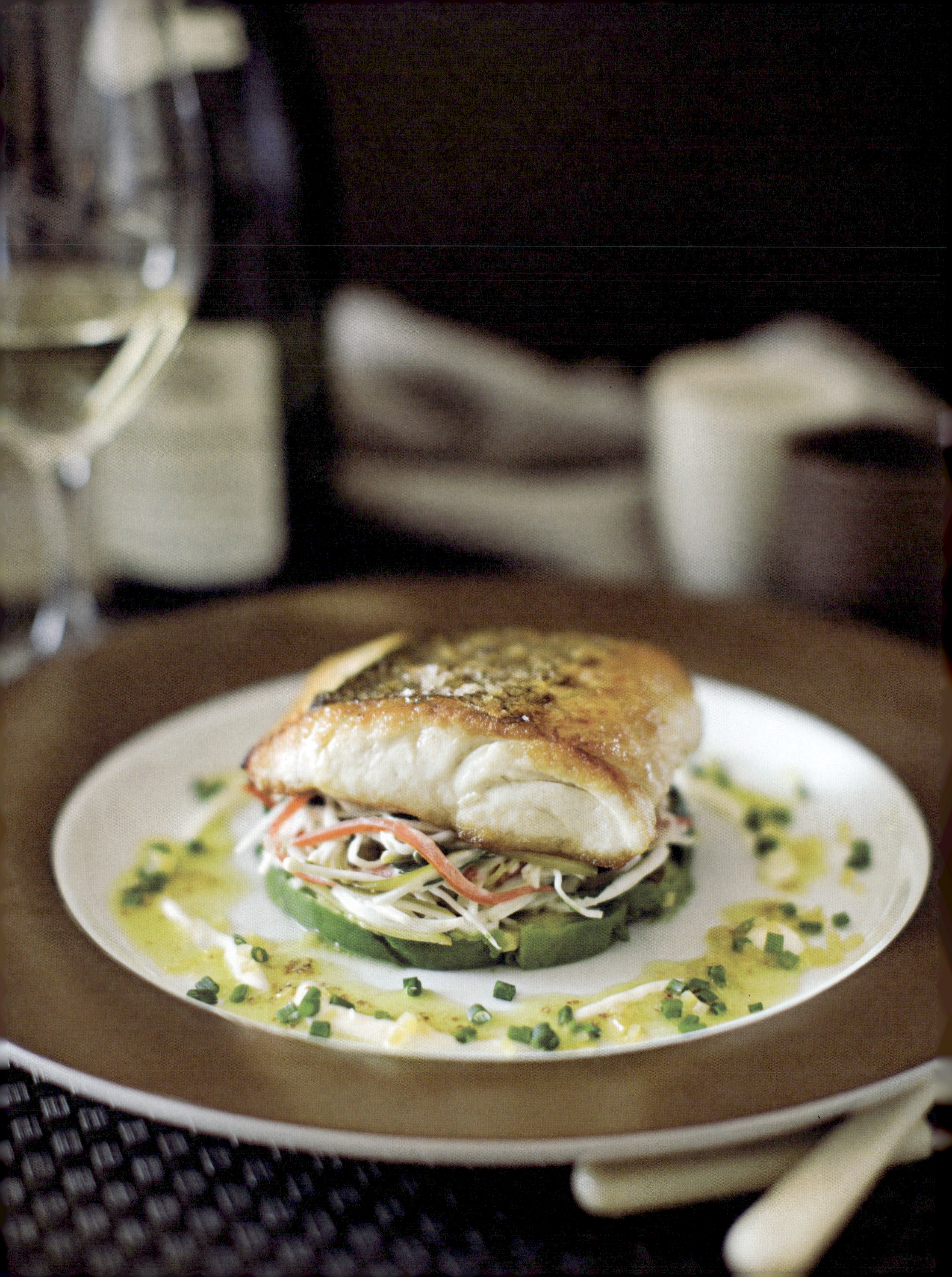

Grilled Shark Bay Pink Snapper, Celeriac & Avocado Remoulade & Preserved Lemon Vinaigrette

Kieran and Tory Wardle run the homestead on Dirk Hartog Island, near Shark Bay. In 2008, for their wedding, we created a shared table banquet for forty-five people and served snapper that Kieran had caught the day before.

6 pink snapper fillets, with skin
Sea salt & freshly ground black pepper
Olive oil
40 g unsalted butter
1 tablespoon freshly chopped chives
Preserved lemon vinaigrette (p. 268)

Remoulade
3 small avocados
$\frac{1}{2}$ a head of celeriac, peeled and washed
Small amount of shaved red onion
2 Roma tomatoes, de-seeded and cut into long thin strips
1 tablespoon tiny capers
1 tablespoon cornichons, thinly sliced
2 tablespoons flat-leaf parsley, coarsely chopped
Remoulade dressing (p. 268)

SERVES 6

Preheat the oven to 220°C.

Heat a large (preferably non stick) frypan over medium-high heat.

Season the snapper on both sides, drizzle a little oil into the pan and drop in the butter; when the butter is foaming place the fish skin side down into the pan. Allow the fish to cook for 1 to $1\frac{1}{2}$ minutes, then carefully turn over. Reduce the heat and cook for a further 1 minute.

Place the fish on an oven tray into the oven and cook about 2 minutes; it will be cooked when the fish flakes when pressed with the tip of your finger.

Cut the avocados in half, slice and lay on the centre of each plate.

Toss the remaining remoulade ingredients together, lightly coat with the dressing, reserving some to drizzle onto the plates.

Place the remoulade salad on top of the avocado. Drizzle the preserved lemon dressing around the plate, follow with a little of the remoulade dressing.

Place the cooked fish on top of the remoulade and serve immediately.

Domaine Raveneau Premier Cru 'Montee de Tonnere'
Chablis, France

This is a fabulous, authentic Chablis that is taut and restrained when young, but builds into a glorious chardonnay with some age. It is a fine match with the fish, and its complexity (wet-stones) pairs well with the earthy, celeriac remoulade.

Domaine Matassa Cuvee Nouge
Côtes de Catalanes, France

You could go with a safe option here, but why not consider something quirky? Domaine Matassa has hints of burnt matchstick, minerals and earthiness, but then the palate opens with layers of flavour – citrus, musk and melon. It's a very interesting combination with bouillabaisse.

Bouillabaisse

Every household in the Bouches de Rhône region will have their own rules for making bouillabaisse, and even though my recipe steers a little away from the traditional version, it has the key flavours that remind me why it is one of the greatest fish dishes on the planet. If you have difficulty purchasing small fish, cut larger fish into chunks or tranches, keeping the bone intact.

Preheat the oven to 170°C.

To make the rouille, wrap the garlic cloves in foil and bake in the oven for approximately 10 minutes until soft; allow to cool.

Squeeze the soft garlic into the bowl of a food processor with the mustard, cayenne and sea salt.

Moisten the saffron with a little hot water and add to the mixture.

Blitz to incorporate the ingredients. Then, with the machine running, slowly drizzle in the olive oil to emulsify with the other ingredients. This should take about 1 minute. Test for seasoning and reserve in refrigerator.

Pour the olive oil into your largest saucepan and place over low-medium heat. Add the onion, fennel, leek, garlic and fennel seeds and cook slowly for 10 minutes, until the vegetables have softened.

Add the potatoes, saffron, tomatoes and herbs and cook for a further 5 minutes stirring constantly. Add 1 or 2 cups of fish stock and bring to a rapid boil. Cook for another 6 minutes.

Break the crabs in half and add to the stock, together with the larger fish you have sourced. Cover the seafood with more fish stock and bring to the boil.

After 1 to 2 minutes add the smaller fish and mussels, ensuring they are covered by the stock, and bring the broth to a rapid boil. Once the fish is cooked and the mussels have sprung open, sprinkle in the Pernod and turn off the heat.

Carefully lift the seafood out of the broth and into a large soup tureen.

Check the broth for seasoning, adding a little salt if required then ladle it, together with the cooked vegetables, into the tureen.

Place a piece of baguette onto the base of each serving plate and ladle the broth, fish and vegetables over the top. Serve with a dish of rouille on the side, ready to dob on top at the last moment.

Rouille
4 garlic cloves, unpeeled
2 egg yolks
1 teaspoon Dijon mustard
$\frac{1}{2}$ teaspoon ground cayenne pepper
1 teaspoon sea salt
$\frac{1}{2}$ teaspoon saffron strands, lightly roasted, crushed
1 cup olive oil

Bouillabaisse
90 ml extra virgin olive oil
1 brown onion, chopped into 1 cm dice
1 head of fennel, chopped into 1 cm dice
1 leek, white part, chopped into 1cm dice
3 garlic cloves, minced
$\frac{1}{4}$ teaspoon crushed fennel seeds
500 g potatoes, peeled & cut into 1cm cubes
1 teaspoon saffron strands
400 g tomatoes, peeled & chopped
1 large sprig of thyme
1 bay leaf
1 strip of orange rind
2 litres fish stock (p. 275)
2 or 3 raw blue manna crabs, carapace & dead man's fingers removed & washed
2.5 kg of four or five varieties of small fish, scaled, gilled & gutted
A handful of mussels, scrubbed & de-bearded
30 ml Pernod

1 baguette, sliced & brushed with olive oil & garlic & crisped in the oven
A sprinkle of flat-leaf parsley

SERVES 6–8

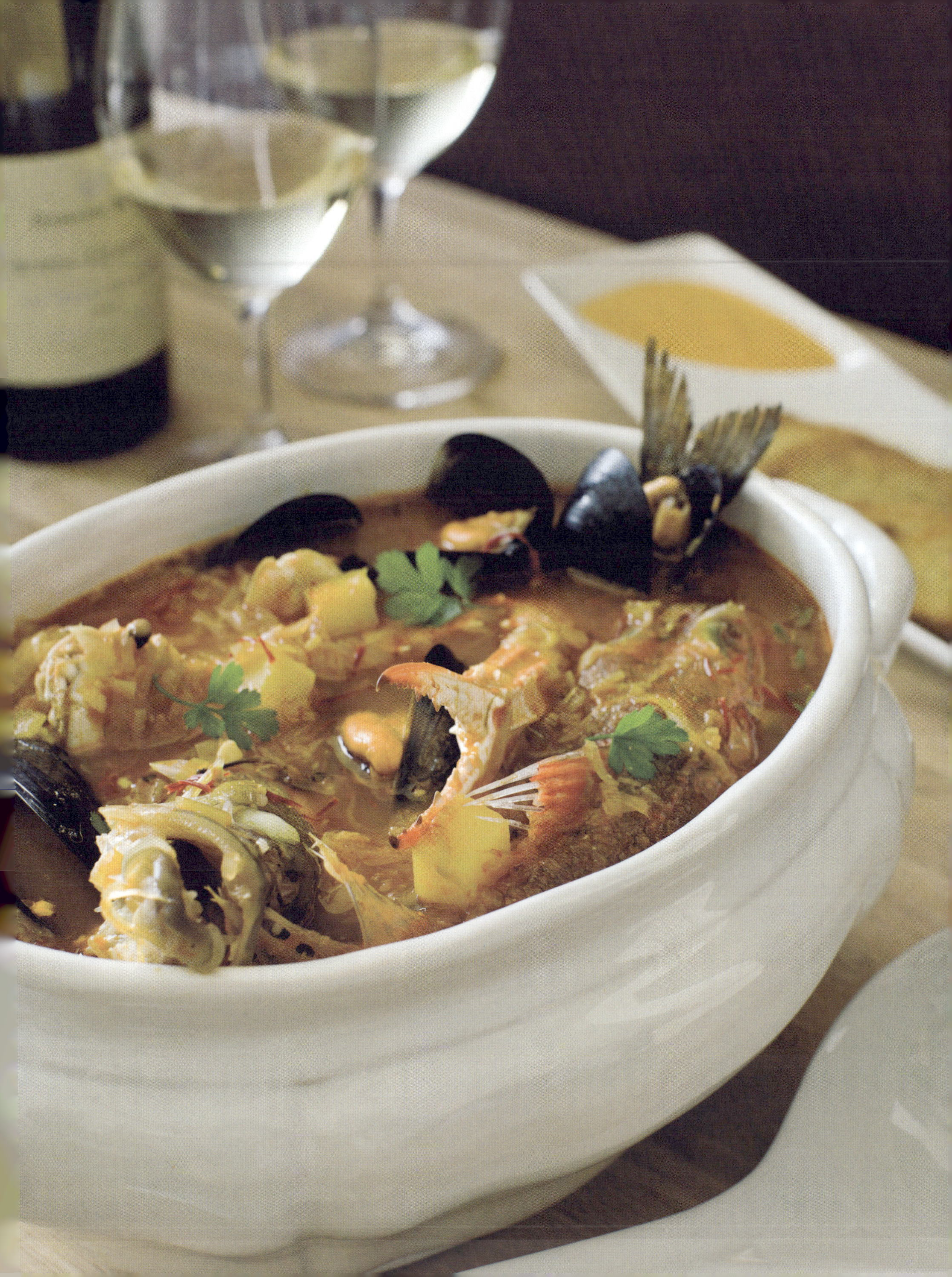

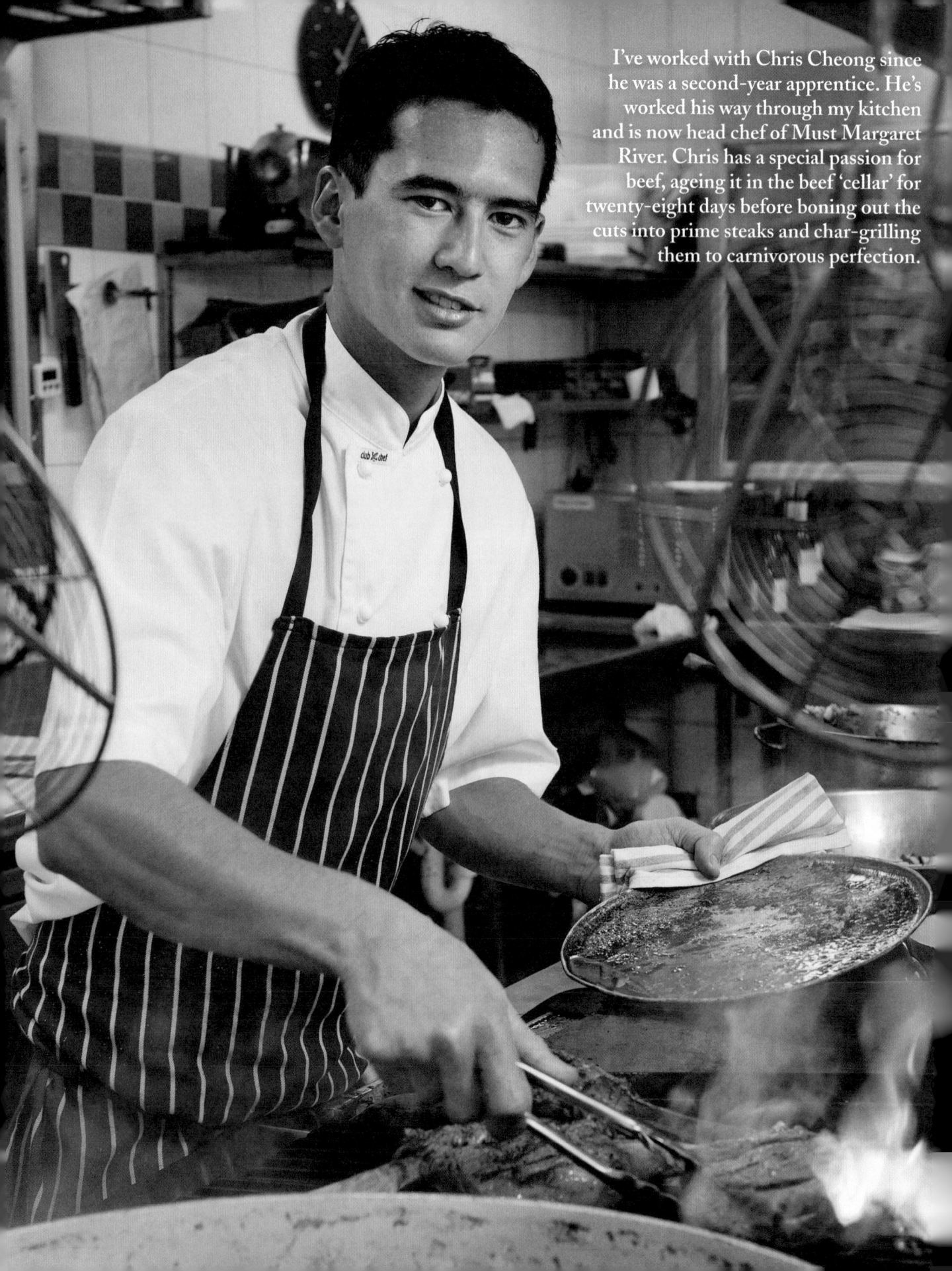

I've worked with Chris Cheong since he was a second-year apprentice. He's worked his way through my kitchen and is now head chef of Must Margaret River. Chris has a special passion for beef, ageing it in the beef 'cellar' for twenty-eight days before boning out the cuts into prime steaks and char-grilling them to carnivorous perfection.

Paella

It was September 1986 and I had parked our Kombi in the ancient part of Valencia, Spain, and walked with Tam (my then girlfriend, now wife) to a renowned paella restaurant off the Plaza de la Reina.

We entered through fortress-like wooden doors into a dining room with rough stone walls, dotted with colourful ceramic plates.

We ordered seafood paella and, while it was being prepared, I pushed open the kitchen door to sneak a look.

There were three matronly women in headscarfs slaving over a massive cast-iron hob that radiated immense heat. They were working on eight paellas, which were scattered over the stove at different stages of cooking – the pans near the centre were rapidly bubbling away, while a couple rested on the side, almost ready to be sent to the dining room. It smelled amazing. While I was taking in the scene, the head cook spotted me. She began hurling a stream of obscenities at me, waving her arms in an effort to shoo me out of the kitchen and leaving me in no doubt that I wasn't welcome there.

When the dish came, our seafood, which included prawns, mussels, cuttlefish and clams of some type, tasted briny and sweet. The rice had soaked up the flavours from the stock and the seafood had formed a crust on the base of the pan that was caramel-luxe.

We got back to the Kombi close to midnight to find it had been ransacked of cameras, backpacks and port. Truly an unforgettable evening.

I've nurtured the following recipe for years in an effort to re-create the flavours of the paella we ate that night and often cook it for friends. Indeed, cooking paella is a social event and a performance of sorts as the ingredients are slowly, individually, added to the bubbling pan. It is the perfect party food.

Marques de Murrieta Reserva Capellania White
Rioja, Spain

The option here is a sturdy white wine or a lighter, fleshy red. The selection has some age and has oak-infused fruit, which provides a wealth of flavour to complement the seafood, chicken, chorizo and spices. The wine has weight to contrast and balance the grainy, rustic mouth-feel.

My Seafood Paella

I always take my paella pan and burner to Margaret River over summer and cook this recipe for our friends. To make it you will need a 40–45 cm wide paella pan, though you can scale down the pan size and quantities when cooking for fewer people.

Bring 2 cups of shellfish stock to a rolling boil in a large pot, tip in mussels, fit lid, and steam for 1 minute until they open.

Strain the mussels through a colander, reserving the cooking liquor. Cool the mussels.

Separate the flesh from three-quarters of the mussels and discard these shells. Use the flesh and remaining shells, discarding any that have failed to open.

Cook the spring onion, garlic and chilli in olive oil over moderate heat in paella pan until onion is translucent. Add the chicken and chorizo sausage, cook for 1 minute. Add the capsicum and cook for a further 2 minutes, stirring all the time with a wooden spoon.

Add the saffron and paprika then the rice, chickpeas, salt and pepper and cook for a further 2 minutes, stirring all the while until the rice becomes slightly transparent.

Add the mussel cooking liquor and enough shellfish stock to cover the rice; stop stirring. After another 4 to 5 minutes, start to add the remaining ingredients.

Sprinkle over the peas and press the vongole into the top of the rice; sprinkle over the cuttlefish then press the prawns into the rice. Finally, add the scallops, followed by the mussels, lightly pressing the shells into the rice. You may need to place a large tray or a layer of foil over the pan to help steam the seafood that is lying on the top.

Reduce the heat and listen for the faint crackle of the rice 'catching' on the base of the pan. Check that the rice is almost fully cooked, turn off the heat and allow the pan to sit for another 5 minutes.

Serve at the centre of the table sprinkled with flat-leaf parsley and wedges of lemon.

2 litres hot shellfish stock (p. 275)
500 g local mussels, scrubbed & de-bearded
60 ml extra virgin olive oil
1 spring onion, chopped into 1 cm chunks (include the green bits!)
4 garlic cloves, chopped finely
1 red chilli, de-seeded & chopped finely
400 g boneless, skinless free-range chicken thighs, cut into 2 cm chunks
1 x 15 cm chorizo sausage, cut into 5 mm slices
1 small red capsicum, chopped into 1 cm x 2 cm chunks
1 small yellow capsicum, chopped into 1cm x 2 cm chunks
1 teaspoon quality Spanish saffron, lightly roasted
1 teaspoon sweet smoked Spanish paprika
1 kg Arroz Calasspara (Spanish rice for paella)
1 cup cooked chickpeas
2 teaspoons sea salt & freshly ground black pepper
1 cup of freshly podded peas, sugar snap peas or beans
250 g of fresh vongole, rinsed
250 g fresh cuttlefish or squid tubes, cut into thick strips
12 large whole raw prawns, peeled & de-veined
12 plump scallops
$\frac{1}{4}$ cup flat-leaf parsley, chopped
2 fresh lemons, cut into wedges

SERVES 10–12

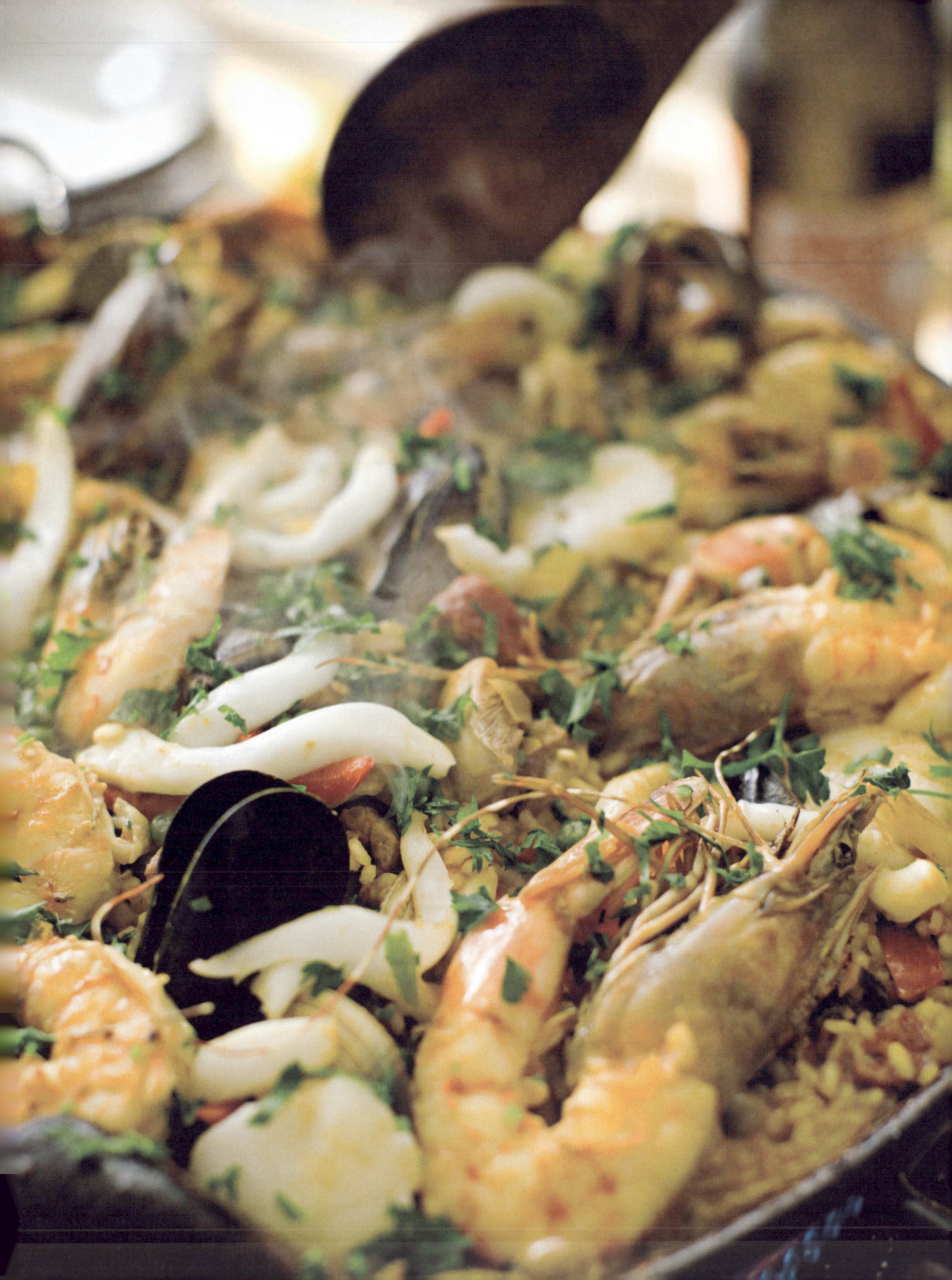

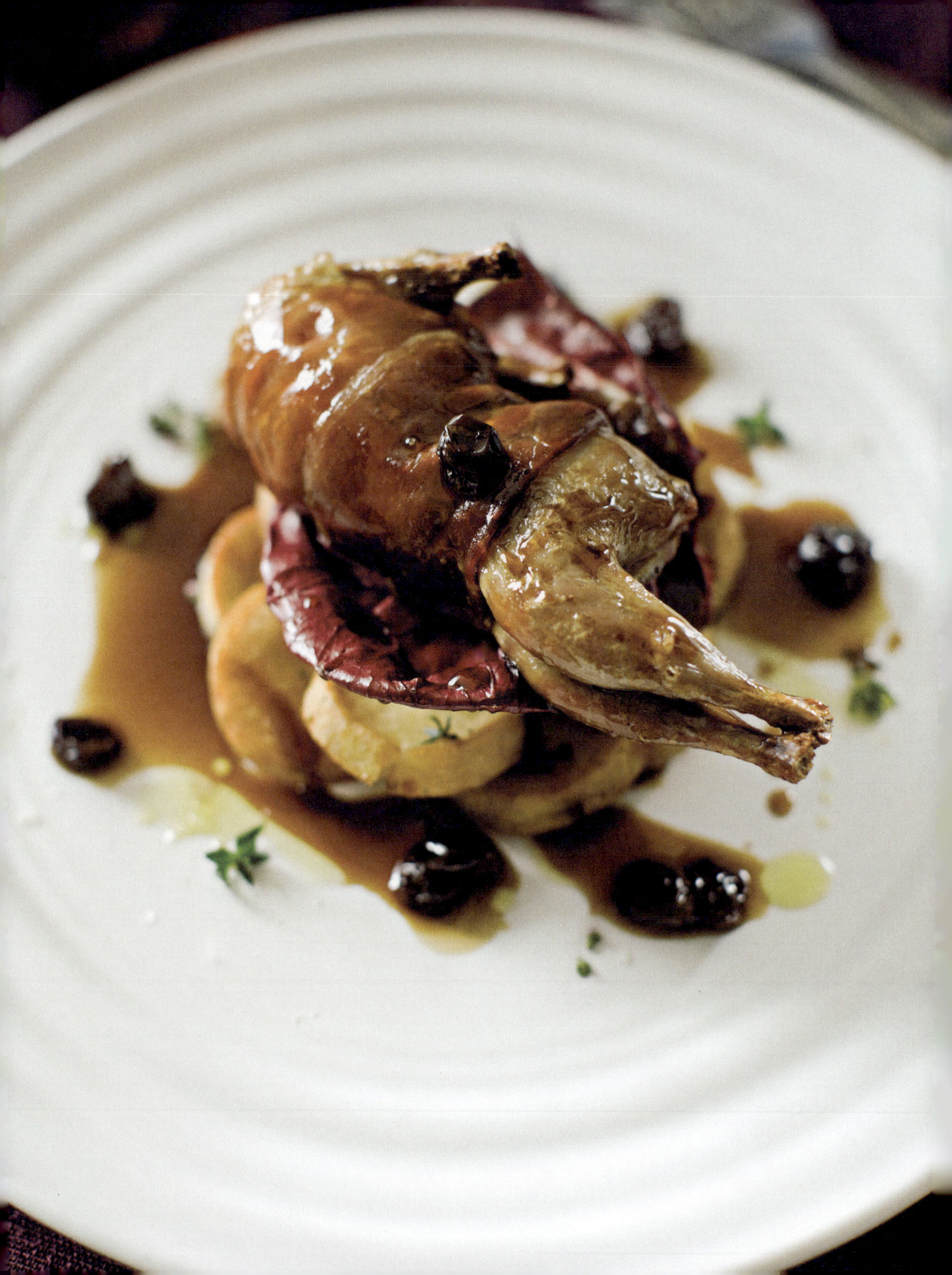

Rotisserie Quail Wrapped in Pancetta, Warm Potato & Radicchio Salad & Sherry-Soaked Flame Grape Jus

This dish combines salty, bitter and sweet – crisp pancetta, radicchio and grapes with quail – into a dish that sings! The secret to the amazing potatoes is to cook them in duck fat. I always reserve duck fat if I roast one at home, storing it in the fridge for cooking potatoes this way.

750 g baby potatoes
Sea salt & freshly ground black pepper
2 garlic cloves, minced
A few sprigs of thyme
50 ml duck fat, melted
150 g radicchio lettuce
Red wine vinaigrette (p. 268)
6 'jumbo' quails, approximately 180 g each
Finely grated zest of 1 lemon
24 fine slices flat pancetta
Butcher's twine
30 ml extra virgin olive oil

Flame Grape Jus

Madeira jus (p. 281)
6 tablespoons sherry-soaked flame grapes (p. 284)

SERVES 6

Preheat the oven to 220°C.

Peel and slice the potatoes 5 mm thick. Place into a bowl, season well and add the minced garlic, a few sprigs of fresh thyme and the melted duck fat. Toss well and place potatoes onto a large baking tray then into the oven for 15 minutes until golden brown and cooked through. Keep warm.

Tear the radicchio lettuce into a bowl, drizzle in a little vinaigrette, toss and reserve.

Partially de-bone the quails, splitting them down the backbone and removing all bones but leaving the leg, thigh and wing bones intact. Place the flattened quails on the bench, season both sides and sprinkle with thyme and lemon zest.

Fold the quails back to their original position and lay four slices of pancetta in a long strip and wrap around the breast section of each quail. Tie the quails with twine, once around the leg-thigh area and and once around the wing-breast area, and rotisserie at approximately 220°C for 8 minutes. Alternatively, seal the quails on both sides for 30 seconds in a hot, non-stick frying pan moistened with a little olive oil, then place onto a tray into the oven at 220°C for 8 minutes. Rest the quails in a warm place for another 3 to 4 minutes while you prepare the plates.

Place some warm potatoes on the base of the plate, top with radicchio, balance a quail on top, sprinkle some of the soaked flame grapes into the Madeira jus and drizzle over the quail.

Muga Rosé
Rioja, Spain

There is lots going on with this quail dish and it requires a wine that will accentuates and contrast both the flavours and textures. This Spanish rosé has fresh stonefruit, a dry mouth-feel that's almost salty, and offers a restrained mineral undertone that complements the darker quail meat, salty prosciutto and sweetness in the salad.

Franz Haas Pinot Nero
Alto Adige, Italy

Richness is essential with this lobster dish and a full-bodied white wine such as Antinori 'Bramito del Cervo' chardonnay from Umbria in Italy would be a safe selection. A red wine option is more challenging, but this pinot nero still combines with the flavours of the white meats, the softness of the risotto and the texture of the beurre blanc.

Mount Barker Chicken & Lobster Boudin Blanc, Caramelised Onion Risotto, Broad Beans & Vermouth Beurre Blanc

Taking the time to make a boudin blanc may seem a daunting process, but the end result – a sausage that is rich yet lightly textured and almost sweet with lobster flavour – is well worth the effort.

Pod the broad beans and reserve. Bring a pot of salted water to the boil, drop in the beans and blanch for 30 seconds; drain and cool the beans in some iced water. Slip the tough skin off the beans and reserve the tender glossy seeds (place in the refrigerator if not using immediately).

To make the risotto, bring the chicken stock to the boil then keep at a slow simmer. Put the oil in a large heavy saucepan over low heat. Add the onion and garlic and sweat until soft. Add the rice and stir to coat in the oil. Start adding the chicken stock, a ladle at a time. Increase the heat to medium and stir for about 5 minutes until the stock has been absorbed. Add the caramelised onion. Continue adding stock, a ladle at a time, stirring constantly until the rice is al dente – this should take about 16 minutes. Stir in the butter and grated parmesan, being careful to ensure the rice is not stodgy (add a little more stock if it is). Season to taste and serve immediately.

To prepare the sausages, heat a saucepan of water to a slow simmer (about 80°C). Drop in the boudins and cook for 10 minutes, making sure the temperature stays at 80°C – any higher and they may burst! Remove boudins from the water to drain; add the broad beans to the water to warm through. Heat a large non-stick frying pan over medium-high heat and add the butter. Once the butter is foaming add the sausages and lightly brown on one side for 20 seconds before turning over. Remove from the pan and keep warm.

Serve the risotto onto large plates, top with a sausage, scatter with broad beans and drizzle with vermouth beurre blanc.

6 chicken & lobster boudins blanc (p. 129)
30 g unsalted butter
1 kg fresh broad beans
Vermouth beurre blanc (p. 284)

Risotto
75 ml olive oil
1 garlic clove, crushed
1 medium onion, chopped finely
300 g Carnaroli rice
750 ml chicken stock
¾ cup caramelised onion (p. 266)
60 g parmesan, grated
40 g butter
Sea salt & freshly ground black pepper

SERVES 6

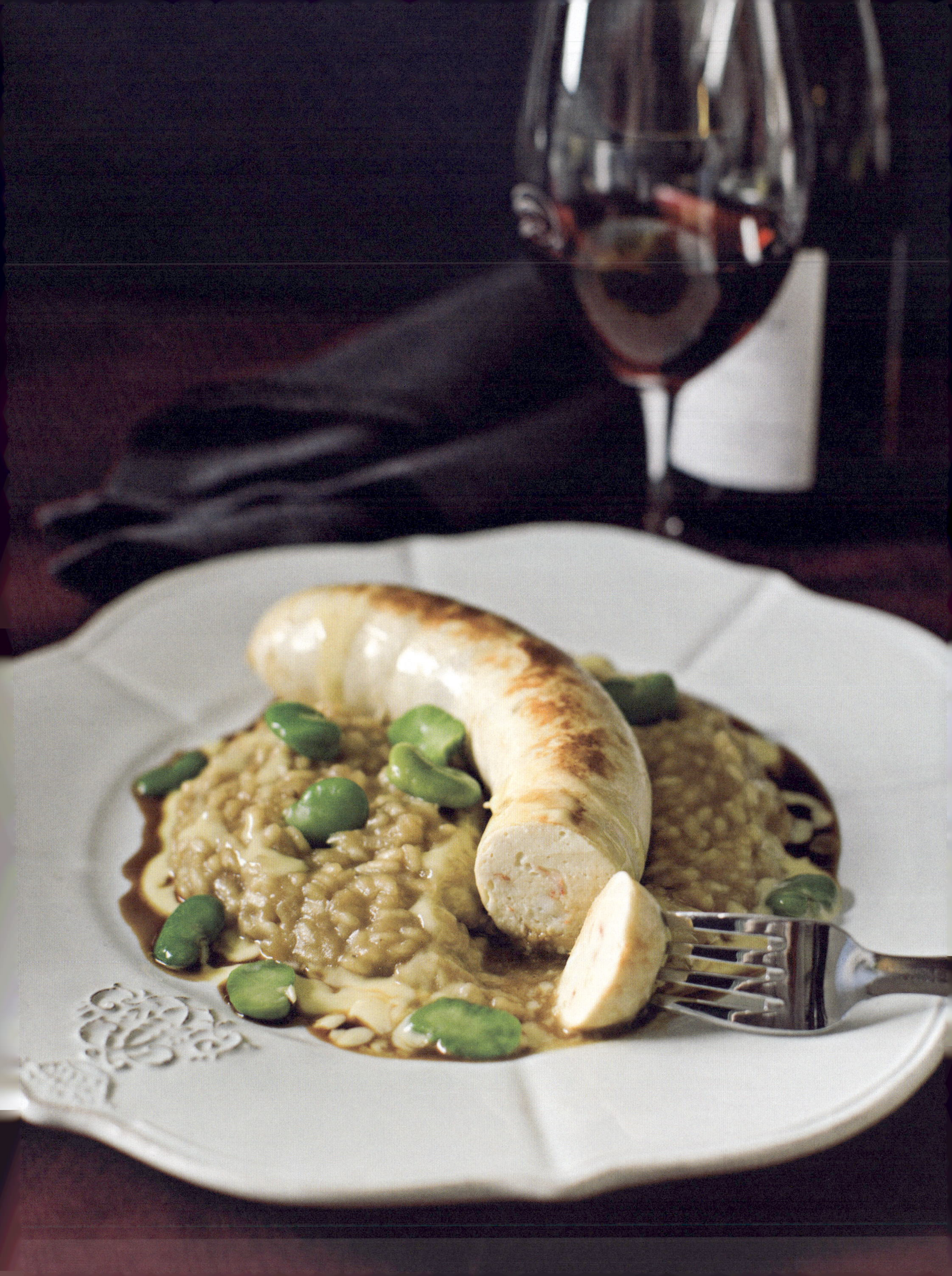

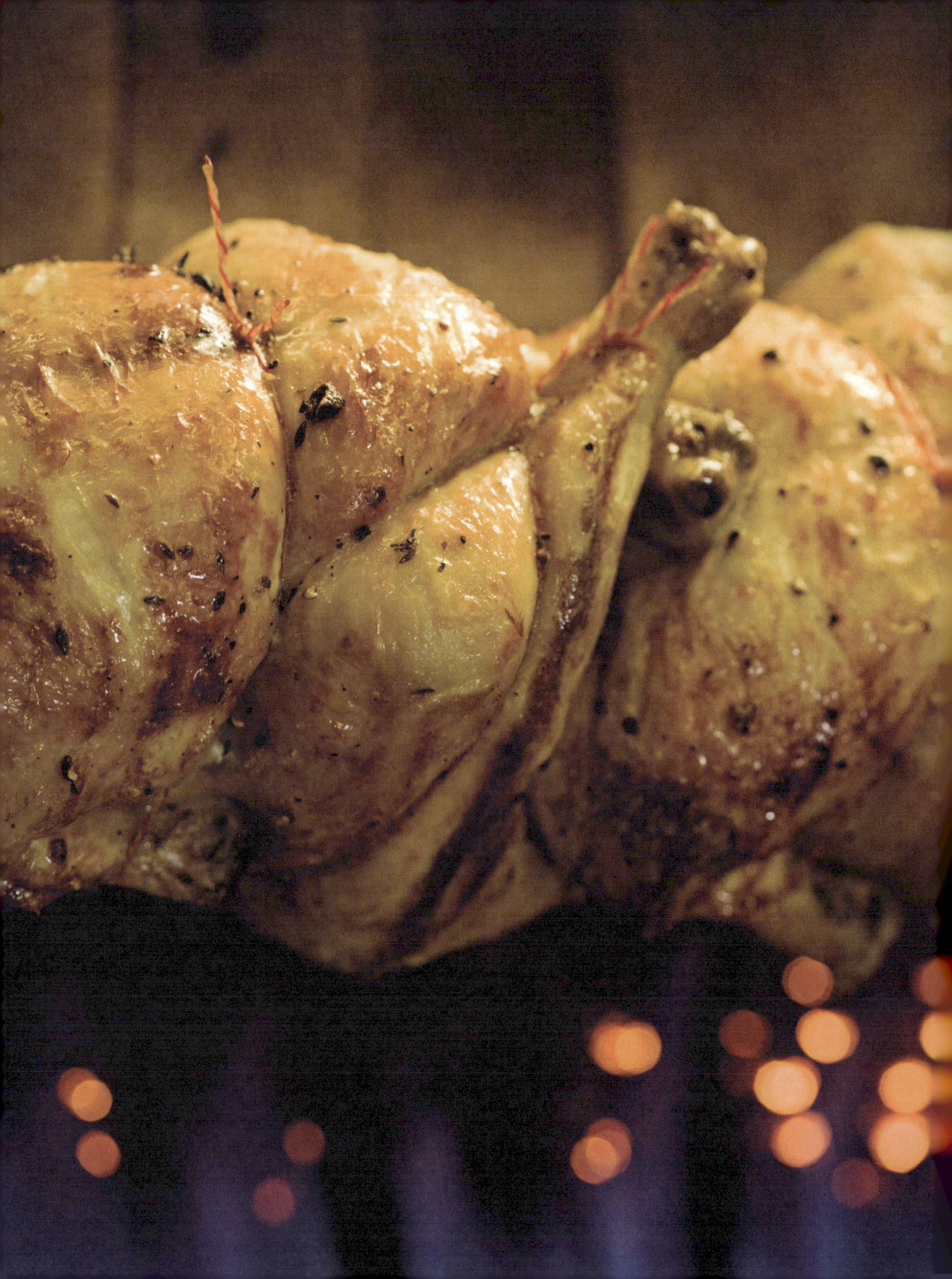

Rotisserie Free-Range Chicken, Minted Green Beans, Paris Mash & Manjimup Black Truffle Jus

French traiteurs often have huge rotisseries facing the street tempting passers-by. At Must Perth we have a rotisserie perched in the kitchen window. We don't overplay the flavours – sea salt, freshly ground black pepper, a sprig of thyme and garlic is all that's needed. When cooking chicken allow around 30 minutes for each 500 grams of chicken at 200°C.

....

1 large free-range chicken (approximately 1.8 kg)
Sea salt & freshly ground black pepper
A sprig of fresh thyme
1 garlic clove, cracked
Minted green beans (p. 200)
Paris mash (p. 203)

Manjimup Black Truffle Jus
Madeira jus (p. 281)
20 g black truffle

SERVES 4-6

Preheat the oven to 220°C.

Rub the chicken generously inside and out with sea salt and pepper, poke the thyme and garlic clove into the cavity and place onto a rack on the roasting tray (or onto your rotisserie skewer) and place in the oven to roast for 20 minutes.

Reduce oven temperature to 200°C and cook a further 1 hour and 20 minutes. Remove from the oven and keep warm.

Serve a dollop of Paris mash on each plate, topped with a sprinkle of beans.

Joint the chicken and place on top of the vegetables. Shave the truffle into the warmed Madeira jus and finish with a generous splash over the chicken.

Chateau les Trois Croix Fronsac
Bordeaux, France

This recipe provides an ideal opportunity to select a red wine. The rotisserie chicken is slightly more robust than other chicken dishes and the wine won't diminish its flavour. The mash helps soften the Bordeaux tannins and contrasts the texture. Green beans and truffle jus also complement the characteristics of this leafy, earthy wine.

Billaud Simon 1ᵉʳ Cru Les Vaillons
Chablis, France

Chicken can work with red or white wine, but poaching the chicken provides softer, more delicate meat and favours white wine. This vivid, shapely, and classically pure wine has the acidity to contrast with the chive buerre blanc.

Poached Mount Barker Chicken Breast, Manjimup Black Truffles, Sweet Corn Purée & Chive Beurre Blanc

Poached chicken has an ideal partner in fresh black truffles and Al Blakers grows some of the finest in Australia. Leave the prepared chicken breasts in the refrigerator overnight to allow the truffle aroma to infuse the flesh. Chicken skin is available from your poultry supplier. The final step in the cooking process is optional – the poached breast is superb whether or not it is coloured in the foaming butter.

Place a large sheet of cling film over the workbench. Stretch a chicken skin on top. Finely shave the truffle over the skin, sprinkle with thyme leaves, season generously and drizzle with extra virgin olive oil. Place a chicken fillet on top and carefully wrap the skin completely around the flesh.

Next, roll the cling film tightly around the chicken, ensuring that there is a 'tail' of cling film at each end of the parcel, and the film has rolled around the breast at least 3 or 4 times. Holding one tail in each hand, roll the parcel tightly on the bench so that the breast firms into a tight cylinder. Tie the tails off tightly with twine and repeat with the rest of the chicken breasts.

Rest the chicken breasts in the refrigerator overnight.

Bring a large pot of water to the boil and reduce to a shimmering simmer (about 90°C). Add the breasts to the hot water, bring the water back up to temperature and cook for 20 minutes. Turn off heat and allow the chicken to rest in the water for another 5 minutes.

Remove the fillets and carefully peel off the cling film and allow to drain while you heat a large non-stick pan over medium-high heat. Add the butter to the pan and, once it's foaming, carefully add the chicken. Allow to colour on one side for about 30 seconds, turn over and repeat, then remove to a cutting board.

Slice the chicken and serve with sweet corn purée, confit baby carrots and a drizzle of chive beurre blanc.

6 large pieces of chicken skin, washed
30 g black truffle
Fresh thyme leaves
Sea salt & freshly ground black pepper
Extra virgin olive oil
6 chicken breast fillets, about 180 g each
Butcher's twine
60 g unsalted butter

Sweet corn purée (p. 285)
Chive beurre blanc (p. 284)
Confit baby carrots (p. 267)

SERVES 6

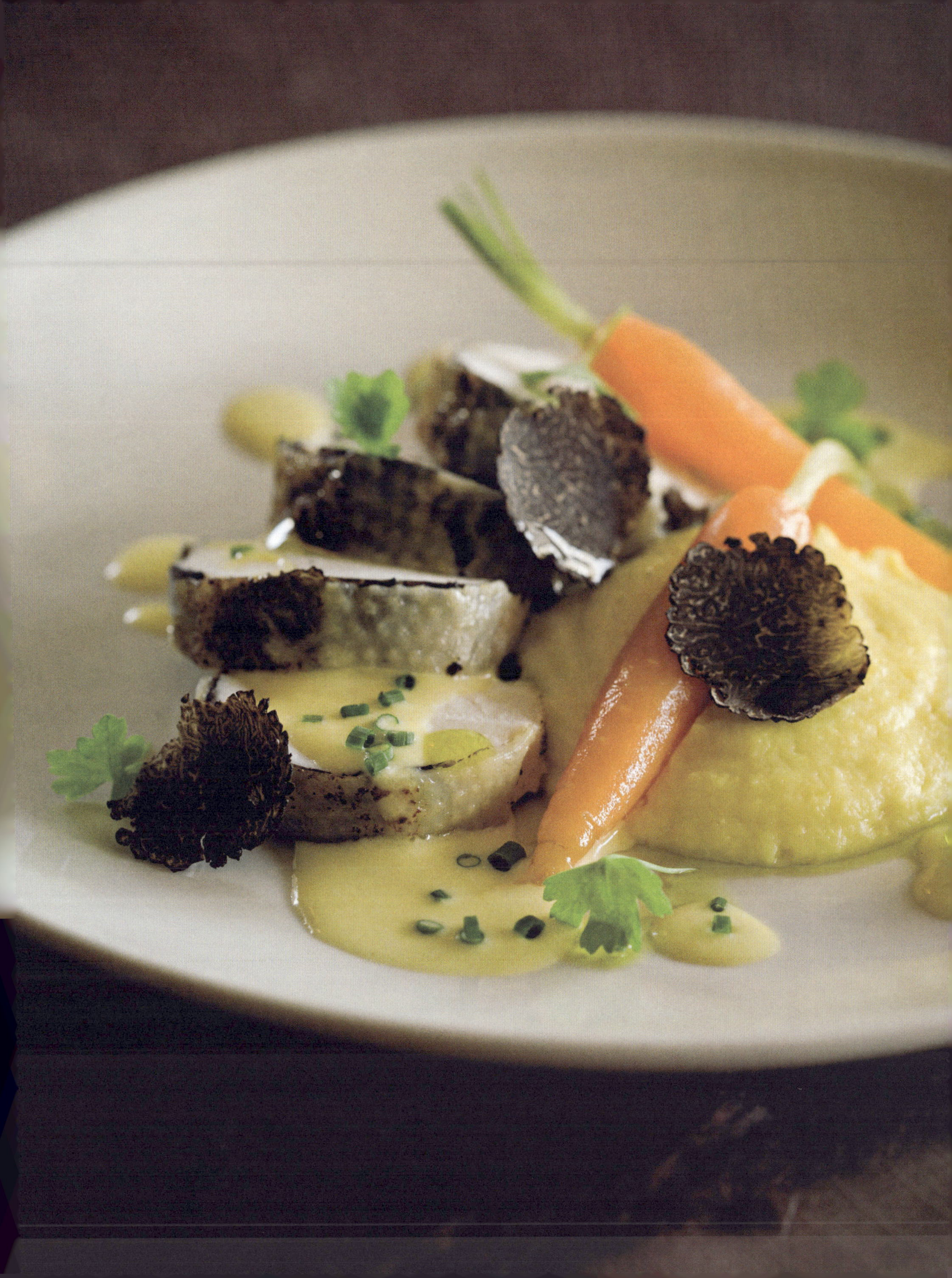

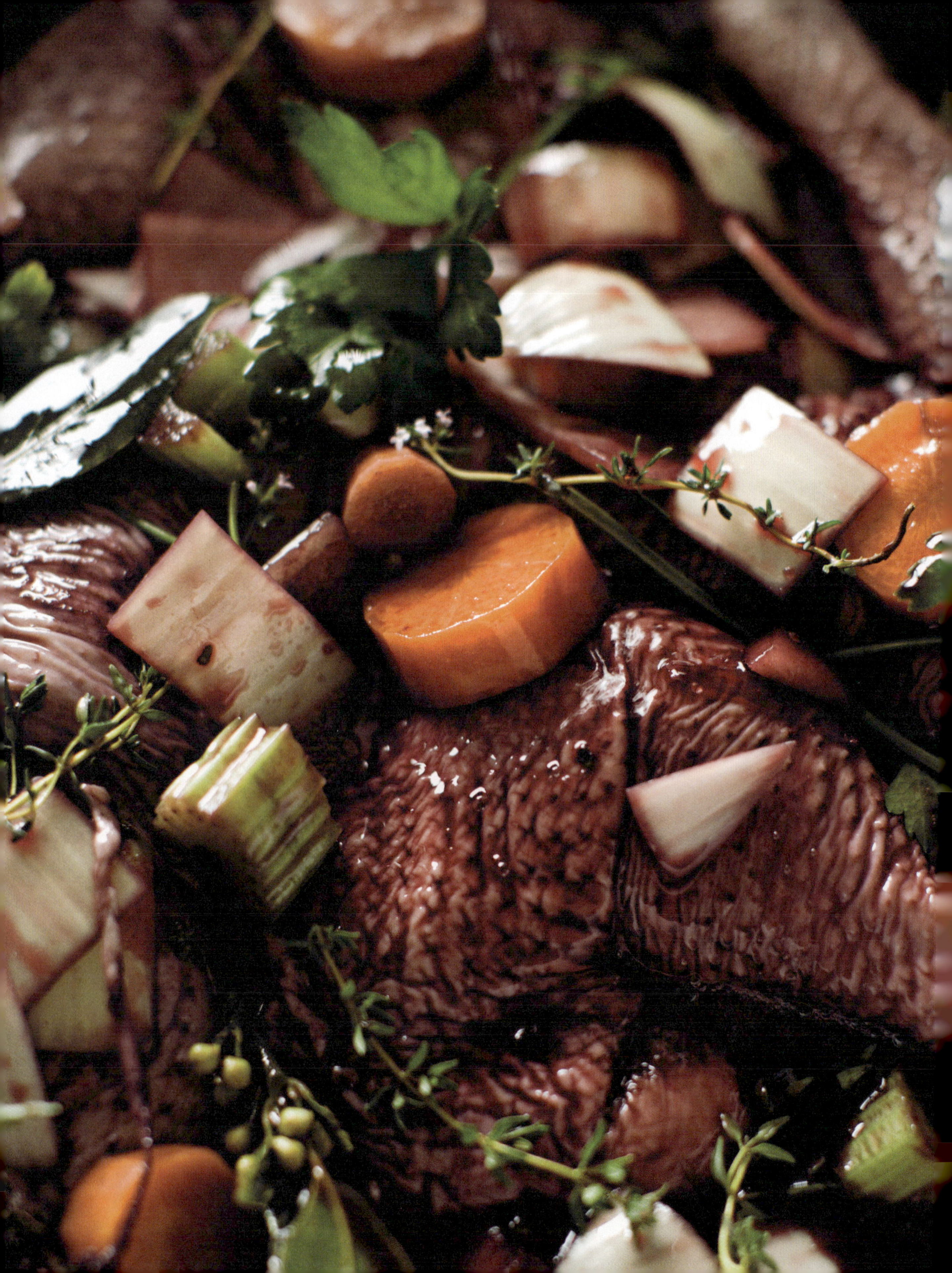

Coq au Vin

I haven't seen the true 'coq' chicken in Australia; a large male chook a couple of years old. The meat from these chickens is tougher and more flavoursome than our delicate free-range hens, and lends itself to long slow cooking until the bone and flesh part company. I use the largest Mount Barker free-range marylands I can get, then marinate them for at least 2 days before cooking.

6 large chicken marylands

Marinade

1 bottle full-bodied red wine
150 ml port
A sprig of thyme
1 bay leaf
A sprig of rosemary
1 onion, chopped
1 carrot, chopped
1 celery stick, chopped
2 garlic cloves
120 g smoked bacon or pancetta
4 black peppercorns

Extra virgin olive oil
1 litre brown chicken stock (p. 274)
Sea salt & freshly ground black pepper
300 g sweet carrots, peeled
50 g unsalted butter
Sea salt & freshly ground black pepper
Paris mash (p. 203)
Mixed leaf salad (p. 211)
Flat-leaf parsley, to garnish

SERVES 6

Place the chicken and marinade ingredients into a large non-reactive bowl and combine. Cover and place in the refrigerator for 48 hours. (Be patient – marinating for this length of time is critical to the flavour of the braise.)

Preheat the oven to 150°C.

Drain and pat the chicken dry, reserving the marinade liquid and vegetables separately.

Heat a large casserole dish over medium heat, add a little olive oil and brown the chicken on all sides; remove and reserve. Add the vegetables and pancetta piece to the casserole dish and sauté until golden brown. Add the reserved marinade liquid, increase the heat and boil, skimming off the foam as it rises. Boil until this liquid has reduced by half.

Add the stock and bring to the boil, then reduce the heat to low; add the chicken, cover and place into oven to cook for about $1\frac{1}{2}$ to $1\frac{3}{4}$ hours. The chicken can also be cooked in a slow cooker on high for about 2 to $2\frac{1}{2}$ hours. When it is cooked the meat will yield when prodded with a finger.

Remove from oven (or slow cooker) and allow to cool for 30 minutes.

Remove chicken and pancetta and cover with foil. Strain cooking liquor, discarding the remaining vegetables. Slip the skin off the pancetta with a knife and cut into lardons (strips the size of your little finger); reserve with the chicken.

Place the peeled carrots into a saucepan of salted water and boil until tender. Drain and blitz to a fine purée in a food processor, reserve, cover with foil to keep warm. Return the cooking liquor to the casserole dish and reduce by about half, until there's about 500 ml remaining.

Stir through a generous amount of carrot purée until the sauce has reached the desired thickness then whisk in the butter. Check the sauce for seasoning then immediately return the chicken and pancetta to warm through.

Serve onto a platter in the centre of the table, sprinkle with parsley leaves. The perfect accompaniments are a steaming bowl of Paris mash and a green leaf salad; or perhaps green beans (p. 200).

Denis Mortet Bourgogne (Rouge)
Burgundy, France

This sublime pinot has the weight and fruit power to complement this classic dish. There is an aromatic harmony (particularly if you have used pinot in the cooking), a textural balance and a similar flavour profile to the special character of this wine-based chicken dish.

Kooyong 'Estate' Pinot Noir
Mornington Peninsula, Victoria

Duck and pinot is always safe, but an unquestioned match. The mash in this dish would suggest a slightly more edgy, tannic Burgundian wine match; but the jus and cabbage is best alongside a fuller Australian pinot, such as this one, which offers pure fruit, a wealth of flavour and gentle sappiness.

Duck Leg Confit, Braised Red Cabbage, Pear Salad & Madeira Jus

Duck confit originated from peasant traditions of preserving meats in salt and fat. These days we cook confit for its complex herbal savoury flavour, crisp skin and delicate moist flesh. Try cooking your confit in a slow cooker with a temperature around 80°C to 95°C for a moist, tender result.

- Place the duck legs into a large non-reactive bowl, sprinkle generously with sea salt and a pinch of pepper, then add the white wine, onion, orange rind, cinnamon stick, bay leaf, some scrunched thyme leaves and garlic. Mix well, cover and refrigerate for 24 hours.
- Heat the duck fat in a slow cooker on a high setting. Wipe excess salt and marinade off the duck legs and place in a slow cooker, ensuring the legs are completely covered with fat. Fit the lid and cook on high for 3 to $3\frac{1}{2}$ hours until the meat gives way when squeezed and the flesh has shrunk back from the knuckle.
- Alternatively, the duck legs can be cooked in a lidded casserole dish in the oven at 120°C for around the same amount of time
- Carefully remove the duck legs from the fat and allow to cool then refrigerate. Once chilled, remove the legs from the fridge and use a large knife or cleaver to trim the knuckle.
- Place the legs in a deep storage container, melt the fat and flood the legs with fat to store. They will keep for a week stored in the fat.
- To reheat, pick the legs out of the fat and wipe off the excess. Place them skin side up on a tray under a medium-high grill and cook for 8 to 10 minutes to crisp the skin and render off the fat. Rest in a warm oven (100°C) for another 5 to 6 minutes before serving.
- Place a dob of mash on the centre of each warm plate. Top with warm red cabbage and a confit duck leg. Slice the pear into matchstick-sized pieces and sprinkle on top of duck, drizzle with Madeira jus and serve.

6 large duck marylands
Sea salt & freshly ground white pepper
150 ml white wine
1 brown onion, chopped
Rind of 1 orange
1 cinnamon stick
1 bay leaf
Fresh thyme leaves
3 garlic cloves, minced
1.5 litres duck fat

Braised red cabbage (p. 196)
Madeira jus (p. 281)
Paris mash (p. 203)
1 beurre bosc pear
Extra virgin olive oil

SERVES 6

mains

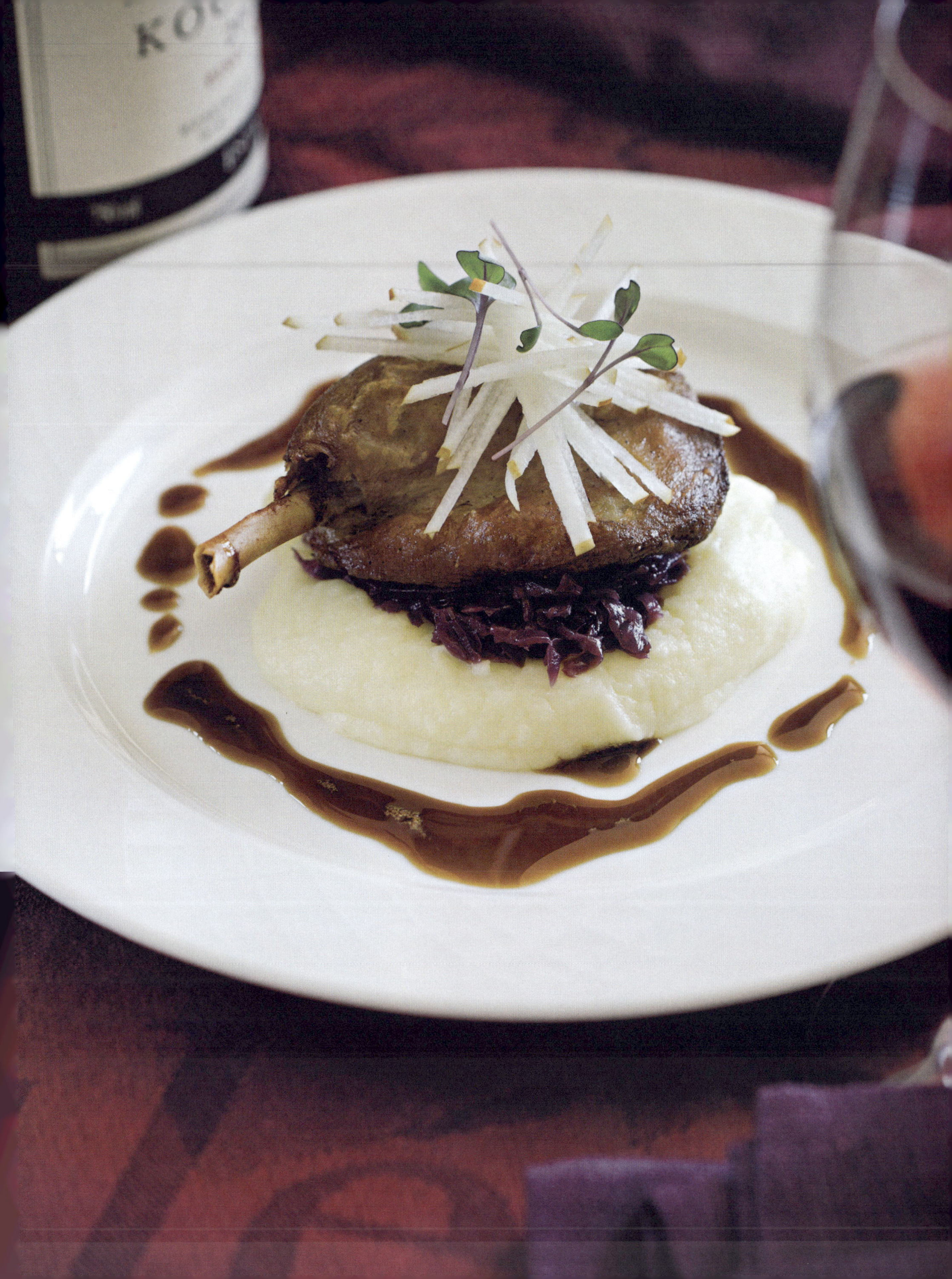

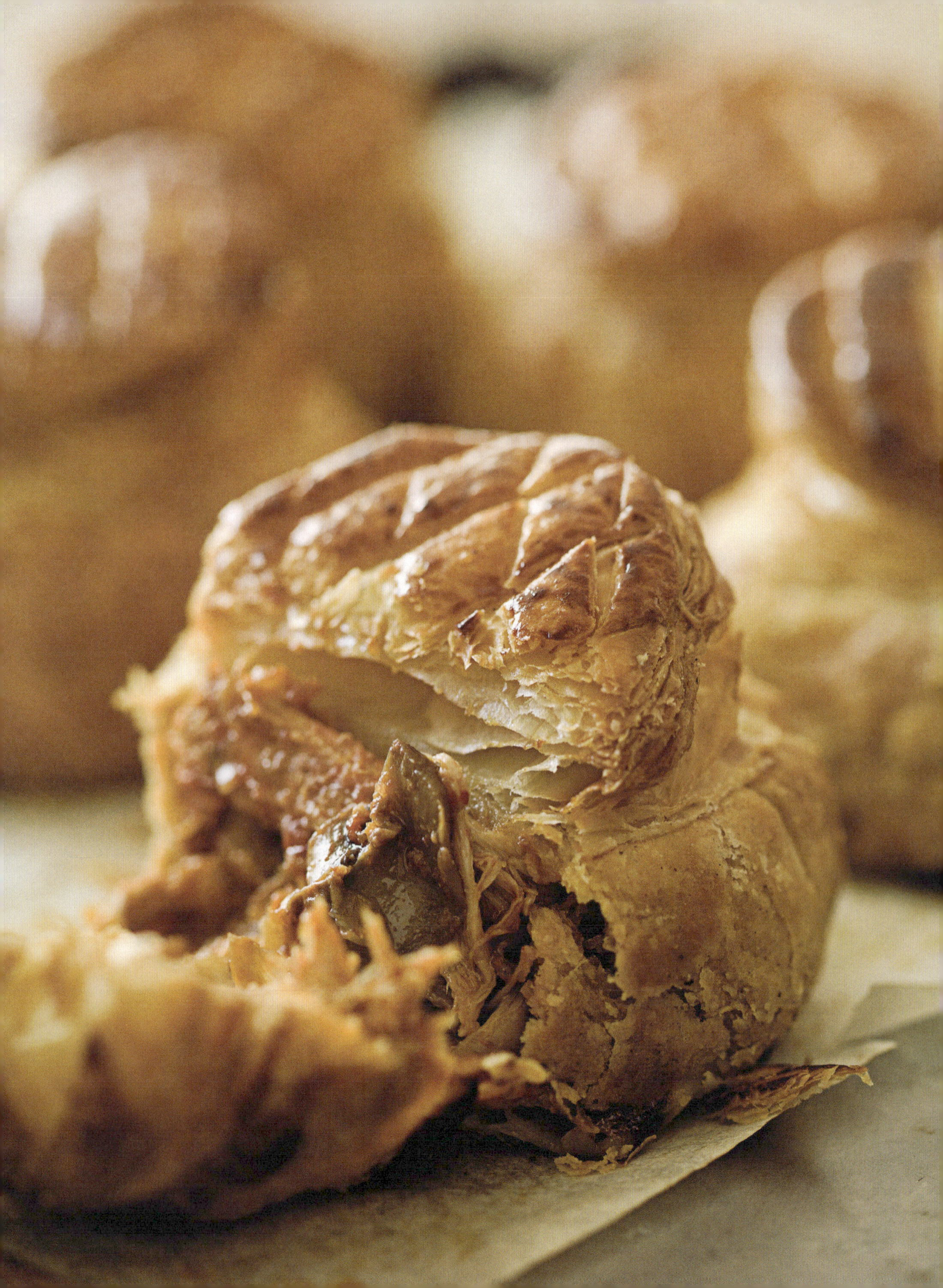

Rabbit & Portobello Mushroom Pie

This is the same ragoût that we serve with our gnocchi. Make sure you reserve the reduced cooking liquor to add back to the pie mix and to use as a sauce. This pie tastes terrific with celeriac purée (p. 197) and the avocado, Over the Moon feta, spinach and pine nut salad (p. 210).

....

50 g butter

600 g Portobello mushrooms, sliced

Rabbit ragoût (p. 83)

Sea salt & freshly ground black pepper

Pâté brisée (p. 285) or use Carême sour cream flaky pastry

1 egg

60 ml milk

SERVES 6

Melt the butter in a large frying pan over high heat and sauté the mushrooms for 3 to 4 minutes. Reduce the heat and cook for a further 3 minutes until the moisture from the mushrooms has evaporated.

Add 1 cup of the rabbit cooking liquor to the pan and simmer over low heat until the mixture has thickened. Incorporate the rabbit flesh and stir well, then season and place the mixture in the refrigerator to cool.

Onto a floured bench roll out the brisée pastry to a 3 mm thickness. Cut six large circles of pastry and sit inside 6 cappuccino-style coffee cups, so that the excess pastry flops over the edge of the cup.

Place some filling in each cup, fold the top over, brush with the beaten egg and milk and press another small circle of pastry onto the top. Etch a grid pattern onto the 'lid' with a knife and brush with egg wash. Place the cups in the refrigerator to set.

Once cooled, tip out the pies and place onto an oven tray lined with baking paper.

Preheat the oven to 220°C.

Bake the pies in oven for 18 minutes or until golden brown and puffed.

Serve the pies with a little of the rabbit sauce, truffle potato mash (p. 203) and rocket and parmesan salad (p. 213).

Isole e Olena Chianti Classico
Tuscany, Italy

This pie immediately brings Italian wine to mind, and Chianti Classico is an ideal option. Isole e Olena is high up in the Tuscan hills and has defined fruit, earthy, savoury flavours and grainy tannins, that contrast and complement this rustic pie.

Luke Lambert Syrah
Yarra Valley, Victoria

The lighter veal meal and the savoury purée pairs best with full-bodied whites or medium-bodied reds. This lighter style syrah (shiraz) is textural and lively with floral, juicy fruit, which balances the lighter meat. The taut, finely knit texture is a perfect match for the softness of the mash and crispy crepes.

Pan-Fried White Rocks Veal Shin Crepe, Rocket Purée & Organic Grape Glaze

Sona and Harry Toutikian make the organic muscat grape glaze in this recipe. Sona crushes the homegrown grapes and boils the must down to a dark, glossy syrup. Vincotto is a good alternative and can be found in specialty food stores.

To make the rocket purée, place all ingredients into a food processor bowl and blitz until smooth. Store in an airtight jar in the refrigerator for up to 2 days.

Preheat oven to 150°C. Cut the veal into large chunks, approximately 4 cm square.

Heat the olive oil in a large saucepan, add the veal and brown on all sides, then remove and place into a large casserole dish. Into the saucepan, add the onion, carrot, celery, pancetta and herbs, cook over medium heat to colour, about 5 minutes, then add to the casserole. Pour in the wine; boil rapidly until reduced by half. Add the stock, bring to the boil then pour over the veal and vegetables in casserole.

Cover and place into the oven for 3 to $3\frac{1}{2}$ hours – the veal should be meltingly tender. (It can also be cooked in a slow cooker on high for 4 to $4\frac{1}{2}$ hours.) Cool in the stock for 30 minutes.

Remove the veal and pancetta and drain. Tear the veal flesh apart with your fingers, dice the pancetta with a knife and place into a bowl.

Strain the cooking liquor, skim off the fat and reduce over high heat to a viscous syrup – there should be about three-quarters of a cup remaining; return this to the veal mixture, mix well, check seasoning and refrigerate for 10 to 15 minutes.

Lay out the crepes, place a generous amount of veal filling on each, fold in the sides and roll to a neatly shaped cylinder. Place in the refrigerator until required.

Preheat the oven to 180°C.

Heat the butter in a non-stick frying pan, add the crepes and cook for about 1 minute until golden brown, then carefully flip over. Remove to a baking tray and place into oven for about 12 minutes.

Place a dob of celeriac purée onto serving plates, top with two crepes, drizzle rocket purée and organic grape glaze around the plates and serve.

Rocket Purée
150 g rocket washed and picked
50 g flat-leaf parsley
Juice of 1 lemon
80 ml extra virgin olive oil
1 tablespoon pine nuts, roasted
Sea salt & freshly ground black pepper

1.2 kg veal shin or shank flesh
40 ml extra virgin olive oil
1 large onion, coarsely chopped
1 celery stick, chopped
2 carrots, chopped
150 g pancetta, skinned and cut into 1 cm 'lardons'
600 ml white wine
1 litre beef (or brown chicken) stock (p. 274)
1 bay leaf
A sprig of fresh thyme
Sea salt & freshly ground black pepper
50 g unsalted butter

12 large crepes (p. 286)
Celeriac purée (p. 197)
Organic grape glaze

SERVES 6

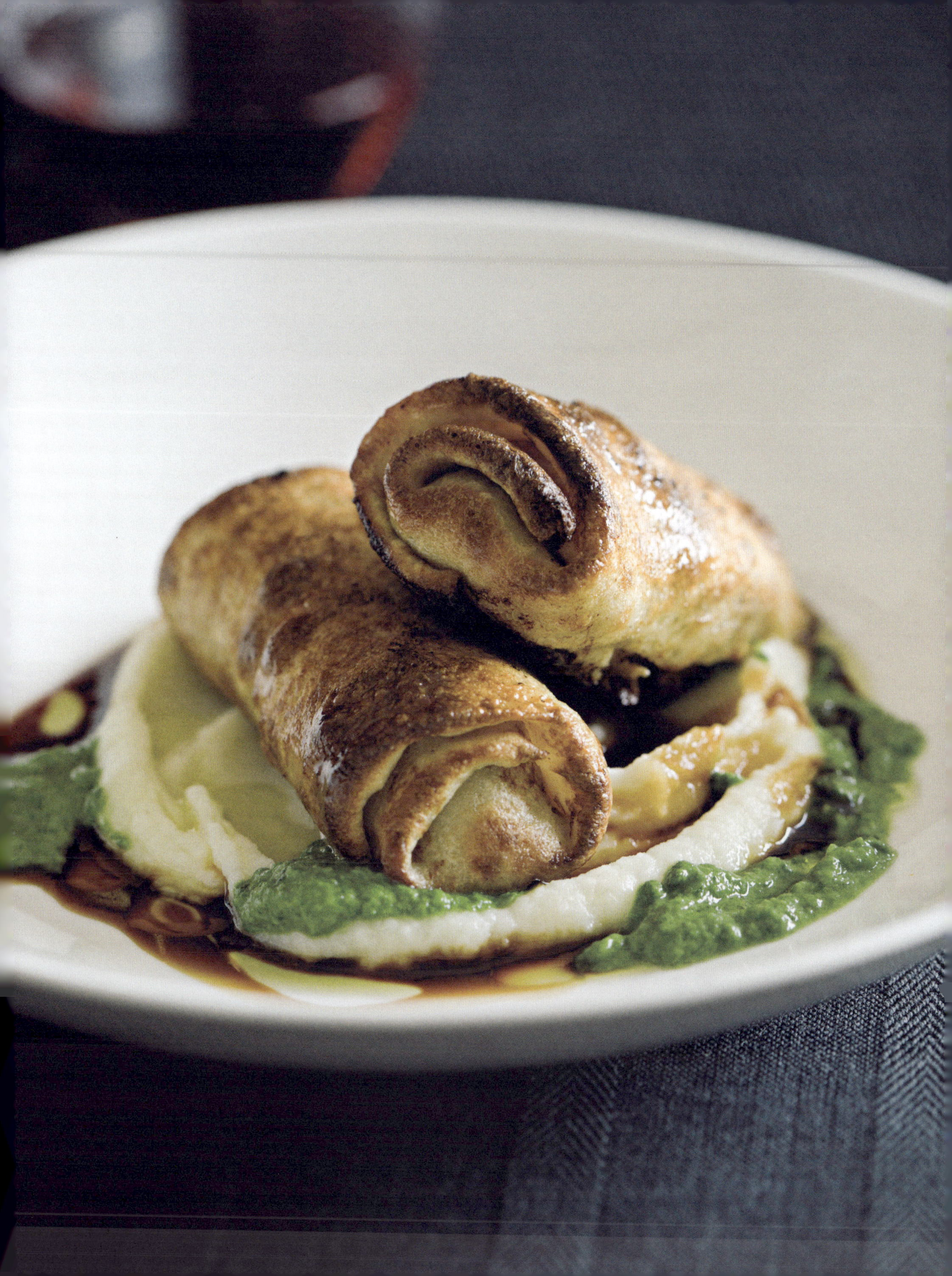

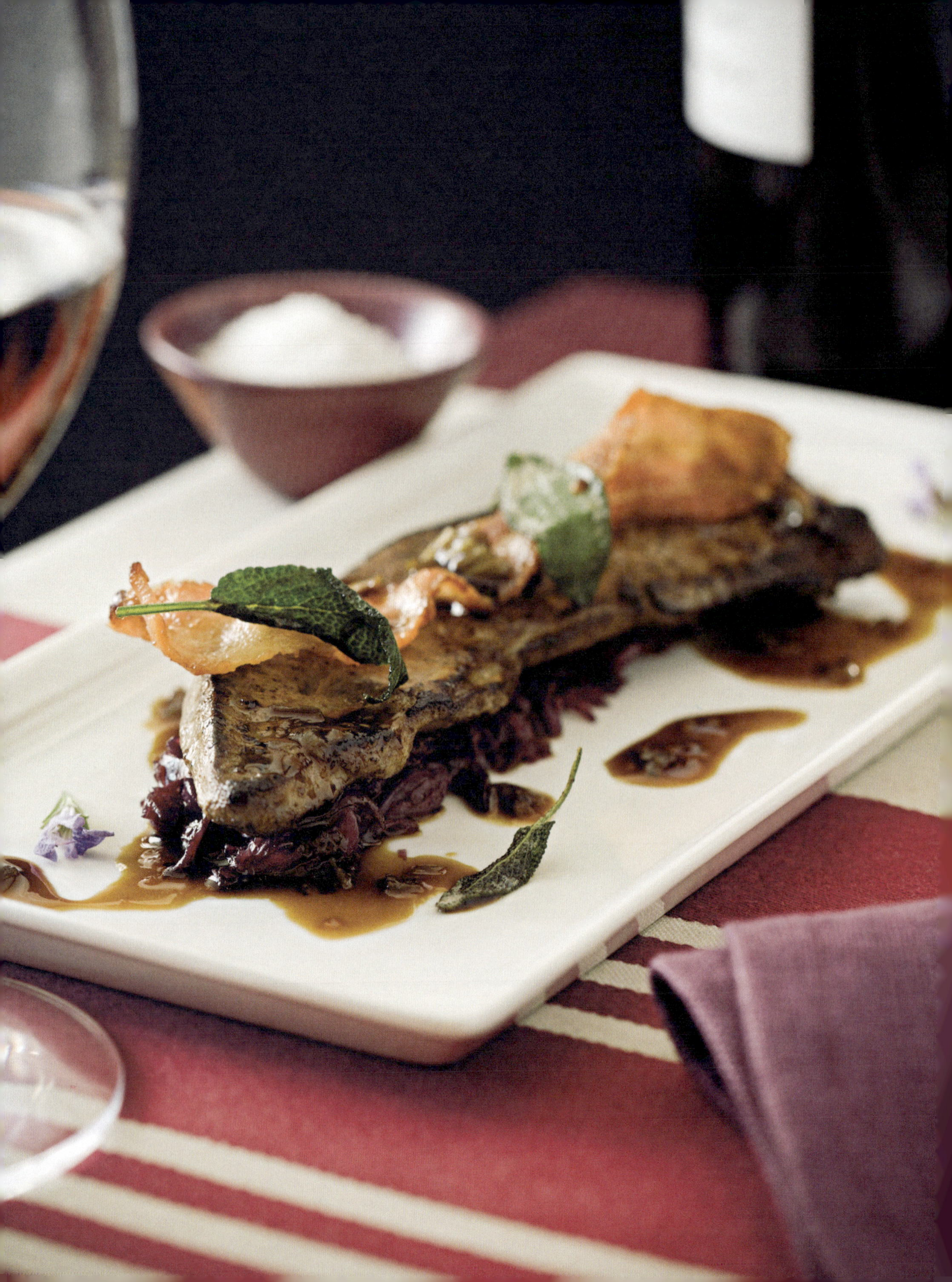

Seared White Rocks Veal Liver, Crisp Pancetta, Red Cabbage Braise & Sauce Diable

When I worked at London's Dorchester Hotel in the 1980s, I was exposed to the finest ingredients the world had to offer: crates of fresh wild mushrooms from Italy, glistening foie gras and aromatic truffles from France and pale, sweet Dutch veal. When I returned to Australia I was surprised to find the most amazing veal being produced 160 kilometres south of Perth. White Rocks is the finest veal available in Australia.

....

6 thin slices flat pancetta

Olive oil

Fresh sage leaves

1 kg White Rocks veal liver, cut into 6 thin slices

60 g unsalted butter

Sea salt & freshly ground black pepper

Red cabbage braise (p. 196)

Sauce diable (p. 283)

SERVES 6

Preheat the oven to 200°C.

Lay a sheet of baking paper onto an oven tray, place the pancetta slices on top, and top with another sheet of paper and another baking tray. Place into the oven for 6 to 10 minutes, until the pancetta is golden brown and crisp. Remove and allow pancetta to cool to room temperature on the tray.

Heat a little olive oil in a small saucepan to about 190°C. Add the sage leaves and cook for 10 seconds until they are deep green and crisp. Remove and drain on kitchen paper.

Season liver slices with salt and pepper. Heat a large non-stick frying pan over high heat, drizzle with olive oil and place the liver into the pan to sear for 30–45 seconds. When the cooked side is a rich, dark brown colour, flip it over and cook for another 30–45 seconds. Drop the butter into the pan to foam and coat the liver.

Remove and serve on the red cabbage braise, topped with the pancetta, a sprinkle of crisp sage leaves, drizzled with sauce diable.

Paris mash (p. 203) or celeriac purée (p. 197) are excellent accompaniments.

Benevelli Piero 'Bric du Succ' Dolcetto d'Alba
Piedmont, Italy

The combination of strong flavoured liver and salty pancetta requires a confident wine, but not one too overpowering. This dolcetto has a generous appeal, which is calming, rather than confronting, and perfect with the texture and earthy flavours of meat.

La Spinetta Barbaresco 'Starderi'
Piedmont, Italy

Confit of goat is strongly flavoured and needs a high-toned wine with strength, firmness and acid-tightness. This very serious wine, balances the rustic flavours and slices through the meat fat. It's a winner.

Confit Goat Shank, Braised Flageolet Beans, Tomatoes & Mint Pistou

As well as duck, we confit salmon, goose, pork – even carrots and potatoes. Long, slow cooking in duck fat always results in a tender and flavoursome dish. Goat shanks may be hard to find, so pre-order from a good continental butcher or consider using full legs and carving the flesh after the second cooking.

- Ensure the shanks have been trimmed at the base and tip to approximately 12 cm in length. Season the shanks with salt, garlic, thyme and pepper and marinate overnight in the refrigerator. Soak the beans in plenty of water overnight.
- Preheat the oven to 140°C. Wipe any excess ingredients from the shanks and place into a casserole dish. Cover with a lid and cook for 2 to 2½ hours, until the flesh is tender to the touch. Remove from the oven and allow the shanks to cool in the fat. Increase the oven temperature to 180°C.
- Alternatively, the shanks can be cooked in a slow cooker for about the same amount of time.
- While the goat is cooking, place the drained flageolet beans into a large saucepan, cover with cold water and bring to the boil. Reduce the heat to a simmer and cook for about 45 minutes until the beans are tender.
- Drain the beans and return to the saucepan, add the stock and herbs, simmer over the lowest heat until the beans have soaked up most of the cooking liquid. Remove the herbs, stir through the butter and season the beans. Keep warm.
- Combine all the ingredients for the pistou in the bowl of a blender or food processor. Process for approximately 45 seconds, until the ingredients are still slightly grainy (don't over-process). Reserve in a cool place.
- Remove the shanks from duck fat (strain and reserve the duck fat in the refrigerator to re-use).
- Place on a metal tray and heat in the oven until the shanks are warmed through.
- Spoon the bean braise onto each plate, sprinkle with confit tomatoes, top with the goat shanks and finish with a dob of mint pistou.

12 baby goat (capretto) shanks
1 tablespoon sea salt
1 garlic clove, crushed
A few sprigs of thyme
¼ teaspoon ground white pepper
300 g flageolet (navy) beans
2 litres duck fat (can be re-used)
1 litre brown chicken stock (p. 274)
A sprig of rosemary
1 bay leaf
50 g butter
Sea salt & freshly ground black pepper
Confit tomatoes (p. 267)

Mint Pistou

40 g flat-leaf parsley, washed and picked
70 g mint, washed and picked
85 g pistachio kernels, roasted lightly
½ teaspoon roasted cumin powder
50 ml lemon juice
130 ml extra virgin olive oil
50 g parmesan, grated finely
Sea salt & freshly ground black pepper

SERVES 6

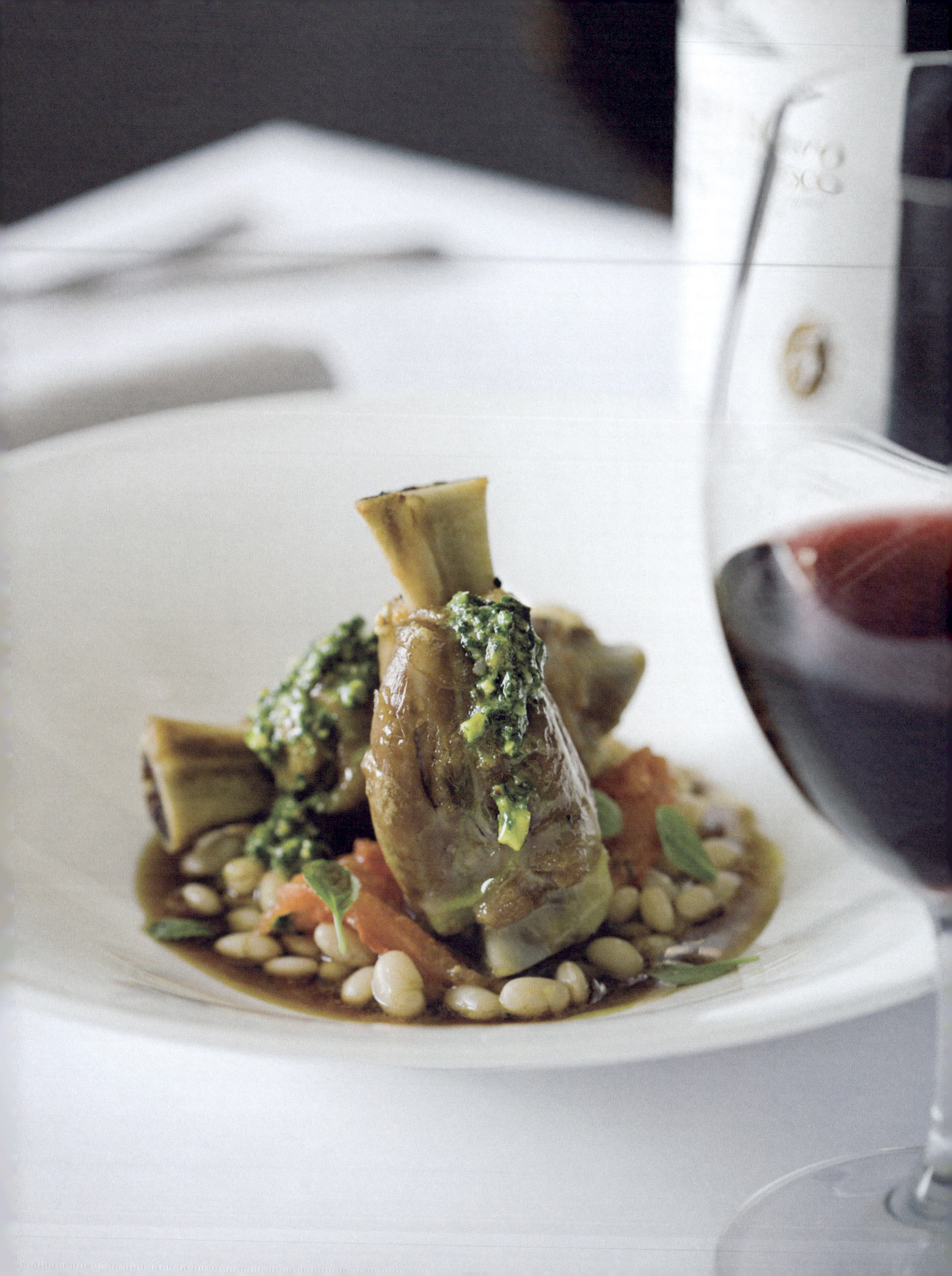

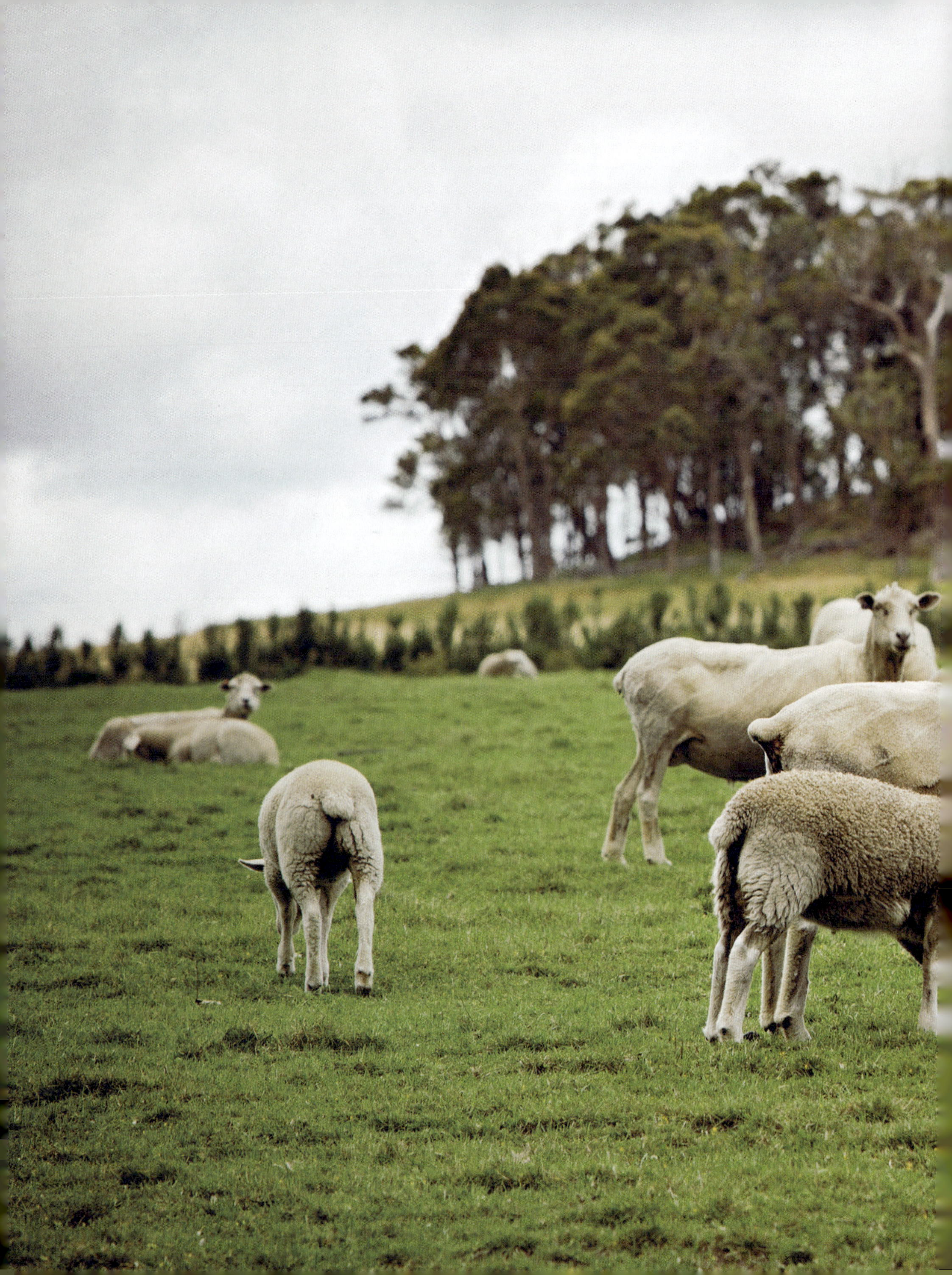

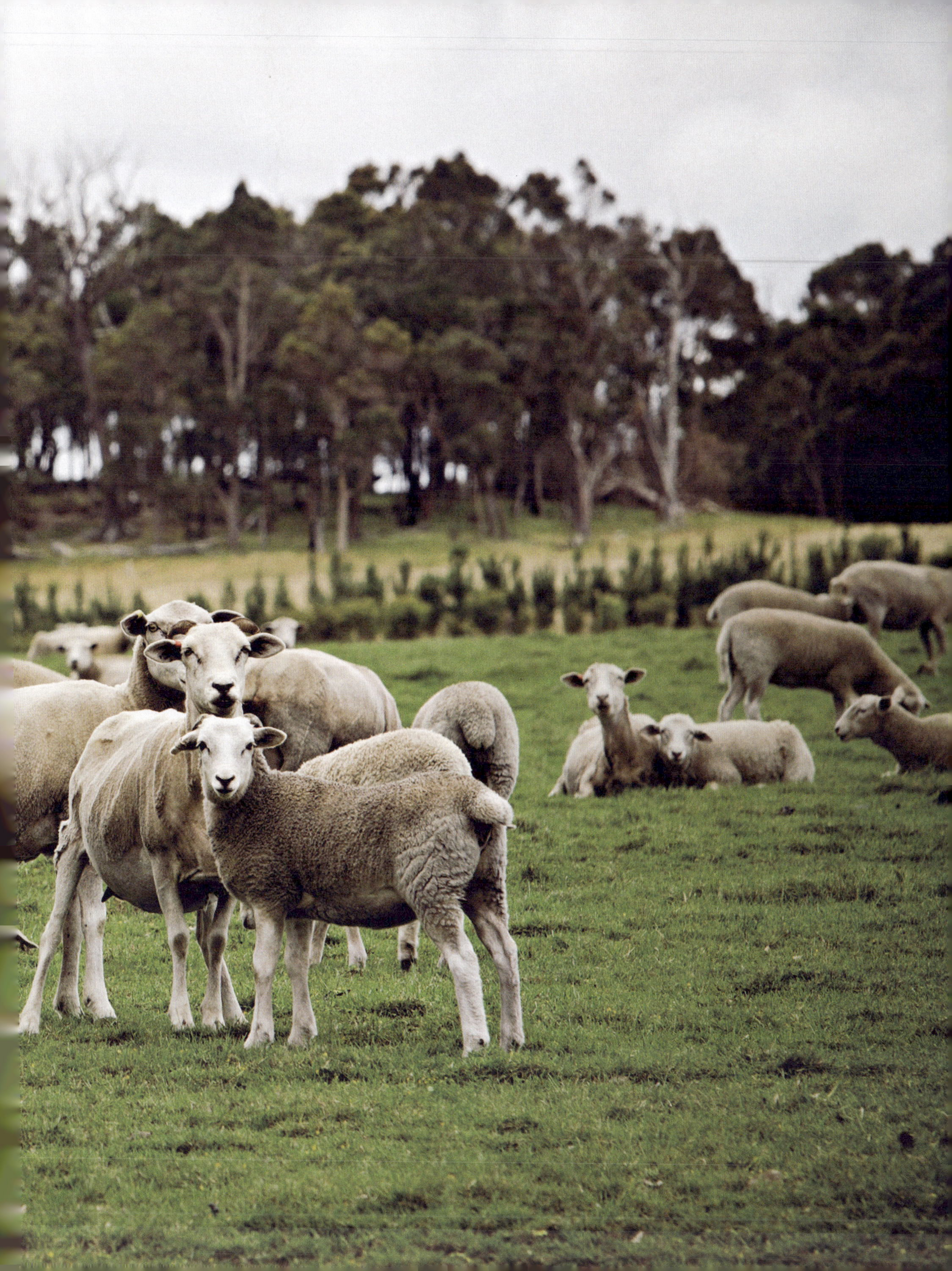

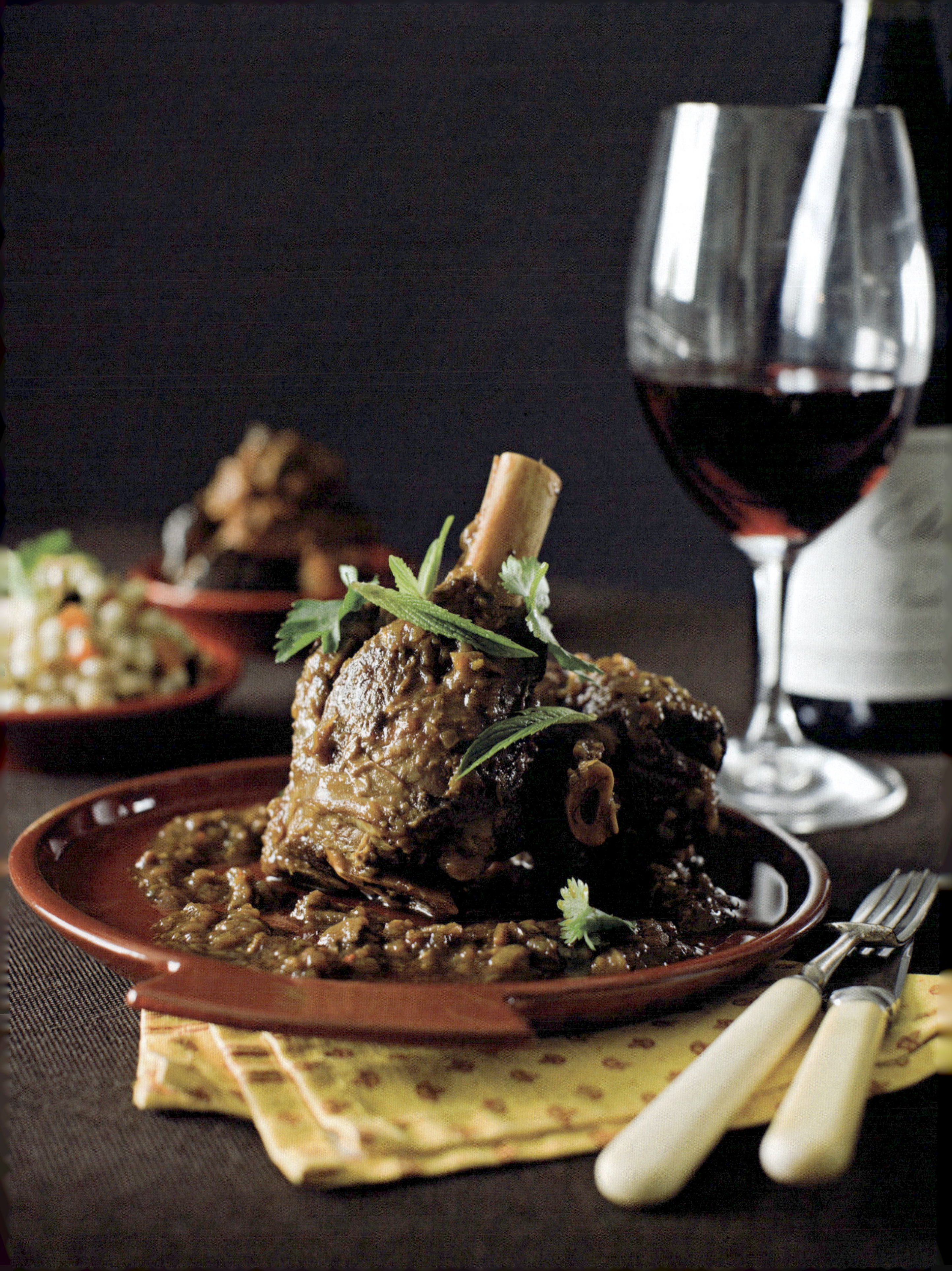

Middle Eastern Spiced Lamb Shank Braise, Mograbieh Salad & Eggplant Tagine

A few years ago I worked with Melbourne chef and author Greg Malouf at the West Cape Howe winery. We produced a Middle Eastern feast including spiced roast baby lamb, wood-fired Lebanese pizzas, harira (the traditional soup of Morocco) and grilled haloumi. This recipe reminds me of the warm spices and flavours that Greg thoroughly understands.

50 g butter

6 lamb shanks (ask your butcher to 'tip' them – that is, to take the top off the bone so the meat shrinks back along it)

2 brown onions, chopped finely

3 garlic cloves, crushed

1 heaped teaspoon each of ground cumin, coriander, ginger powder and black pepper

1 small red chilli, de-seeded and chopped

1 cinnamon stick or 1 heaped teaspoon cinnamon powder

A few strands of saffron soaked in the juice of 1 lemon

1 can (400 g) crushed tomatoes

125 g pitted dates, chopped

500 ml brown chicken stock (p. 274)

Sea salt to taste

Coriander, parsley and mint leaves
Mograbieh salad (p. 212)
Eggplant tagine (p. 199)

SERVES 6

Preheat the oven to 160°C.

Heat the butter in a large heavy saucepan or casserole. Place the lamb shanks into the pan and brown on all sides. Remove the shanks; keep aside while you make the sauce.

Add the onions and garlic to the pan and lightly sauté until they are translucent. Place a lid on the pan and cook for 10 minutes until the onions have softened. Add the spices, chilli, garlic, cinnamon stick and lemon, stir well and cook for a further minute. Add the crushed tomatoes, chopped dates and stock, bring to the boil and simmer for 10 minutes.

Place the lamb shanks into a lidded casserole or ovenproof dish and pour the sauce over the top. Place the lid on and pop into the oven to cook for $2\frac{1}{2}$ to 3 hours.

Alternatively, you could use a slow cooker on high for about 3 to 4 hours. When the lamb falls off the bone when prodded with your finger, it's ready.

Season the braise, then serve sprinkled with herbs, the mograbieh salad and a dish of eggplant tagine.

Etude Pinot Noir
Carneros, California, USA

This Californian pinot has attractive weight and mid-palate texture that pairs well with, and soothes the warmth of, the chilli and Middle Eastern spices. It also balances the sweetness from the dates and complements the tender lamb shanks.

Bruno Giacosa Barbera 'Falletto'
Piedmont, Italy

Bruno Giacosa Barbera has earthy, wild berry fruit, with crisp acidity – bracing rather than broad – with just the right weight and balance for the pork. It also contrasts with the textural mash and has a savoury feel that matches the herb salad.

Rotisserie Pork Fillet, Pumpkin & Parmesan Mash & Torbay Herb Salad

At Must we use Plantagenet free-range pork from the Great Southern, and Jarrahdene free-range pork from Margaret River.

Preheat the oven to 180°C.

To make the mash, place pumpkin chunks into a baking tray, cover with foil and bake in oven for around 1 hour until soft. Remove, scoop out the flesh and reserve in the bowl of a food processor.

Meanwhile, wrap the garlic cloves in foil and bake for about 15 minutes until soft. Squeeze out the flesh from the skin and combine with the pumpkin.

Bring the cream to the boil in a small saucepan, cook for 3 to 5 minutes until it thickens and add to the food processor bowl. Season with sea salt and freshly ground white pepper and blitz the pumpkin to purée and combine the ingredients. Add the parmesan and stir through before serving.

Trim the pork fillets of sinew and fat, removing the butt and tail ends. Season with salt and pepper and sprinkle with thyme leaves.

Lay 5 to 6 slices of pancetta in a long strip on the benchtop, place a pork fillet on top and roll to wrap the fillet tightly. Then wrap the fillet with cling film and place in the fridge until required. Repeat the process with the remaining fillets.

If you have a rotisserie, preheat the oven to 220°C. Unwrap the pork and thread onto the rotisserie skewer. Cook the pork fillet for 10 minutes then rest, covered in a warm place, for 5 to 8 minutes.

Alternatively, heat a large, non-stick frypan on medium-high heat, add olive oil, unwrap the fillets and seal on all sides to a light golden brown. Place in a 220°C oven for about 6 minutes, reduce the heat to 150°C and cook for a further 4 minutes. Remove and rest, covered, in a warm place for 5 to 8 minutes.

Serve a dob of pumpkin and parmesan mash onto warm serving plates, top with slices of pork and a sprinkle of microherbs, drizzle with olive oil and Madeira jus.

3 pork fillets
Sea salt & freshly ground black pepper
300–400 g flat pancetta, sliced finely
Fresh thyme leaves
Extra virgin olive oil
Fresh microherbs
Madeira jus (p. 281)

Pumpkin Mash

1 kg butternut or Japanese pumpkin, cut into large chunks
3 garlic cloves
75 ml cream
60 g butter
Sea salt & ground white pepper
3 tablespoons parmesan, grated

SERVES 6

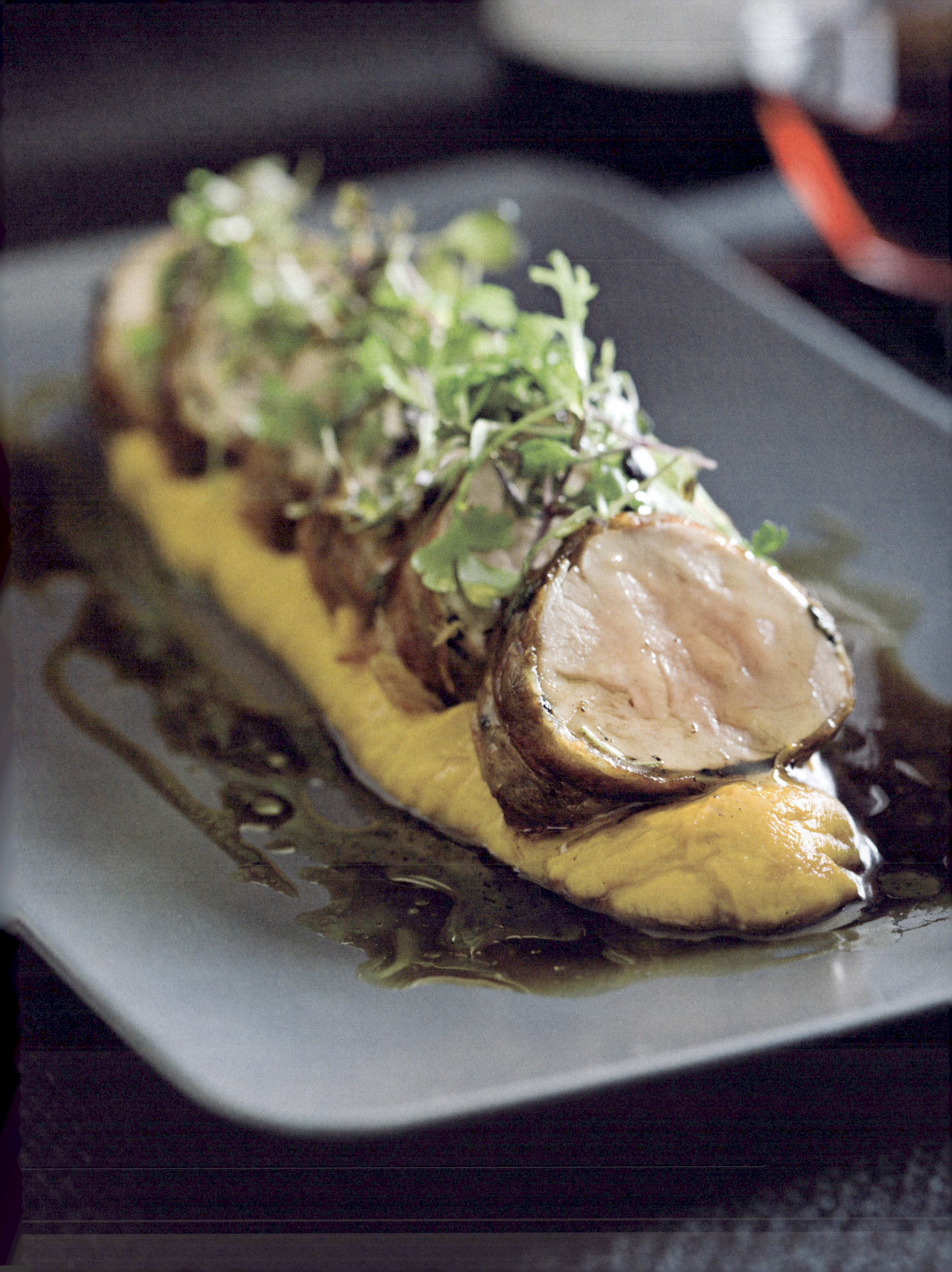

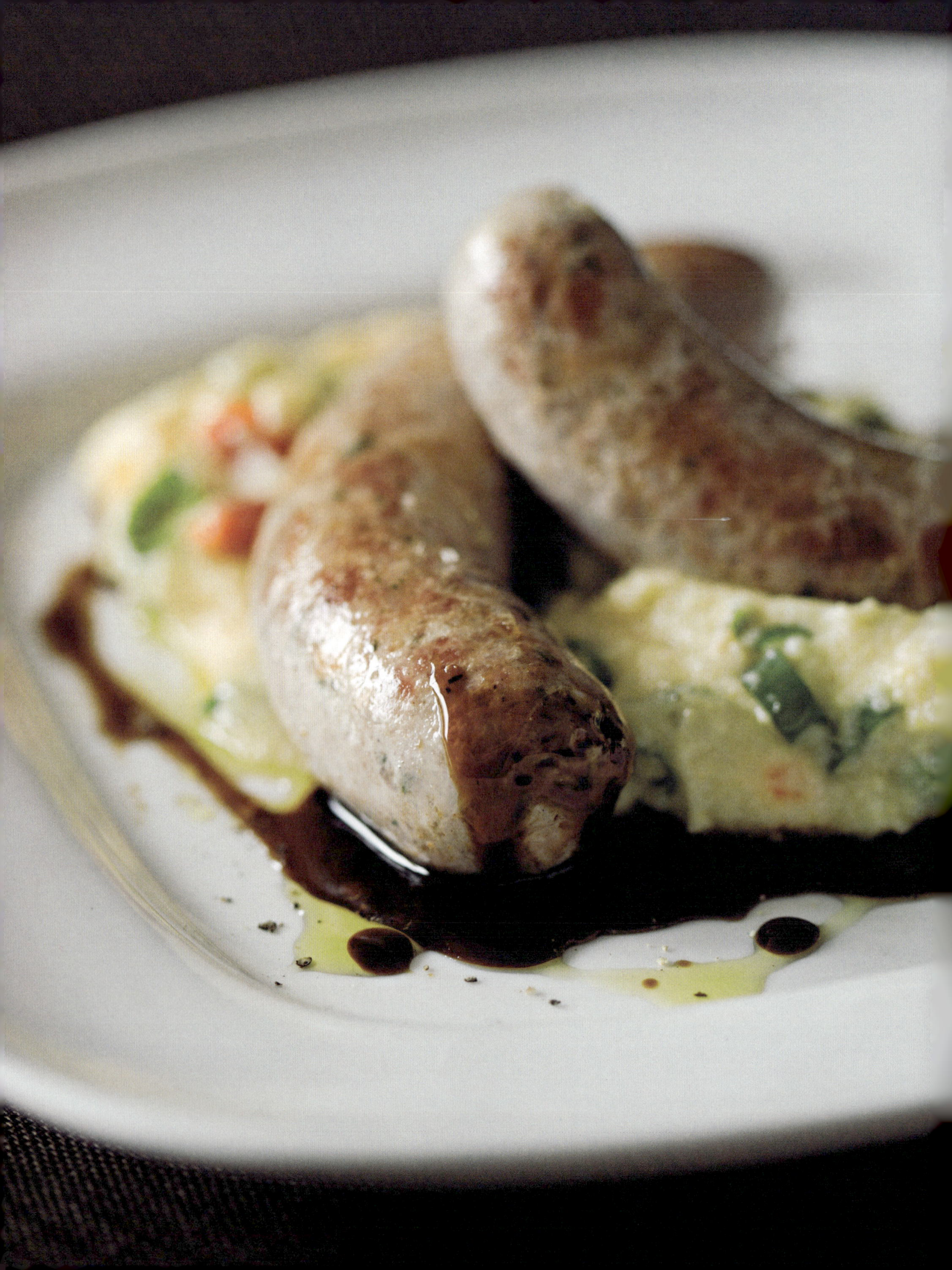

Must Pork Sausage, Soft Polenta & Roast Capsicum

'Instant' polenta is simple to use and speedy for the home cook. At Must we often use a traditional polenta, Moretti Bramata, which is raw. If you prefer to use this, increase the milk and water to 300 ml each for the same quantity of polenta, and cook it gently for at least 40 minutes. The result will be rich, nutty and sublime.

Must pork sausages (p. 132)
Olive oil
Red wine jus (p. 283)

Soft Polenta

250 ml milk
250 ml water
65 g fine instant polenta
60 g parmesan
40 g butter
Sea salt & freshly ground black pepper
1 red capsicum, roasted, peeled and cut into strips
2 spring onions, sliced finely

SERVES 6

Preheat the oven to 180°C.

To make the soft polenta, combine the milk and water in a large saucepan over high heat; once simmering, reduce the heat and drizzle in the polenta, stirring constantly. Cook for approximately 5 minutes, stirring frequently.

Remove from the heat; add the cheese, butter and seasoning, stirring well to combine. Stir through the roast capsicum and spring onions; keep warm.

Heat a little oil over medium heat in a large non-stick frying pan, add the sausages and cook for 1 minute on each side to lightly brown.

Place the sausages on a baking tray and into the oven. Cook for 6 minutes, then serve on top of the warm polenta, drizzled with a little red wine jus.

Remelluri Joven (Tinto)
Rioja, Spain
Remelluri is flavoursome and medium-bodied with enough weight and flavour to pair with pork. It also has the freshness and brightness to cut through the fatty sausage texture and solidness of the polenta.

Domaine Du Vieux Lazaret Chateauneuf-du-Pape
Rhône, France

This blended red from the southern Rhône has typical earthy, roasted red fruit that glides over the palate like satin. It is finely knit with gentle, warming fruit and a suggestion of floral savouriness. There is fullness and opulence, but it doesn't diminish the pork belly flesh and copes with the PX edge. The sturdy tannins, volume and earthiness is a perfect match for the braise.

Crispy Pork Belly, Pedro Ximénez Lacquer, Chickpea & Chorizo Braise

Making crispy-skinned pork belly is a two-day job. Day one is spent cooking the belly to a silken texture, using a slow steaming process to render out some of the fat. After cooking, the belly is then pressed overnight in the fridge to set its shape, ready for a high temperature 'flash bake' to crisp the skin. The lacquer provides a sweet–sour glaze that accompanies the rich meat perfectly.

....

Preheat the oven to 180°C.

Score the fat side of the pork belly only to skin depth using a crisscross pattern – about 30 to 40 scores each way. Sit the pork belly on an oven rack over the sink, skin side up. Boil a large pot of water and ladle the boiling water over the skin so that it lightens in colour and bunches up. Pat the skin dry and rub the belly all over with a generous quantity of sea salt and pepper.

Place the vegetables, garlic and thyme into the base of a deep oven tray, pour over the wine and place the pork on top. Place a sheet of baking paper over the pork skin then cover the tray tightly with foil and place into the oven to cook for about 3 hours.

To check if the pork is cooked, press the skin with your finger, the flesh will feel soft and will yield. Re-seal and cook a little longer if the pork is not tender.

Remove the foil and baking paper, return to the oven and increase the heat to 220°C. Cook the pork, uncovered, for another 30 to 40 minutes, frequently basting the skin with the pan juices. Remove the pork from the oven once the skin is golden, place onto a large tray, placing another tray or chopping board on top weighed down with several heavy tins so the pork will flatten. Chill in the refrigerator overnight.

Preheat the oven to 260°C.

Place the pork on an oven tray, skin side up. Brush olive oil over the skin and season with a good sprinkle of sea salt and place in the oven for 20 minutes until the skin is puffed, crisp and golden. Remove and rest on a chopping board.

Serve the chickpea braise onto serving plates, carve off a chunk of the belly, place on top and drizzle with Pedro Ximénez lacquer.

1.5 kg pork belly
Sea salt & freshly ground black pepper
1 large brown onion, chopped
2 celery sticks, chopped
1 large carrot, peeled and chopped
1 head of garlic, smashed
A good handful of fresh thyme
500 ml white wine
Extra virgin olive oil

Pedro Ximénez lacquer (p. 282)
Chickpea & chorizo braise (p. 198)

SERVES 6

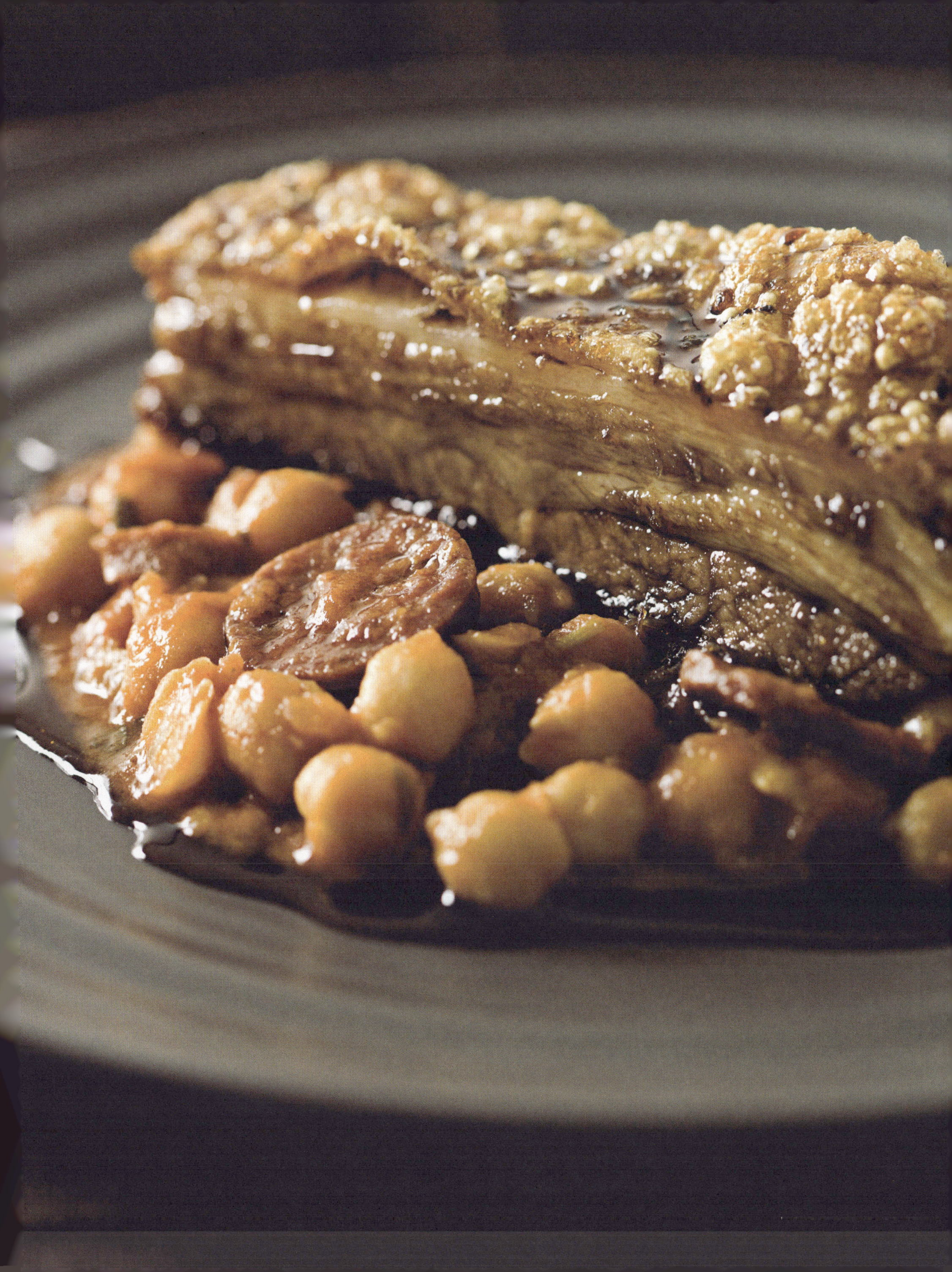

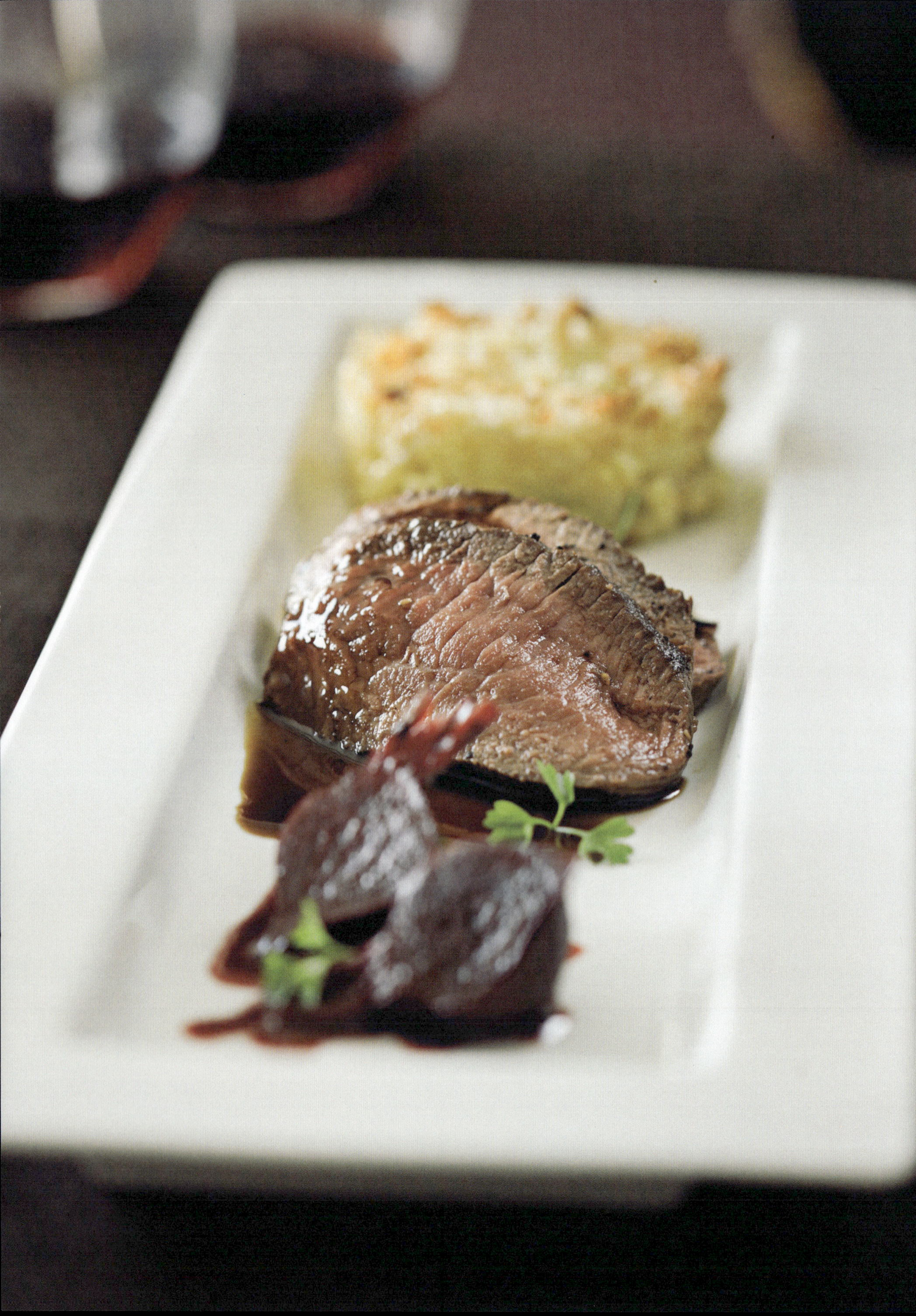

Seared Margaret River Venison, Glazed Baby Beetroot & Potato Gratin

Venison is a naturally lean meat that will dry if overcooked. The key is a short cooking time (to rare) with a long rest to allow the muscle to relax and the juices to re-absorb.

Glazed Baby Beetroot

12 baby beetroot, peeled
200 ml red wine vinegar
200 ml red wine
100 ml port
500 ml water
100 g muscovado sugar
A pinch of salt

1.2 kg venison loin, denuded
Olive oil
Sea salt & freshly ground black pepper
Potato gratin (p. 205)
Port wine jus (p. 283)

SERVES 6

To cook the beetroot, combine the vinegar, wine, port, water, sugar and salt in a large saucepan, bring to the boil and simmer for 1 minute. Add the beetroot, reduce heat and cook for approximately 30 minutes until tender.

Remove the beetroot, return the saucepan to the stove and boil the liquid until it has reduced to a syrupy consistency.

Return the beetroot to the liquor to warm through before serving.

Preheat the oven to 220°C.

Season the venison, heat a large frying pan over high heat, drizzle in some olive oil and sear venison for about 30 seconds on each side, remove and place on a baking tray. (Keep the frying pan aside to make the sauce.)

Place the venison into the oven for about 8 minutes to cook to rare. Remove and rest the meat in a warm place, covered with aluminum foil, for 5 to 6 minutes.

Slice the venison and lay onto warm plates, scoop on some potato gratin, spoon on some beetroot and drizzle with jus.

Torbeck 'Natural Wine Project' Grenache
Barossa Valley, South Australia

The gamey venison meat is best matched with a fresh, pure-fruited wine – not too heavy or strongly flavoured. Torbreck has crafted this juicy grenache, which is shapely, svelte, and doesn't dominate the venison. It also pairs with the sweet, earthy beetroot.

Paul O'Meehan runs a feedlot on his family property near Borden at the foot of the Stirling Ranges northeast of Albany. He selects eight- to ten-month-old Angus crossbred cattle from quality farmers in the region, and holidays the cattle on his feedlot. Paul fattens the cattle on his special grain 'muesli' (using cereals grown on his farm) for eighty to 100 days, before selecting those he intends to brand as 'Butterfield'.

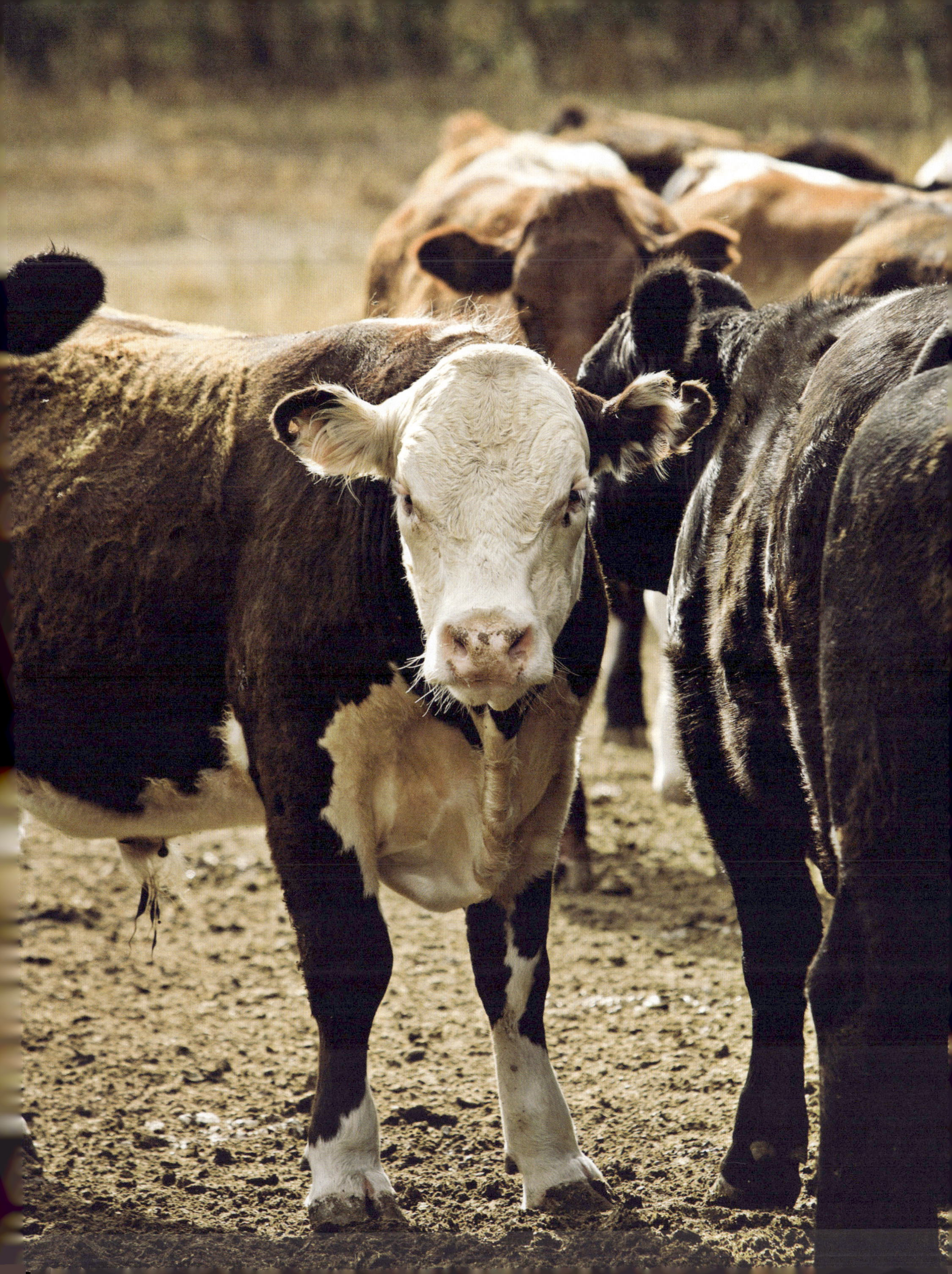

Spinifex Esprit
Barossa, South Australia

Aged beef is easily matched with a range of red wines, so it's worth selecting one that is dense, vibrant and seamless. Spinifex Esprit has the power to stand alongside this beef steak – it is concentrated, not heavy; fruity, not sweet. The texture will also contrast with the salty frites.

Char-Grilled Dry-Aged Sirloin Steak, Frites & Béarnaise

Peter Stocker, our butcher in Perth, hangs our steaks on the bone for twenty-eight days, which results in a old-fashioned flavour reminiscent of my childhood and makes the ultimate bistro meal – steak, frites and Béarnaise.

··

Preheat the oven to 220°C.

Warm a chargrill pan or barbecue grill to the highest temperature, drizzle the steaks with oil and season on both sides.

Place onto the chargrill or barbecue and cook for about 1½ minutes on each side to mark the steaks with a grill pattern.

Place the steaks on a baking tray and place in the oven until almost cooked (if using the barbecue, reduce the heat to low and drop the lid), then remove, dob the steaks with butter and rest for 5 minutes in a warm place (the steak will finish cooking in this time).

Toss the watercress with a little red wine dressing, place the rested steaks onto serving plates and add a handful of pommes frites and a generous serve of Béarnaise sauce.

6 x 250–300 g sirloin steaks (ask your butcher to trim most of the fat from the top of the steaks)
Extra virgin olive oil
Sea salt & freshly ground black pepper
250 g watercress
50 g unsalted butter
Béarnaise sauce (p. 278)
Red wine vinaigrette (p. 268)
Pommes frites (p. 203)

SERVES 6

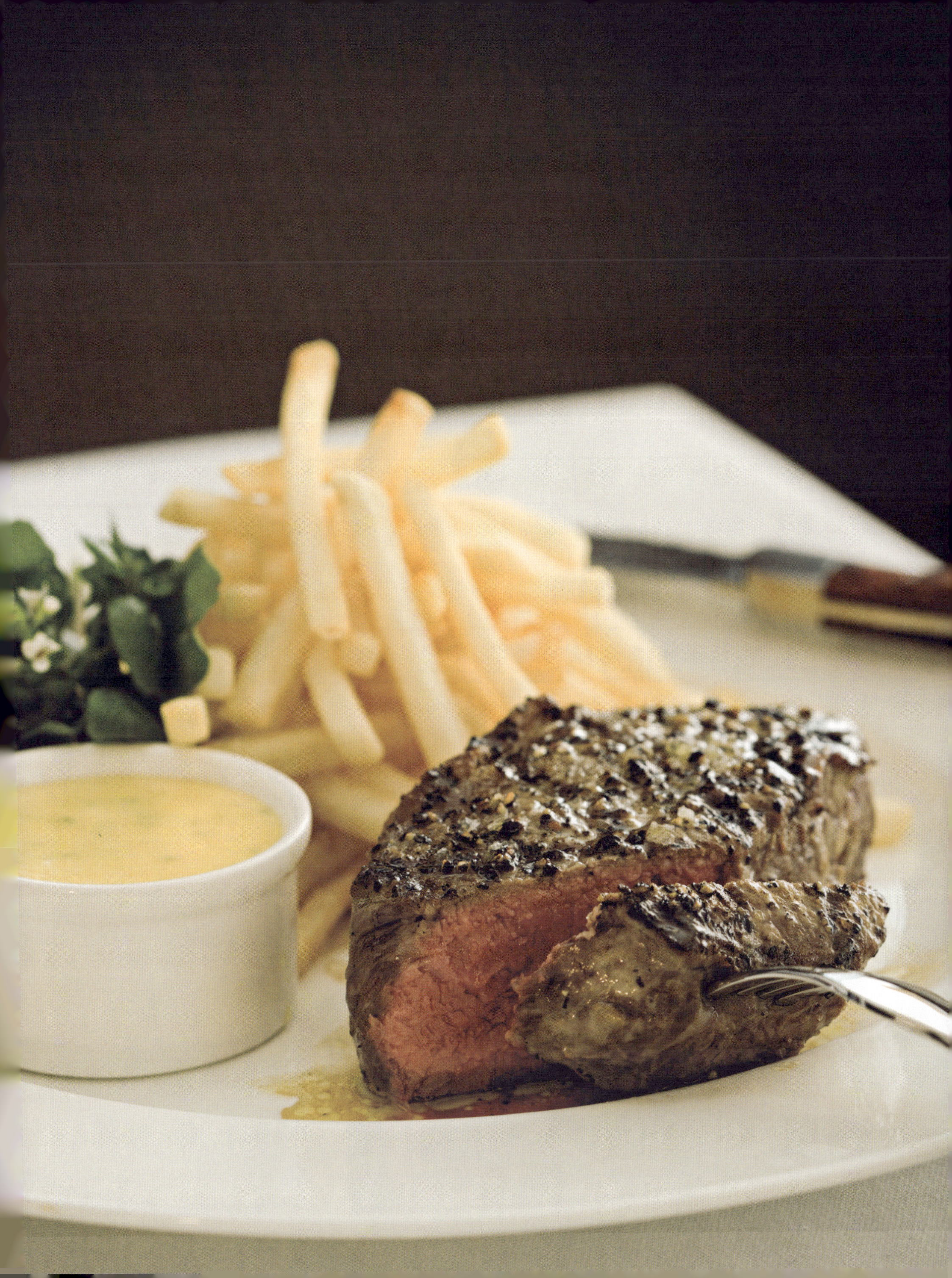

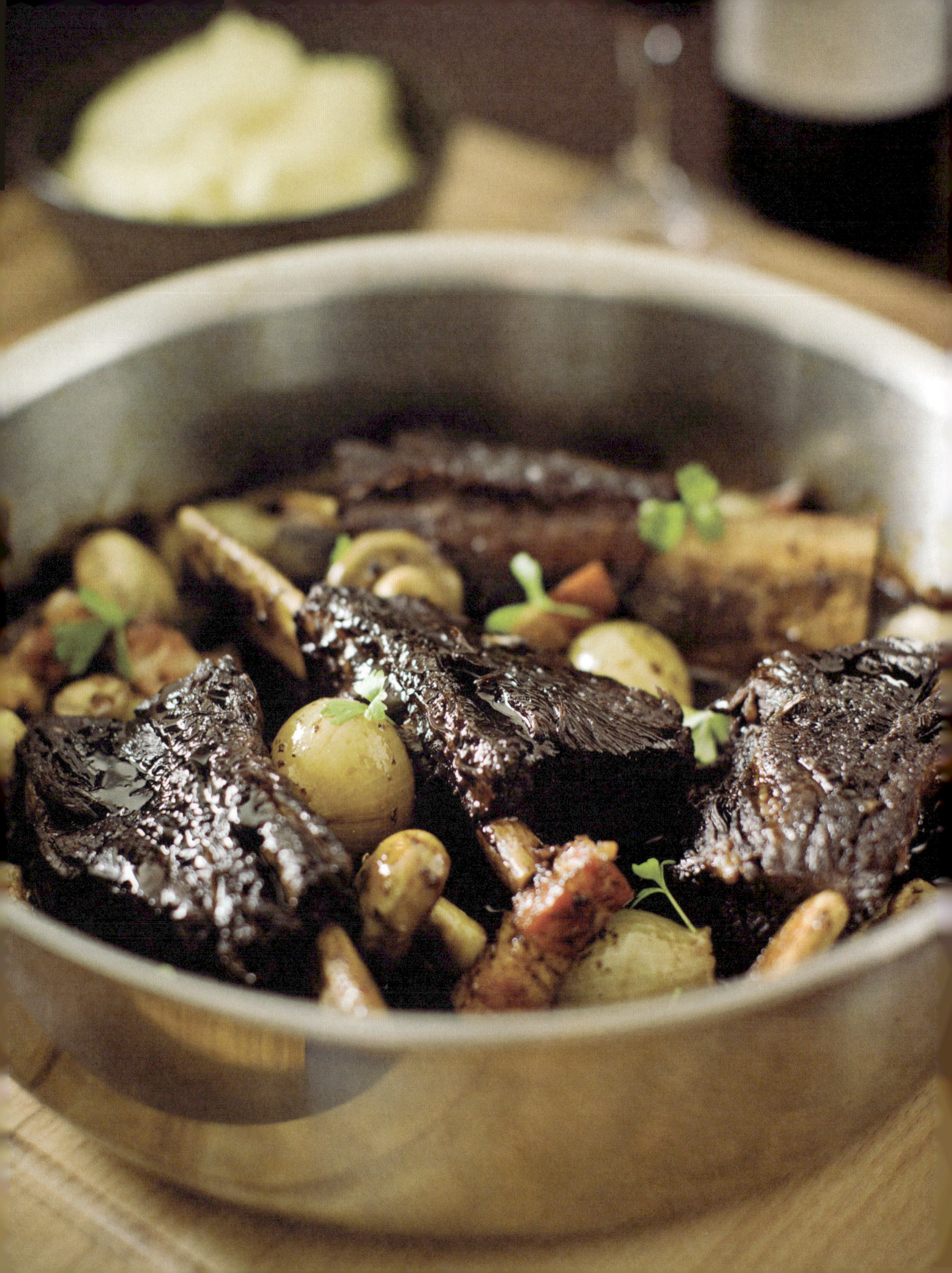

Beef Rib Bourguignon with the Classic Garnishes

A few years ago my butcher Peter brought me a sample of fleshy Butterfield beef spare ribs, taken from the neck end of the forequarter. I marinated and braised them as I would for beef cheeks, cooking the meat long and slow on the bone. A gelatinous and tasty success, these ribs have featured on our menus ever since.

Chateau Haut-Batailley
Pauillac, France

This serious, classic Pauillac has the concentration and structure (tannins and acid) that is calmed and softened by the meaty, succulent flesh of the beef. Importantly, the earthy, red-berried flavours are strongly defined and not overwhelmed by the meat.

---

6 fleshy beef ribs, about 15 cm long
2 litres beef stock (p. 274)
12 small button onions, peeled
24 small pickling mushrooms, washed and stalks trimmed
100 g piece of smoked streaky bacon, cut into 1 cm 'lardons'
50 g unsalted butter
Sea salt & freshly ground black pepper
Flat-leaf parsley, to garnish

Marinade

1 bottle full-bodied red wine
150 ml port
1 sprig of thyme
1 bay leaf
1 sprig of rosemary
1 onion, chopped
1 carrot, chopped
1 celery stick, chopped
2 garlic cloves
4 black peppercorns

SERVES 6

Place the beef ribs and marinade ingredients into a large bowl and combine. Cover and marinate in the fridge for 48 hours (be patient, marinating for this amount of time is critical to the flavour of the braise).

Preheat the oven to 150°C.

Drain and pat the beef dry, reserving the marinade liquid and vegetables separately.

Heat a large casserole dish over medium heat, adding a little olive oil then brown the beef on both sides, remove and reserve.

Tip the vegetables into the casserole and sauté until golden brown. Add the reserved marinade, increase the heat to high and boil, skimming off the foam as it rises. Boil until this liquid has reduced by half.

Add the stock and boil for 15 to 20 minutes until it has reduced by at least one-quarter. Reduce the heat to low, add the beef, cover and place into the oven to cook for about 3 to 3½ hours, until the flesh easily separates from the bone.

Alternatively, the beef can also be cooked in a slow cooker on high for about the same amount of time.

Remove from the oven (or turn off slow cooker) and allow to cool for 30 minutes with the lid off.

Remove the ribs and cover with foil so they don't dry out. Strain the cooking liquor, discarding the remaining vegetables.

Return the cooking liquor to a casserole dish or large saucepan, add the onions and cook over moderate heat for about 30 minutes until the onions are tender.

Remove the onions and the increase heat until the sauce has reduced to a viscous syrupy consistency – there should be about 400–500 ml remaining.

Remove from the heat, whisk through the butter and keep warm.

Drizzle a little oil in a non-stick frying pan and sauté mushrooms with bacon for 4 to 5 minutes until golden, add to sauce.

Place the sauce back over low heat, check for seasoning and place the ribs back in to warm through.

Place the beef and sauce onto a deep serving dish in the centre of the table, sprinkled with parsley leaves.

Serve with Paris mash (p. 203) or pan-fried gnocchi (p. 77).

Domaine L'Oratoire St Martin Réserve des Seigneurs Cairanne *Côtes-du-Rhône, France*

Cassoulet is typically rustic French fare and pairs well with similar style wines. This is a serious Côtes-du-Rhône with plenty of power, weight and earthy, roasted red fruit, which is softened by the texture and harmonised by the flavour of this dish. It's a little funky, too, which works with the earthy cassoulet flavours.

Cassoulet

Like any classic bistro dish, true cassoulet has many regional variations, though every recipe has white haricot or navy beans. Long soaking is important to plump the beans with water before cooking, so give them a swim for at least a day (2 days is better). A slow cooker is ideal for this recipe, though it will extend the cooking time – I have cooked the beans for 8 hours on a low setting with the most amazing result!

Cook the navy beans in plenty of water until they are al dente (around 40 minutes), drain and reserve.

Prepare the cooking liquor: drop the duck fat in a large heavy saucepan over low heat, add the shallot, garlic, leek, carrot and celery and sweat for 10 minutes.

Add the tomato purée, white wine, herbs and beef stock and bring to the boil, skimming off any impurities.

Add the pork rind, spare ribs and bacon, reduce heat to a slow simmer and cook for about 1 to 1½ hours until the meats are tender.

Remove the ribs and bacon and allow to cool.

Add the beans to the stock and cook at a slow simmer for 2½ to 3 hours until the beans are meltingly tender and have soaked up much of the stock. Check the beans for seasoning and cool.

Preheat the oven to 150°C.

Place your largest casserole dish on the bench and start to build the dish. Spoon in some of the beans with their juice, then top with the duck legs, which have been drained of any fat, and sausage. Cut the bacon into large chunks and scatter across the dish, and separate the pork spare ribs and poke among the other meats. Spoon over the remainder of the beans, add a little more juice from the pot to moisten, sprinkle with breadcrumbs then place into the oven for at least 3 hours.

While the cassoulet is cooking, break the surface crust several times with the back of a spoon to stop it from getting too dry.

Serve from the centre of the table, with a green leaf salad (p. 211) and as much fanfare as you can muster!

1 kg white haricot (navy) beans, soaked for 24 to 48 hours
50 g duck fat or butter
100 g shallot, chopped finely
3 garlic cloves, minced
½ leek, white part only, chopped finely
1 carrot, diced finely
1 celery stick, diced finely
½ cup tomato purée
200 ml white wine
1 bay leaf
A sprig of fresh thyme
2 litres beef stock (p. 274)
75 g pork rind, finely chopped
1 strip US-style pork spare ribs
250 g smoked continental bacon
6 duck confit legs (p. 160)
6 small Toulouse sausages; blanched (we add red wine and extra garlic to the sausage recipe on p. 132, or you could purchase coarse Italian sausages)
Sea salt & freshly ground black pepper
50 g breadcrumbs

SERVES 6

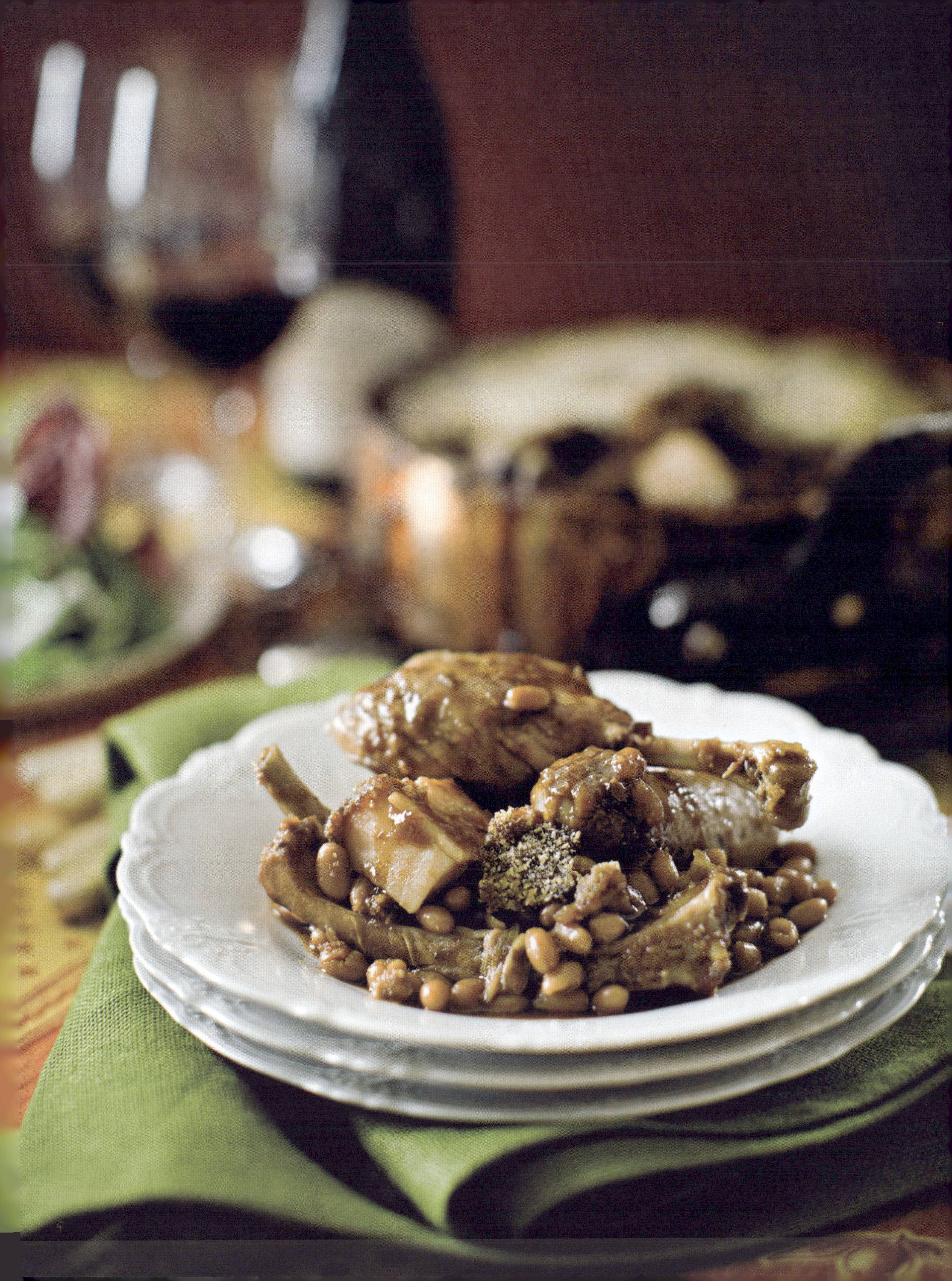

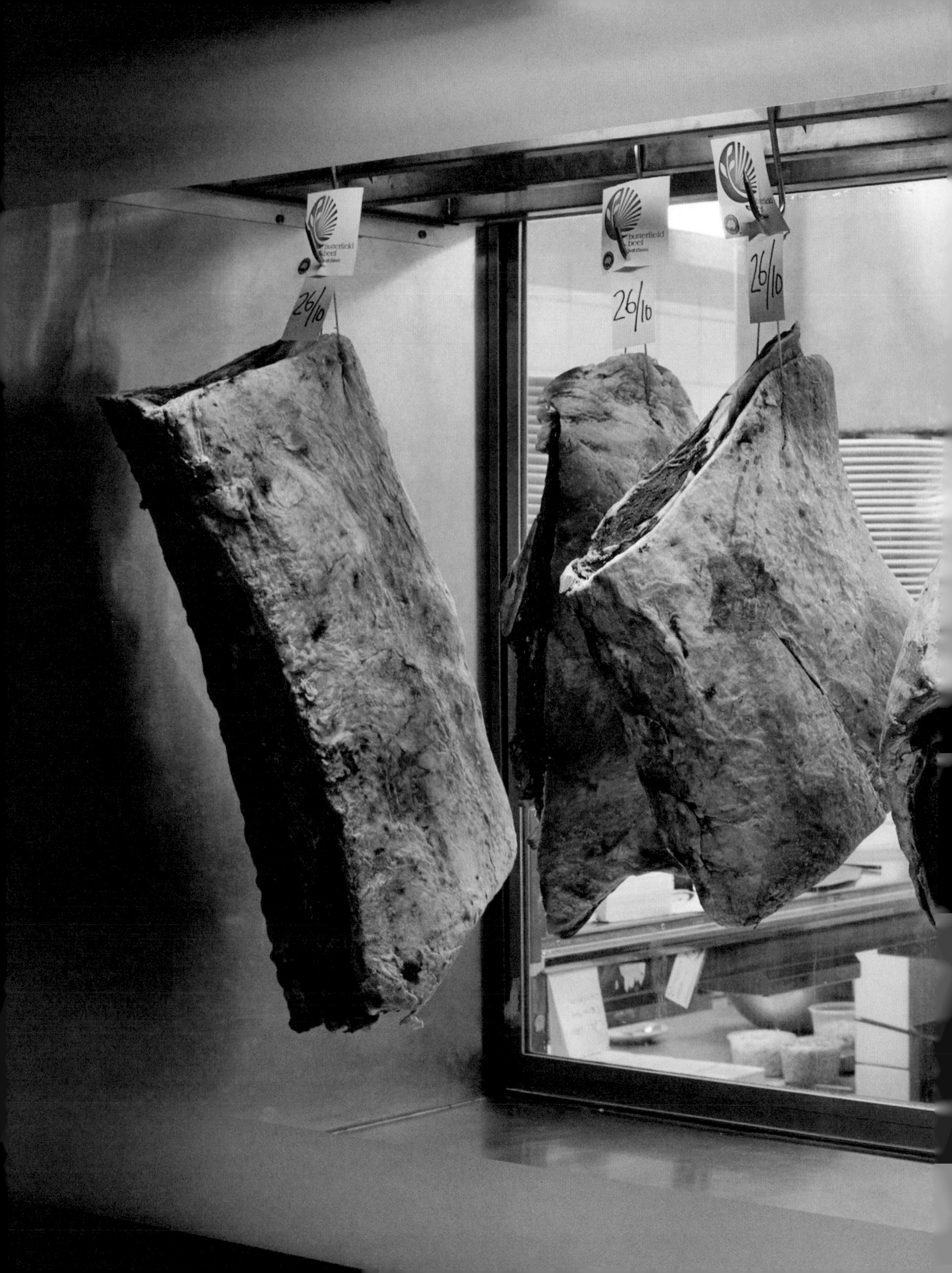
butterfield beef
26/10
butterfield beef
26/10
26/10

butterfield
beef
just classic
26/10
butterfield
beef
just classic
26/10
26/10

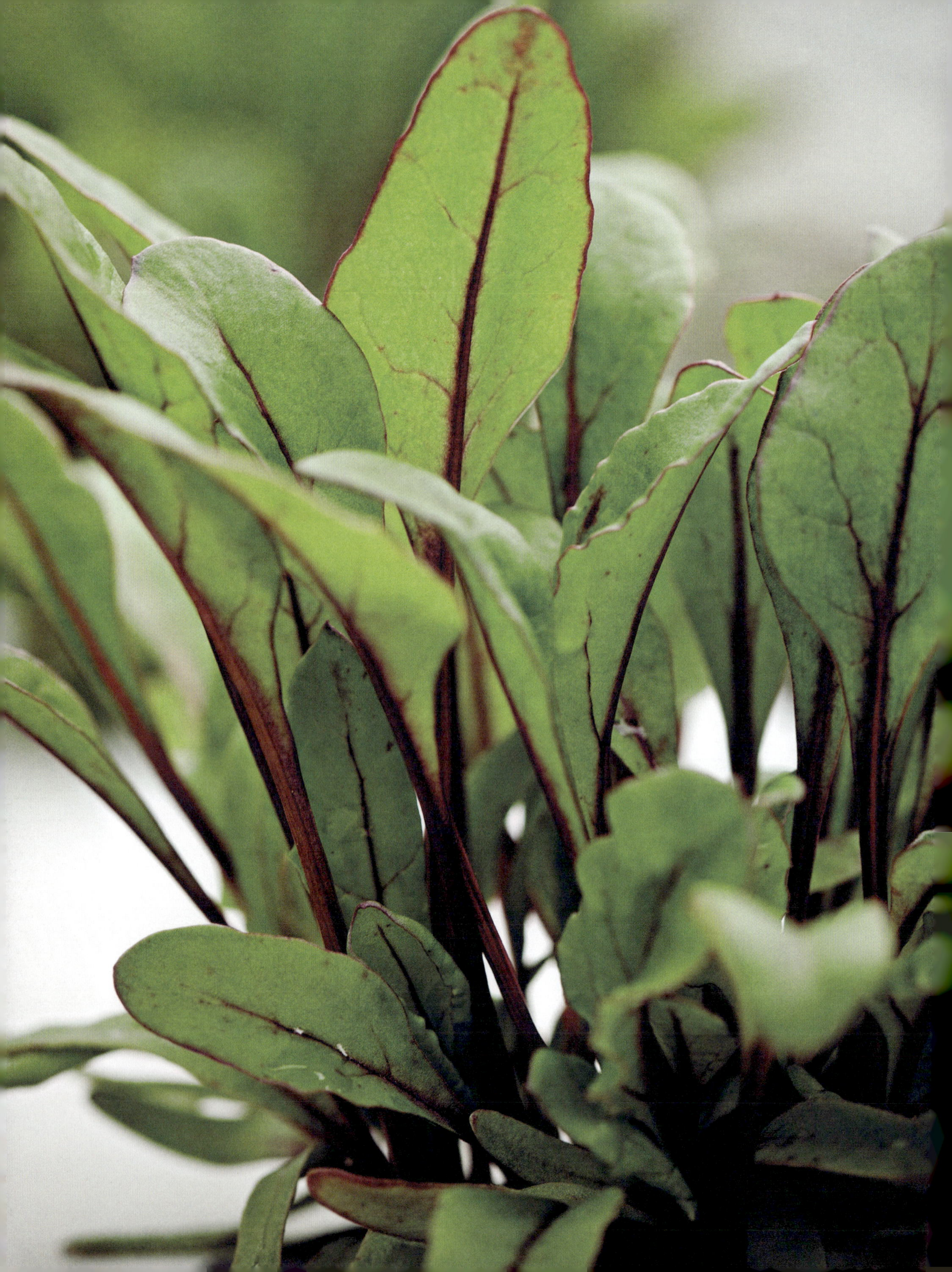

sides

Allan Hill is a long-time surfer who moved to the Margaret River region from the Gold Coast in 1988 and then became interested in growing hydroponic herbs and lettuces. Today, Al relishes taking his teenage son Jake out to catch some waves on weekends. Recently Jake was caught inside a break. 'He copped a few big ones on the head, but he toughed it out and paddled out for more.'

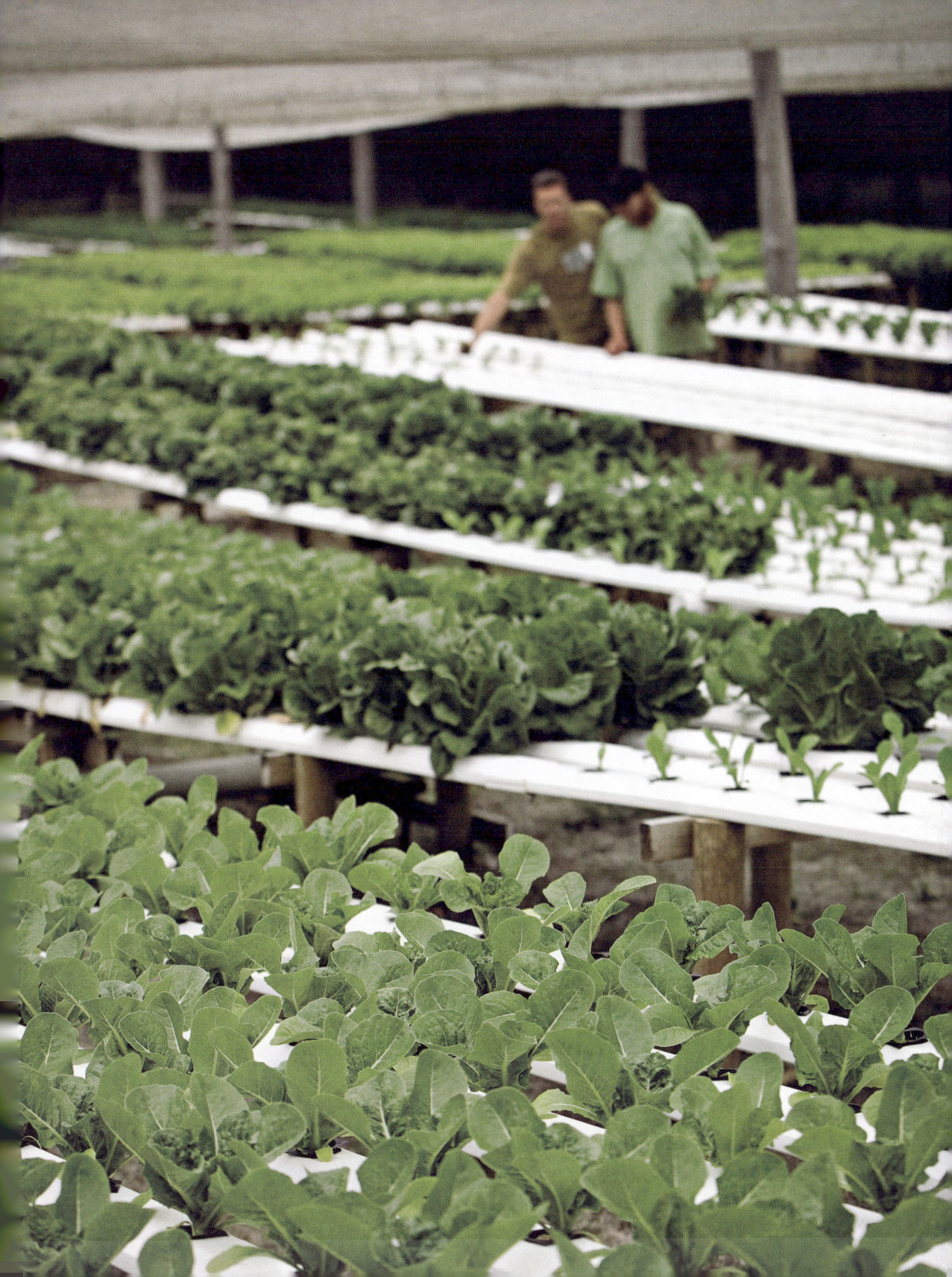

Braised Red Cabbage

I have a weekly ABC radio food segment, 'Lazy Susan', with the amiable and entertaining Russell Woolf. I gave this recipe to a caller named Harry. He called back the next week to say that the dish didn't work, so I invited him to Must for a cabbage cooking 'masterclass'. He learned that 'steaming' the cabbage for 25 to 30 minutes with the lid on softens the vegetable and helps incorporate the flavours.

Cut the cabbage into quarters, remove the core and slice finely.

Heat the olive oil in a large lidded saucepan over low heat. Add garlic and onion and sweat for about 3 minutes.

Add the vinegar and sugar, stirring for 1 minute to dissolve the sugar. Add the cabbage, place the lid on the pot and steam over low heat for 25-30 minutes until the cabbage is tender.

Season with salt and pepper; cook for another 10-15 minutes with the lid off until most of the liquid that has come out of the cabbage has evaporated. The cabbage will be richly purple coloured and tender with a sweet and sour flavour.

Braised red cabbage will keep well stored in the refrigerator for 3 to 4 days.

1 red cabbage
A little olive oil
2 garlic cloves, minced
1 brown onion, chopped finely
150 ml red wine vinegar
150 g brown sugar
Sea salt & freshly ground black pepper

SERVES 6 - 8

Celeriac Purée

Celeriac is at its best from autumn into winter. Its ugly exterior hides a creamy flesh with a sweet, nutty flavour. It works beautifully when combined with potato as in this purée, or finely shaved in a classic remoulade (p. 143). A word of caution: do not combine potatoes with the celeriac when using the food processor or you'll have a gluey mess.

1 head of celeriac (about 300 g), peeled and chopped roughly

600 g royal blue potatoes, peeled and chopped roughly

80 ml cream

50 g butter

Sea salt & ground white pepper

SERVES 6

Wash the celeriac then place into a large pot of salted cold water. Bring to the boil and cook until tender.

Cook the potatoes separately 10 minutes after the celeriac, they will take less time to cook.

Drain the celeriac and give it a few minutes to 'steam off' (for excess moisture to evaporate) before placing into a food processor bowl. Blitz to a fine purée.

Drain the potatoes, allow to 'steam off' for a few minutes then pass through a ricer or press through a fine sieve and add to the celeriac.

Bring the cream and the butter to a simmer, until butter melts, then remove from the heat. Pour the mixture over the purée and season well. Whip with a whisk to incorporate. Serve warm.

Chickpea & Chorizo Braise

This dish is adapted from a recipe by Joanne Weir and has been served many times at our Tapas Long Table lunches. It makes a remarkable hearty dish on its own, or pairs well with the Pedro Ximénez lacquered crispy pork belly (p. 178).

Place the chickpeas in a saucepan with the onion, clove powder, cinnamon stick, bay leaf, thyme and parsley and cover with chicken stock.

Simmer, uncovered, until the chickpeas are tender, about 40 minutes, then discard the onion and herbs and keep the chickpeas in the cooking liquor.

Heat the oil over low heat in a large frying pan. Add the garlic and onion and sweat for 5 minutes. Add the chorizo sausages; increase the heat to medium and cook for a further 5 minutes.

Add the chickpeas and their cooking liquor and simmer slowly until the liquid has almost vanished, this will take 30 to 40 minutes or so. Add seasoning if required.

Remove the chorizo from the pan, slice, then return to the chickpeas. Serve warm.

2 cups dried chickpeas, soaked overnight then drained
1 onion, halved
A pinch of clove powder
1 cinnamon stick
1 bay leaf
A few sprigs of thyme
A few sprigs of flat-leaf parsley
1 litre brown chicken stock (p. 274)
3 tablespoons extra virgin olive oil
1 medium onion, chopped finely
3 garlic cloves, minced
360 g chorizo sausages
Sea salt & freshly ground black pepper

SERVES 6

Eggplant Tagine

This spicy, sweet and sour eggplant begs to be partnered with braised or barbecued lamb or grilled chicken.

2 eggplants

90 ml extra virgin olive oil

1 brown onion, chopped finely

3 garlic cloves, minced

1 red chilli, de-seeded and chopped

1–2 cm knob of fresh ginger, chopped finely

2 teaspoons cumin powder

1 teaspoon coriander powder

½ teaspoon ras el hanout

700 ml tomato purée

50 g brown sugar

½ teaspoon saffron strands moistened with a little lemon juice

60 ml white wine vinegar

Sea salt & freshly ground black pepper

SERVES 6

Cut the eggplant into 2 cm square chunks and drizzle with a little oil. Heat a barbecue or chargrill pan cook the pieces of eggplant on all sides until marked; reserve.

Combine the remaining olive oil, the onion, garlic, chilli and ginger in a large deep saucepan over low to medium heat and cook for 10 minutes, stirring constantly, until the onion is tender.

Add the spices, increase the heat and cook for 2 to 3 minutes to release their aromas.

Add the tomato, sugar, saffron and vinegar; simmer over medium heat for 20 minutes until the sauce has thickened.

Add the eggplant, reduce the heat and place a lid on the saucepan, then cook for 10 to 15 minutes until the eggplant is tender. Season and serve.

Green Beans, Mint, Extra Virgin Olive Oil & Sea Salt

Pick crisp, fine, brightly coloured beans for this recipe.
The sprinkle of mint just before serving adds a delightfully fragrant note.

Place a large pot of salted water on to boil.

Remove any stalks from the beans, drop into the boiling water and blanch for three minutes.

Meanwhile, heat the olive oil in a large frying pan over medium-high heat and finely shave the mint leaves.

Strain the cooked beans, allow to drain for a few seconds then add to the frying pan, tossing through the shaved mint and seasoning.

Serve immediately, pouring any oil left in the pan over the beans.

600 g green stringless beans
A handful of fresh mint leaves
50 ml extra virgin olive oil
Sea salt & freshly ground black pepper

SERVES 6

Macaroni Cheese

My eldest son Alex wrote this recipe for a cookbook as part of a school project. It makes a great companion to an aged beef steak.

500 g elbow macaroni
960 ml full-cream milk
100 ml cream
100 ml evaporated milk
60 g unsalted butter
60 g plain flour
250 g Comté cheese (or gruyère), grated, plus 100 g extra to sprinkle on top
Sea salt & freshly ground black pepper
A pinch of freshly ground nutmeg

MAKES 6 SIDE SERVES

Cook the macaroni according to the packet instructions then refresh in cold water and drain.

Preheat the oven to 180°C.

Butter a 2 litre casserole dish.

Pour the milk, cream and evaporated milk into a saucepan and warm over medium heat.

Melt the butter over low to medium heat in a heavy saucepan. Add the flour and stir over heat for 3 minutes. Gradually add the warm milk mixture and stir constantly over low heat for 5 minutes. Add the grated cheese, seasoning and nutmeg. Add the macaroni and stir well to combine.

Pour the mixture into the buttered casserole dish and sprinkle with the extra cheese. Bake for about 30 minutes until the top is golden brown and the macaroni is bubbling.

Pickled Beetroot

This sweet-savoury accompaniment to the Wagyu beef carpaccio (p. 108) also works well with roasted duck and seared venison dishes. It can also be stored in the refrigerator for up to two weeks.

Peel the beetroot and cut into julienne (fine strips).

Place all the other ingredients in a small saucepan and bring to the boil over high heat. Add the sliced beetroot, return to the boil then turn off heat and allow to cool.

Chill before using.

One large beetroot (about 250 g)
90 ml red wine vinegar
160 g light brown sugar
50 ml water
1 bay leaf
½ teaspoon sea salt
2 black peppercorns
1 clove

MAKES ABOUT 200 G

Paris Mash

This reminds me of the wonderfully decadent mash I have so often eaten in Paris bistros. At Must we add the liquid to the potatoes once they are cooked, which makes them easier to pass through a fine strainer. To make a truffled potato mash simply add 20 grams of shaved Manjimup black truffles and drizzle the warm mash with a little truffle oil.

....

1 kg royal blue potatoes, peeled
300 ml milk
100 g unsalted butter
1 tablespoon sea salt
A pinch of ground white pepper

SERVES 6-8

Place the potatoes in a large pot of salted water, bring to the boil and simmer until tender. Drain and allow to 'steam off' (for the excess moisture to evaporate) for 2 minutes.

Meanwhile, put the butter and milk in a saucepan and bring to the boil. Remove the milk mixture from the heat, season, then add the potatoes. Stir once or twice with a wooden spoon to combine the potatoes and the liquid.

Pass the mixture through a fine strainer or potato ricer into a mixing bowl. Once passed, test for seasoning; if the consistency is a little stiff add a little more boiled milk then serve.

Pommes Frites

An essential side dish, classic French fries are a bistro staple and remain ever-popular. You will need a thermometer for this dish.

....

1 kg royal blue potatoes
500 ml vegetable oil
for deep-frying
Sea salt

SERVES 4-6

Peel the potatoes and cut into long 'sticks' (with 6 mm sides) and pat dry with a kitchen towel.

Heat the oil to 130°C in a large heavy saucepan.

Blanch the potatoes in the oil in batches until they are just tender. Drain and refrigerate until required.

Increase the temperature of the oil to 190°C, add the blanched potatoes in batches and fry until golden and crisp. Drain on kitchen paper and season with sea salt. Serve immediately.

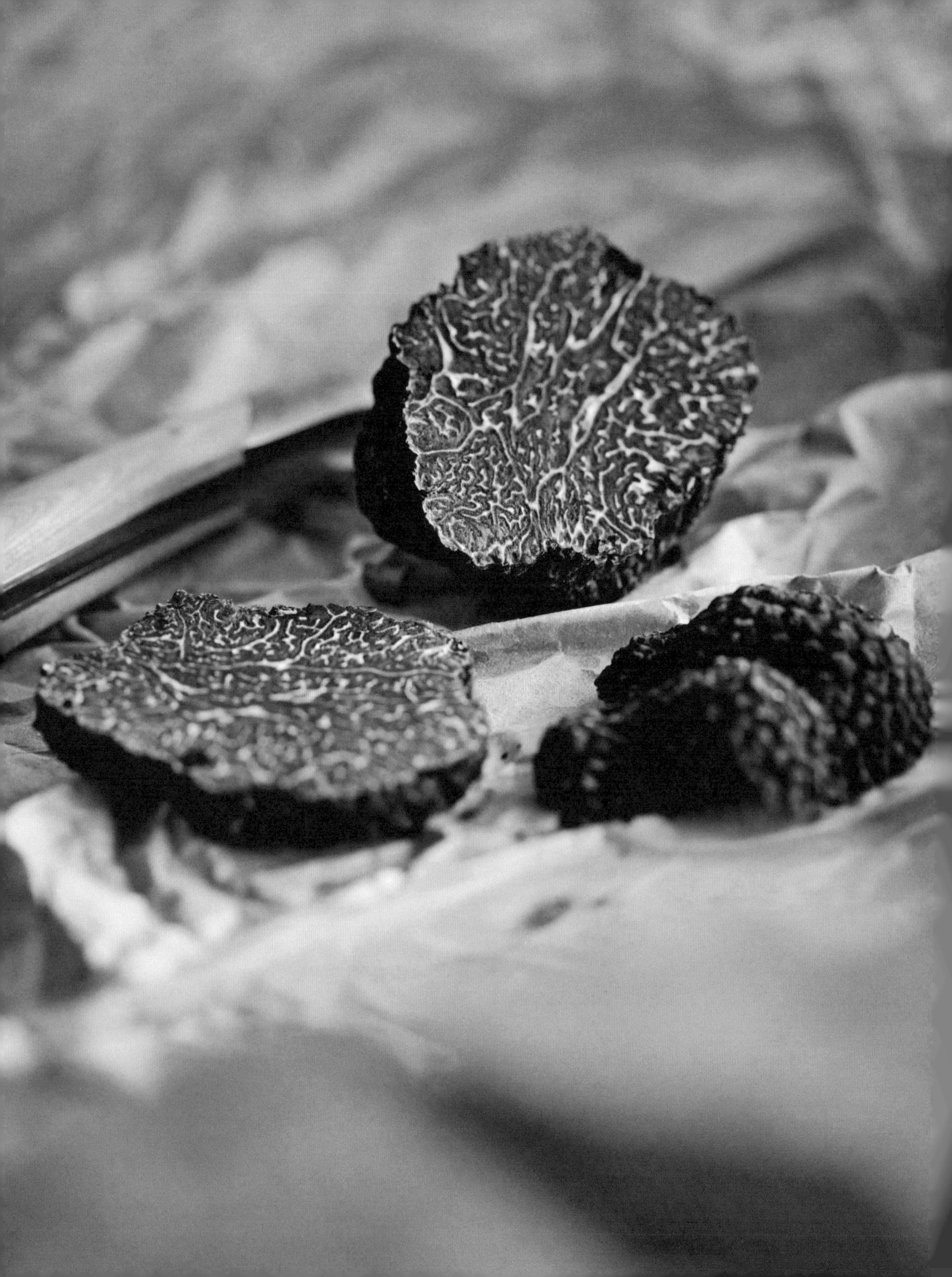

Potato Gratin

Gratin dauphinois is a classic dish of French haute cuisine. Potatoes are usually sliced, flooded with cream and cooked in the oven until they form a crust. My version is a little lighter. I cut the potatoes into strips and salt them to drain some of the excess starch. Then I cook the potatoes on the stovetop, before setting in a baking dish to crisp up with a topping of gruyère cheese and breadcrumbs.

1 kg royal blue potatoes, peeled
1 tablespoon table salt
20 g unsalted butter
1 small brown onion, chopped finely
2 garlic cloves, minced
A sprig of fresh thyme
1 bay leaf
375 ml milk
475 ml cream
A pinch of ground white pepper
A pinch of ground nutmeg
100 g gruyère cheese, grated finely
Breadcrumbs

SERVES 6

Use a mandolin to cut the potatoes into fine matchstick-sized strips. Place onto a tray and sprinkle with the salt, mixing well to coat the potato. Leave for 5 to 10 minutes to 'weep'.

Melt the butter in a large heavy saucepan, add the onion and garlic and sweat over low heat for 10 minutes until the onion is tender.

Add the herbs, milk, cream, white pepper and nutmeg. Bring to the boil and cook for another 5 to 6 minutes – the liquid will thicken a little as it boils.

Meanwhile, squeeze the potatoes firmly to remove excess moisture. Sprinkle the potatoes into the pot, reduce the heat to low and cook for about 15 minutes; stir gently to ensure the potatoes do not 'catch' on the bottom of the pan. Test that the potatoes are tender and check the seasoning. Transfer to a buttered casserole dish, removing the bay leaf and thyme stalks.

Preheat the oven to 180°C

Sprinkle the potatoes with the gruyère and breadcrumbs and pop into the oven for 25 to 30 minutes, until golden brown and bubbling.

Ratatouille Provençale

After requests from Claude, my restaurant manager at the time, we put this dish on the menu. It works really well with lamb, fish and chicken, and is a simple meal in itself served with grilled sourdough bread.

Dice all the vegetables into $1\frac{1}{2}$ cm pieces; keep separate.

Heat a little of the olive oil in a deep heavy saucepan over low heat, add the onion and garlic and sweat for 8 to 10 minutes until softened.

Add the tomato, thyme and bay leaf, increase heat to medium and cook for a further 10 minutes. Add the diced capsicum and cook for another 10 to 12 minutes until they have softened a little.

Lower the heat and add the zucchini, and cook a further 8 to 10 minutes until it has softened a little.

Meanwhile, heat the remaining oil in a large frying pan over medium-high heat, add the eggplant and cook for 6 to 8 minutes, stirring frequently until golden and softened. Add to the other vegetables.

Cook the ratatouille over low heat for another 10 to 15 minutes, gently stirring the pot regularly. Check that the vegetables have all cooked to a velvety tenderness, season and reserve until required.

Ratatouille improves in flavour after 24 hours and will keep in the refrigerator for three days.

120 ml extra virgin olive oil
1 large brown onion
400 g tomatoes, blanched & peeled
1 red capsicum, de-seeded
1 yellow capsicum, de-seeded
1 eggplant
1 zucchini
2 garlic cloves, minced
A sprig of thyme
1 bay leaf
Sea salt & freshly ground black pepper

SERVES 6

Torbay Asparagus, Extra Virgin Olive Oil & Sea Salt

Torbay is a hotspot for growing asparagus in Western Australia. Its rich, loamy soils, abundant rainfall and the warming influence of the nearby ocean create perfect growing conditions.

18 large asparagus spears
50 ml extra virgin olive oil
Sea salt & freshly ground black pepper

SERVES 6

Check the base of the asparagus talks when you buy them – if the cut face is dry or the base is shrivelled it isn't fresh.

Hold the spear at the tip (be careful, fresh asparagus will be brittle) and peel the thick bottom section of the spear, up to 5 cm from the base.

Bring a large pot of salted water to boil, drop in the asparagus spears, cook for 2 to 3 minutes, remove and drain.

Heat the olive oil in a large frying pan over high heat. Toss in the asparagus, season and toss to coat.

Serve immediately, pouring any oil left in the pan over the asparagus.

Mike Skivinis always runs a slasher through his avocado orchard before sending in the workers to pick. I know why. I visited the farm café with our kitchen team for lunch last summer, and found Mike stuffing a large dugite snake that he had just beheaded into a jar. The long kikuyu grass and abundant water on the property are like heaven – for snakes.

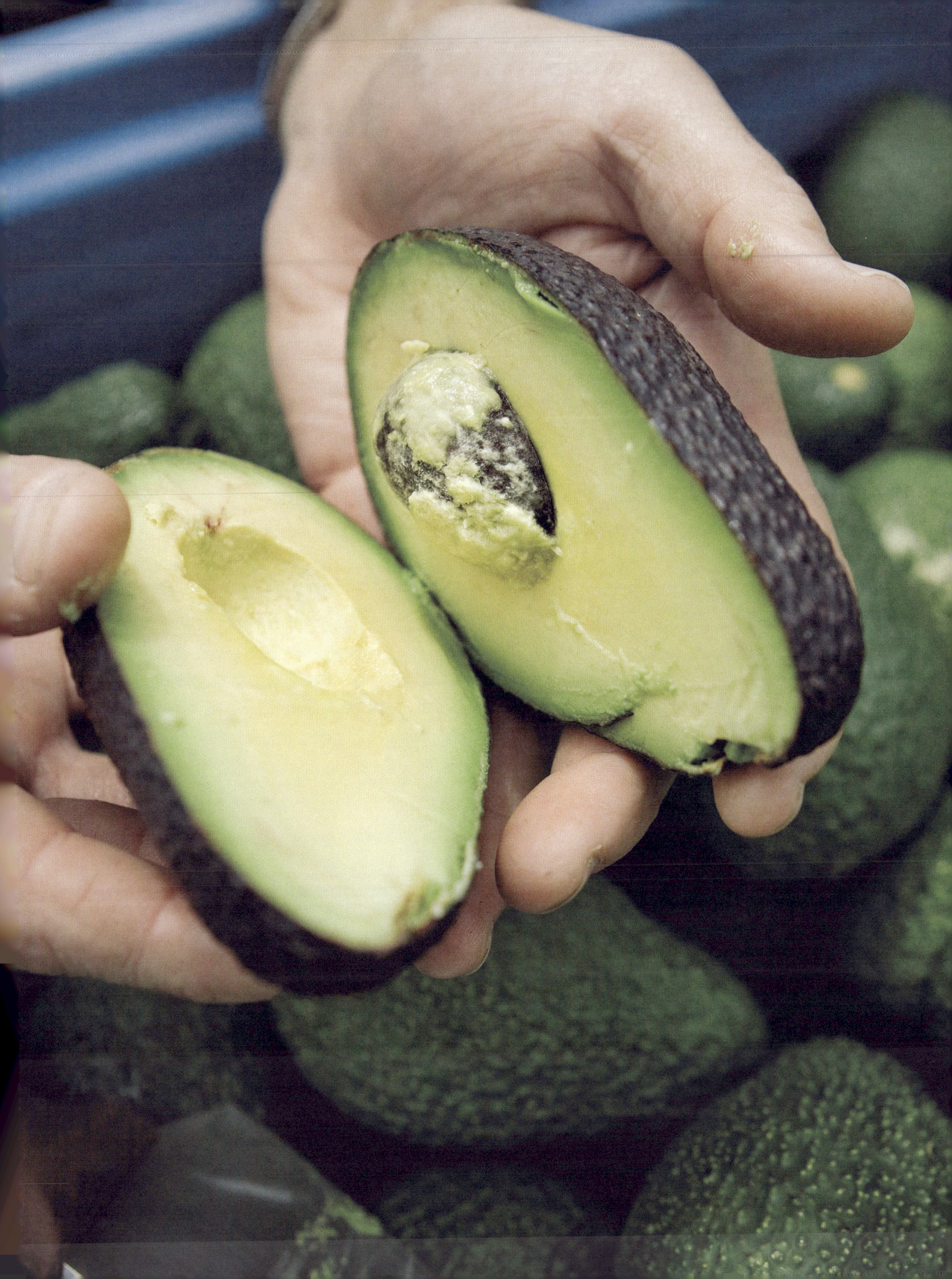

Avocado, Over the Moon Feta, Spinach & Pine Nut Salad

The organic feta produced by Over the Moon Organics dairy in Redmond (on Western Australia's south coast) is soft, decadently rich and tastes like the milk of the Jersey cows from which it is produced.

Ensure the spinach is well washed and dry. Place in a large bowl, add sliced avocado, pine nuts and crumbled feta.

Combine the red wine vinegar and oil in a bowl and whisk together with a little sea salt and pepper.

Gently toss the dressing through the salad, and serve immediately.

200 g baby spinach, washed
2 ripe avocados
$\frac{1}{4}$ cup pine nuts, lightly roasted
100 g Over the Moon feta

Dressing

25 ml red wine vinegar
75 ml extra virgin olive oil
Sea salt & freshly ground black pepper

SERVES 6

Mixed Leaf Salad

When it comes to green or mixed leaf salads, I like the lettuce to star, and a great dressing to tie the different textures and tastes of the lettuces together.

200–250 g mixed lettuce leaves
Red wine vinaigrette (p. 268)

SERVES 6

Tips to a great leafy salad:

* Use fresh lettuces, which are just about the easiest thing to grow in a home garden. The secret ingredient is water.

* Use a variety of lettuces: butter lettuce is soft and sweet; rocket, peppery; radicchio is crunchy with a bitter edge; while young frisée is savoury with a crisp bite.

* Gently wash and spin the lettuces dry before storing them in a sealed container in the fridge to chill before serving. This restores the crunch to the lettuce.

* Use about 2 to 3 tablespoons of dressing for every 4 cups of lettuce.

* Use a ridiculously oversized bowl when tossing the lettuce, and use your hands not salad spoons. Gently lift the leaves with open fingers, evenly distributing the dressing. Transfer from the mixing bowl to the serving bowl once tossed.

Moghrabieh Salad

Lebanese moghrabieh is like giant couscous grains, in fact it's often known as pearl couscous. It's made from semolina and flour and toasted until dry and is available from Mediterranean and gourmet food stores.

Bring $2\frac{1}{2}$ litres of salted water to the boil, drop in the moghrabieh and simmer slowly for 20 to 25 minutes until tender.

Drain and place into a large bowl, add the oil, lemon juice and seasoning. Allow to cool to room temperature.

Add the remaining ingredients and toss well. Cover and keep at room temperature if using within an hour or two.

300 g mograbieh

2 tablespoons extra virgin olive oil

Juice of one lemon

Sea salt & freshly ground black pepper

1 red chilli, de-seeded and chopped finely

4 spring onions, sliced finely

1 red capsicum, roasted, peeled and cut into strips

A handful of currants, soaked in hot water and drained

$\frac{1}{4}$ cup almond slivers, roasted

$\frac{1}{3}$ cup each of flat-leaf parsley, coriander and mint, washed and picked

SERVES 6

Rocket & Parmesan Salad

Don't be heavy-handed when shaving the parmesan. Use a light touch and the shavings will be paper thin, perfect to melt in the mouth when eating this salad. Biemme parmesan is made in a small family-run dairy in the Po River Valley of northern Italy.

200 g baby rocket or wild rocket leaves

60 g Reggiano or Biemme Grana Padano parmesan, shaved with a vegetable peeler

Dressing

70ml extra virgin olive oil

50 ml aged balsamic vinegar

A pinch of sea salt

SERVES 6

Ensure the rocket is well washed and dry. Place into a large bowl.

Add the dressing ingredients into a bowl and whisk lightly to combine. Sprinkle the dressing over the rocket, tossing to coat the leaves lightly.

Sprinkle in the shaved parmesan and toss through.

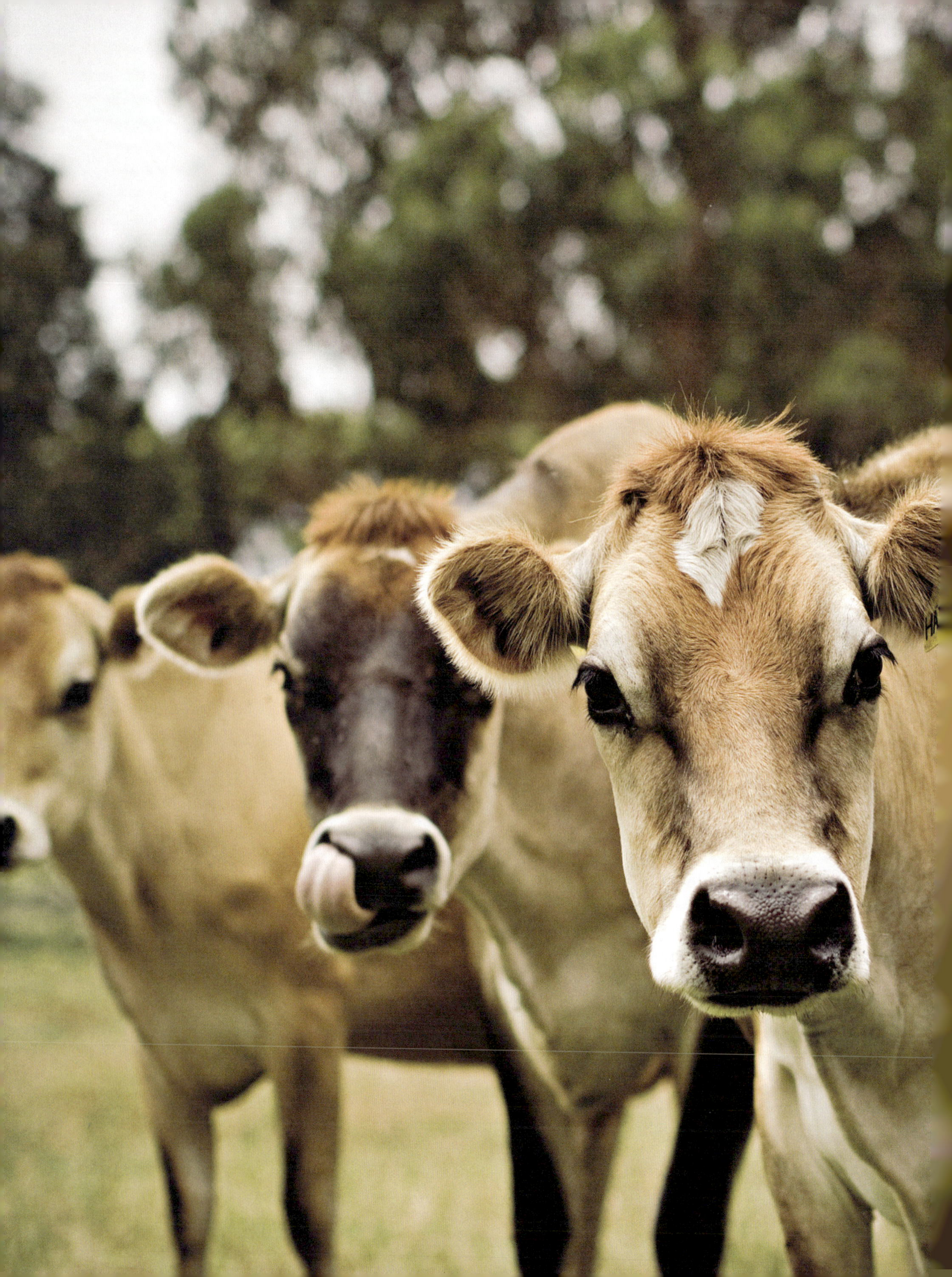

desserts

Juliet Bateman and her partner David Schober bought their dream property near Redmond in the Great Southern, which then inspired horse-mad Juliet to buy another horse. David said that would be too expensive, so Juliet settled on a Jersey cow, which then led to a small herd of milking cows, and then to the Over the Moon Organics dairy and cheese factory!

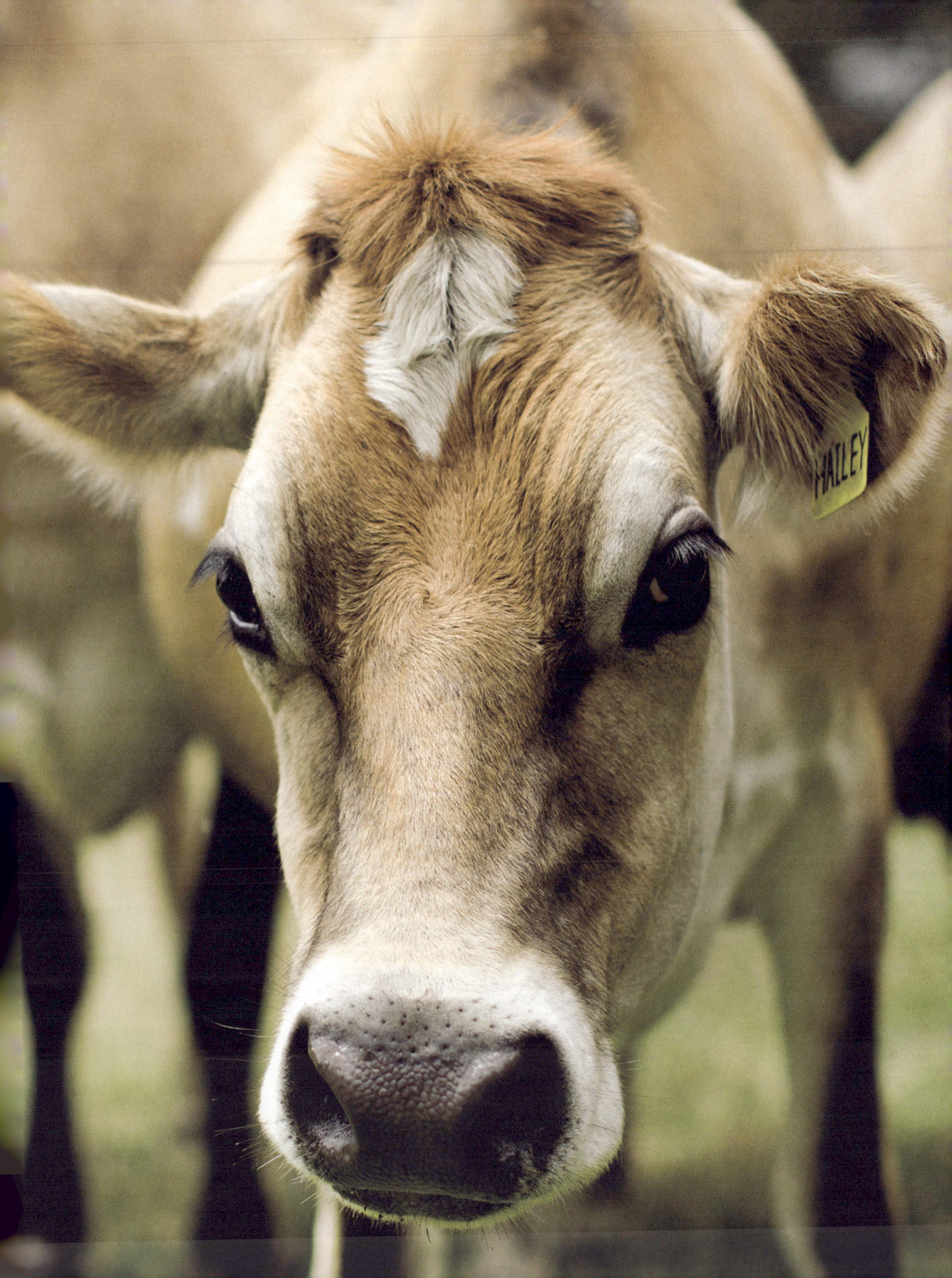
HAILEY

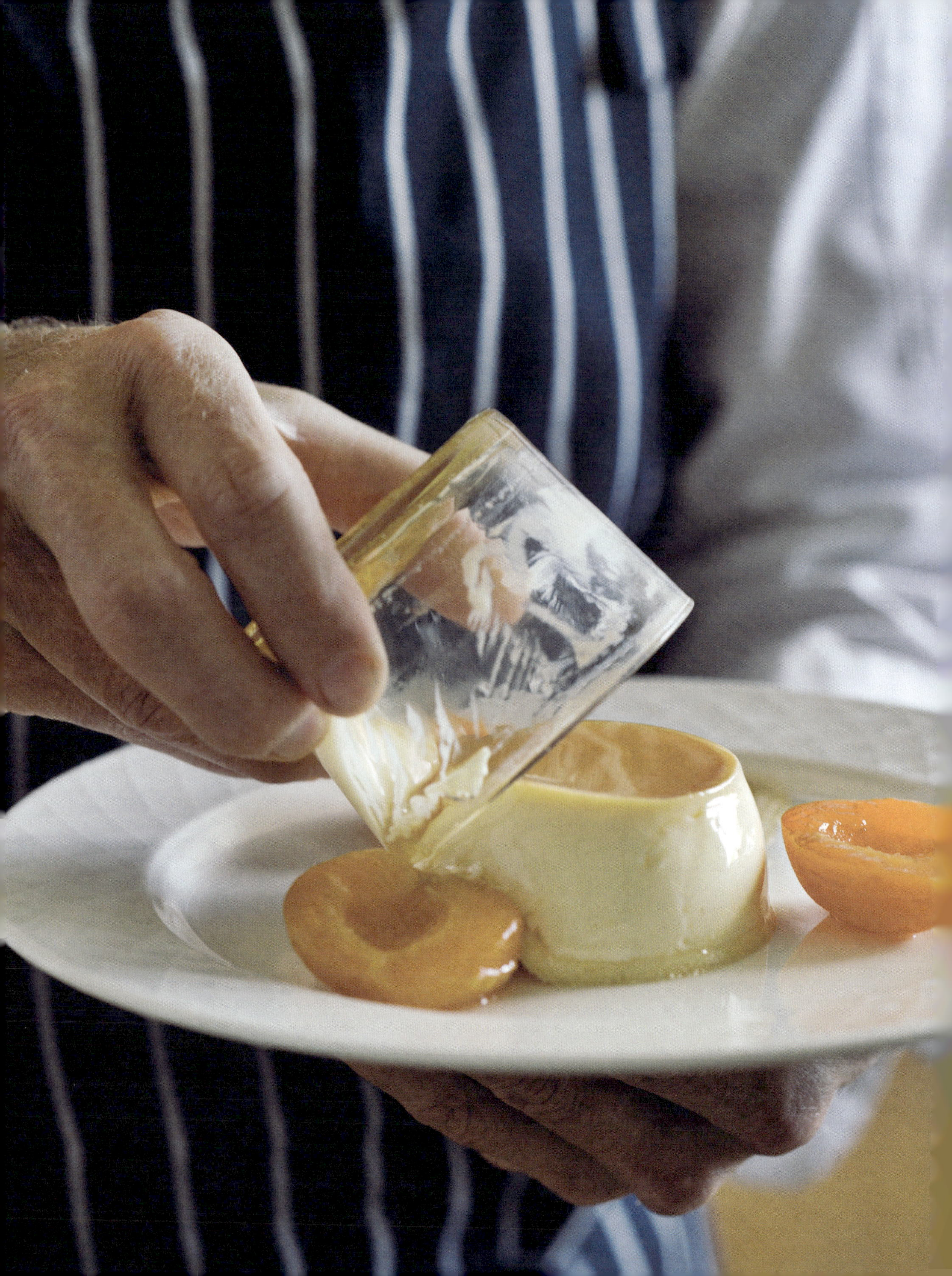

Crème Caramel & Poached Apricots

Crème caramel is a classic bistro dessert. Fresh apricots, simply poached, add acid and colour, but if you miss them during their short season try blood oranges, grapefruit or oranges.

---

4 x 50 g eggs
4 egg yolks
100 g caster sugar
700 ml milk

Caramel
330 g (1½ cups) caster sugar
185 ml water
125 ml water, extra

Poached apricots (p. 289)

SERVES 6

To make the caramel, put the sugar and 185 ml water in a saucepan and bring to the boil. Brush the sides of the pan with a wet pastry brush to avoid sugar crystals forming on the sides of the pan.

When the syrup is a deep caramel colour, remove from the heat and tip in the extra water (be careful, it will spit). Return to the heat for a few moments, stirring to ensure the caramel is completely dissolved.

Pour the caramel into 6 small dishes (approximately 180 ml), rotating the dishes to line the sides with the hot liquid. Allow to cool.

Preheat the oven to 160°C.

Combine the eggs, yolks and sugar in a large bowl, then whisk lightly.

Boil the milk and add to the egg mixture while whisking.

Strain the mixture into a bowl then ladle into the caramel-lined dishes. Place the dishes into a large baking dish, then carefully pour boiling water into the dish to a height of 2 cm to create a water bath. Bake for approximately 30 minutes until the custard has just set.

Remove the custard dishes and cool to room temperature, refrigerate overnight.

Place the crème caramel dish into a bowl of hot water for 10 to 15 seconds to soften the caramel around the sides. Press lightly on the edge of the custard to create an air pocket that will release when the caramel is turned out on the serving plate.

Arrange poached apricots around the plates and serve with whipped cream.

Chateau Suduiraut Sauternes
Bordeaux, France

The custard texture of this crème caramel is best paired with a wine that has both elegance and strength. Suduiraut is a lustrous gold colour and full of 'apricot' botrytis, which is ideal with the poached apricots. The textural feel of both the wine and dessert is soothing and unctuous.

Domaine Disznoko Aszû 5 Puttonyos
Tokaji, Hungary

Tokaji has wonderful botrytis, orange-quince flavours, piercing acidity and varying levels of sweetness. The disznoko has a fresh, energetic style with intense sweetness and searing acidity, which balances the sweetness and creaminess of the crème brûlée and pairs with the sesame snap. The sorbet is always a challenge, but this has the acid and sweetness to survive.

Orange & Vanilla Crème Brûlée, Raspberry Sorbet & Sesame Snap

On my last trip to France with my family, we travelled from the Alps to Provence to stay at a friend's hotel on the outskirts of Nîmes. We dined at L'Annexe, the second restaurant of Michelin-starred local chef Olivier Douet. This recipe is a re-creation of the remarkable dessert I was served that night. You will need a blowtorch to finish this dessert successfully.

Preheat the oven to 150°C.

Split vanilla pod lengthways and scrape the seeds into a saucepan with the cream and orange zest. Heat over medium heat until the cream starts to simmer, remove from heat and steep for 10 minutes.

Whisk the egg yolks with the sugar and Grand Marnier. Strain the warm milk over the eggs and whisk lightly to combine. Skim off any froth from the custard, and pour into shallow dishes (I use 10 cm square by 2.5 cm deep dishes).

Place into a deep baking tray, then surround the filled dishes with 1 cm of boiling water. Carefully cover the tray with foil and bake for 20 to 23 minutes. Check that the brûlées are cooked by shaking one of the dishes – the custard will barely wobble when it has set. Refrigerate until cold.

Increase the oven temperature to 170°C.

To make the sesame snaps, pour the milk into a small saucepan and bring to the boil; reduce the heat to low, then add the butter and glucose. When the butter has melted add the sugar and stir well to dissolve. Remove from the heat and add the seeds, stirring well. Refrigerate until cool.

Line an oven tray with baking paper, place spoonfuls of the cooled mixture onto the paper and spread flat with a palette knife. Bake for about 6 to 8 minutes until crisp and lightly golden. Remove from the oven, cool for 1 minute, then press a 5 cm cutter into the cooked snap to form a perfect disc. Cool to room temperature and store in a sealed airtight container until required.

Remove the brûlées from the fridge and onto a metal tray. Sprinkle one with a thin, even layer of sugar then caramelise with a blowtorch. Repeat for the remaining brûlées. Place onto serving plates with a scoop of sorbet on top, then finish with a sesame snap.

Crème Brûlée

1 vanilla bean
1 teaspoon orange zest, grated finely
575 ml cream
6 x 59 g egg yolks
120 g sugar
25 ml Grand Marnier

Caster or cassonade sugar for the brûlée crust
Raspberry sorbet (p. 290)

Sesame Snap

50 ml milk
150 g unsalted butter
50 g glucose
150 g caster sugar
300 g sesame seeds
50 g poppy seeds

SERVES 6

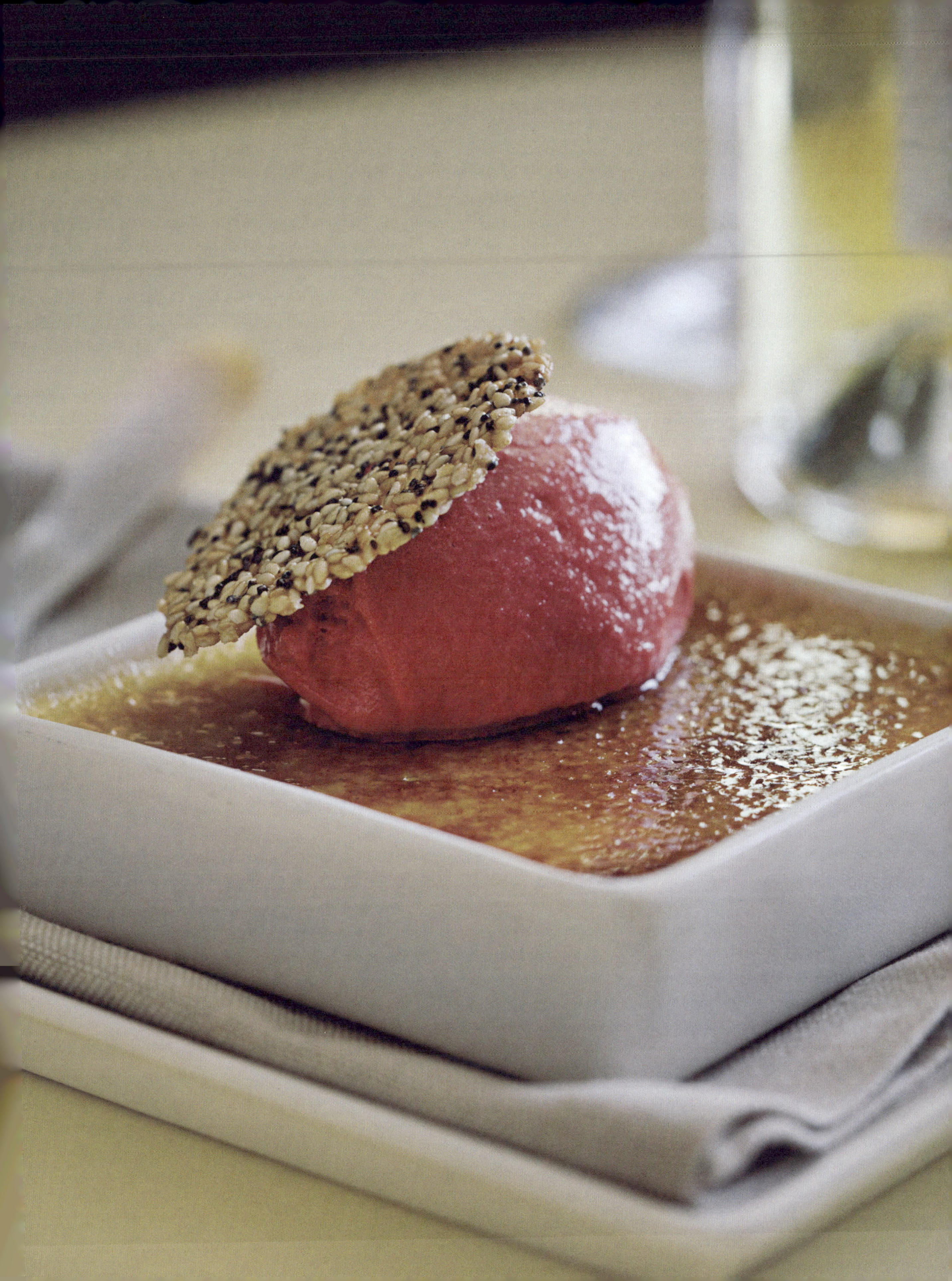

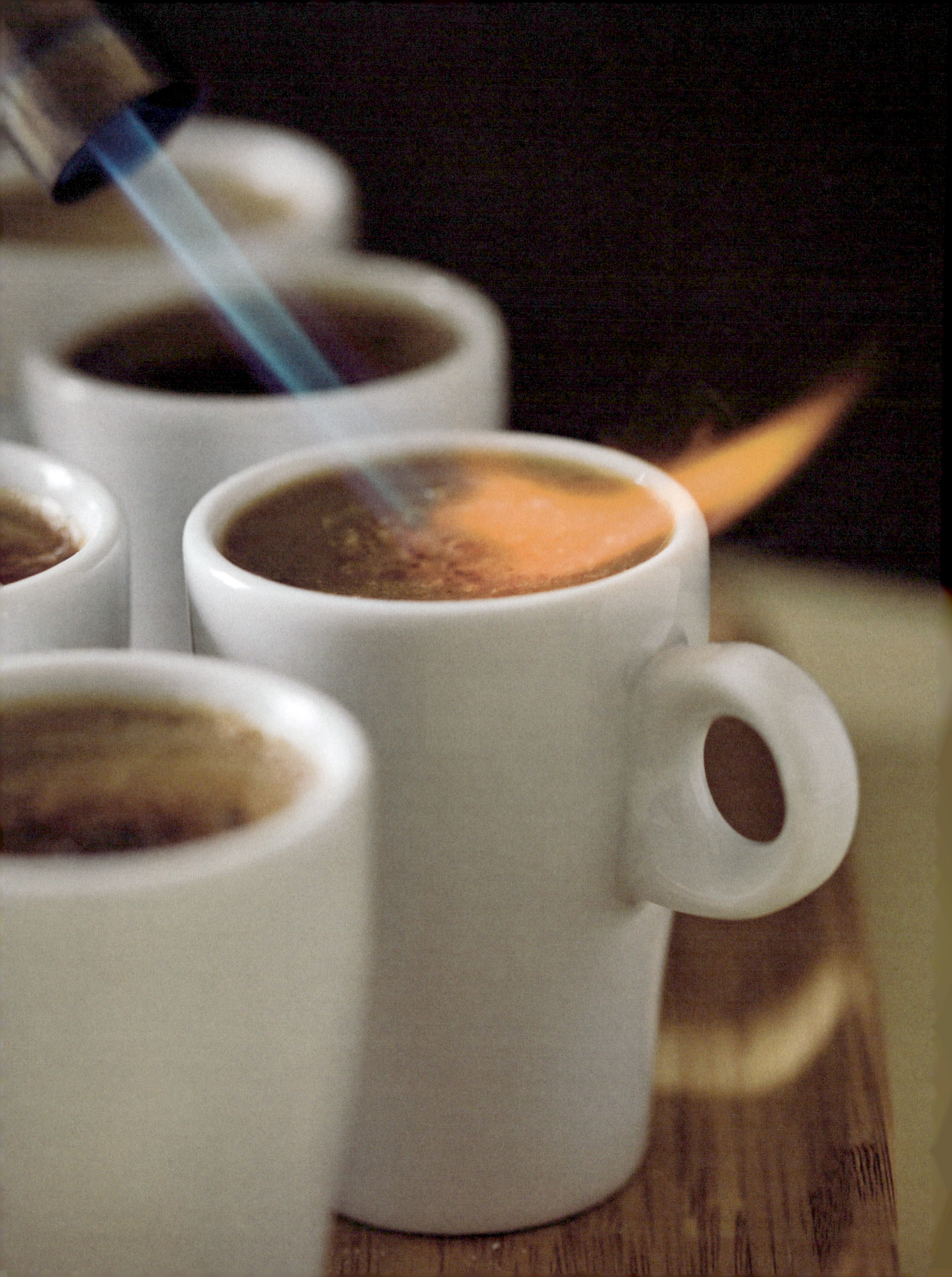

Espresso Crème Brûlée

Tiny crèmes brûlées are a great option for a stand-up dessert, and demitasse cups, which hold about 60 ml, are the perfect size. If you don't have tequila, substitute it with Kahlua or even sambucca.

500 ml cream
60 ml espresso coffee
6 x 59 g egg yolks
75 g light muscovado sugar
40 ml Patrón XO Café Tequila
Caster or cassonade sugar for the brûlée crust

SERVES 9–10

Preheat the oven to 150°C.

Place the cream into a saucepan over medium heat; on simmering, remove from heat, add coffee and put to one side.

Whisk the egg yolks with the sugar and preferred liqueur. Strain the warm milk over the eggs and whisk lightly to combine. Skim off any froth from the custard and pour into demitasse cups.

Place into a deep baking tray, then surround the filled cups with 2 cm of boiling water. Carefully cover the tray with foil and place into the oven to cook for 12 to 15 minutes. Check that the brûlées are cooked by shaking one of the cups – the custard will only just wobble when it has set. Refrigerate until cold.

Remove the brûlées from the fridge and place onto a metal tray. Sprinkle one with a thin even layer of sugar then caramelise it using a blowtorch.

Repeat for all the brûlées and serve immediately.

Isole e Olena Vin Santo
Tuscany, Italy

Espresso crème brûlée doesn't require a fortified wine, which will detract from the creamy custard and crisp caramel edge. Vin Santo is variable in style, but this one is often nutty and smoke-infused with an orange, caramel tone. The viscosity and opulence matches the texture and weight of the dish. The espresso influence provides an additional challenge, but combines with the wine's nutty caramel and smoky elements.

Phillip Marshall not only grows asparagus, he also nurtures raspberries in greenhouses on his farm at Torbay, dodging the thorns to pick the berries over the summer months. Most are sold to locals at the weekly Albany farmers' market, the finest of its type in Australia.

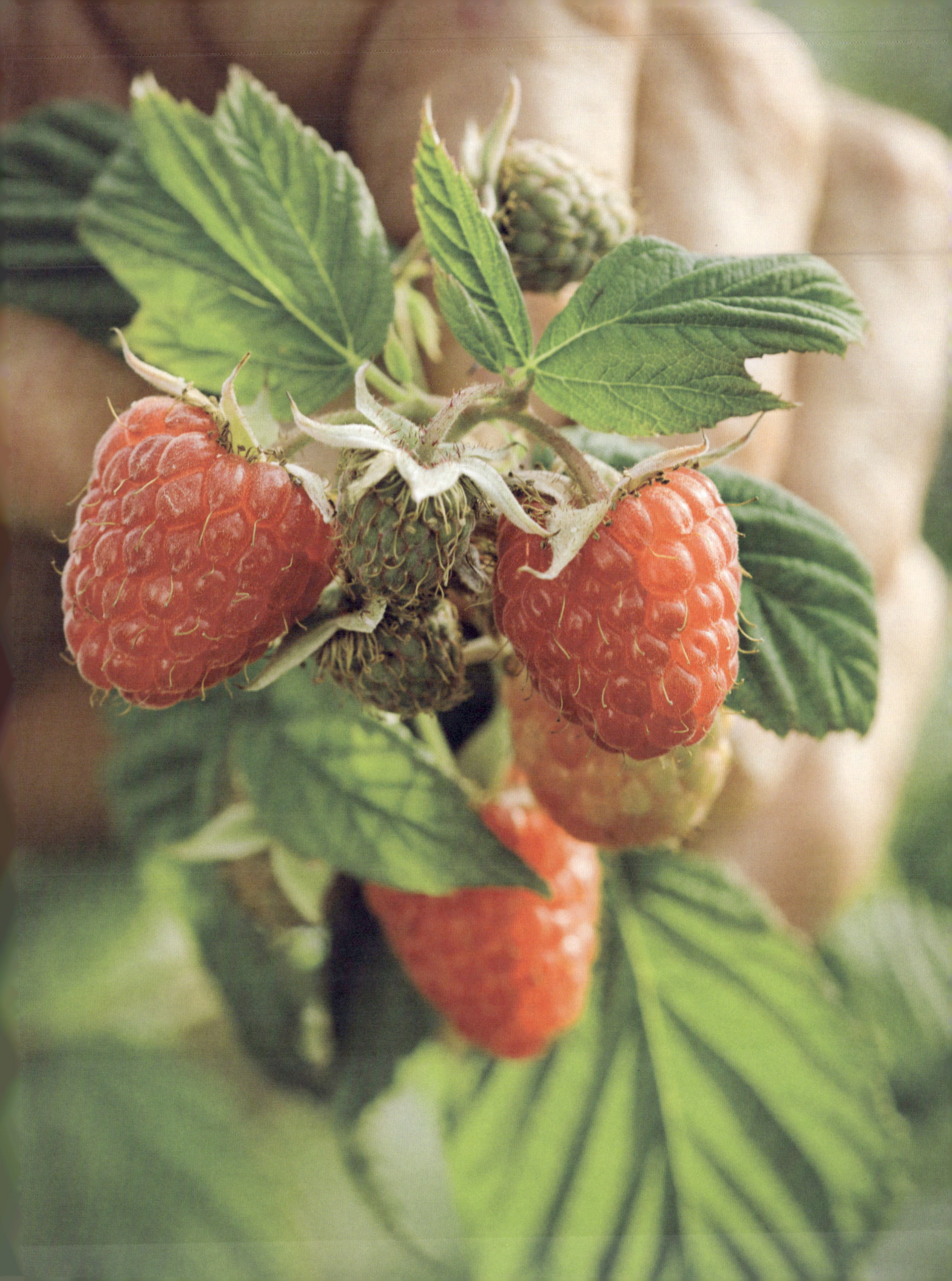

Willi Opitz Beerenauslese Pinot Noir
Neusiedlersee, Austria

Willi is the 'king' of Austrian sweet wine, and this pinot noir is attractively pink, with a cherry-strawberry opulence that perfectly complements the bright berry dessert. The weight of flavour, alcohol viscosity and sweetness sit comfortably with the creamy fruit compote.

Summer Berries, Crème Chantilly, Meringue & Crème de Fraises

This is inspired by the classic Eton mess, first created as a picnic dish in exclusive English schools in the 1930s. I've 'messed' with the recipe, adding ice cream, blueberries and raspberries, and a splash of a strawberry liqueur from Boudier, one of the finest producers in France.

Hull and slice the strawberries, place half of them into a food processor and process to a purée. Pass through a fine strainer to remove the seeds; reserve.

Whip the cream and sugar together to light peaks.

Mix the sliced strawberries, blueberries and raspberries in a bowl and add the crème de fraisis.

Crush the meringue and add to the cream. Gently fold through some strawberry purée.

Serve into glasses alternating the meringue-cream mixture with ice cream and the berries.

1 punnet of strawberries
250 ml whipping cream
40 g caster sugar
1 punnet of blueberries
250 ml whipping cream
1 punnet of raspberries
60 ml Gabriel Boudier crème de fraises
6 pieces of crisp, ready-made meringue
6 scoops vanilla ice cream (p. 290)

SERVES 6

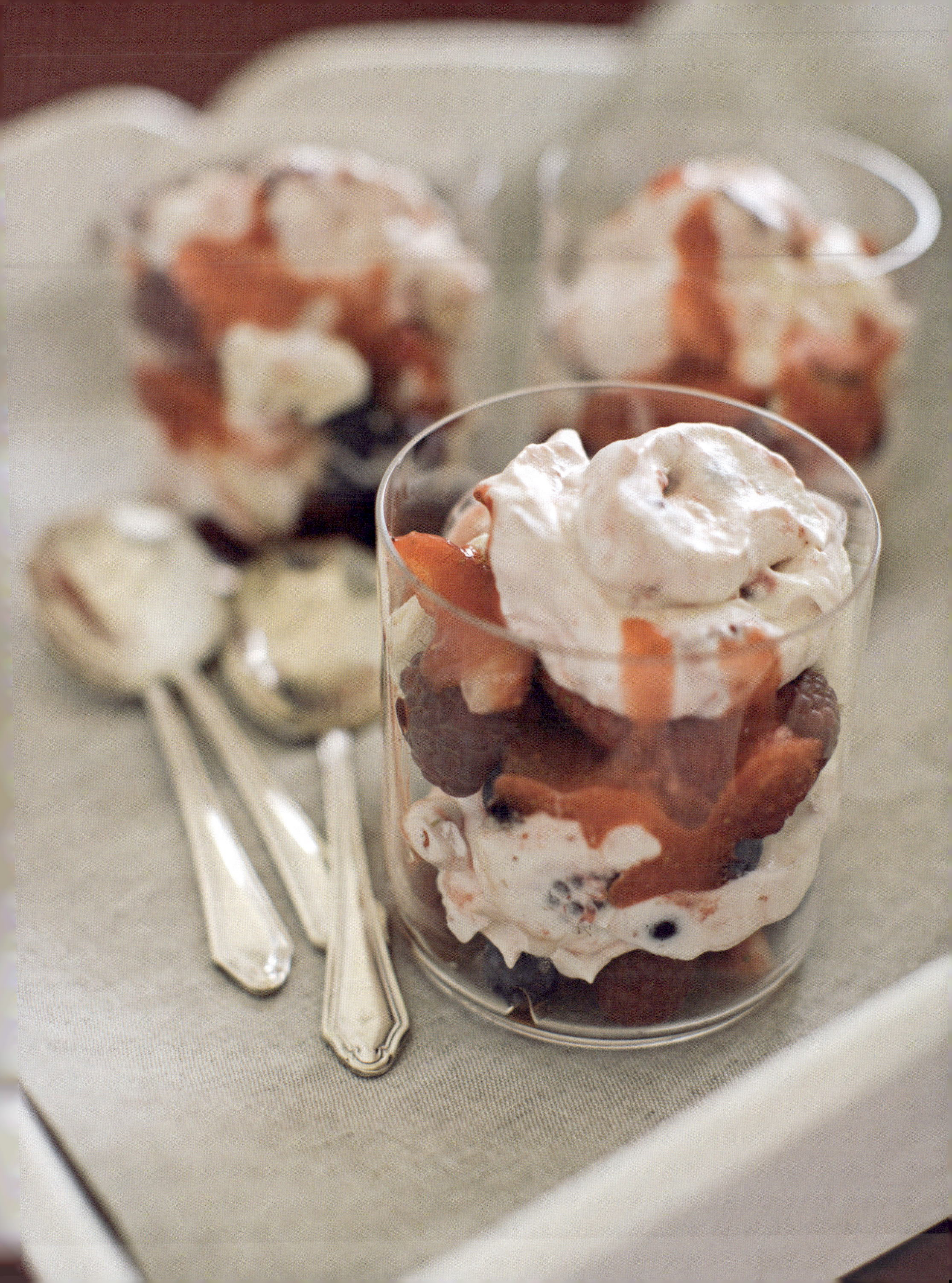

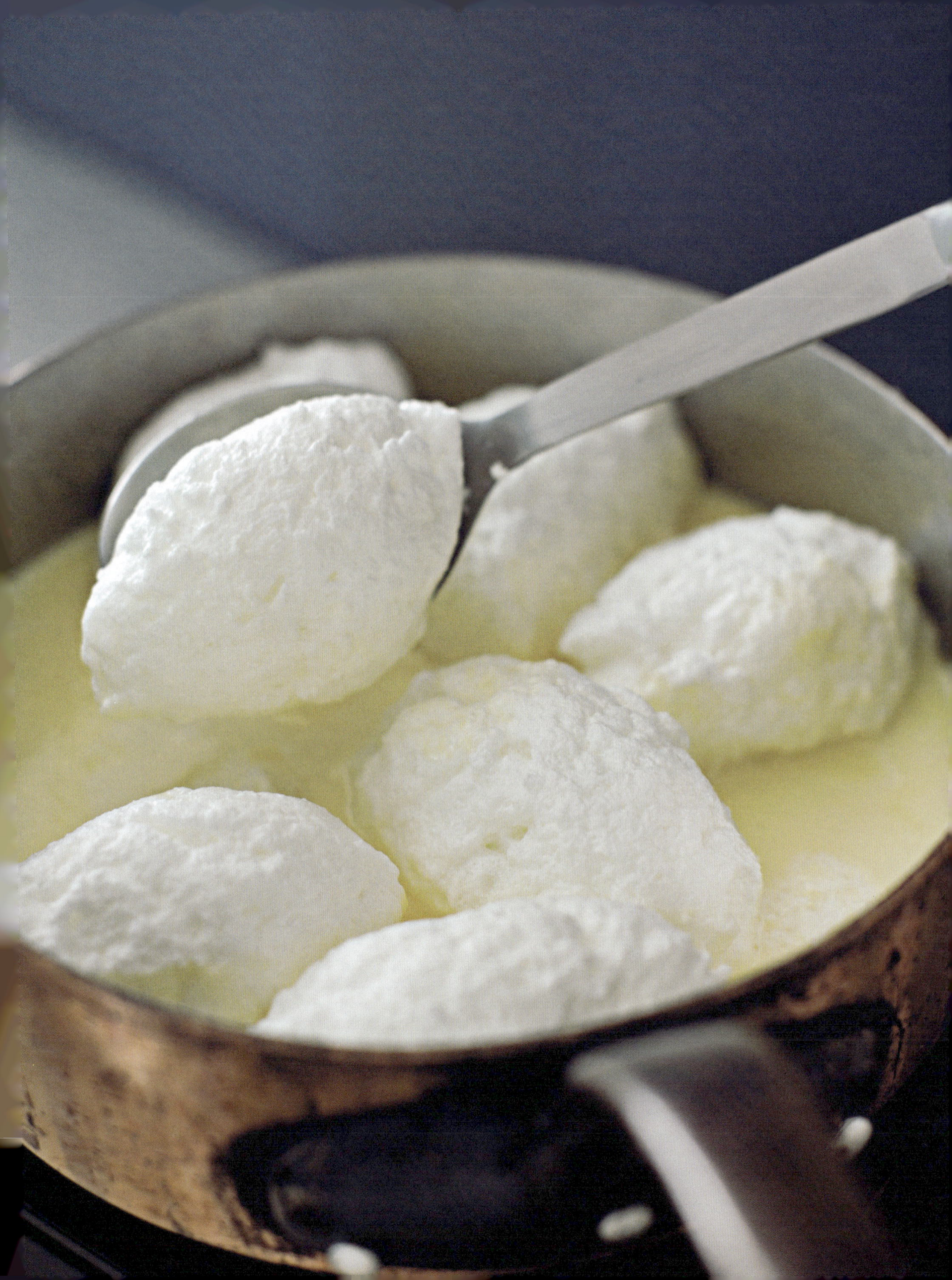

Iles Flottantes

No self-respecting French bistro chef wastes egg whites. These 'floating islands' of poached meringue were popular in the 1970s, but good food, like good fashion, is timeless.

Floating Islands

6 egg whites
A pinch of salt
100 g caster sugar
½ teaspoon vanilla extract
500 ml milk
1 vanilla bean, split lengthways
6 egg yolks
85 g caster sugar
60 ml Cointreau

Orange Caramel

½ cup finely strained orange juice
125 g caster sugar
Extra 70 ml strained orange juice

Pistachio praline (p. 234)

SERVES 6

Beat the egg whites with salt to soft peaks. Gradually add 100 g of sugar and the vanilla extract, constantly beating the whites to firm, glossy peaks.

Bring the milk to a boil in a large wide pan then reduce to a slow simmer.

Using 2 spoons, shape the meringue into large 'footballs' and slip into the milk to cook, turning twice, for approximately 2 minutes each side. Set aside.

When the islands are cooked, strain the milk into the bowl and add the vanilla bean.

Whisk the yolks and the remaining sugar. Pour the milk onto the egg mixture, whisking to combine. Pour into a clean saucepan.

Place the saucepan on low heat, stirring constantly until the custard coats the back of a spoon. Strain, add Cointreau and cool in the refrigerator.

To make the caramel, combine the orange juice and sugar together in a small saucepan and cook over high heat, brushing the sides of the saucepan with a wet pastry brush to avoid the mixture crystallising.

When the colour has darkened to a deep gold, remove from the heat, pour in the extra orange juice, stirring well to combine and cool immediately.

To serve, place an 'island' in the centre of each plate, flood the plate with custard, drizzle with orange caramel and sprinkle with pistachio praline.

Royal Tokaji Blue Label Aszû 5 Puttonyos
Tokaji, Hungary

The influence of orange in this dessert is best matched with a wine of similar flavour. The heavily botrytis Royal Tokaji has rich marmalade-apricot tones, generous weight and unctuous texture, with a brisk edge that balances the sweet, creamy meringue and combines with the custard and orange caramel.

Lemon Mascarpone Tart

Dr Loosen Beerenauslese Riesling, *Mosel-Saar-Ruwer, Germany*

This lemon mascarpone tart, with its creamy texture and richness, is nicely contrasted by Dr Loosen's intense, but delicate Beerenauslese riesling. The crisp acidity cuts through the substance of the tart and provides a fresh, zesty lemon finish.

This recipe is a departure from the standard baked custard. Somewhat richer, it is made as a sabayon, whipped over a bath of hot water until it is cooked, poured into a sweet pastry shell then refrigerated to set.

Preheat the oven to 190°C.

Mix the butter and icing sugar together to a smooth paste. Add the egg yolks and salt and mix well. Add the flour and work until the ingredients have combined.

Roll the dough into a ball and refrigerate for 1 hour.

Flour your workbench and roll out the dough to about a 3 mm thickness.

Lightly butter a 28 cm (3 cm deep) loose-based flan tin, and line with the pastry. Let the edges of the pastry overlap the sides to be trimmed off after the flan is cooked.

Prick the base of the tart with a fork, line with baking paper and fill with baking beads. Blind bake for 15 minutes. Remove the paper and filling, reduce the heat to 180°C and bake for a further 15 to 20 minutes until golden.

Remove from the oven, trim off the excess pastry edges with a serrated knife and allow to cool.

To make the filling, combine the egg yolks, lemon juice and sugar in a large bowl. Place over a saucepan of simmering water, ensuring the base of the bowl does not touch the surface of the water, and whisk until thick and lightly coloured. Remove from the heat and whisk butter into the mixture then cool to room temperature.

Fold in the mascarpone and pour the mixture into the cool flan base. Allow to set for 2 to 3 hours in the refrigerator, and serve with double cream.

Pastry

300 g unsalted butter, at room temperature
150 g icing sugar, sifted
3 x 59 g egg yolks
A pinch of salt
375 g plain flour, sifted

Filling

8 x 59 g egg yolks
100 ml lemon juice
250 g caster sugar
250 g unsalted butter (at room temperature), cubed
150 g mascarpone cheese

SERVES 6-8

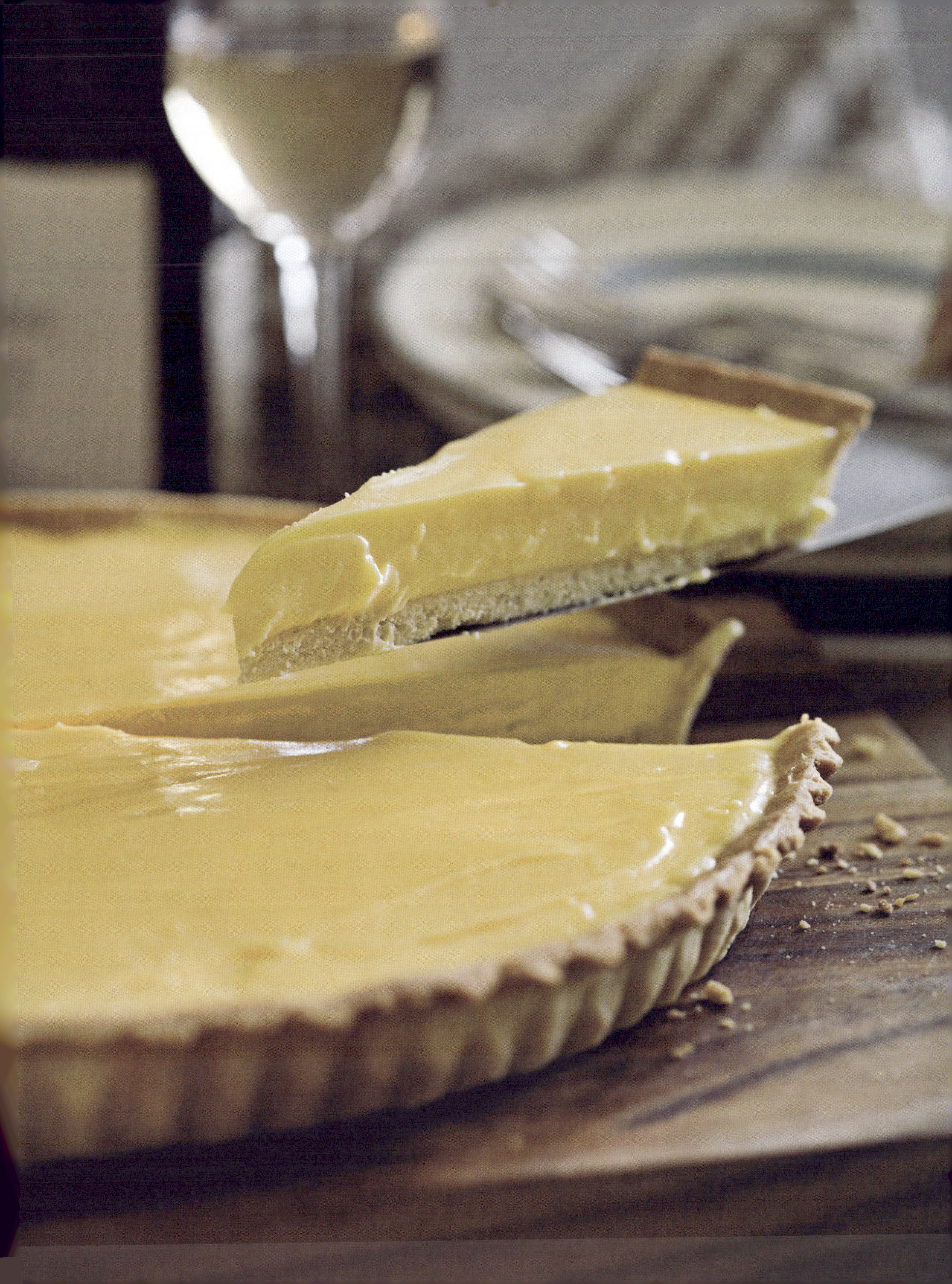

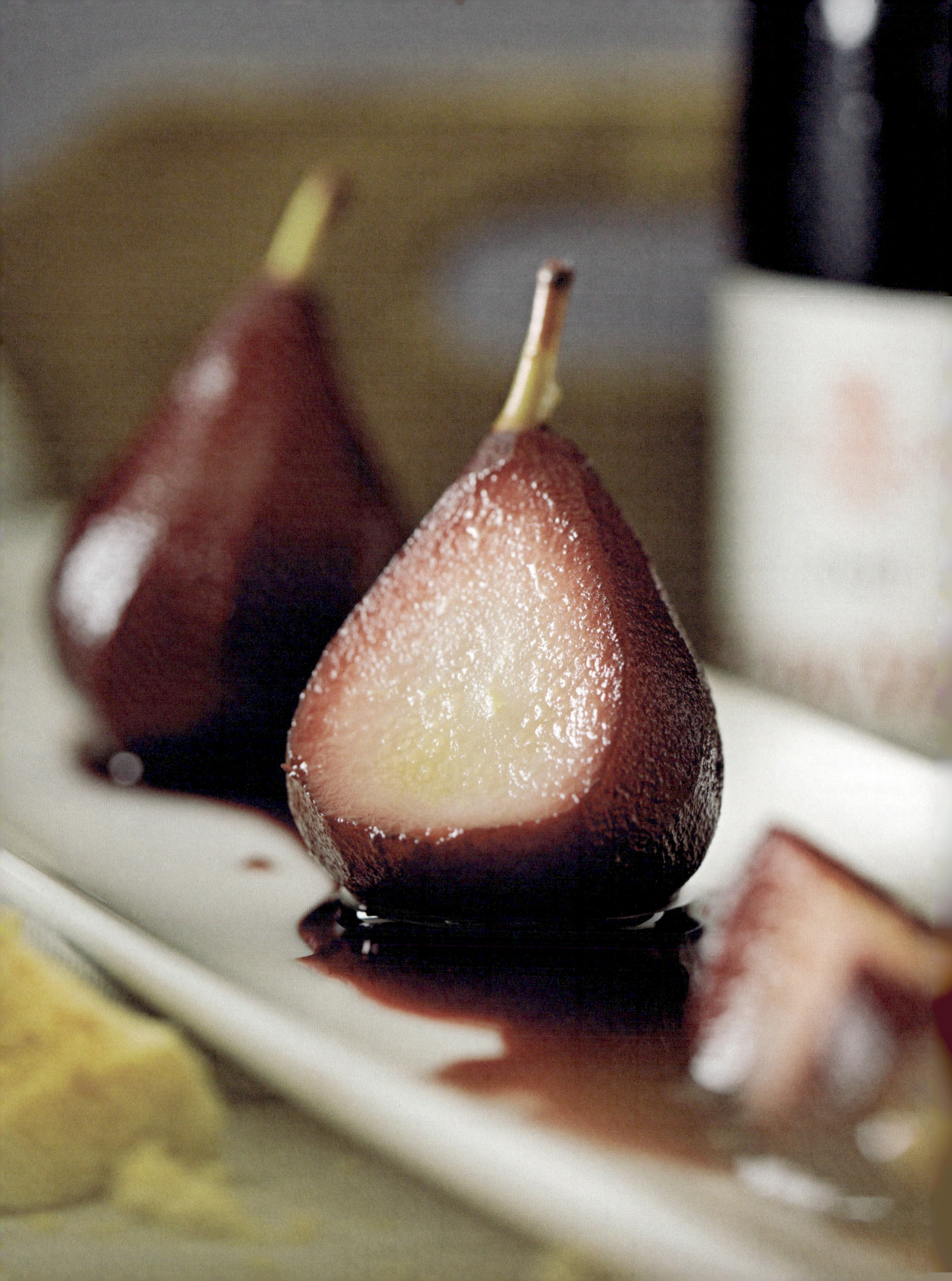

Pears Poached in Shiraz, Honeycomb & Crème Fraîche

A pear poached in red wine is another timeless French bistro dessert. These small, sweet red-blush pears are grown in Donnybrook from mid to late summer. They are perfect for poaching as they hold a firm texture and happily soak up the colour of the wine.

12 red blush pears

Poaching Liquid

1 bottle Côtes-du-Rhône (or shiraz grenache blend)
250 g caster sugar
Juice of 1 lemon
1 vanilla bean
2 cloves
1 star anise
8 allspice berries
2 cinnamon sticks
250 ml water

Honeycomb

160 g sugar
25 g honey
60 g glucose
30 ml water
7 g bicarbonate of soda

Crème fraîche, for serving

SERVES 6

To make the honeycomb, line a deep baking tray with baking paper and reserve. Combine the sugar, water, honey and glucose in a large saucepan over medium heat. Stir to dissolve the ingredients together then increase the heat and cook until the syrup has reached 150°C.

You need to move quickly now. Remove from the heat, use a tea strainer to sift the bicarbonate of soda into the caramel, whisking to combine. The honeycomb will rapidly expand so be careful.

Pour the honeycomb into the lined tray. Do not move or touch the tray after you have poured in the honeycomb as it may 'flop'.

Let the honeycomb cool to room temperature, then store in an airtight container until required.

To poach the pears, combine all the poaching liquid ingredients in a large saucepan and bring to a simmer for 5 or 6 minutes.

Peel the pears and add to the poaching liquid. Place a sheet of baking paper weighed down with a saucer to hold the pears under the liquid as they cook.

Cook the pears on a very slow simmer for 45 minutes to 1 hour, until they are firm but cooked through.

Take the saucepan from the heat and allow the pears to cool in the liquid. Refrigerate the pears in the liquid until required.

To serve, strain some of the liquid into a small saucepan, simmering it to reduce by about two-thirds to a rich, deeply coloured glaze; allow to cool.

Core the pears if you wish, place onto serving plates and drizzle with syrup. Serve with crème fraîche and crumbled honeycomb.

Clos des Paulilles 'Rimage' Banyuls
Roussillon, France

The sweet pears with reduced berried syrup will require a strongly sweet, warming alcohol-based wine. The lively red-berried flavours of this fresh-style banyuls have the same varietal base as the fruit flavours of the pears and poaching liquid.

Chateau Orignac Pineau des Charentes
Cognac, France

This is a rustic sweet fortified wine (it can be red or white) from France that complements the Cognac-infused chocolate mousse. It is viscous in texture, sweet-fruited, with deep layers of flavour, and is Cognac-laced (from ageing in Cognac barrels). Importantly, this red (red wines may still be deeply yellow or tawny) has the synergy and presence to work with this food combination.

Valrhona Chocolate & Cognac Mousse, Blood Plum Sorbet & Pistachio Praline

This dessert is silky and decadent. If, however, you prefer a lighter version, whip the egg whites and fold them through with the whipped cream.

Preheat the oven to 170°C.

To make the pistachio praline, place the pistachios onto a sheet of baking paper and lightly roast in the oven for 5 minutes. Cool, keeping them on the oven tray.

Place the sugar and water in a small saucepan and stir gently over low heat until the sugar dissolves. Increase the heat to high and boil, without stirring, for about 5 minutes. Use a brush dipped in hot water to 'wash' down the inside of the pan to prevent the sugar from burning.

When the sugar has turned a golden colour remove from the heat and allow to cool for 30 seconds, then pour the caramel over the pistachios.

Allow to cool, then break the sheet of toffee into pieces and blitz in the bowl of the food processor to a coarse crumble. Keep in an airtight container until required.

To make the mousse, chop the chocolate into even-sized pieces and place in a bowl over hot (not boiling) water to melt without allowing the base of the bowl to touch the water.

In a separate bowl, mix the egg yolks with the warmed milk, Cognac and glucose syrup, then stir through the melted chocolate.

Whip the cream lightly with the icing sugar (it should easily dollop off a spoon), and fold gently into the chocolate mixture.

Pour into serving glasses, and refrigerate for at least 1 hour before serving.

Serve topped with blood plum sorbet, sprinkled with crushed pistachio praline.

Pistachio Praline

75 g pistachios, shelled
115 g caster sugar
100 ml water

250 g Valrhona Guanaja chocolate
60 ml milk
4 eggs, separated
1 tablespoon glucose syrup
40 ml Cognac (I use Hennessy VSOP)
250 ml whipping cream
50 grams icing sugar

Blood plum sorbet (p. 290)

SERVES 6

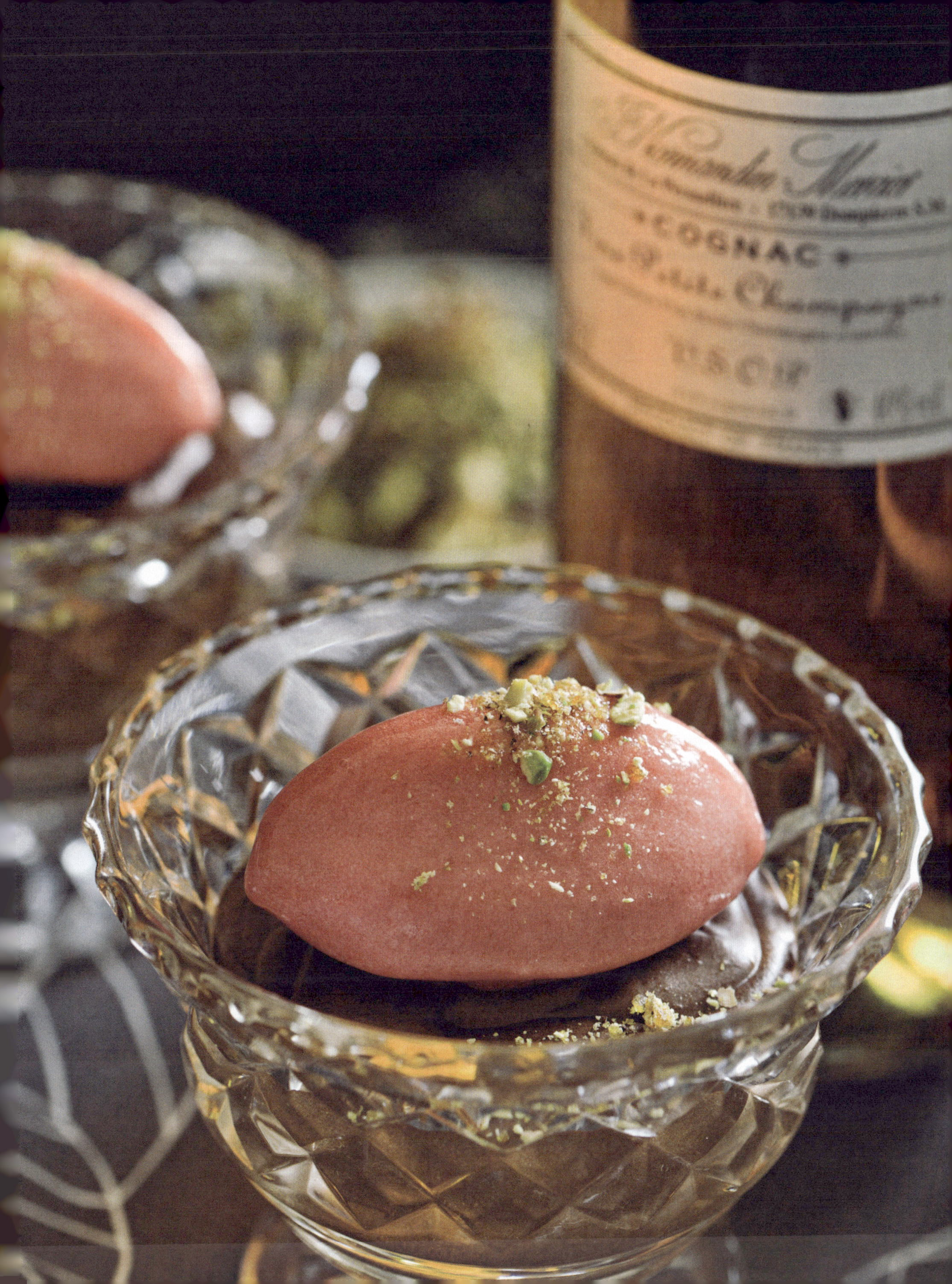
COGNAC

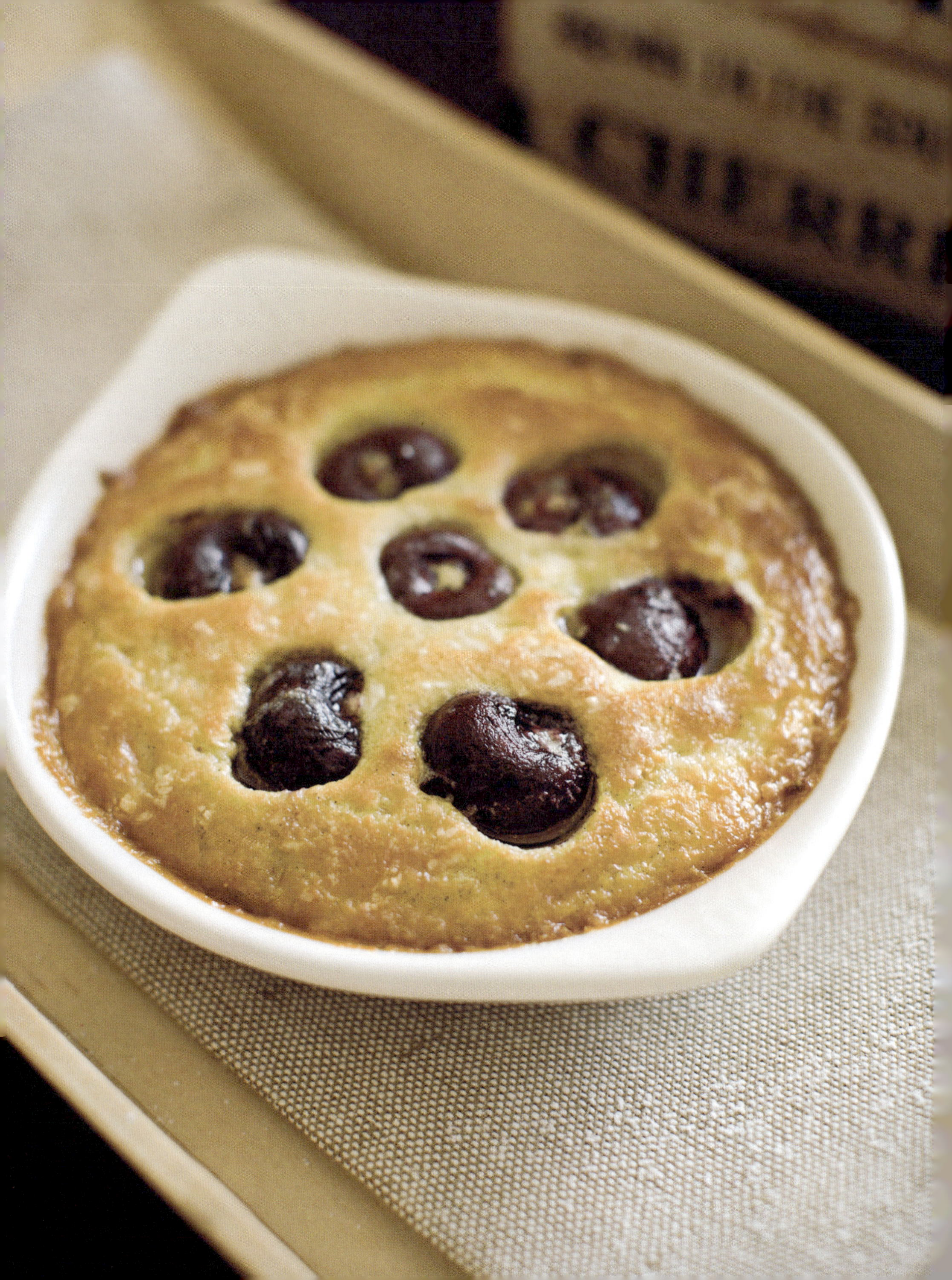

Cherry and White Chocolate Clafoutis

One freezing January day I went in search of Le Comptoir, Chef Yves Camdebordes' 25-seat bistro on the Left Bank. On arriving, I spied a large ceramic dish of Limousin cherry clafoutis, but when I ordered a serve at the end of the meal it had all gone. Here's my version of a dish I didn't get to taste.

---

4 eggs

130 g caster sugar

2 vanilla beans

Finely grated zest of ½ lemon

360 ml milk

1 tablespoon Massenez Crème de Griottes (cherry liqueur)

120 g melted butter, plus butter to grease the clafoutis dishes

130 g plain flour

60 g white chocolate, grated

600 g fresh cherries, pitted

SERVES 6

Preheat the oven to 180°C.

Combine the eggs and sugar.

Split the vanilla beans lengthways and scrape out the seeds into the eggs together with the lemon zest. Beat together lightly.

Add the milk and cherry liqueur and beat while pouring the melted butter into the mix. Gradually beat in the flour to make a smooth batter.

Brush six 16 cm clafoutis dishes with a little melted butter, drop cherries into the base of the dishes and pour batter over the top.

Sprinkle the chocolate over the top and place into the oven to cook for approximately 25 minutes. The clafoutis will be cooked when a skewer inserted into the centre of the pudding comes out clean.

Serve warm with double cream or vanilla ice cream (p. 290).

Stella Bella Pink Muscat
Margaret River, Western Australia

Light in alcohol, sweetly aromatic and lightly 'frizzante', this very energetic wine is perfect to contrast the soft clafoutis and balance the sweet and sour cherries. It's easy to become enthusiastic and generous when drinking these half-bottles!

Henriques & Henriques Madeira 'Malmsey' 15 Years Old
Madeira, Portugal

Malmsey is the sweetest style of Madeira. It has textural viscosity and rich alcoholic warmth. The classic oxidised caramel tone with acid cut is perfect with the caramelised appley tart and creamy vanilla ice cream.

Tarte Tatin

A century ago, Loire Valley hotelier Stephanie Tatin was caught in a pre-lunch rush. Too busy to finish an apple pie, she placed the pastry on top of the apple-sugar-butter mix and turned out the tarte to be served hot. Yet it took the actions of celebrated chef Louis Vaudable of Maxims in Paris, who covertly travelled to the region, tasted the dish and re-created it, thus ensuring its position as a culinary classic.

Preheat the oven to 220°C.

Peel the apples, core and quarter them, then 'wash' them with lemon juice to stop them from browning.

Place the butter into a large, heavy ovenproof frying pan to melt, add the apples and cook over medium heat for 3 minutes. Sprinkle the sugar over the apples and stir gently and evenly to coat.

Increase the heat and cook for approximately 15 minutes until the sugar and butter forms a golden caramel and the apples have softened.

Remove the pan from the heat, check that the butter caramel mix is still liquid and not burnt (it may require a little water to loosen it up).

Grease six 10 cm by 2½ cm deep mini tart tins with butter then pour some of the caramel into the little tins; tightly arrange the apples pieces on top (cut side facing up).

Allow to cool to room temperature (this can all be completed before your dinner, ready to cook).

To finish, cut six circular sheets of puff pastry that snugly fit on top of the tins, 'dock' the pastry by pricking it with a fork and lay on top of the apples.

Place into the oven for approximately 20 minutes until the pastry is golden brown and puffed. Remove from the oven.

Now comes the tricky bit so do protect your hands with kitchen gloves or tea towels. Sit a serving plate on top of a tin and flip over so that the tarte turns out onto the dish. Repeat the process for the other five tartes.

Serve immediately with vanilla ice cream and drizzle with the remainder of the caramel.

12 Granny Smith apples
Juice of 1 lemon
120 g butter, extra for greasing tins
220 g caster sugar
375 g butter puff pastry (I use Carême puff pastry)
Vanilla ice cream (p. 290)

SERVES 6

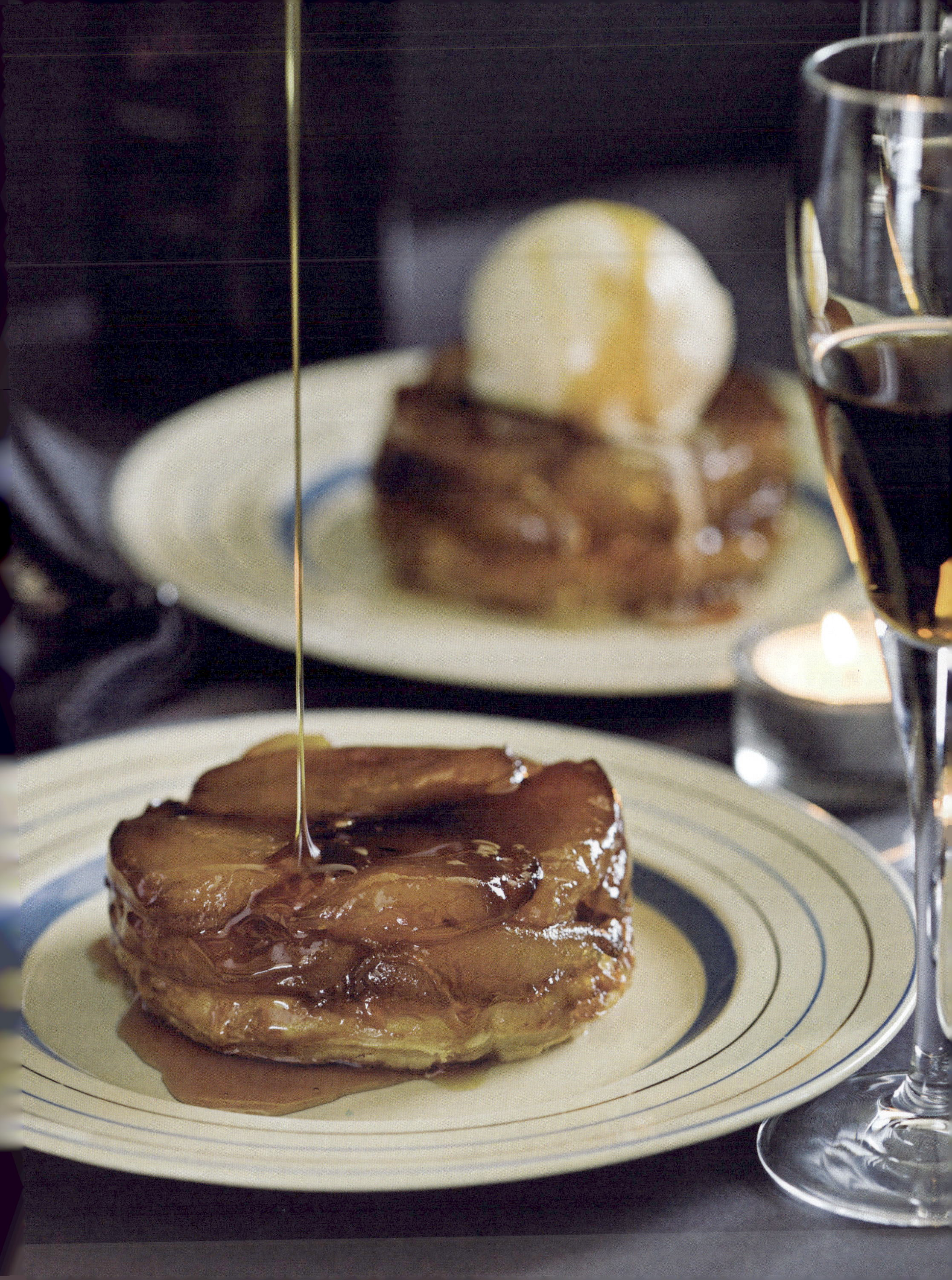

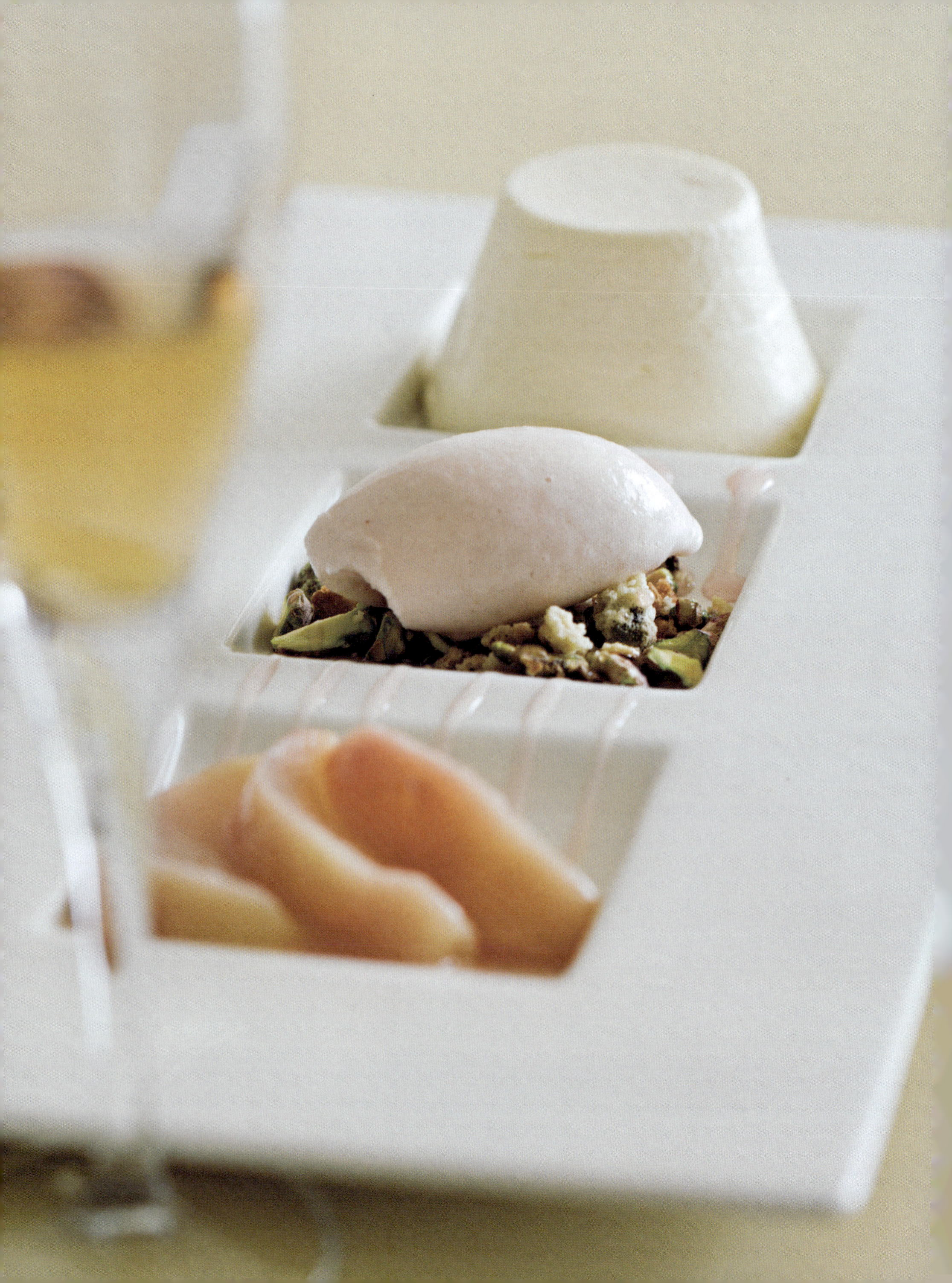

White Peach & Mascarpone Cheesecake

Lisa Schreurs completed her apprenticeship at Must and still works here; she also assisted with the food photography in this book. This dessert is her dish, a combination of some recipes of mine together with her own ideas and deft touch.

---

1 leaf Gelita 'gold' gelatine

50 g caster sugar

1 egg

15 g extra caster sugar

165 g mascarpone cheese

85 g white peach purée

35 ml Gabriel Boudier Crème de Peches (peach liqueur)

200 ml cream

120 g pistachio cantucci biscuits (I use Simon Johnson's)

Poached white peaches and peach syrup (p. 289)

White peach sorbet (p. 290)

SERVES 6

Soak gelatine leaf in some cold water.

Place the sugar and 40 ml of water into a small saucepan and boil until the temperature reaches 120°C.

Whisk the egg with 15 g of sugar until fluffy, slowly adding the hot water-sugar syrup. Keep whisking the mixture as it cools.

Squeeze the excess water from the gelatine and melt in a bowl over hot water, then add it to the egg mixture, whisking until it has cooled.

Fold the mascarpone through the beaten egg, being careful not to overmix, as the mixture may separate. Then fold the peach purée through the mixture.

Whip the cream to soft peaks and also lightly fold through.

Pour the mixture into oiled, flexible 6 cm moulds, and refrigerate until set, about 4 or 5 hours, preferably overnight.

Turn out the cheesecakes onto six serving plates. Slice the poached peaches and arrange on each plate then crush the cantucci biscuits, placing a small pile on each plate topped with the white peach sorbet. Drizzle over a little peach syrup and serve immediately.

La Spinetta 'Bricco Quaglia' Moscato d'Asti
Piedmont, Italy

This cheesecake is textural and creamy but not heavy or overtly sweet. Moscato d'Asti is low in alcohol, sweet and sparkling, with the brightness of fruit to lift the peach flavours and freshen and contrast the texture. A lovely match.

Taylor's 20 Year Old Tawny Port
Oporto, Portugal

The butterscotch and caramel flavour of this dessert balances the oxidative 'rancio' (nutty) character of Taylor's Tawny Port. It has sweetness to cope with the ice cream, delicacy of flavour and the added weight and generosity of 20 per cent warming alcohol to contrast with the chill.

Butterscotch Pudding & Dulce de Leche Ice Cream

The smoothest texture I have ever tasted in an ice cream comes from slow-simmered condensed milk. The French call it 'confiture de lait', but I prefer the Spanish name, which sounds as it tastes – sweet, rich and indulgent. If you don't have an ice-cream churn just dob the cooked 'dulce' on top of the pudding with crème fraîche or double cream.

....

To make the sauce, place the sugar and cream into a saucepan over medium heat, bring to the boil, simmer for 1 minute then remove from heat. Whisk in butter and reserve.

Preheat the oven to 190°C.

Butter six pudding dishes, approximately 8 cm wide by 7 cm deep, then line each bottom with a circle of buttered baking paper.

Place the butter and sugar into the bowl of a mixer and beat for 5 to 7 minutes until lightly coloured and very fluffy. Reduce the speed to medium, adding the egg yolks one at a time. Once the eggs are combined, add the sifted flour and cinnamon bit by bit, alternating with the wet ingredients, mixing slowly.

In a separate bowl, beat the egg whites to soft peaks; stir one-third of the whites into the pudding batter, then gently fold in the remainder.

Fill the prepared dishes two-thirds full with the batter, and place into a deep baking tray flooded with 1 cm of boiling water. Seal the baking tray tightly with aluminium foil.

Place into the oven and cook for 25 to 35 minutes, until a skewer inserted comes out clean. Serve topped with a scoop of dulce de leche ice cream and a drizzle of butterscotch sauce.

Pudding

120 g unsalted butter, at room temperature (plus butter for brushing dishes)

180 g brown sugar

3 eggs, separated

270 g self-raising flour, sifted

15 g cinnamon powder

190 ml milk

$\frac{1}{2}$ teaspoon vanilla extract

60 ml butterscotch schnapps or dark rum

Butterscotch Sauce

150 g light muscovado sugar

175 ml cream

35 g butter

Dulce de leche ice cream (p. 287)

SERVES 6

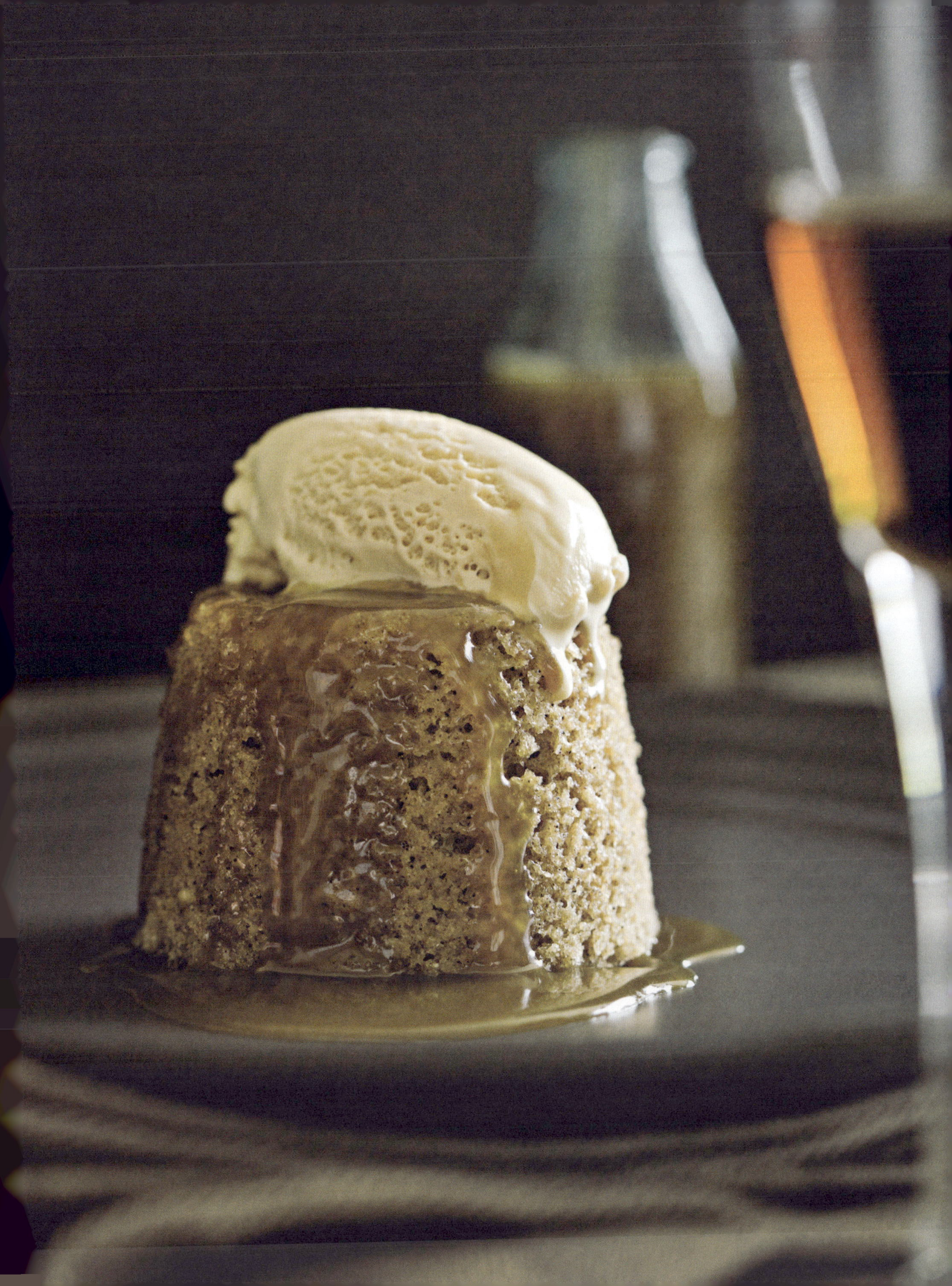

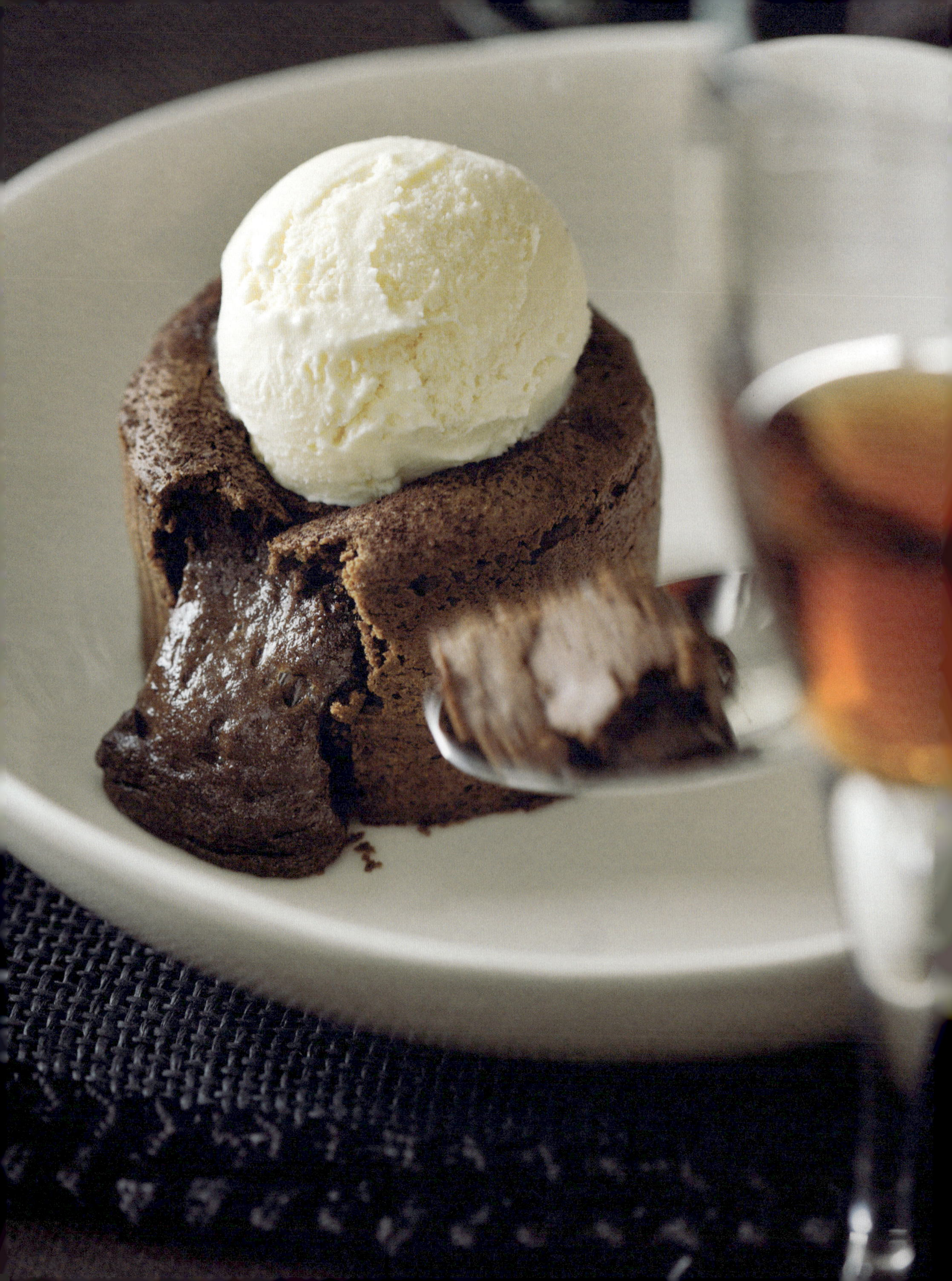

Chocolate Moelleux

A rich chocolate pudding with a molten centre is perhaps best known as a chocolate fondant. Whether fondant or moelleux, the result is the same – pure decadence! We use metal cylinders as moulds to cook our moelleux, but they can also be cooked in small dishes and served in the dish.

185 g Valrhona Guanaja chocolate

185 g unsalted butter

4 x 59 g eggs

4 large egg yolks

90 g caster sugar

150 g plain flour

25 g Valrhona cocoa powder, plus extra for dusting

Vanilla ice cream (p. 290)

SERVES 6

Line six individual baking rings (approximately 8 cm wide by 5 cm high) with baking paper, and place onto a baking tray.

Chop the chocolate and butter into even-sized pieces and place in a bowl over hot (not boiling) water to melt without allowing the base of the bowl to touch the water.

Beat the whole eggs, yolks and sugar until pale and fluffy. Fold the melted chocolate and butter into the egg mixture, then sift the flour with the cocoa and carefully fold through.

Fill the prepared moulds to three-quarters with the mixture, place in refrigerator to set for at least 2 hours.

Preheat the oven to 200°C.

Bake the moelleux for 8 minutes, remove from the oven, dust with cocoa and place onto serving plates.

Slide the pudding from the mould and top with a scoop of vanilla ice cream.

Warre's Vintage Port 1994
Douro, Portugal

Chocolate will destroy many sweet wines, so it's important to match this dessert with a wine that is both robust and sufficiently comparable in flavour profile. Hot climate fortified wines have a warm chocolate-like character and power of alcohol (even if fortified) to pair with chocolate desserts. Warre's 1994 has both the richness of fruit, warmth of vintage and confidence to complement this dessert.

desserts

Alvear Pedro Ximénez
Jerez, Spain

This dessert requires a wine of equivalent sweetness and weight. Using the same sherry that is drizzled over the ice cream harmonises the topping and dessert flavours. The warming alcohol is a delicious contrast to the freezing ice cream.

Vanilla & Valrhona Chocolate-Chip Ice Cream & Pedro Ximénez Sherry

It is a Spanish tradition to pour the sweet elixir made from Pedro Ximénez grapes over vanilla ice cream. I add another dimension of flavour and texture by stirring Valrhona chocolate shavings through our vanilla bean ice cream.

....

Chop the chocolate into even-sized pieces and place in a bowl over hot (not boiling) water to melt.

Lay a large sheet of baking paper onto the benchtop and spread the chocolate as thinly as possible onto the paper.

Place into the refrigerator to cool.

Crumble cold chocolate into chards and stir through freshly churned vanilla ice cream.

Allow to set in the freezer for at least 2 hours.

Scoop into cold serving glasses and allow your guests to pour over the chilled Pedro Ximénez sherry.

90 g Valrhona Guanaja chocolate
Vanilla ice cream (p. 290)
Pedro Ximénez sherry

SERVES 6

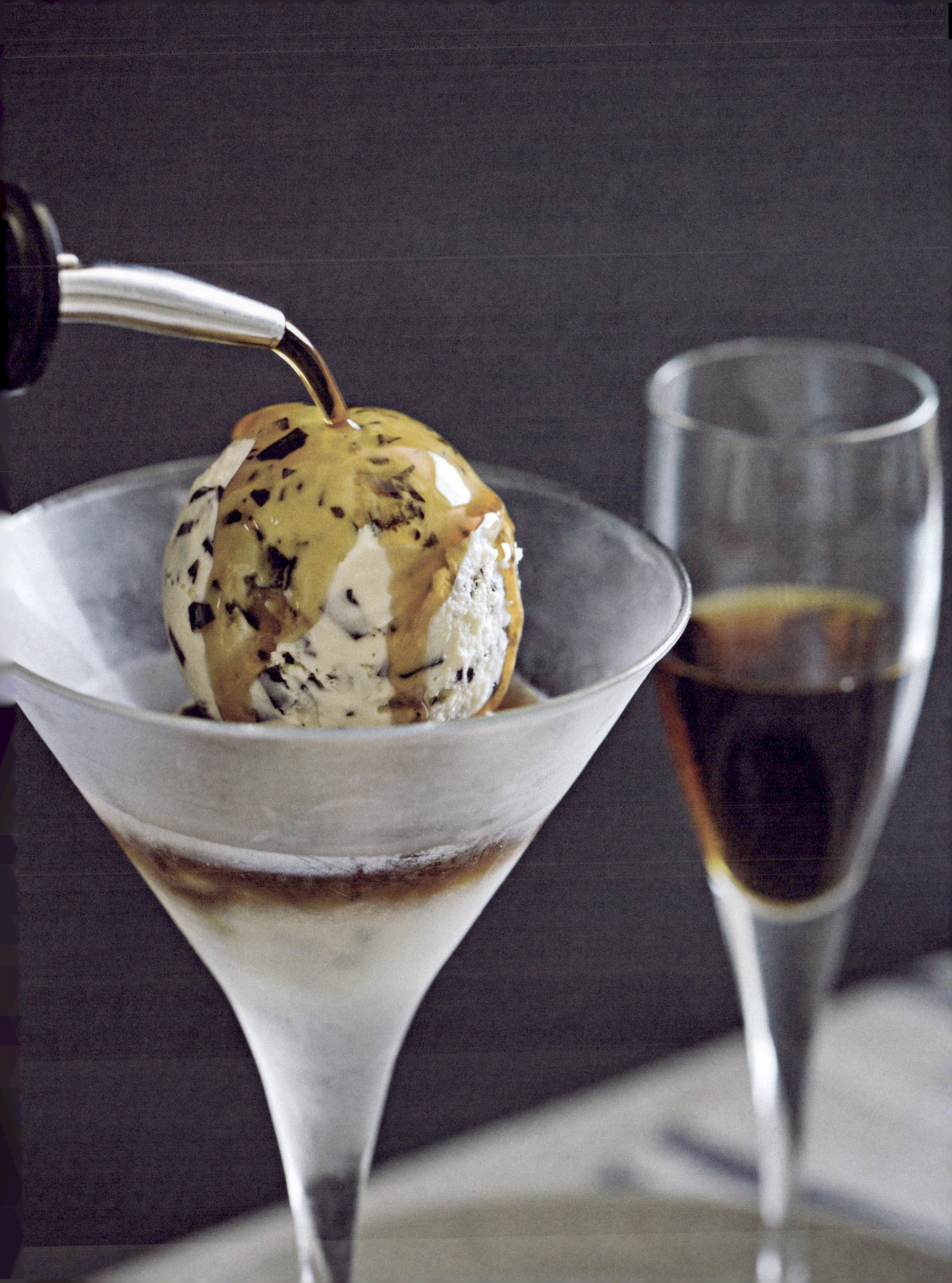

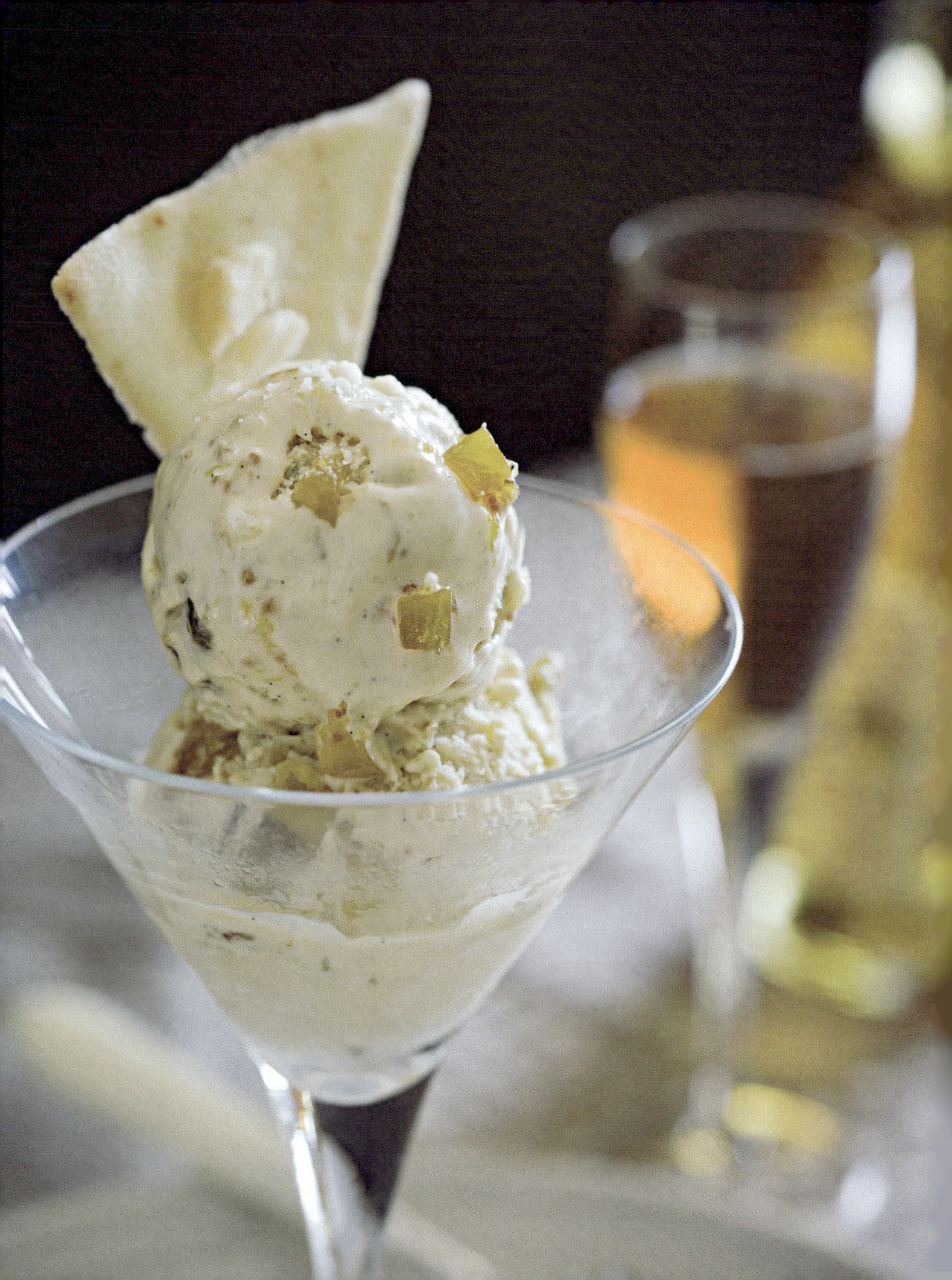

Olive Oil, Fig & Macadamia Ice Cream & Macadamia Tuile

In 2001 I was judge of the Great Southern regional cooking competition. The winning dish, from Debra Hartmann, was a perfect ice cream made with a grassy, local extra virgin olive oil, dried figs and macadamias. I asked Debra if we could use the recipe at Must and she obliged. Soon after, the dish was listed as one of *Gourmet Traveller*'s top 50 dishes of the year!

75 g roasted macadamia nuts, chopped

90 g glacé figs, chopped

130 ml extra virgin olive oil (use a full-flavoured oil)

230 ml milk

85 g dark muscovado sugar

1½ cinnamon sticks

1 vanilla bean, split lengthways

7 egg yolks

80 g raw brown sugar

300 ml cream

SERVES 6

Place the chopped macadamia nuts and figs into the freezer to chill.

Combine the oil, milk, muscovado sugar, cinnamon and vanilla in a saucepan and bring to the simmer, making sure all the sugar has dissolved. Remove from the heat and leave to infuse for 1 hour.

Whisk the yolks and raw sugar with an electric mixer until thick and pale.

Strain the milk mixture into the egg yolks and whisk together, then return to the saucepan.

With the saucepan on low heat, stir constantly with a wooden spoon until the custard thickens and coats the back of the spoon. Remove from the heat, stir in the cream and cool in the refrigerator.

Churn the mixture in an ice-cream machine; add the chopped figs and macadamias when the mixture is almost set. This recipe makes 1 litre of ice cream.

Store in the freezer. Serve with macadamia tuiles (p. 288).

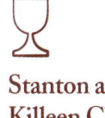

Stanton and Killeen Classic Tokay
Rutherglen, Victoria

This fortified wine is warm, attractively plush and generously sweet. It complements and balances the cinnamon and fig, with the slight nutty, tea-like oxidative character providing a contrasting savoury edge.

Barbera Respizo Di Vigna Ca'd Carussin
Piedmont, Italy

I tasted this wine in a smart London restaurant called Arbutus. It is a rare find, but it has everything that works with this dessert: it is vibrant and red (perfect with the berried fruit), it is driven by acidity, which contrasts with the sponge-based, softly textured bombe Alaska and it is lightly fortified, which helps balance the sweetness.

Must Bombe Alaska, Strawberry Syrup & Sebbes Road Blueberries

Each summer the blueberry farm on Sebbes Road, south of Margaret River, opens its gates to the public for 'pick your own' harvesting. Our boys Sam and Alex are bemused that there is still an 'honour system' in use there. It reminds me of my childhood in this dairy community, when we never locked our farmhouse and the safest place for Dad to leave the Holden keys was in the ignition.

Preheat the oven to 200°C.

Mix the egg yolks and salt with one quarter of the sugar and whisk lightly to combine.

Whip the egg whites with the remaining sugar to firm peaks. Fold in the yolks, lemon zest and sifted flours.

Gently scoop the sponge mix out onto an oven tray (lined with baking paper) to about a 7 mm thickness. Cook for approximately 5 minutes until a light golden colour.

Remove and cool.

Cut the sponge to fit the six square moulds you intend to use, which will be around 7 cm in size. Line the bases of the moulds with baking paper, then insert a piece of sponge. Fill the mould with vanilla ice cream, top with another piece of sponge and place into the freezer to set.

Wash and hull the strawberries, blitz in food processor with the icing sugar and pass through a fine strainer.

To serve, remove the bombes from their moulds, pipe Italian meringue on top, then brown the meringue with a blowtorch.

Serve with a drizzle of the strawberry coulis and a sprinkle of fresh blueberries.

Sponge

4 egg yolks
50 grams caster sugar
2 egg whites
Zest of $\frac{1}{2}$ lemon
A pinch of salt
38 g plain flour
12 g (1 tbsp) cornflour

Vanilla ice cream (p. 290)
1 punnet fresh strawberries
10 g icing sugar
Italian meringue (p. 289)
1 punnet fresh blueberries

SERVES 6

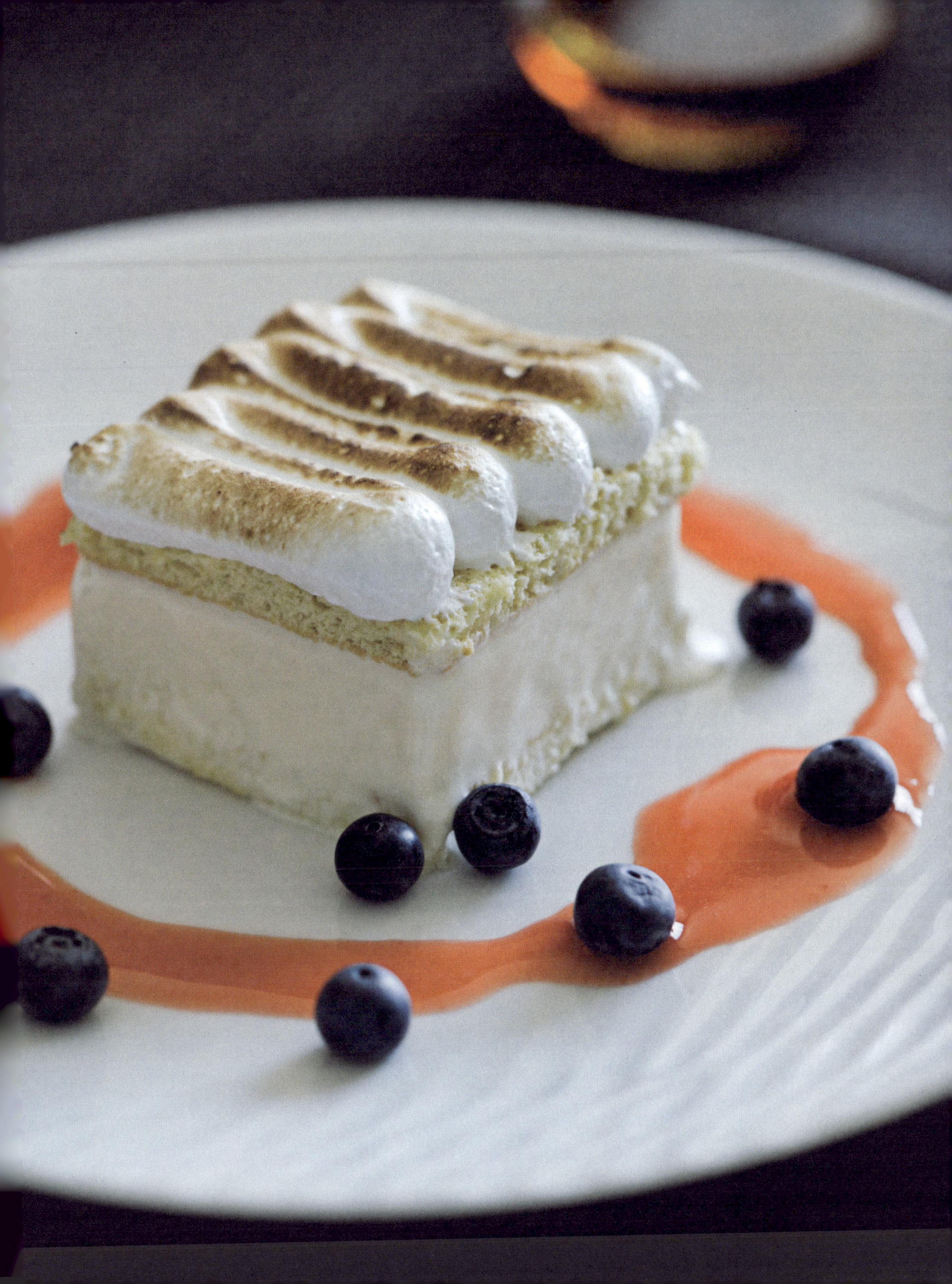

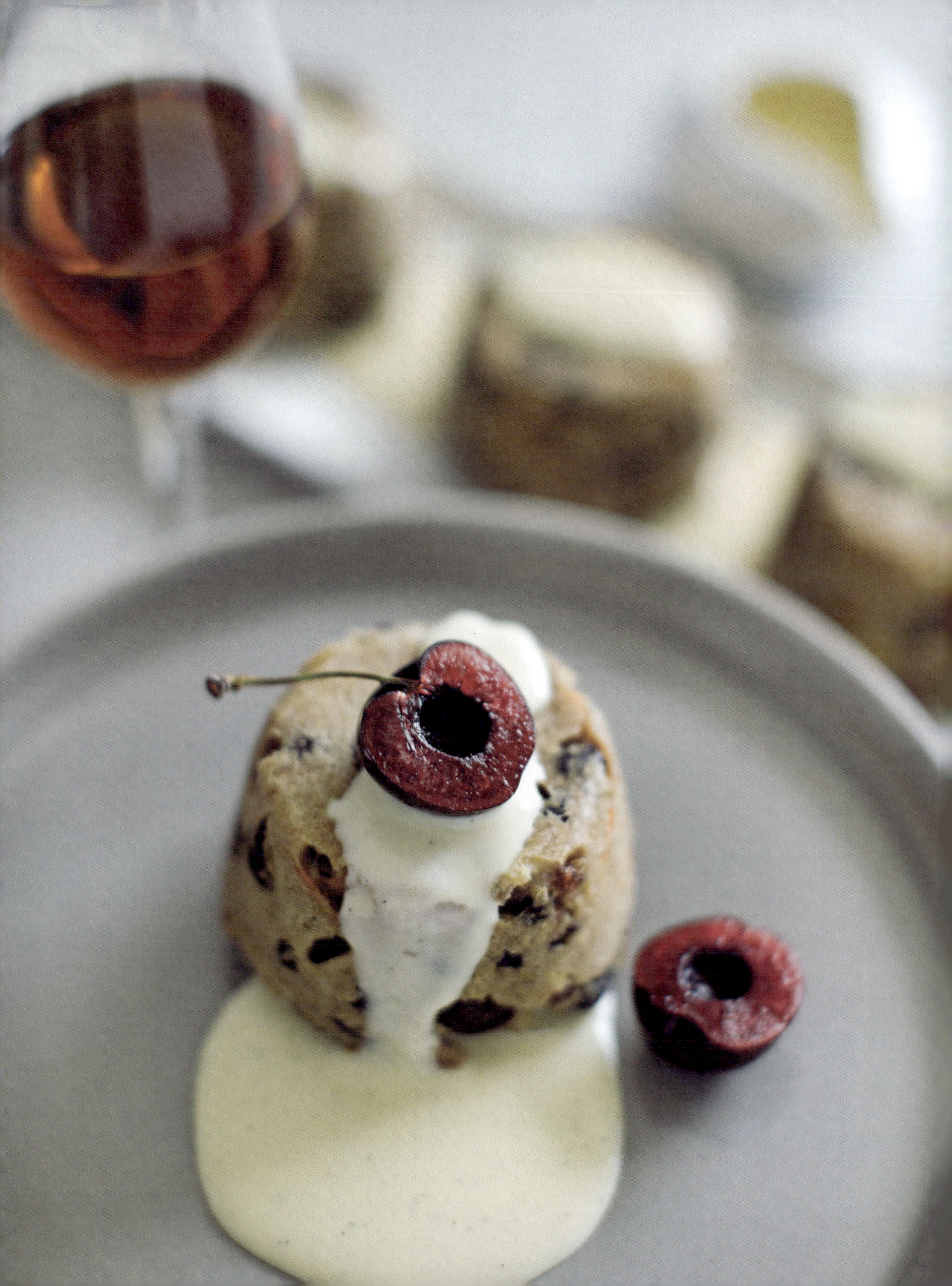

Must Christmas Pudding

Every Christmas we bake hundreds of these, which are so much lighter than a traditional plum pudding – perfect for our summer. If made in advance they will also keep for a month.

Talijancich Liqueur Muscat
Swan Valley, Western Australia

This wine is beautifully poised and decadent – shapely and raisined with layers of opulent sweet flavours. It harmonises and integrates with the compote of dried fruit flavours and the undertone of citrus in the pudding. Importantly, it has the alcohol strength to avoid being overpowered by the fruit flavours. The texture of the crème topping combined with the viscosity of the muscat is almost sinful.

---

100 g cranberries
120 g raisins
120 g currants
100 g sultanas
80 g mixed peel
100 ml Cognac (I use Hennessy VS)
Juice of 1 orange
180 g light muscovado sugar
180 g unsalted butter, at room temperature
Zest of 1 orange, grated finely
3 eggs
120 g plain flour
$\frac{1}{2}$ teaspoon ginger powder
1 teaspoon mixed spice powder
$\frac{1}{2}$ teaspoon cinnamon powder
40 g slivered almonds
40 g carrot, finely gratedt
80 g breadcrumbs
60 ml milk
Juice of 1 orange

MAKES 12 PUDDINGS

Mix the dried cranberries, raisins, currants, sultanas and mixed peel together with the Cognac and orange juice and leave in the refrigerator to marinate for up to 3 days.

Preheat the oven to 180°C.

Place the muscovado sugar, butter and orange rind together in the bowl of a mixer and whip until pale and fluffy; this may take 10 minutes or longer.

Add the eggs one by one, beating to combine.

Sift the flour and spices together and add to the bowl, but do not mix. Add the almonds and grated carrot, moisten the breadcrumbs with the milk and orange juice and, together with the steeped fruit, add to the bowl. Carefully fold the ingredients until combined.

Spoon the mixture into buttered dariole-shaped moulds, approximately 7cm wide by 6 cm deep. Loosely cover with cling film and place into a large, deep oven tray. Flood the tray with 1 cm of boiling water, tightly seal the top, firstly with cling film then aluminium foil and place into the oven.

Cook for approximately 1 to $1\frac{1}{4}$ hours, until a skewer inserted comes out clean. Add more hot water to the baking tray if it evaporates away during the cooking process. Remove from the oven and cool. They are ready to be served now.

If serving the puddings at a later date, ensure they are loosely covered with cling film, place in a steamer tray over a large saucepan of simmering water and steam until heated through, approximately 30 to 45 minutes.

Serve warm with Cognac crème Anglaise (p. 287) and fresh cherries.

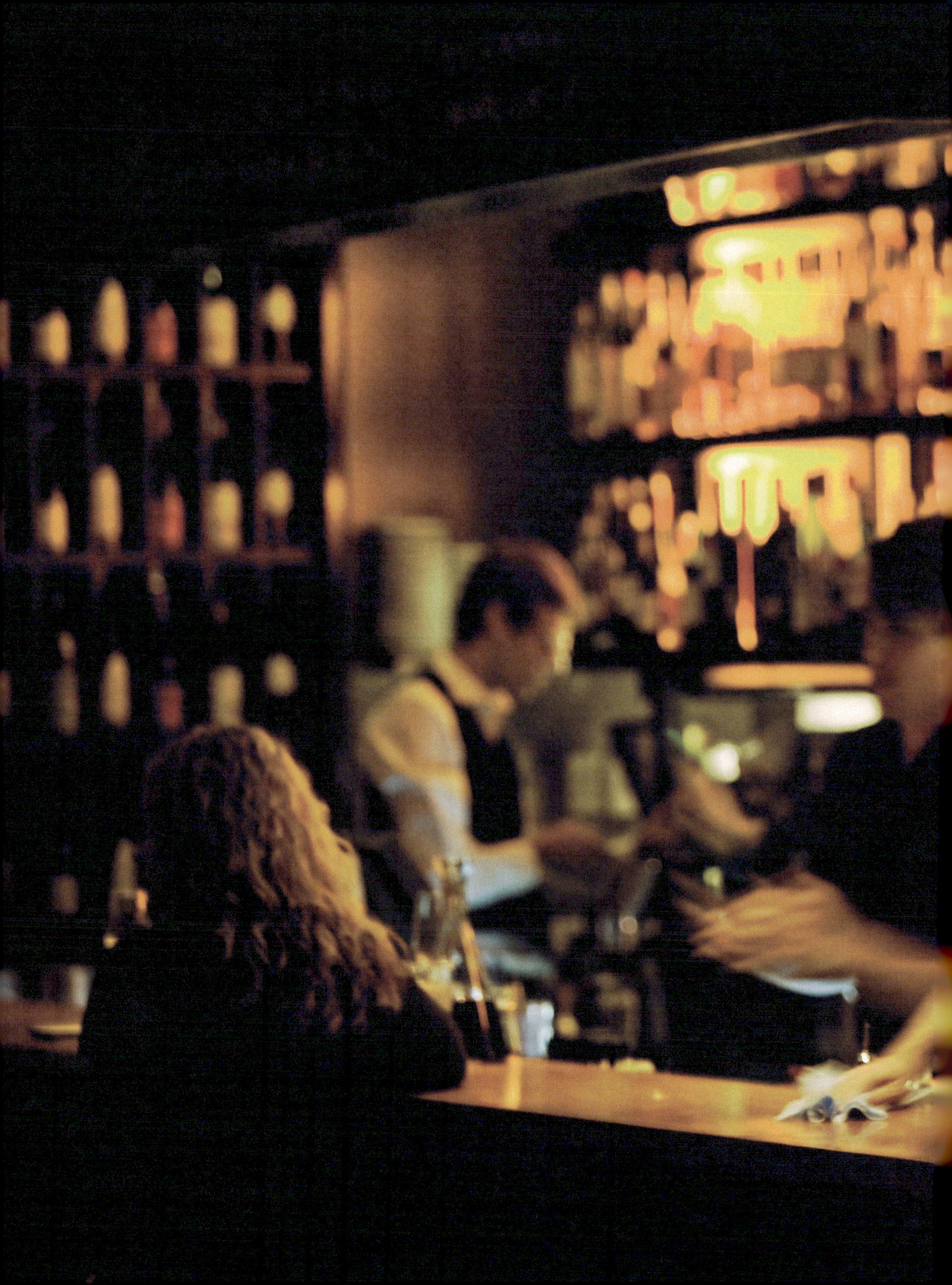

cocktails

French-Oaked Rob Roy

60 ml Glenmorangie Burgundy Wood Finish scotch whisky
8 ml Noilly Prat dry vermouth
8 ml Noilly Prat sweet vermouth
A dash of Free Brothers whiskey barrel-aged bitters
A fresh cherry, to garnish

Stir ingredients over ice. Strain into chilled martini glass. Garnish with a whisky-soaked cherry.

French Martini

Stir vodka over ice with a drizzle of pineapple juice.
Strain into martini glass rinsed with Chambord.
Garnish with pineapple sliver.

60 ml Cîroc vodka
10 ml Chambord black raspberry liqueur
Freshly crushed pineapple juice
Sliver of pineapple, to garnish

Monte-Negroni Equal measures of Campari, gin and Amaro Montenegro Serve short on the rocks with a twist of orange.

Garden of Eden

Muddle the pear and sage.
Add vodka, absinth and juice.
Shake well over ice.
Strain into a tumbler filled with ice.

¼ green anjou pear
5 fresh sage leaves
30 ml Wyborowa pear vodka
5 ml La Fée Bohemian absinth
30 ml cloudy apple juice

Lilletini

15 ml Lillet blanc
45 ml gin
15 ml Grand Marnier
Juice of ½ blood orange
A dash of orange bitters

Shake ingredients over ice and double strain.
Garnish with a rose petal.

Apricot Crusta

Shake ingredients over ice.
Strain into sugar-rimmed glass.
Garnish with a twist of orange.

45 ml Paul Giraud Vieille Réserve Cognac
15 ml Massenez Liqueur d'Abricot
Juice of 1 lemon
A dash of Angostura bitters
Powdered sugar and orange rind, to garnish

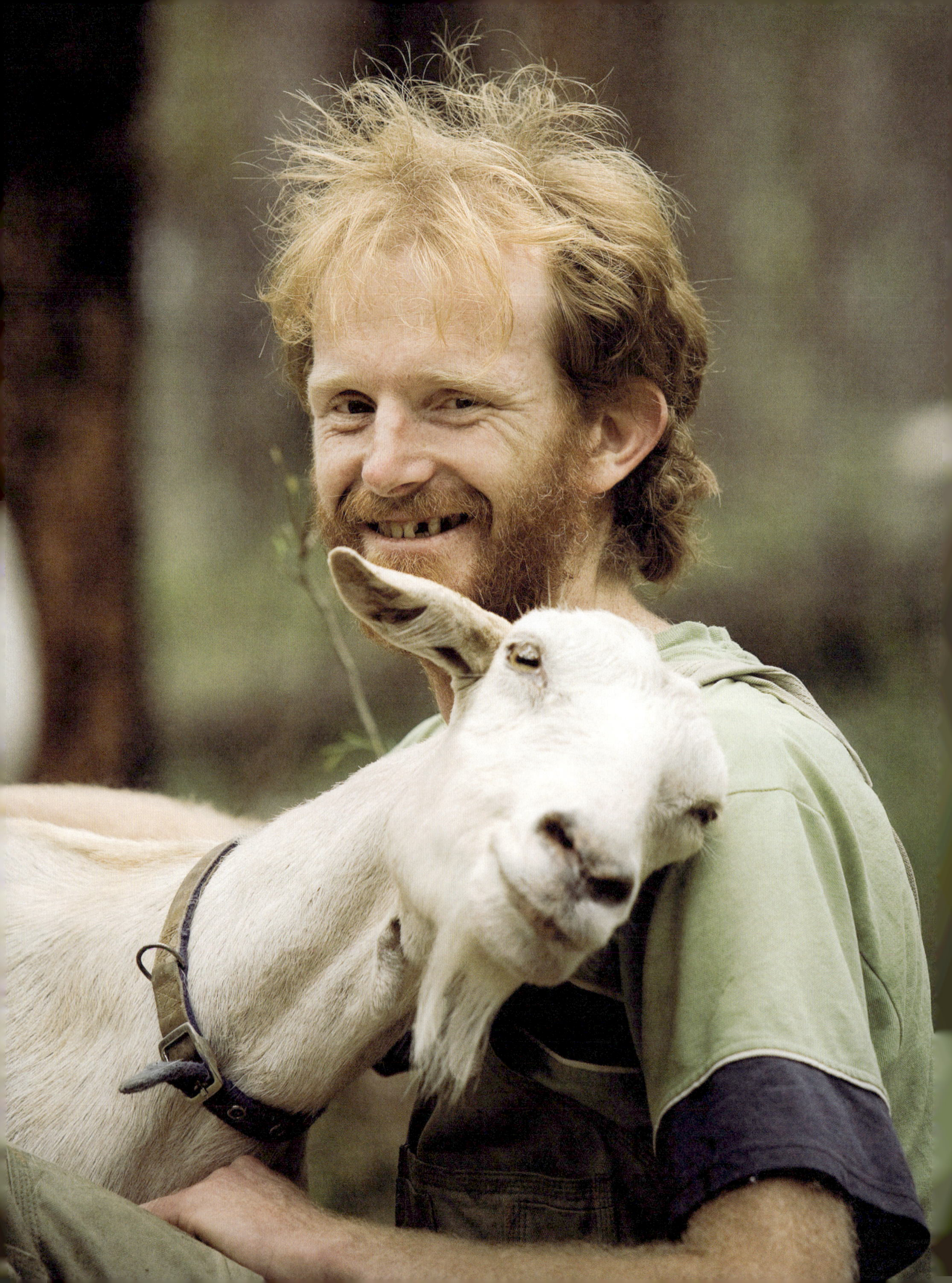

basics

When John Saunders calls 'Picnic time!' his goats stand to attention, ready to file out of the milking shed door. They then scoot up the gravel path and browse happily in the banksia and jarrah bushland on the Ringwould property. The goats adore John and nuzzle him frequently. The animals eat everything – new shoots, old bark, dead leaves, even blackberry and tea tree – and yet Ringwould goat's cheese is always so delicious!

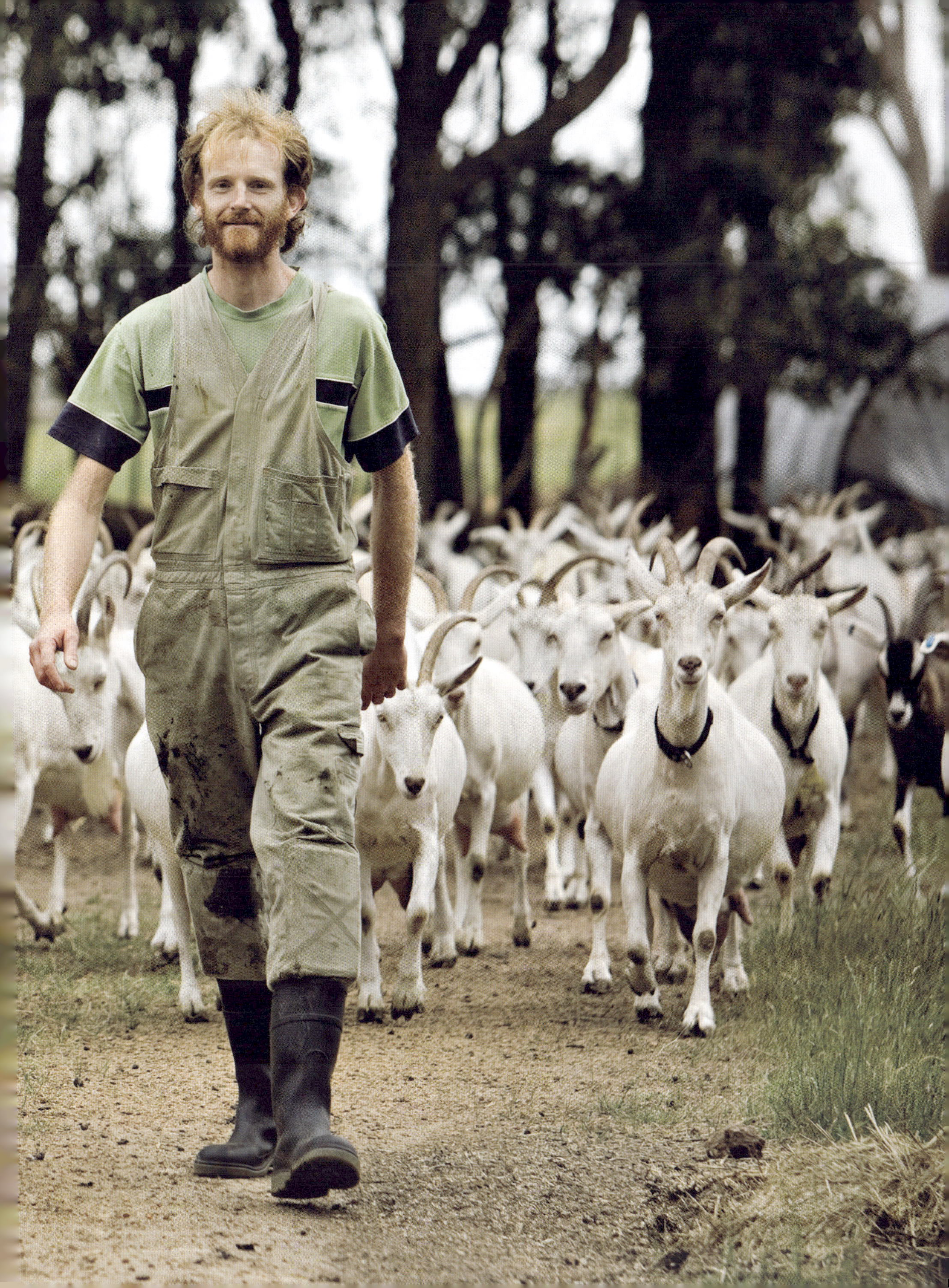

Caramelised Onion

A great base for so many dishes – stir it through a risotto, top a puff tart with it before adding goat's cheese, or toss it through pan-fried gnocchi.

Melt the butter in a large heavy saucepan over moderate heat.

Add the onions, sugar and salt and stir the onions as they start to brown for about 5 minutes. Reduce the heat to low and cook for about 30 to 45 minutes, stirring regularly. As they cook the onions will become golden brown and moist.

During the cooking process, let the onions 'catch' lightly on the bottom of the pan, then stir to combine the rich colour and flavour. Add a little water if needed to help lift the pan stickings. Once they are a rich golden colour remove and cool.

The onions will keep for up to 5 days in a sealed jar in the refrigerator.

60 g unsalted butter
1–1.2 kg brown onions, peeled and finely sliced
1 teaspoon sugar
1 teaspoon salt

MAKES ABOUT 1 CUP (250 ML)

Chilli Tomato Jam

Be careful to stir this jam regularly, particularly towards the end of its cooking, as it may burn.

Combine all of the ingredients in a large heavy saucepan and cook over moderate heat for approximately 45 minutes to 1 hour. The finished jam will have thickened and darkened a little.

Remove the bay leaf, clove and thyme sprig. Place into bowl of a food processor and blitz to a fine purée.

Store in the refrigerator for up to 2 weeks.

500 g fresh tomatoes, peeled, de-seeded and chopped
1 fresh red chilli, chopped (de-seeded if you don't like it too hot)
A sprig of fresh thyme
1 clove
½ bay leaf
110 g sugar
110 ml red wine vinegar
1 teaspoon sea salt

MAKES ABOUT 500 ML

Confit Baby Carrots

These melt-in-the-mouth carrots are terrific matched with slow-braised beef dishes. I've also partnered them with the poached chicken breast (p. 156). You can substitute extra virgin olive oil for the duck fat if you wish.

750 ml duck fat

2 bunches medium-sized (10-12 cm) baby carrots

Sea salt & freshly ground black pepper

SERVES 6 AS A SIDE

Heat duck fat in a large saucepan to 90°C.

Peel, trim and wash the carrots, drop into the warm fat and cook for approximately 35 to 45 minutes until tender.

Remove and drain the fat from carrots. Season and serve while still warm.

Confit Tomatoes

Confit tomatoes can be used in so many recipes – they add a burst of intense flavour and velvety texture to almost any dish.

3 large Roma tomatoes

1 garlic clove, chopped finely

A few sprigs of thyme

Sea salt & freshly ground black pepper

Extra virgin olive oil

Preheat the oven to 160°C.

Blanch and peel the tomatoes. Cut into quarters lengthways, de-seed and place onto a sheet of baking paper on an oven tray. Sprinkle with chopped garlic, thyme sprigs and salt and pepper.

Drizzle with the olive oil and place in oven for approximately 30 minutes. The tomatoes will have intensified in flavour and dehydrated a little. Remove and reserve at room temperature if you are using them on the same day.

The tomatoes will keep for 3 to 4 days if kept in the refrigerator; take them out to warm to room temperature before using.

Dressings

For the following recipes, simply whisk the ingredients together and reserve.

Cabernet Vinaigrette
40 ml hazelnut oil
70 ml extra virgin olive oil
30 ml Forum cabernet vinegar
Sea salt & freshly ground black pepper

Citrus Vinaigrette
25 ml lemon juice
25 ml white wine vinegar
1 heaped teaspoon wholegrain mustard
1 teaspoon thyme leaves
125 ml extra virgin olive oil
Sea salt & freshly ground black pepper

Hazelnut Vinaigrette
40 ml hazelnut oil
70 ml extra virgin olive oil
20 ml Bouquet de Figues vinegar
10 ml red wine vinegar
Sea salt & freshly ground black pepper

Pomegranate Dressing
2 tablespoons pomegranate molasses
1 teaspoon Dijon mustard
1 tablespoon lemon juice
2 tablespoons extra virgin olive oil

Preserved Lemon Vinaigrette
25 ml lemon juice
25 ml white wine vinegar
1 heaped teaspoon wholegrain mustard
1 teaspoon thyme leaves
125 ml extra virgin olive oil
1 tablespoon preserved lemon rind, finely diced
Sea salt & freshly ground black pepper

Red Wine Vinaigrette
110 ml extra virgin olive oil
30 ml Forum cabernet vinegar
1 garlic clove, smashed
Sea salt & freshly ground black pepper

Remoulade Dressing
100 ml mayonnaise
½ tablespoon Dijon mustard
1 tablespoon lemon juice
1 tablespoon water
1 tablespoon fennel or dill tops, chopped coarsely
Sea salt & freshly ground black pepper

Sherry-Walnut Vinaigrette
40 ml walnut oil
70 ml extra virgin olive oil
30 ml sherry vinegar
Sea salt & freshly ground black pepper

Walnut Vinaigrette
40 ml walnut oil
70 ml extra virgin olive oil
20 ml Bouquet de Figues vinegar
10 ml red wine vinegar
Sea salt & freshly ground black pepper

Black Truffle Vinaigrette

We make this dressing each winter during Manjimup Truffle Week. Drizzle it over lightly steamed or grilled shellfish or chicken, or dress a mixed leaf salad for a special mid-winter meal.

1 heaped teaspoon Dijon mustard
40 ml white truffle oil
70 ml light oil
20 ml sherry vinegar
1 tablespoon lemon juice
15–20 g Manjimup black truffle

Whisk the dressing ingredients together, then shave in the truffle, whisking well.

Dill & Grain Mustard Dressing

This is the perfect accompaniment to cured ocean trout (p. 62), with the honey providing a sweet balancing act to the acid kick of vinegar and lemon.

1 egg (separate and keep the yolk)
1 tablespoon wholegrain mustard
1 tablespoon Dijon mustard
Juice of 1 lemon
1 tablespoon champagne vinegar
½ tablespoon honey
100 ml grapeseed oil
100 ml extra virgin olive oil
1 tablespoon fresh dill, chopped finely
Sea salt & freshly ground black pepper

Place the yolk, mustards, lemon, vinegar and honey in a large mixing bowl and whisk well. Add the grapeseed oil and whisk to combine, then add the olive oil, whisking again. Add the dill; season to taste.

Egg Pasta Dough

Kids love to help make pasta and it is more than worth the effort. This is the basic recipe, but if you want a firmer bite try substituting one-third of the flour with fine durum semolina.

- Sift the flour into a mound on the benchtop.
- Mix the eggs together in a bowl with 2 tablespoons of cold water. Make a well in the centre of the flour and pour in the egg mixture. Gradually mix the ingredients from the centre of the well with your fingertips – work from the centre, drawing in more flour bit by bit.
- Once the ingredients have been worked together, knead the dough for 1 minute, roll into a ball, cover with cling film and rest in the refrigerator for at least 1 hour.
- Divide the dough into 5 or 6 even portions, keeping the unused portions wrapped.
- Press the dough flat with the heel of your hand, lightly dust with flour and roll through the largest setting of your pasta machine. Gradually reduce the setting on the machine and repeat until you have rolled the dough to the second-last setting, which is about 1.5 mm thick.
- For ravioli, run the sheets through the same setting a second time to prevent them from shrinking.
- For angel hair pasta, run through the finest setting before cutting with your finest cutter.

500 g bread-making or '00' flour
2 x 50 g eggs plus 8 egg yolks
30 ml olive oil
1 tablespoon salt
Fine semolina, for dusting

MAKES 700 G OF DOUGH

Green Olive Tapenade

Try this unusual combination of piquant green olives, capers and cornichons with pale meats and chicken.

15 g salted capers
200 g green olives, pitted
40 g cornichons
1 tablespoon finely chopped flat-leaf parsley
Juice of ½ lemon
1 teaspoon fresh thyme leaves
50 ml extra virgin olive oil
1 boiled egg (8 minutes), peeled
Freshly ground black pepper

To make the tapenade, soak the salted capers for 30 minutes in warm water, then rinse in cold water and drain.

Combine all the ingredients except the boiled egg in the bowl of a food processor and blitz to a coarse paste. Remove and place into a bowl.

Rub the egg through a fine grater into the mix and stir through to combine; check seasoning and reserve in the refrigerator.

Baharat

mixed spices

Baharat means 'spices', the blends of which vary widely across the Middle East. I use my version of this sunny spice mix to season lamb meatballs (p. 26). I also sprinkle it on lamb cutlets with some olive oil, garlic and lemon zest before barbecuing.

30 g ground black pepper
6 g cardamom powder
30 g cinnamon powder
8 g clove powder
18 g coriander powder
30 g cumin powder
18 g nutmeg powder
30 g paprika powder

MAKES 170 G

Sift the spices into a large bowl, mix well and store in an airtight container for up to 6 months.

Before using, roast the mix in a 180°C oven for 1 to 2 minutes to release the flavours.

basics 271

mixed spices

Quatre Èpices

This is the key four-spice spice blend in our pork sausages (p. 132).

Sift the spices into a large bowl, mix well and store in an airtight container for up to 6 months.

30 g allspice powder
10 g clove powder
30 g ginger powder
30 g nutmeg powder

MAKES 100 G

Ras el Hanout

This fragrant mix of spices forms the aromatic base for many Middle East and North African foods. Sprinkle onto quail or chicken before grilling.

Sift the spices into a large bowl, mix well and store in an airtight container for up to 6 months.

27 g allspice powder
30 g ground black pepper
20 g cardamom powder
23 g cinnamon powder
20 g ginger powder
10 g mace powder
33 g nutmeg powder
3 sachets saffron powder
13 g turmeric powder

MAKES 180 G

Peach or Pear Chutney

We make this chutney with peaches in summer when they are abundant and we use pears during the colder months.

400 g caster sugar

160 g Granny Smith or bramley apples, peeled and grated

160 g brown onion, chopped finely

1½ tablespoons orange rind, grated coarsely

Juice of 3 oranges

1 tablespoon salt

30 g fresh ginger, grated

½ teaspoon nutmeg powder

1 pinch saffron stamens

A pinch of cayenne pepper

400 ml white wine vinegar

1.2 kg peaches or pears

350 g ripe tomatoes

80 g sultanas

MAKES 1 KG

Place the sugar, apples, onion, orange rind and juice, salt, spices and vinegar in a large, heavy non-aluminium saucepan. Simmer over medium heat, stirring frequently for about thirty minutes, until thick and syrupy.

De-seed the peaches and chop roughly. Blanch the tomatoes in boiling water, peel, de-seed and chop, then add both to the saucepan, along with the sultanas.

Cook until the mixture has thickened and sets when cooled (test a little on a small plate in the refrigerator). Pour into sterilised jars and seal while hot.

Store in the pantry for up to 6 months.

Beef Stock

This is the base for hundreds of sauces in the French kitchen. Substitute the beef bones for veal bones for a 'sweeter' stock. Make sure you use a combination of meaty bones and marrow bones; the fleshy bits give flavour to the stock and the marrow provides body and richness.

Preheat the oven to 220°C.

Place the bones and/or meat into a large roasting pan and roast, stirring occasionally, for 30 to 45 minutes until deep brown.

Add the carrot, celery, onion and leek, toss through the bones, reduce heat to 200°C and roast a further 10 minutes.

Use a slotted spoon to lift the bones and vegetables out of the roasting pan into a large deep saucepan. Pour off any remaining fat and add the red wine to the roasting pan, simmering over high heat to loosen the pan-stickings and reduce wine by half.

Pour mixture into the saucepan, add the remainder of the ingredients and pour 4 litres of cold water to liberally cover the bones.

Bring to the boil, skim, then reduce the heat to a slow simmer, uncovered, over lowest heat for at least 4 hours, and up to 8 hours; the longer and slower you cook, the more flavoursome it will be. Make sure you regularly skim the stock while it is cooking.

Strain through a fine sieve into another saucepan, place over high heat and reduce the stock, still skimming regularly, to about 2 litres. This reduced stock is now ready to use as a sauce base.

Cool, cover and refrigerate for 5 days or freeze for up to 1 month.

Lamb Stock

Lamb bones can also be used to make a great stock and a beautiful reduced jus for roast meat.

Substitute at least half of the beef bones with lamb, being careful not to over-brown them as they tend to make a very dark stock. Drop a vanilla pod into the stock as it is cooking; this secret ingredient matches the fuller flavour of lamb perfectly.

Chicken Stock

Replace the beef bones with chicken carcasses and/or wings, and use white wine instead of red to deglaze the roasting pan. This stock will only take about 4 hours to cook.

---

2 kg meaty beef bones or oxtail, ribs, chuck or brisket, chopped

1 kg beef marrow bones, chopped

Olive oil

2 carrots, diced

2 celery sticks, diced

2 onions, diced

1 leek, diced

300 ml red wine

3 garlic cloves, smashed with the back of a knife

100 g mushrooms (Portobello or field), chopped

400 g fresh tomatoes, chopped

4 black peppercorns

1 bay leaf

A sprig of fresh thyme

A few parsley stalks

MAKES 2 LITRES

Fish Stock

Use only white fish bones for this stock; dhufish, pink snapper, red emperor, coral trout and whiting are ideal. If you use the head, remove the eyes and gills or they will discolour the stock.

Olive oil
1 brown onion, chopped
1 leek, white part only, chopped
2 celery sticks, chopped
1 small fennel bulb, chopped
1 bay leaf
A sprig of fresh thyme
4 white peppercorns
250 ml white wine
2 kg white fish bones

MAKES 2 LITRES

Put the olive oil into a large saucepan over low heat. Add the vegetables, herbs and peppercorns and sweat for 5 minutes to release their flavours. Add the wine, increase the heat and simmer for 1 minute.

Rinse the fish bones, drop into the pot, add enough cold water to cover the bones and increase the heat to a simmer. Cook for about 30 minutes, skimming regularly. Remove from the heat and cool for 5 minutes.

Strain the liquid through a fine sieve, then reserve in the refrigerator for 2 days, or in the freezer for up to 1 month.

Shellfish Stock

For a rich and sweetly flavoured stock I use whole crabs, but raw lobster or prawn shells will give a similar result.

2 kg raw blue manna crabs
1 large brown onion, chopped
2 carrots, chopped
2 celery sticks, chopped
Olive oil
A sprig of fresh thyme
1 bay leaf
4 white peppercorns
250 ml white wine
1 x 400 g tin crushed tomatoes

MAKES 2 LITRES

Flip up the triangular flap on the underside of the crabs and pull off the carapace. Remove the dead man's fingers, take the flesh from the claws and reserve in the fridge for another dish. Repeat for each crab.

Put the olive oil in a large saucepan over medium heat. Add the vegetables and sweat for 5 minutes to release their flavours. Add the herbs, peppercorns, wine and crab shells and increase the heat to high. Cook for 5 to 6 minutes until the shells become orange in colour. Crush the shells with a heavy wooden spoon then add the tomatoes and 2 litres of water. Bring to the boil and simmer for 45 minutes, skimming constantly.

Strain the liquid through a fine sieve. The stock will keep in the fridge for up to 3 days, or freeze for up to 1 month.

stocks & sauces

Vegetable Stock

Use this as a substitute for white chicken stock.

Place the vegetables in the bowl of a food processor and blitz to a coarse rubble. Heat a little olive oil in a large heavy saucepan over low to medium heat, add the vegetables and sweat for 15 minutes, stirring regularly.

Barely cover the vegetables with water, bring to the boil, skim then reduce heat to a slow simmer, uncovered, over lowest heat for about $1\frac{1}{2}$ hours.

Strain the liquid through a fine sieve and cool. The stock will keep in the fridge for up to 3 days, or freeze for up to 1 month.

3 celery sticks, diced
3 onions, diced
2 leeks, white part only, diced
3 carrots, diced
1 head of fennel, diced
4 garlic cloves, peeled
4 large tomatoes
100 g champignon mushrooms
Extra virgin olive oil
1 bay leaf
A sprig of fresh thyme
4 white peppercorns
A few parley stalks

MAKES 1.5 LITRES

'White' Chicken Stock

Every week at Must we make hundreds of litres of this sweet, rich stock to flavour soups and to use as a base for risotto.

3 kg chicken carcasses/wings
2 celery sticks, diced
2 onions, diced
2 leeks, white part only, diced
3 garlic cloves, smashed
4 white peppercorns
1 bay leaf
A sprig of fresh thyme
A few parsley stalks

MAKES 2 LITRES

Rinse the chicken bones with cold water and place into a large saucepan. Add the remaining ingredients then just cover with cold water.

Bring to the boil, skim, then reduce the heat to low and simmer, uncovered, for 2 hours until it is richly flavoured. Make sure you regularly skim the stock while it is cooking.

Strain the liquid through a fine sieve and cool. The stock will keep in the fridge for up to 3 days, or freeze for up to 1 month.

Béarnaise Sauce

Béarnaise is the perfect sauce for steak. It differs from hollandaise – the classic butter emulsion sauce – in that it is based on a tarragon vinegar reduction rather than white wine vinegar, and is flecked with flavoursome tarragon leaves. To make a good Béarnaise means using clarified butter.

To produce about a cup (250 ml) of clarified butter, put 300 g of unsalted butter into a small saucepan over low heat and bring to the boil. Remove from the heat, skim off the froth that rises to the surface and decant the clear 'clarified' butter from the milky dregs at the bottom of the pan.

To make sauce, strip the leaves from the bunch of tarragon, keeping them aside.

Put the tarragon stalks, vinegar, water, peppercorns, shallots and bay leaf in a small saucepan and bring to the boil until reduced by half. Strain and reserve.

Measure 25 ml of the reduction into a metal bowl together with 25 ml of water, then add the egg yolks and whisk together.

Place the bowl over a steaming double boiler filled with just-simmering water and whisk until thick and creamy.

Remove the bowl from the double boiler and whisk for another 30 seconds or so. Slowly pour in the warm clarified butter, whisking constantly; this should take about 2 minutes. Add the lemon juice and season with a little salt if required.

If the consistency of the sauce is too thick, whisk in a little warm water; if it's too thin whisk in a little more clarified butter.

Keep the sauce covered in a warm place for a maximum of 2 hours before serving.

Finely chop the tarragon leaves and add to the finished sauce.

1 bunch tarragon, washed
60 ml white wine vinegar
120 ml water
1 shallot, chopped finely
4 white peppercorns
1 bay leaf
3 egg yolks
220 g clarified butter
1 tablespoon lemon juice
Table salt

MAKES 250 ML

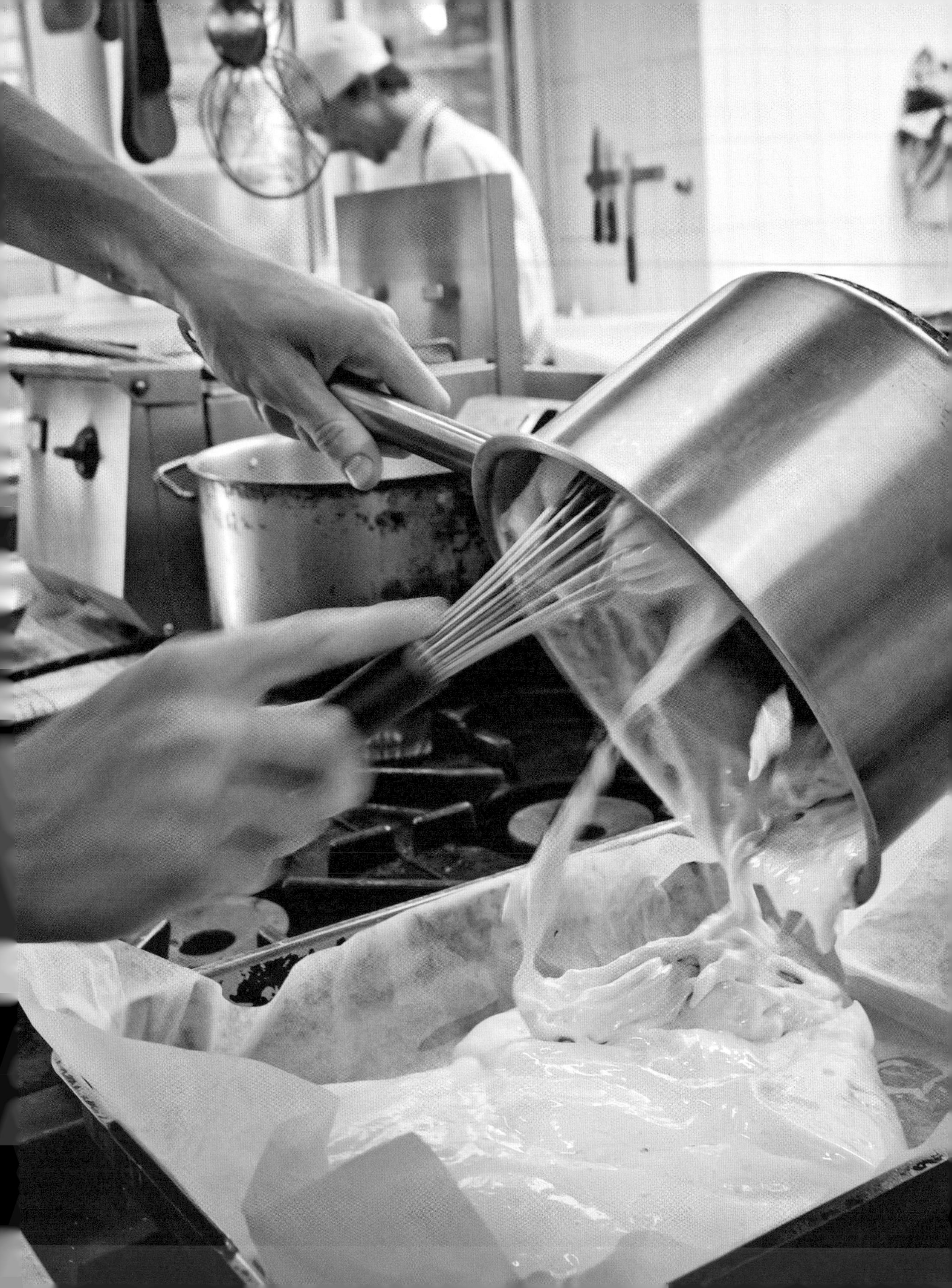

Hollandaise Sauce

Hollandaise is perhaps the best known of French sauces and it is the doting mother to a family of derivatives, including choron sauce.

Put the vinegar, water, peppercorns, shallots and bay leaf in a small saucepan and bring to the boil until reduced by half. Strain and reserve.

Measure 25 ml of the reduction into a metal bowl together with 25 ml of water, then add the egg yolks and whisk together.

Place the bowl over a steaming double boiler filled with just-simmering water and whisk until thick and creamy.

Remove the bowl from the double boiler and whisk for another 30 seconds or so. Slowly pour in the clarified butter, whisking constantly; this should take about 2 minutes. Add the lemon juice and season with a little salt if required.

If the consistency of the sauce is too thick, whisk in a little warm water; if it's too thin whisk in a little more clarified butter.

Keep the sauce covered in a warm place for a maximum of 2 hours before serving.

Choron Sauce

Add 1 tablespoon of tomato paste (at room temperature) to the hollandaise recipe and whisk well to combine. Keep the sauce covered in a warm place for a maximum of two hours before serving.

60 ml white wine vinegar
120 ml water
1 shallot, chopped finely
4 white peppercorns
1 bay leaf
3 egg yolks
1 tablespoon lemon juice
220 g clarified butter (p. 278)
Table salt

MAKES 250 ML

Madeira Jus

Madeira lends a sweet richness to this sauce, but if it is difficult to find use a sweet Oloroso sherry.

2 litres brown chicken stock (p. 274)
300 ml Madeira
60 g unsalted butter
Sea salt & freshly ground black pepper

MAKES ABOUT 500 ML

Pour Madeira into a large saucepan and reduce over high heat to about 150 ml. Add the stock and keep boiling until the sauce has reduced by about three-quarters, skimming regularly; the stock should be quite glossy and a little viscous. Remove from the heat and add the butter to the sauce, whisking constantly to combine. Once all the butter has been added, check the seasoning and keep warm.

Mayonnaise

Take the time to make your own mayonnaise, it's quite easy and the end product more than rewards the effort.

2 egg yolks, at room temperature
1 tablespoon Dijon mustard
1 tablespoon lemon juice
1 tablespoon white wine vinegar
1 teaspoon sea salt
A pinch of ground white pepper
250 ml light oil (grapeseed or sunflower)

MAKES 250 G

Mix the yolks, mustard, lemon juice, vinegar, salt and pepper together in a bowl. Constantly whisk by hand, adding the oil in a fine drizzle. Once the oil has been added, give a 'supercharged' whisk to thicken the mayonnaise. Store in the refrigerator for up to 3 days.

Pedro Ximénez Lacquer

I love the rich sweet caramel flavour of Pedro Ximénez sherry. This lacquer is brushed or drizzled on pork, chicken or even around a warm tart.

Place all the ingredients in a saucepan and boil until reduced to a thick, glossy consistency.

Cool and seal in an airtight jar in the refrigerator until required; it keeps indefinitely.

375 ml Pedro Ximénez sherry
150 ml red wine vinegar
1 star anise
Rind ½ orange
40 g muscovado sugar

MAKES ABOUT 150 ML

Ravigote Mayonnaise (Must Tartare Sauce)

Ravigote sauce is traditionally a vinaigrette served with rich offal dishes. I decided to whisk the components through a mayonnaise, giving a piquant-creamy result that moistens the jambon persillé (p. 126) on our charcuterie plate. The result also doubles as our tartare sauce, the only sauce to serve with crumbed oysters (p. 43).

Soak the capers in cold water for 1 hour then wash, strain and chop finely. Whisk all of the ingredients together well.

25 g salted capers
15 g cornichons, chopped finely
10 g chives, chopped
5 g flat-leaf parsley, chopped
25 g red onion, chopped finely
100 g mayonnaise
Juice of ½ lemon
Sea salt & freshly ground black pepper

MAKES ABOUT 200 G

Red Wine Jus

stocks & sauces

The classic sauce for red meats.

....

300 ml red wine
100 ml tawny port
2 litres beef stock (p. 274)
60 g unsalted butter
Sea salt & freshly ground black pepper

MAKES ABOUT 500 ML

Pour the red wine and port into a large saucepan and reduce over high heat to about 150 ml.

Add the stock and keep boiling until the sauce has reduced by about three-quarters, skimming regularly; the stock should be quite glossy and a little viscous.

Remove from the heat and add the butter to the sauce, whisking constantly to combine. Once all the butter has been added, check seasoning and keep warm.

Port Wine Jus

Use the red wine jus recipe above as a base, but instead of the red wine use 500 ml of tawny port.

Sauce Diable

Sauce diable means 'devil's sauce', and it has a real tang from its intense reduction of vinegar and white wine spiced with white pepper. Serve it with rich meats or offal.

....

200 ml white wine
150 ml white wine vinegar
12 white peppercorns
100 g shallots, chopped finely
1 garlic clove
A sprig of thyme
1 bay leaf
2 litres beef stock (p. 274)
60 g unsalted butter
Sea salt

MAKES ABOUT 500 ML

Put the wine, vinegar, peppercorns, shallots, garlic, thyme and bay leaf into a saucepan and boil until the liquid has almost evaporated. Add the stock and keep boiling until the sauce has reduced by about three-quarters, skimming regularly; the stock should be quite glossy and a little viscous.

Remove from the heat and add the butter, whisking constantly to combine. Once all the butter has been added, check the seasoning, strain and keep warm.

basics 283

Vermouth Beurre Blanc

Beurre blanc is a classic partner for fish and white meats. For a sauce with a little more tang, make a lemon beurre blanc, substituting the vermouth for an extra 100 ml lemon juice.

....

Place the shallot, vermouth, peppercorns and lemon juice into a small saucepan and boil for about 5 minutes until almost completely reduced. Add cream and simmer for about 4 minutes until thickened. Season with a pinch of sea salt.

Remove from the heat; slowly add knobs of the butter, whisking the sauce constantly until the butter has been completely combined. Strain the sauce and keep covered in a warm place for a maximum of 2 hours before serving.

Chive Beurre Blanc
Substitute the vermouth for white wine, and stir through 2 tablespoons of finely chopped chives just before serving.

1 shallot, chopped finely
150 ml vermouth
2 white peppercorns
Juice of 1 lemon
100 ml cream
100 g unsalted butter, at room temperature
Sea salt

MAKES ABOUT 250 ML

Sherry-Soaked Flame Seedless Grapes

Alf Edgecombe supplies us with the most beautiful dried flame grapes from his property in the Swan Valley. He dries them on the vine by snipping the cane to each bunch, then lets the grapes sun-dry and shrivel, which concentrates the sugars. We soak them in syrup for our cheese platters, and use them in a delicious sauce for quail.

....

Put the sugar, sherry and water into a saucepan, bring to the boil then simmer for 5 minutes. Add the grapes, return to a simmer then turn the heat off. Allow the grapes to cool in the syrup.

Store in a sealed container in the refrigerator for up to 1 month.

250 g dried flame seedless grapes
80 g white sugar
125 ml dry sherry
250 ml water

Sweet Corn Purée

Sweet corn is most abundant in the warmer months.
Frozen sweetcorn can be used instead when fresh is out of season.

4 ears sweet corn
80 ml cream
20 g butter
Sea salt & freshly ground white pepper

MAKES 400 G, SERVES 6

Bring a large saucepan of salted water to the boil.

Peel off the husk and remove the silk from the sweet corn. Drop into the water and simmer for about 7 minutes until the kernels are tender, then drain.

Once cool enough to handle, slice the kernels from the cob, placing them into the bowl of a food processor.

Place the cream in a small saucepan, bring to the boil and simmer for 1 minute. Add the butter and, when melted, add to the food processor together with sea salt and a pinch of pepper. Blitz the corn to a very fine purée and serve warm.

Pâté Brisée

This is the rich, flaky pastry that forms the base for all the savoury tarts we make at Must.

250 g unsalted butter
350 g plain flour
1 teaspoon salt
3 egg yolks
125 ml ice-cold water

MAKES ABOUT 700 G, ENOUGH FOR 1 LARGE TART (OR 40 MINI TARTS)

Cut the butter into small cubes and place onto a tray in the freezer for 10 to 15 minutes. Sift the flour and salt onto the butter and toss into the bowl of a food processor. Pulse the flour and butter together until the mixture resembles breadcrumbs, then tip into a large bowl. Add the yolks and gently combine with your fingertips, slowly adding the water so that the dough begins to hold together. Once the dough has incorporated, stop adding water. Knead with the heel of your hand four or five times until the dough is smooth.

Keep refrigerated for 2 days, or freeze for up to 1 month.

sweets & pastry

basics

sweets & pastry

Crepes

When I think of crepes I think of Paris. On a recent visit there was a streetside crêperie near our hotel in Saint Germain. The aromas of sweet cinnamon and chocolate infused with the cooking crepes were irresistible, and we stopped several times for a bite on our way to a gallery or museum.

....

Place all of the ingredients except the clarified butter into the bowl of a food processor and blitz until smooth. Strain into a bowl and refrigerate for at least 1 hour to settle.

Once rested the mix should be the consistency of pouring cream – if it is a little too thick add a splash of milk and whisk through.

Heat a 30 cm non-stick pan over medium-high heat, brush with clarified butter and ladle in enough crepe mixture to coat the base of the pan evenly, swirling and tilting the pan to ensure the mixture is as thin as possible.

Return to the heat until the crepe dulls and firms then turn it over – to do this use a spatula or your fingers to loosen an edge (or flip it if you dare!).

Finish cooking the crepe, another 20 to 30 seconds, then slide it out of the pan and onto a clean tea towel. Repeat until the mixture is finished.

Once the crepes are cool they should be stacked onto a large plate and covered with cling film in the refrigerator until required.

Crepes should be eaten on the day they are cooked.

250 g flour
1 teaspoon table salt
4 x 50 g eggs
25 ml olive oil or melted butter
1 teaspoon caster sugar
600 ml milk
Clarified butter (p. 278)

MAKES 12–15 LARGE CREPES

Cognac Crème Anglaise

sweets & pastry

This classic dessert custard takes on the flavour of the Cognac so use the best you can afford.

....

500 ml milk
6 egg yolks
100 g sugar
60 ml Cognac

MAKES ABOUT 700 ML

Pour the milk into a saucepan bring to a simmer over medium heat.

Whisk the eggs and sugar together until quite light and fluffy.

Pour the hot milk over the eggs, whisking constantly.

Return the mixture to low heat, stirring constantly with a wooden spoon until the mixture coats the back of the spoon. Immediately strain then add the Cognac and whisk through.

Keep warm to use straight away or refrigerate for up to 2 days to use cold.

Dulce de Leche Ice Cream

Cooking condensed milk for hours gives this ice cream an indulgent texture, like eating frozen velvet!

....

395 g tin Nestlé condensed milk
500 ml cream
5 egg yolks
50 g sugar

MAKES ABOUT 1 LITRE

Place the can of condensed milk into a large saucepan of water, bring to the boil and simmer for $3\frac{1}{2}$ hours. Keep watch over the saucepan taking care not to let the water evaporate as the can may explode. Remove from the water and allow to cool.

Pour the cream into a saucepan and bring to a simmer.

Whisk the egg yolks and sugar in a large bowl, add the hot cream, whisking to combine; return to the saucepan. With the saucepan on low heat, cook gently, stirring constantly with a wooden spoon until the custard thickens and coats the back of the spoon.

Remove the pan from the heat, stir in the cooked condensed milk, pass through a fine strainer and cool in refrigerator.

Churn the mixture in an ice-cream machine (following the manufacturer's instructions).

This ice cream is best eaten on the day it is made, but it can be stored in the freezer for up to 1 month.

basics

Macadamia Tuiles

Tuiles are crispy wafers, or tiles, of pastry. The pastry is quite wet and requires time to set in the refrigerator, and it is spread rather than rolled out. They are best eaten on the day they are cooked, though you can store them for up to a week in an airtight container.

....

Make your tuile 'template' using a plastic lid from an ice-cream container. On it, draw a 6 to 8 cm triangle and cut the shape out with a trimming knife. Cut the excess plastic away from the edge of the template and it is ready to use.

Preheat the oven to 180°C.

Cream the sugar and butter together in a mixer or food processor. Gradually add the egg whites, with the machine set on low speed. Add the salt, orange zest and flour, mix lightly until combined.

Allow the tuile batter to rest in the fridge for at least 1 hour (the batter can be kept refrigerated for 2 to 3 days).

Butter a large sheet of baking paper and place onto a baking tray. Place a teaspoon of the batter onto the tray, then place the shaping template over the batter and, using a palette knife, spread the mixture evenly so that it fills the template in a thin even layer; remove the template and repeat until the baking tray is filled. Sprinkle each tuile with some crushed macadamia nuts.

Place the tuiles into the oven and bake for 8 to 10 minutes until golden brown on the edges – they will be a little lighter in the centre. Allow to cool on the tray and carefully remove.

100 g caster sugar
60 g unsalted butter, at room temperature
2 egg whites
A pinch of salt
Zest of 1 orange
70 g plain flour
90 g macadamia nuts, crushed

MAKES 12–15

Italian Meringue

Italian meringue can be made in advance and stored in the freezer for up to 1 month. You will need a digital thermometer for this recipe.

....

150 g caster sugar
15 g liquid glucose
40 ml water
3 egg whites

Place sugar, glucose and water into a saucepan over medium heat until the sugar is dissolved. Increase the heat and boil until syrup reaches 120°C.

Whisk the egg whites in a mixer to soft peaks. Slowly trickle the hot syrup into the whites, whisking constantly. Continue whisking after all the syrup has been added, until the egg whites have cooled to room temperature; they should have firm, glossy peaks.

Poached Peaches or Apricots

My mother would pick peaches and apricots from the fruit trees surrounding our farmhouse in Cowaramup. Simply poached in a light syrup, they formed the basis for dessert at many family feasts.

....

750 g caster sugar
1 litre water
250 ml white wine
2 cinnamon sticks
2 cloves
2 cardamom pods, cracked
Juice of 1 lemon
6 firm, ripe large peaches or 12 apricots

Combine the sugar, water, wine, spices and lemon juice in a large saucepan and bring to a simmer. Reduce the heat and cook for 5 minutes to allow the flavours to blend.

Bring the liquid back to the boil; cut the peaches or apricots in half and discard the stones.

Drop the fruit into the boiling liquid, reduce the heat to a slow simmer and cook the peaches for 8 to 9 minutes or the apricots for 4 to 5 minutes. Make sure you do not overcook the fruits – they should not be mushy.

Remove the saucepan from the heat and place a sheet of baking paper on the top of the liquid to seal the surface. Allow the fruit to cool in the liquid, then refrigerate.

A portion of the liquid can be reduced to a syrup; return the strained liquid to a saucepan and cook over medium heat to reduce by about two-thirds.

Vanilla Ice Cream

Freshly churned ice cream is a real delight. This recipe has a silky richness that is the perfect end to a relaxed bistro meal and will keep in the freezer for up to 1 month. You will need an ice-cream maker for this recipe.

Split the vanilla bean lengthways and scrape out seeds, placing bean and seeds with the milk in a large saucepan over medium heat; heat to a shimmer (almost boiling).

Whisk egg yolks and sugar and pour on the hot milk, whisking to combine. Return to the saucepan and place on low heat; cook gently, constantly stirring with a wooden spoon until the custard thickens and coats the back of the spoon. Remove from the heat and stir in the glucose then cream. Pass through a fine strainer and cool in the refrigerator.

Churn the mixture in an ice-cream machine and freeze.

1 vanilla bean
500 ml milk
6 egg yolks
200 g sugar
15 g liquid glucose
500 ml cream

MAKES 1.25 LITRES

White Peach Sorbet

I always look forward to the first peaches in spring. To protect the sweet-smelling perfume of the white peaches, as used here, I don't cook them before I purée them. Do try and eat this delicious sorbet on the day that it's made.

Combine the sugar, water and glucose in a large saucepan over medium heat and stir until the sugar has dissolved. Bring the syrup to the boil and remove from the heat. Add the lemon juice and peach purée, pass through a fine strainer and chill in the refrigerator.

Once cold, churn in an ice-cream machine and store in the freezer for up to 1 month.

Raspberry Sorbet / Blood Plum Sorbet

Use the same recipe as above but substitute raspberry purée or blood plum purée for the peach.

225 g caster sugar
190 ml water
50 g glucose syrup
Juice of 1 lemon
600 g white peach purée

MAKES 1 LITRE

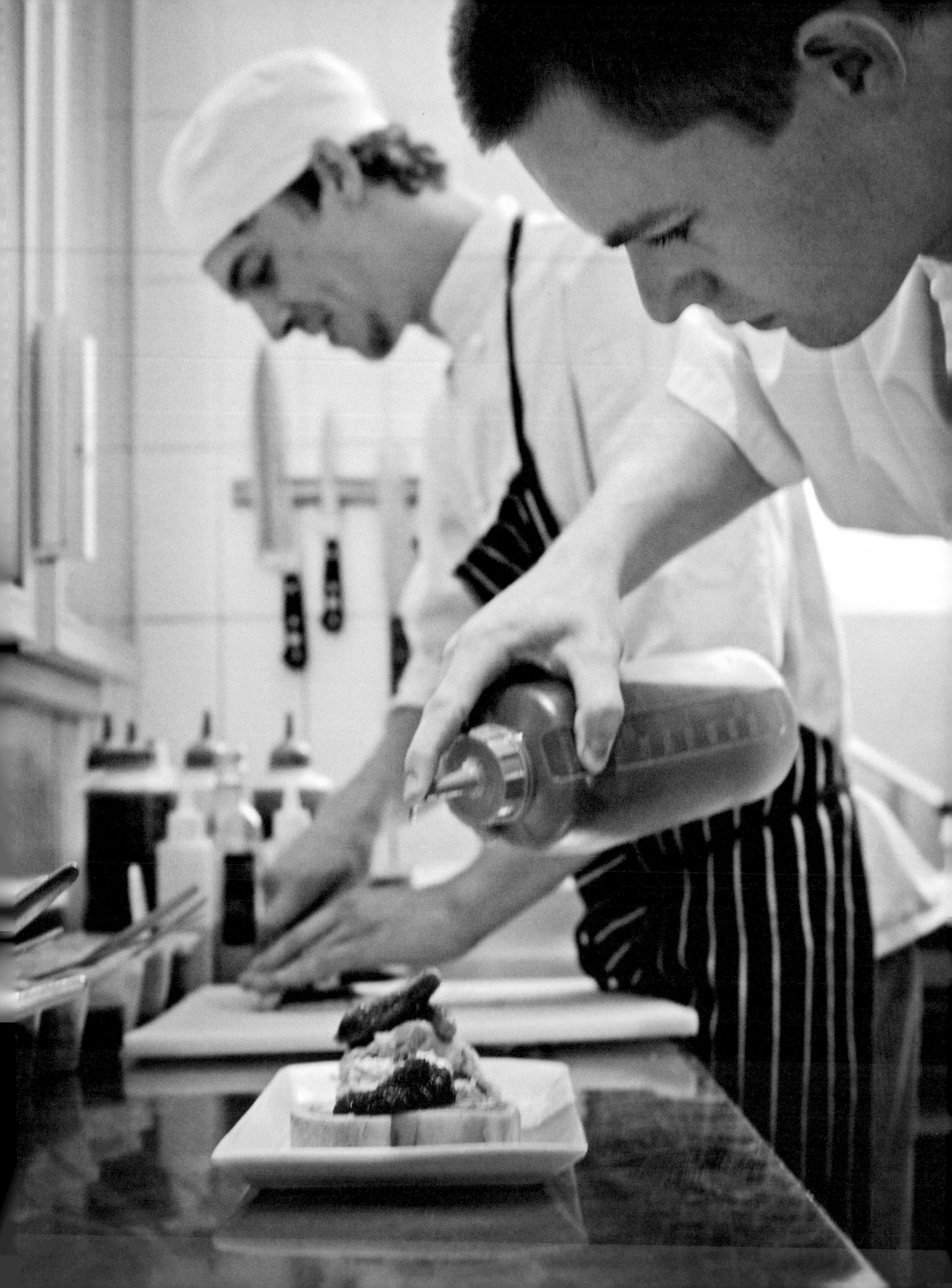

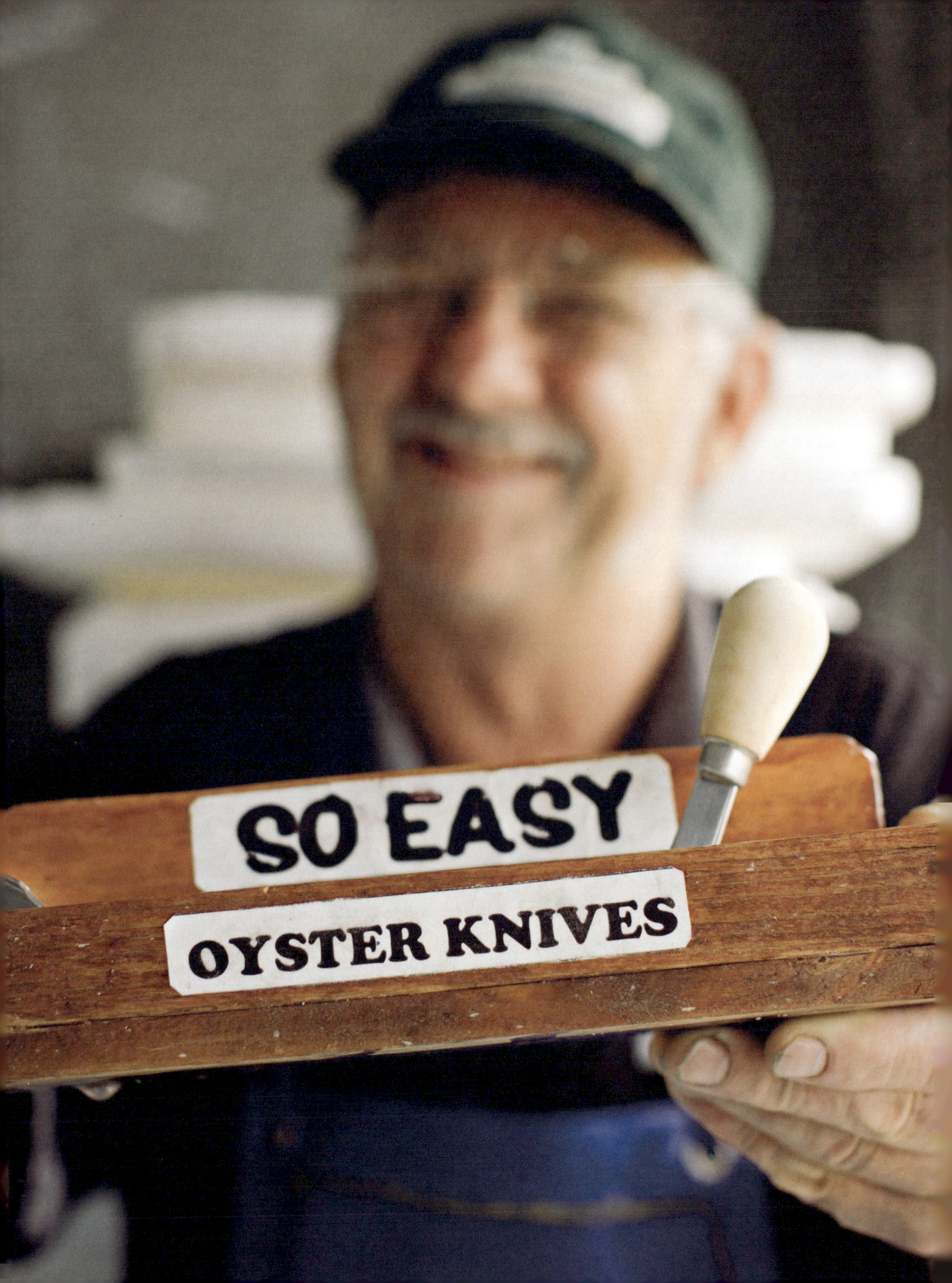
SO EASY
OYSTER KNIVES

Select Bibliography

Stephanie Alexander, *The Cook's Companion*, Viking, Melbourne, 1996.
Colman Andrews, *Flavours of the Riviera: Discovering the Real Mediterranean Cooking of France and Italy*, Grub Street, London, 2000.
Pepita Aris, *Recipes from a Spanish Village*, Conran Octopus, London, 1990.
Antonio Carluccio, *Antonio Carluccio's Vegetables*, Headline, London, 2000.
Claire Clark, *Indulge*, Whitecap, Vancouver, 2007.
Carole Clements, *A Flavour of Normandy*, Cornstalk, Sydney, 1996.
Linda Dannenberg, *Paris Bistro Cooking*, Clarkson Potter, New York, 1991; *Paris Boulangerie-Patisserie*, Clarkson Potter, New York, 1994.
Serge Dansereau, *The Bathers' Pavilion: Menus and Recipes*, ABC Books, Sydney, 2006.
Stefano de Pieri, *Modern Italian Food*, Hardie Grant Books, Melbourne, 2004.
Fiona Dunlop, *New Tapas*, Mitchell Beasley, London, 2002.
Jenny Ferguson, *Cooking for You and Me*, Methuen Hayes, Sydney, 1987.
Jean Claude Frentz, Michel Poulain & Anne Sterling, *Charcuterie Specialties*, Cicem Paris/John Wiley, New York, 1996.
Jane Grigson, *Charcuterie and French Pork Butchery*, Grub Street, London, 2001.
Ian Hemphill, *Spice Notes*, Macmillan, Sydney, 2000.
Diana Henry, *Crazy Water, Pickled Lemons: Enchanting Dishes from the Middle East, Mediterranean and North Africa*, Mitchell Beasley, London, 2002.
Philip Johnson & Kris Riordan, *e'cco: Recipes from an Australian Bistro*, Random House, Sydney, 1999.
Thomas Keller, *Bouchon*, Artisan, New York, 2004; *Under Pressure: Cooking Sous Vide*, Artisan, New York, 2008.
Jean-Jacques Lale-Demoz, *Jean-Jacques' Seafood*, Nelson, Melbourne, 1986.
Bruno Loubet, *Bistrot Bruno*, Macmillan, London, 1995.
Greg Malouf & Lucy Malouf, *Arabesque*, Hardie Grant Books, Melbourne, 1999; *Saha*, Hardie Grant Books, Melbourne 2005.
Jacques Médecin, *Cuisine Niçoise*, Grub Street, London, 2002.
Sean Moran, *Let It Simmer*, Lantern, Melbourne, 2006.
Gordon Ramsay, *Gordon Ramsay's Just Desserts*, Quadrille, London, 2001.
Michel Roux, *Eggs*, Quadrille, London, 2005; *Pastry*, Quadrille, London, 2008; *Culinaria France*, Könemann, Cologne, 1999.
Guy Savoy, *La Cuisine de mes bistrots*, Hachette, France, 1998.
Jane Sigal, *Backroad Bistros, Farmhouse Fare*, Pavilion, London, 1995.
Nigel Slater, *The Kitchen Diaries*, Fourth Estate, London, 2005.
Martha Stewart, *Pies and Tarts*, Clarkson Potter, New York, 1985.
Christian Teubner & Sybil Gräfin Schönfeldt, *Great Desserts*, Hamlyn, London, 1989.
Joanne Weir & Carin Alpert, *From Tapas to Meze: Small Plates from the Mediterranean*, Ten Speed Press, San Francisco, 2004.
Patricia Wells, *Bistro Cooking*, Workman, New York, 1989; *The Paris Cookbook*, Kyle Cathie, London, 2001; *At Home in Provence*, Kyle Cathie, London, 1997.
Marco Pierre White, *Canteen Cuisine*, Ebury Press, London, 1995.
Paula Wolfert, *The Cooking of South West France: A Collection of Traditional and New Recipes from France's Magnificent Rustic Cuisine*, Wiley, New Jersey, 2005.

Debra Hartmann's olive oil ice cream recipe (p. 249) was adapted from a recipe by **David Tsirekas**, of Perama Greek Restaurant, Sydney.

First published in 2010
by UWA Publishing
Crawley, Western Australia 6009
www.uwap.uwa.edu.au

UWAP is an imprint of UWA Publishing, a division of The University of Western Australia.

THE UNIVERSITY OF WESTERN AUSTRALIA
Achieving International Excellence

This book is copyright. Apart from any fair dealing for the purpose private study, research, criticism or review, as permitted under the Copyright Act 1968, no part may be reproduced by any process without written permission. Enquiries should be made to the publisher.

Copyright © Russell Blaikie 2010

Photography © Craig Kinder 2010
Wine notes © Paul McArdle 2010

The moral right of the author has been asserted.

National Library of Australia Cataloguing-in-Publication entry:

Blaikie, Russell.
Must eat / Russell Blaikie ;
photography by Craig Kinder.

ISBN 9781921401671 (pbk.)
Cookery, French.
Food and wine pairing.

641.5944

Design and typeset by
Anna Maley-Fadgyas
Photographic artwork and CMYK colour conversion by Henrik Tived
Printed by Imago

Cover photograph: Must Winebar,
519 Beaufort St, Highgate WA 6003
www.must.com.au

Thanks

Kate McLeod from UWAP, who asked me to write this book in 2008. Kate also worked her magic, styling the photos and applying her 'cool filter' to the wonderful food images.

Craig Kinder, who has made this project so enjoyable. He's made my food look so naturally edible with his awesome ability behind the lens, and his sense of humour and easy-going personality made the many long days cheerful.

Paul McArdle, for his wonderful ability to match wines from around the globe with my food, and his special ability to convert what he tastes into words.

Lisa Schreurs from my kitchen in Perth was my assistant with the recipe preparation and styling. She constantly amazed me with her fine attention to detail and commitment.

The teams past and present at Must Winebar in Perth and Must Margaret River, particularly my head chefs Andre Mahe and Chris Cheong who provided incredible support, allowing me to spend precious time away from the kitchen to work on this book. And Aaron Commins, for weaving his magic with the cocktails in this book, and his commitment to making Must a great place to be.

I've written about the many suppliers and producers of food and wine who have provided Must with so much of our success, but a special thanks, too, to the many others who have not been named here.

Thanks also to Graham Clarkson, Alana Blowfield, Alex England, Annika Kristensen, Kellie Willcock and my partners in wine – Garry, Stefan, Carlo and David Gosatti, Grant and Wendy Mason, and Tony and Loretta van Merwyk. Also thanks to Lisa, Cherida and Melissa for maintaining sanity in the office.

A final thank you to my family – Tamara, Alex and Samuel – for bearing with me when in writing mode and Melanie Ostell, publisher, Zoe Harpham, recipe editor, Anna Maley-Fadgyas, designer, and Henrik Tived, touch-ups, who have helped transfer my vision for this book onto the page.

Index

Angel hair pasta (to make) 270; with crab, chilli, tomato & basil 84
Apricots crusta (cocktail) 261; poached 289
Asparagus 207; asparagus, feta & macadamia crumble 55
Avocado avocado & celeriac remoulade 143; avocado, feta, spinach & pine nut salad 210; salsa 44

Baharat 271
Baharat lamb meatballs 26
Barramundi & potato, artichoke, olive & truss tomato salad 140
Beans cassoulet 188; flageolet beans 168; green beans with mint 200
Béarnaise sauce 278
Beef beef cheek ravioli 90; beef rib bourguignon 187; carpaccio 108; slow-braised Wagyu beef shin 80; spicy beef skewers 29; sirloin steak (grilled) 184; steak sandwich 30
Berries & crème Chantilly 226
Beetroot glazed baby beetroot 181; pickled beetroot 202
Black olive tapenade 10
Beurre blanc chive 284; lemon 284; vermouth 284
Blood plum sorbet 290
Blueberry & strawberry syrup 250
Bombe Alaska 250
Bouillabaisse 144
Butterscotch pudding & sauce 242

Cabbage, red braised 196
Cabernet vinaigrette 268
Caramel 219; orange caramel 229
Cassoulet 188
Celeriac celeriac & avocado remoulade 143; celeriac purée 197

Champagne crème 107
Cheese (savoury) comté tart 94; fontina cheese toasties 6; Fourme D'Ambert tart 95; gougères 14; gruyère toasties 9; parmesan & pumpkin mash 174; pumpkin ravioli & goat's cheese 86; taleggio soufflé 105; tomme de chévre gnocchi 79; (sweet) lemon mascarpone tart 230; white peach & mascarpone cheesecake 241
Cherry and white chocolate clafoutis 237
Chicken chicken liver parfait 119; chicken & lobster boudin blanc (to make) 129, (to cook) 152; coq au vin 159; paella 148; poached chicken breast & black truffles 156; rotisserie chicken 155; stock (brown) 274, (white) 277
Chickpeas chickpea & chorizo braise 198; chickpea batter 44; chickpea-battered oysters 44; hummus 65
Chilli tomato jam 266
Chive beurre blanc 284
Chocolate chocolate & Cognac mousse 234; chocolate chip ice cream 246; chocolate molleux 245; white chocolate & cherry clafoutis 237
Choron sauce 280
Christmas pudding 253
Clarified butter 278
Cocktails apricot crusta 261; French martini 257; French-oaked Rob Roy 256; Garden of Eden 259; Lilletini 260; Monte-negroni 258
Cognac crème Anglaise 287
Comté tart 94

Confit baby carrots 267; duck leg 160; goat shank 168; tomatoes 267
Court bouillon 66
Crab bisque 70
Crème brûlée orange & vanilla 220; espresso 223
Crème caramel 219
Crème Chantilly 226
Crepes 286; veal shin 164
Croutons 74
Crumbed oysters 43
Cured ocean tout 62

Dressings & vinaigrettes black truffle vinaigrette 269; cabernet vinaigrette 268; citrus vinaigrette 268; dill & grain mustard dressing 269; hazlenut vinaigrette 268; red wine vinaigrette 268; preserved lemon vinaigrette 268; pomegranate dressing 268; preserved lemon vinaigrette 268; red wine vinaigrette 268; remoulade dressing 268; shallot vinegar 40; sherry-walnut vinaigrette 268; verjuice dressing 55; walnut vinaigrette 268
Duck cassoulet 188; duck leg confit 160; duck rillettes 122

Egg pasta dough 270
Egg wash 43, 94
Eggplant tagine 199
Espresso crème brûlée 223

Fish bouillabaisse 144; crispy-skinned barramundi 140; nannygai fillets (steamed) 139; ocean trout (cured) 62; paella 148; pink snapper (grilled) 143; snapper brandade tartlets 22; stock 275; whiting fillets (crumbed) 59

the end 295

French martini (cocktail) 257
French onion soup 73
French-oaked Rob Roy (cocktail) 256

Garden of Eden (cocktail) 259
Glazed baby beetroot 181
Gnocchi (to make) 77; pan-fried with tomme de chèvre 79; pan-fried with beef shin 80; pan-fried with rabbit ragoût 83
Goat shank (confit) 168
Gougères 14
Green beans & mint 200
Green olive tapenade 271
Green salad 211

Ham & parsley terrine 126
Hazelnut salad 94
Honeycomb 233
Horseradish crème 108
Hummus 65

Ice cream chocolate chip 246; Dulce de Leche 287; olive oil, fig & macadamia 249; vanilla 290
Iles flottantes (poached meringue) 229
Italian meringue 289

Jambon persillé (ham & parsley terrine) 126

Lamb baharat-spiced meatballs 26; Middle Eastern spiced lamb shank braise 173; stock 274
Leek & potato soup 74
Lemon beurre blanc 284; mascarpone tart 230; preserved lemon vinaigrette 268
Lilletini (cocktail) 260

Macadamia crumble 55; tuiles 288
Macaroni cheese 201

Mayonnaise 281
Meringue Italian 289; poached 229
Mint pistou 168
Minted green beans 200
Mixed leaf salad 211
Mograbieh salad 212
Monte-negroni (cocktail) 258
Moorish beef skewers 29
Mouclade (curried mussel soup) 69
Mushrooms cepe mushroom tart 100; mushroom crème 91; mushroom & rabbit pie 163

Olives olive, potato, artichoke & tomato salad 140; olive tapenade (black) 10, (green) 271
Onion caramelised 266; caramelised onion risotto 152; French onion soup 73
Orange caramel 229; vanilla crème brûlée 220
Oysters natural 40; crumbed 43; chickpea-battered 44; Rockefeller 47

Paella 148
Parsley purée 80
Pasta angel hair pasta with crab, chilli, tomato & basil 84; egg pasta dough 270; macaroni cheese 201; see ravioli
Pâté brisée (pastry) 285
Peaches chutney 273, poached 289; sorbet 290; white peach & mascarpone cheesecake 241
Pears chutney 273; pear & walnut salad 95; pear salad 160; poached in shiraz 233
Pedro Ximénez lacquer 282
Pickled beetroot 202
Piperade 139
Pistachio praline 234

Polenta bites 13; soft polenta 177
Pomegranate dressing 268
Pork cassoulet 188; ham & parsley terrine 126; ham & scallop croquettes 18; pancetta-wrapped tiger prawns 17; pancetta-wrapped quail 151; pork belly & Pedro Ximénez lacquer 178; rotisserie fillet 174; pork, pistachio & date terrine 121; sausages (to make) 132, (to cook) 177
Port wine jus 283
Potatoes dauphine potatoes 139; see gnocchi; gratin 205; leek & potato soup 74; Paris mash 203; pomme frites 203; potato & radicchio salad 151; potato, artichoke, olive & truss tomato salad 140
Prawn cocktail 66; soufflé 107; wrapped in pancetta 17
Pumpkin pumpkin ravioli & goat's cheese 86; pumpkin & parmesan mash 174
Purées celeriac 197; parsley 80; rocket 164; sweet corn 285

Quail quail egg benedict tartlets 21; rotisserie quail wrapped in pancetta 151
Quatre èpices (four spices) 272

Rabbit ragoût 83; rabbit & mushroom pie 163
Ras el hanout 272
Raspberry sorbet 290
Remoulade avocado & celeriac remoulade 143; dressing 268
Ratatouille Provençale 206
Ravioli (to make) 270; beef cheek ravioli 90; pumpkin ravioli 86

Red cabbage braise 196
Red capsicum coulis 87
Rocket purée 164; rocket & parmesan salad 213
Rouille 144

Sage butter 87
Salads asparagus, feta & macadamia crumble 55; caramelised fig, jamón ibérico & Roquefort salad 56; caramelised pear & walnut salad 95; frisée & hazelnut salad 94; mixed leaf salad 211; mograbieh salad 212; pancetta & rocket salad 100; pear salad 160; potato & radicchio salad 151; potato, artichoke, olive & truss tomato salad 140; rocket & parmesan salad 213; tomato & citrus salad 58
Sauces (savoury) Béarnaise 278; beurre blanc 284; champagne crème 107; choron (tomato hollandaise) 280; diable 283; hollandaise 280; horseradish crème 108; Madeira jus 281; mayonnaise 281; meatball sauce 26; mint pistou 168; Moorish marinade 29; mushroom crème 91; Pedro Ximénez lacquer 282; piperade 139; red capsicum coulis 87; red wine jus 283; sherry-soaked flame grape jus 151; tartare (ravigote) 282; truffle jus 155; tomato coulis 79; verte 17; (sweet) butterscotch sauce 242; Cognac crème Anglaise 287
Sausages cassoulet 188; chicken & lobster boudin blanc 129, 152; pork 132, 177
Scallops seared scallops & hummus 65; scallop & jamón croquettes 18

Sesame snap 220
Seafood bouillabaisse 144; crab bisque 70; crab, chilli, tomato & basil pasta 84; lobster & chicken boudin blanc 152; mouclade 69; oysters 40–47; paella 148; prawn cocktail 66; scallop & jamón croquettes 18; seared scallops 65; stock 275; tiger prawn soufflé 107; tiger prawns with pancetta 17; yabby, confit tomato & chive tart 101
Sherry-soaked flame seedless grapes 284; grape jus 151
Sorbet blood plum 290; raspberry 290; white peach 290
Soufflés taleggio 105; tiger prawn 107
Soups crab bisque 70; bouillabaisse 144; French onion soup 73; leek & potato soup 74; mouclade 69; see stocks
Spices baharat 271; Moorish marinade 29; quatre èpices 272; ras el hanout 272
Stocks beef stock 274; chicken stock (brown) 274, (white) 277; court bouillon 66; fish stock 275; lamb stock 274; rouille 144; shellfish stock 275; veal stock 274; vegetable stock 276
Sweet corn purée 285

Tarts (savoury) cepe mushroom tart 100; comté tart 94; Fourme D'Ambert tart 95; snapper brandade tartlets 22; yabby, tomato & chive tart 101; (sweet) lemon mascarpone tart 230; tarte Tatin 238
Tarte Tatin 238
Tomatoes chilli tomato jam 266; confit 267; coulis 79; tomato & citrus salad 58; tomato, potato, artichoke & olive salad 140; yabby, tomato & chive tart 101
Truffles black truffle & fontina cheese toasties 6; black truffle vinaigrette 269; truffle jus 155; truffle potato mash 203

Veal crepe 164; liver & pancetta 167; stock 274
Vegetable stock 276
Venison 181
Verte, sauce 17

Yabby, confit tomato & chive tart 101